Parasitic Orobanchaceae

Daniel M. Joel • Jonathan Gressel •
Lytton J. Musselman
Editors

Parasitic Orobanchaceae

Parasitic Mechanisms and Control Strategies

Springer

Editors
Daniel M. Joel
Department of Weed Research
Agricultural Research Organization
Newe-Ya'ar Research Center
Ramat-Yishay, Israel

Jonathan Gressel
Plant Sciences
Weizmann Institute of Science
Rehovot, Israel

Lytton J. Musselman
Department of Biological Sciences
Old Dominion University
Norfolk, Virginia
USA

ISBN 978-3-642-38145-4 ISBN 978-3-642-38146-1 (eBook)
DOI 10.1007/978-3-642-38146-1
Springer Heidelberg New York Dordrecht London

Library of Congress Control Number: 2013943330

Printed on acid-free paper

Springer is part of Springer Science+Business Media (www.springer.com)

Foreword

Editing a book on a 'hot' subject like parasitic plants is possible only when the knowledge of these plants reached a threshold, with sufficient understanding of phylogenetic trends, physiological processes, structural developments, biochemical pathways, gene expression and ecological interactions, as well as breakthroughs in the integrated management of some of the most pernicious weedy parasitic species in agricultural fields. It was not until 2010 that the significant changes in our knowledge of parasitic mechanisms and in the control of some of the weedy species could be appreciated. This allowed the preparation of the comprehensive book on the Orobanchaceae, which integrates basic and applied aspects of this important plant family.

The publication of the book could not be possible without the excellent cooperation of all chapter authors, leading scientists in their respective fields of research, who contributed both basic and cutting-edge information on all key aspects of the parasitic syndrome and on all major aspects of parasitic weed management. I am pleased to thank the chapter authors and co-authors for their excellent contribution and for their cooperation during the process of editing, which obviously took a long while. Thanks are also due to all experts who have peer reviewed the chapters and helped in ensuring the high scientific standards of the book.

My cordial thanks are particularly due to Jonny Gressel for his dedication and important inputs during the planning and preparation of the book and for carefully editing key chapters. Special thanks are also due to Lytton Musselman for helpful review and editing of other chapters.

I hope that the book, presenting the current knowledge in all key aspects of plant parasitism, is not only a source of important information on the Orobanchaceae, but also a stimulus to further research in both basic and applied aspects of plant parasitism.

Newe-Ya'ar Research Center, Israel — Danny Joel

Preface

Plant parasitism is a fascinating phenomenon of extreme intimate plant-to-plant interactions. The world of parasitic plants includes around 20 families, but the Orobanchaceae are the leading models for research. This is not only because some members of the family are parasitic weeds of great economic importance, but also because this family includes the whole trophic spectrum from non-parasitic autotrophs to obligate holoparasites. Many of these species are relatively amenable to laboratory and field experimentation. Research on the Orobanchaceae has yet to peak, but only recently has there been a surge in research with significant achievements particularly in the understanding of the mechanisms of parasitism, which justifies the publication of a new book on parasitic plants.

The evolutionary origin of plant parasitism is associated with regulatory changes in genes that usually fulfil non-parasitic functions. The specific functions of parasitism evolved following the duplication of genes or genomes and by ectopic expression of genes (see Sect. 4.5). In this way parasitic plants acquired features that are common to many non-parasitic plants, but their mode of expression, the extent to which these features have developed and the combination of the different features are unique. These unique features make fascinating scientific research that is aimed at understanding the parasitic plants at the most basic level. These findings can also be exploited at the applied level in designing sophisticated tools for the control of species that cause damage to agricultural crops.

The most recent example of parasitic plant research that significantly contributed to understanding the physiology of plants is the discovery of a novel family of plant hormones, the strigolactones, which was first identified as a group of germination stimulants for the holoparasites *Striga* and *Orobanche*. The detailed knowledge of the Orobanchaceae, presented in this book, should therefore not only reflect on the understanding of parasitic plants belonging to other families, for which little physiological and molecular information is available, but particularly contribute to understanding many features of plants in general.

The main objective of the book is to provide a comprehensive account of the current knowledge on all aspects of the parasitic syndrome within the Orobanchaceae. For this sake, internationally recognized leading scientists were invited as chapter authors. The organization of the book is modular so that each chapter covering a given topic is self-contained while being indexed and fully cross-referenced to related chapters.

The book includes two parts. The first presents the cutting-edge knowledge of all key aspects of parasitism, and the second part is dedicated to the weedy species and their management, presenting and discussing strategies for parasitic weed control. Aspects of the Orobanchaceae that are not related to the parasitic habit are not presented. The diversity of parasitic families within the plant kingdom is briefly covered in Chap. 1, in order to clarify the position of the Orobanchaceae within the world of parasitic plants.

The core of parasitism is a special organ—the haustorium, a unique plant organ that is homologous to roots and physically connects the parasites to their hosts, allowing the physiological bridging between them. The structure of the haustorium, the signalling mechanisms for triggering its initiation and the manner it invades the host tissues are described and dealt with in Chaps. 2–5.

Following the establishment of the physical connection between the parasite and the host, the coordination between them is facilitated by specific chemical and hormonal signalling, allowing the parasite to act as an effective compatible sink in the overall host plant metabolism. While nutrient transfer and other physiological interactions between the parasite and its compatible hosts are discussed in Chap. 6, the host responses to the parasite are discussed in Chap. 7, including a detailed account of host resistance mechanisms.

Unlike the facultative hemiparasitic Orobanchaceae, the obligate parasites can only germinate in the vicinity of host roots. The unique structure of their seeds, the signalling mechanisms behind the ability of the seeds to identify host roots and the physiological aspects of their germination are dealt with in Chaps. 8–12. The strigolactones, a group of chemicals that are exuded by plant roots and serve as germination stimulants for many obligate parasites, are a major focus of Chaps. 10 and 12. The chemical and genetic aspects of strigolactones activity and the biochemical aspects of their biosynthesis are currently on the cutting edge of plant research.

Many Orobanchaceae species are adapted to parasitize specific hosts, a phenomenon that is particularly evident in the weedy species. Nonetheless, the mechanisms behind the adaptation of these species to changes in the availability of hosts are hardly understood. Given the increasing interest in epigenetics, a speculative chapter (Chap. 13) discusses, for the first time, the possibility that epigenetics is involved in determining host specificity.

The Orobanchaceae is a highly diverse plant family with many genera that have previously been included in other families. Recent molecular studies clearly show the phylogenetic relations between the different genera, on which the taxonomy of the Orobanchaceae is currently based. These phylogenetic relations and evolutionary trends are presented in Chap. 14, together with much taxonomic information regarding the current status of 'problematic' genera. Chapter 15 further presents aspects of the genomic evolution of the Orobanchaceae that appears to be extraordinarily dynamic and includes, between others, the reductive evolution of the plastid chromosome following the loss of photosynthesis.

Most Orobanchaceae species are not dominant in their habitat and may easily be ignored in the field because they look like 'ordinary' plants, though some have showy flowers or lack chlorophyll; nonetheless, certain species are keystone species in their plant ecosystems. The ecological aspects of parasitic Orobanchaceae have mainly been studied with some hemiparasitic model plants, particularly species of *Rhinanthus*. The interaction between parasitic plants and their hosts at the plant community level is presented in Chap. 16, with analysis of the impacts of these interactions on the dynamic structure of plant communities and on the interaction between the plant community and other organisms. The potential role of some hemiparasites as a tool in promoting floristic biodiversity by selectively parasitizing species that are too dominant in these habitats is also discussed in this chapter.

Plant parasitism is not only a case of extreme plant-to-plant interactions that can be useful as a tool in scientific research and in the management of certain habitats. It is also a significant threat to agriculture. Some parasitic species are weedy and damage major agricultural crops, leading to heavy economic losses worldwide and threatening food security, especially in poor countries. Potentially climate change may expand the distribution of the weedy species to geographical areas that are currently un-infested, and some non-weedy species may penetrate cultivated areas and become weedy. An updated description of the species that parasitize agricultural crops is presented in Chaps. 18–26, where the current knowledge on all aspects of parasitic weed management is discussed. These chapters are more fully introduced in Chap. 17.

The book is intended for all people who are interested in this remarkable family of parasitic plants, including students, university lecturers, plant scientists, as well as agronomists and weed specialists, breeders and farmers, extension personnel and experts in tropical and subtropical agriculture. The book is suitable for use in various university and college courses, not only in general plant biology, parasitic plants, plant physiology and plant evolution, but also in weed science, plant protection and host–parasite interaction.

Newe-Ya'ar, Israel — Daniel M. Joel
Rehovot, Israel — Jonathan Gressel
Norfolk, VA, USA — Lytton J. Musselman

Contents

Chapter 1
Introduction: The Parasitic Syndrome in Higher Plants

Henning S. Heide-Jørgensen

1.1 Parasitism in Plants

In vascular plants, parasitism is found only in the eudicotyledonous angiosperms, although the gymnosperm *Parasitaxus usta* may be considered a borderline case, because it develops graft-like attachments with roots of another conifer, rather than haustoria, and has only direct water relations with its host while carbon trafficking from the host is mediated by mycorrhiza fungi (Feild and Brodribb 2005; Heide-Jørgensen 2008). Parasitic plants have been able to adapt to all types of plant communities in all environments where flowering plants occur, except the aquatic environment. Competition for water is one of the main driving forces in the evolution of land plants. In an aquatic environment, water is no limitation to plant growth, and there is no advantage in being a parasite removing water from a host. On the other hand, if a land plant, especially during its establishment, exploits another plant's root system and photosynthetic apparatus, it obtains a clear competitive advantage. It is this advantage that has been exploited by terrestrial parasitic plants and enabled them to be represented in nearly all ecosystems from the high arctic to the driest deserts. This is particularly true for the group of parasites that are dealt with in this book, the Orobanchaceae.

The physical connection organ between parasite and host is called a **haustorium**. The term was introduced by A. P. de Candolle (1813) to describe the connection between *Cuscuta* and its hosts. Since then, it has been used for a variety of structures that are involved in nutrient absorption from species to species or from generation to generation, as exemplified by fungal hyphae, sporophyte of mosses and pteridophytes, and embryo of some seed plants. In parasitic angiosperms, the haustorium is "the essence of parasitism" as Job Kuijt (1969) has put it. At first, the haustorium serves as an attachment organ. Then it develops as an intrusive structure

H.S. Heide-Jørgensen (✉)
University of Copenhagen, Copenhagen, Denmark
e-mail: hhj@viscum.dk

D.M. Joel et al. (eds.), *Parasitic Orobanchaceae*,
DOI 10.1007/978-3-642-38146-1_1,

that penetrates host tissues. It later becomes a water and nutrient absorption organ. Its most characteristic anatomical feature, which can be seen in all parasitic plant families, is a **xylem bridge**, connecting the xylem of the parasite to host xylem. A few parasites also develop **phloem connection** (see Sect. 3.9.3). Haustorial morphology and anatomy varies greatly among families and taxa. In some parasites, such as Rafflesiaceae, no haustorial parts are visible outside the host. The internal parts in these parasites form cellular strands within the host tissues and become so diffuse that it no longer makes sense to use the term haustorium (Heide-Jørgensen 2008). In such cases, the vegetative part of the parasite within the host is often referred to as the **endophyte**, while the external parts, which are sometimes limited to flowers or flowering stems, as the **exophyte**. The haustorial variation among major taxa reflects the generally agreed opinion that parasitic plants have evolved independently about a dozen times (Nickrent 2008; see Chap. 14).

Traditionally, parasitism has been recognized in 20 dicotyledonous families (Fig. 1.1a), but recently Olacaceae and Santalaceae were split into several smaller families based on molecular studies (Malécot and Nickrent 2008; Nickrent et al. 2010). Parasitic plants may now be found in up to 28 dicotyledonous families (see Sect. 1.7). Based on the presence of mature haustoria, all families consist solely of parasites except for Orobanchaceae that includes the non-parasitic genus *Lindenbergia* (see Chap. 14) and Lauraceae and Convolvulaceae where only *Cassytha* and *Cuscuta* are parasitic. The total number of parasitic species is close to 4,500, in 270–275 genera. That is about 1 % of all known 260,000 seed plants (Thorne 2002). The majority or 90 % of all parasites are hemiparasites, and root parasites represent 60 % of all parasitic plants (Fig. 1.1a).

Although some parasitic plants have been known since the days of Theophrastus (372–287 B.C.), botanists did not pay much attention to this life form until the nineteenth century. Some members of Rafflesiaceae, Balanophoraceae, and Cynomoriaceae were classified as fungi or placed in a separate class for bizarre excrescences, named Sarcophytae (Trattinick 1828; Kuijt 1969). It was the introduction of the very harmful witchweed *Striga asiatica* (Fig. 18.3b) to maize fields in the eastern USA in the early 1950s that stimulated research in parasitic plants. The first comprehensive scientific treatment of parasitic plants was published by Job Kuijt in 1969, and more recent thorough treatments of major parasite taxa or aspects of the parasitic syndrome include Kuijt (1977, 2003), Calder and Bernhardt (1983), Bhandari and Mukerji (1993), Weber (1993), Press and Graves (1995), Polhill and Wiens (1998), Geils et al. (2002), Joel et al. (2007), Carlón et al. (2008), Heide-Jørgensen (2008), and Mathiasen et al. (2008). Following a series of scientific meetings on parasitic plants since 1973, the International Parasitic Plant Society (IPPS; http://www.parasiticplants.org/default.asp) was founded in 2001. Besides organizing congresses and symposia on parasitic plants, the society publishes the newsletter "Haustorium" that is an excellent source on recent published literature on parasitic plants.

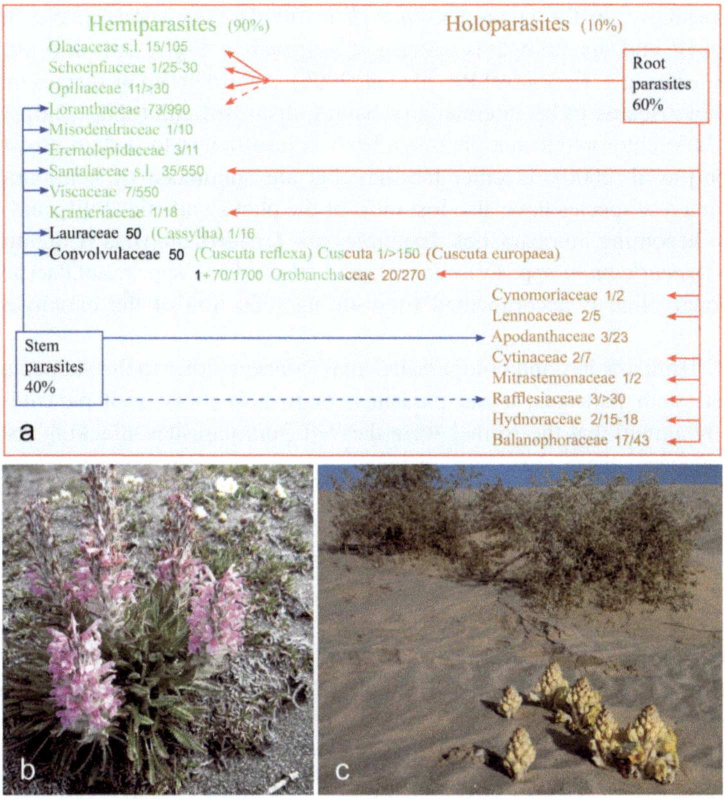

Fig. 1.1 (**a**) Parasitic families arranged according to parasitic types. Family names are followed by a number of genera and species. Broken lines indicate a few exceptions from main type: black, non-parasites; green, hemiparasites; brown, holoparasites; percentages are in relation to the total number of parasitic plants. Apodanthaceae, Cytinaceae, and Mitrastemonaceae used to be in Rafflesiaceae. Orobanchaceae includes the former parasitic Scrophulariaceae and the non-parasitic genus Lindenbergia (modified from Heide-Jørgensen 2008 and 2011). (**b**, **c**) The two main types of parasitic plants. (**b**) The hemiparasitic *Pedicularis lanata* (Orobanchaceae), between plants that serve as its host; high arctic Greenland. (**c**) The holoparasitic *Cistanche tubulosa* (Orobanchaceae) with the host in the background; desert, Qatar (photos: part **b** by Helene Heide-Jørgensen; part **c** by KK Kristiansen)

1.2 Hemi- and Holoparasitism

Two main types of parasitic plants are recognized (Fig. 1.1b, c): (a) **hemiparasites** that are able to photosynthesize although they are not necessarily self-sufficient with carbon and (b) **holoparasites** that have no photosynthetic abilities (dePamphilis and Palmer 1990; Hibberd et al. 1998). Unlike some hemiparasites, when holoparasites have a root system, it is highly reduced and all or the major part of their needs for water and nutrients is derived from their hosts.

Interestingly, in the genus *Cuscuta* (Convolvulaceae), some species have no chlorophyll and are holoparasites (e.g. *C. europaea*; Fig. 1.2a), and others are hemiparasites (e.g. *C. reflexa*; Revill et al. 2005). A third group of species including *C. gronovii* seems to be intermediary having disturbed chloroplast ultrastructure and so little chlorophyll that photosynthesis is insufficient to sustain growth (Van der Kooij et al. 2000). In other families that are dominated by hemiparasites, a single or a few species have also lost most of the photosynthetic ability and may be close to becoming holoparasites. Examples are *Tristerix aphyllus* (Loranthaceae), certain *Arceuthobium* spp. (Viscaceae), and *Phacellaria* spp. (Santalaceae). This evolutionary line is accompanied by a strong reduction of the exophyte (Kuijt 1969).

Both hemiparasites and holoparasites may connect either to the shoot system of the host (stem parasites, aerial parasites) or to host roots (root parasites). It is generally agreed that the former were derived from parasites attacking roots, and in Santalaceae, there are a few species such as *Exocarpos cupressiformis* and *E. pullei* that occur both on roots and on stems (Coleman 1934; Lam 1945). Further, in Orobanchaceae some species may penetrate both roots and rhizomes (Weber 1976, 1993; see Sect. 3.6.1); therefore, the terms root and stem parasites should be used with caution.

Some authors also distinguish between facultative and obligate parasites. **Facultative parasites** may survive without haustorial connection to a host but productivity is better with hosts. Naturally, only hemiparasitic root parasites can be facultative parasites. However, so far no parasitic plant has been documented to complete its lifecycle in a natural environment without haustorial connection to a host. In nature, competition from other plants may eliminate a potential facultative parasite. Therefore, these terms should only be used for parasites grown under artificial conditions (Kuijt 1969). Nonetheless, for a shorter or longer period after germination, some parasites are autophytes nourished by nutrients stored in the seed and/or manufacturing some carbohydrates from photosynthesis in cotyledons. Others at maturity may live for some time from nutrients in storage organs, as suggested for *Nuytsia floribunda* (Fineran and Hocking 1983).

1.3 The Haustorium

Two main types of haustoria are recognized (Kuijt 1969) (see Chap. 3). The **terminal haustorium** (= primary haustorium; Fig. 1.2b, c) develops directly from the apex of the primary root, while **lateral haustoria** (= secondary haustoria; Fig. 1.2d) develop laterally on young lateral or adventitious roots (see Sect. 3.3 for description of terminal and lateral roots of the Orobanchaceae).

In many parasitic plant families, the terminal haustorium is the largest and usually serves as the main functional haustorium throughout the life of the parasite, while lateral haustoria are in most species short-lived and are functional only for a few months or a growing season. In perennials, new haustoria develop each season.

Fig. 1.2 Parasites of the various parasitic plant families. (**a**) Flowering *Cuscuta europaea* (Convolvulaceae) with red achlorophyllous twinning stems; Denmark. (**b**) Terminal haustorium of the hemiparasitic stem parasite *Erianthemum ngamicum* (Loranthaceae) forming a woodrose; most of the tissue is produced by the host *Burkea Africana*; Shakati Nature Reserve, South Africa. (**c**) Tubercle of the holoparasite *Orobanche hederae* (Orobanchaceae) with base of three flowering stems attached to host root (light colour) by terminal haustorium; the host root is wilting distally to the haustorium. (**d**) Rhizome of the holoparasite *Hydnora visseri* (Hydnoraceae) with three lateral haustoria attached to short adventitious roots occurring in rows between buds; lighter host root is *Euphorbia dregeana*; SE of Port Nolloth, South Africa. (**e**) Coiling stems of the hemiparate *Cassytha pubescens* (Lauraceae) with several lateral haustoria attaching its host *Pavonia praemorsa*; Botanical Garden, Copenhagen. (**f**) Self-parasitism in the hemiparasitic stem parasite *Viscum album* (Viscaceae); two young plants have established on a parasite internode after dispersal by the bird *Sylvia atricapilla*. (**g**) Terminal haustorium (asterisk) of the stem parasite *Plicosepalus kalachariensis* (Loranthaceae) and two epicortical roots with lateral haustoria (arrows); South Africa. (**h**) Directional explosive buds and open flower of bird-pollinated *Agelanthus gracilis* (Loranthaceae); Shakati Nature Reserve, South Africa (photos: parts **b**, **c**, **e**, **f**, **h** by HS Heide-Jørgensen; part **d** by LJ Musselman; part **g** by C Calvin)

The numbers of lateral haustoria may amount to several thousands per plant, especially in hemiparasites (Fineran 1963a). Some parasites have both types of haustoria, while others only one. *Cassytha* (Fig. 1.2e) and *Cuscuta* are exceptional by developing haustoria laterally from the stems (see Chap. 3 for detailed description of Orobanchaceae haustoria).

The structure of the mature haustorium varies greatly among families. Both haustorial types may produce an attachment organ, often named a holdfast or adhesive disc. This is particularly important in aerial parasites, where the parasite seedling is not supported by soil particles. In these plants, its function is to glue the young haustorium to the host by secretion of lipidic or pectic substances (Dobbins and Kuijt 1974; Heide-Jørgensen 1989, 1991). In Santalaceae and Loranthaceae, the holdfast often develops a mantle clasping the host, and in the most extreme cases, the clasping folds meet one another on the opposite side (Weber 1980; Fineran and Hocking 1983; Calladine and Pate 2000).

Along the **interface** (the border between parasite and host cells), which often increases tremendously with the splitting up of the haustorium within the host tissues, the cell walls are often thicker (Dobbins and Kuijt 1973) and in some cases labyrinthine walls may also develop (Gedalovich-Shedletzky and Kuijt 1990; Heide-Jørgensen and Kuijt 1993; Fineran and Calvin 2000).

The xylem bridge was once assumed to be the main transport route for water and nutrients from host to all parasites. However, apoplastic markers demonstrated the existence of an apoplastic continuum along the interface of some parasites (Coetzee and Fineran 1987). Pate et al. (1990) demonstrated that only 1 % of the interface cells of *Olax phyllanthi* are xylem-to-xylem connections, while the other interface cells face host parenchyma cells rather than conductive cells. In *Triphysaria pusilla* (Orobanchaceae), many haustoria have no xylem bridge at all or a bridge with incomplete xylem strands, but all haustoria have a well-established intrusive organ and hence a considerable interface area for apoplastic nutrient translocation (Heide-Jørgensen and Kuijt 1995) (see Chap. 6).

The most advanced haustoria from both an anatomical and physiological points of view are those containing phloem with sieve elements, occurring close to or in connection with host sieve elements. This has been demonstrated with some *Cuscuta* and *Orobanchaceae* species (see Sect. 3.9.3). The presence of phloem with fully differentiated sieve tubes so close to host sieve tubes may explain *Cuscuta* being one of the fastest growing parasites. Phloem is known from haustoria in other taxa as well, including stem parasites of Loranthaceae (Calvin 1967), but it never comes as close to host phloem as in the examples above.

1.4 Dispersal and Germination Strategies

Five different strategies for seed dispersal are recognized in parasitic plants and relate to their parasitic syndrome (Kuijt 1969; Heide-Jørgensen 2008).

- The seeds are sticky and in most cases dispersed by birds eating the fruits. By wiping the sticky seeds off the beak or by defecation, the seeds are often placed directly on a branch of a suitable host (Kuijt 1969). Rodents and marsupials may also participate in such seed dispersal (Amico and Aizen 2000). These seeds are relatively large with enough nutrients to produce a large terminal haustorium, while photosynthesis in the endosperm may provide additional nutrients until a vascular connection is established with the host (Kuijt 1969). This strategy is common in stem parasites in Santalaceae *s.l.*, Loranthaceae, and Viscaceae. Dispersal of mistletoes by birds is thus strongly correlated with the behaviour of the birds, which prefer free-standing trees, hedges, and wood edges but avoid the interior of woods.
- Fruits of *Misodendrum* (Misodendraceae) are dispersed by wind and by hygroscopic movements. Long hairy setae (Fig. 1.3b) secure some fruits to the host (Hooker 1847).
- *Arceuthobium* (Viscaceae) has explosive fruits and some seeds may land on suitable hosts 20 m away (Hinds et al. 1963). Explosive fruits are also known from a few species of *Korthalsella* (Santalaceae) (Sahni 1933).
- The parasite seeds contain enough nutrients to sustain the seedling for some weeks. For example, *Cuscuta gronovii* may live for 7 weeks on seed reserves and reach a length of 35 cm before parasitizing a host (Spisar 1910). The young *Cuscuta* stem is guided by volatile oils as shown for *Cuscuta pentagona* (Runyon et al. 2006). This strategy also applies to *Cassytha*, to root parasitic members of Santalales, and to several hemiparasitic Orobanchaceae.
- The seeds are small and numerous with very little nutrient reserves, as in most holoparasitic root parasites. In some holoparasitic Orobanchaceae, seed output per plant is often in the range of 10,000–1,000,000 (Molau 1995). The seeds germinate only when receiving a chemical signal from an adjacent host root (see Chap. 8).

1.5 Host Range

Most parasitic plants, hemiparasitic root parasites in particular, have a wide host range (Kuijt 1979; Gibson and Watkinson 1989; Nilsson and Svensson 1997). Some parasites, like hemiparasitic Orobanchaceae, may attach to several hosts simultaneously. This may provide an ecological advantage since different hosts supply the parasite with different types and amounts of nutrients as shown for *Odontites verna* (Govier et al. 1967). Some stem parasites also have many hosts, such as *Dendrophthoe falcata* (Loranthaceae) that is known from about 400 different host species (Narasimha and Rabindranath 1964). As noted by Kuijt (1979), the relatively few examples of narrow host range (high host specificity) are found among parasites having a terminal haustorium only. *Viscum cruciatum* is known mainly from *Olea europaea*, while the closely related *V. album* is known from more than hundred genera. **Epiparasitism**, which occurs when one parasitic species

Fig. 1.3 Parasites of various parasitic plant families. (**a**) *Psittacanthus calyculatus* (Loranthaceae) with large showy bird-pollinated flowers; Yucatan, Mexico. (**b**) Female inflorescences of *Misodendrum* cf *oblongifolium* (Misodendraceae) on *Nothofagus antarctica* with persistent staminodes used for wind dispersal; Alumine, Patagonia. (**c**) *Comandra umbellata*, a perennial root parasite of the Santalaceae; Albany Pine Bush, New York. (**d**) The tree *Pseudotsuga menziesii* heavily damaged by *Arceuthobium douglasii* (Viscaceae), Oregon, USA; Inset: fruiting female *Arceuthobium* exophytes between needles of the host tree (photos: part **a** by H Adsersen; part **b** by V Thomsen; part **c** by C Gracie; part **d** by HS Heide-Jørgensen)

parasitizes another parasite species, is most common among members of Santalales (Kuijt 1969; Calvin and Wilson 2009). **Self-parasitism**, occurring when haustoria form between different parts of the same plant, is also known (Fig. 1.2f), mainly in *Cuscuta*, *Cassytha*, and the Orobanchaceae (see Sects. 3.4.2 and 3.5).

1.6 Geographical Distribution

Parasitic plants occur in all climatic zones from northern Greenland to Tierra del Fuego and on all continents except Antarctica. Some plants are resistant to certain parasites (see Chap. 7). However, if a parasite is not found on certain plant species, they are not necessarily resistant and the species may still be an acceptable host. The absence of the parasite may also have ecological causes such as the lack of a suitable dispersal agent (e.g. birds), or the light conditions may be insufficient for the parasite. Some plant groups such as ferns, water plants, and orchids are rarely or never parasitized. Genetics and tissue incompatibility determine the maximum number of acceptable hosts, but in practice, host range is mainly influenced by geographical (host distribution) and ecological (dispersal biology and environmental factors) relationships.

1.7 The Parasitic Plant Families (Fig. 1.1a)

Families recently revised based on molecular studies are treated as *sensu lato* (s.l.) (Nickrent 2010). The families are arranged in about 12 orders, indicating that parasitism has evolved independently several times. All families are illustrated including distribution maps in Heide-Jørgensen (2008) and Nickrent (2010).

Santalales

This plant order comprises at least the following eight families:

Olacaceae s.l. is a tropical–subtropical family of root parasitic shrubs, trees, or lianas. A terminal haustorium has not been observed, and the family is considered the most primitive in Santalales. *Olax* is the largest genus and its lateral haustoria may serve as a model for haustoria in the order. For characteristic haustorial features such as the mantle, collapsed zones, interrupted zones, and graniferous tracheary elements, see Fineran (1985, 1991) and Fineran et al. (1987). According to Nickrent (2010) Erythropalaceae, Strombosiaceae, Coulaceae, Octoknemaceae, Ximeniaceae, and Aptandraceae, which are related to the Olacaceae, are independent families. The first three families are assumed to be non-parasitic.

Schoepfiaceae with the single genus *Schoepfia* was earlier included in Olacaceae (Werth et al. 1979; Nickrent and Malécot 2001). Life form, parasitic mode, and distribution are similar to Olacaceae, but the family is not represented in Africa.

Opiliaceae is a small pan-tropical family of root parasitic evergreen trees and lianas (Hiepko 1979, 1982). Lateral haustoria resembling those of Olacaceae are the only type known.

Loranthaceae is the largest family in Santalales with close to 1,000 species of hemiparasitic stem parasites and three root parasites mainly from tropical and subtropical regions. All species are shrubs except the best-known root parasite, the Australian *Nuytsia floribunda*, which is a tree. The holdfast of its numerous

Fig. 1.4 Various holoparasites. (**a**) Inflorescence of *Cynomorium coccineum* (Cynomoriaceae); Algarve, Portugal. (**b**) Exophyte of Apodanthes caseariae (Apodanthaceae) consisting of female flowers with dark stigma on top of fruit; Costa Rica. (**c**) *Cytinus hypocistis* (Cytinaceae) pollinated by a bee (to the left); Southern France. (**d**) *Rafflesia keithii* flower with diameter up to 80 cm (Rafflesiaceae); Sabah, Borneo. (**e**) Flowering *Hydnora johannis* (Hydnoraceae); reproductive parts of the flower are subterranean; South Africa. (**f**) Rhizomes (earlier called pilot roots) of *Hydnora triceps* with short haustorial roots and flower buds in four rows; East of Port Nolloth, South Africa. (**g**) Female inflorescence of *Balanophora latisepala* (Balanophoraceae); Rongla National Park, Thailand (photos: part **a** by FN Rasmussen; part **b** by P Maas; part **c** by HS Heide-Jørgensen; part **d** by P Ø Larsen; parts **e**, **f** by LJ Musselman; part **g** by T Læssøe)

endophyte, and unusual flower construction. In the Mediterranean *Cynomorium* flowers on the succulent axis are so reduced (Fig. 1.4a) that the plants were erroneously considered fungi. There is a perennial rhizome, and lateral haustoria develop from adventitious roots. A terminal haustorium is expected but has not been described (Lanfranco 1960; Kuijt 1969).

Lennoaceae (Boraginales) is mainly Central American and interesting by showing root dimorphism: pilot roots search for host roots and when found they develop short haustorial roots connecting to the host. The species are perennial and succulent. A terminal haustorium is expected (Kuijt 1966, 1969).

Mitrastemonaceae (Ericales) contains South-East Asian root parasites that were previously included in the Rafflesiaceae. The main part of the perennial parasite is embedded in host roots, and only the flowers emerge out of the roots and can be seen above soil. After pollen release, the 16 stamens form a mitre-shaped tube, which is pushed off by the growing pistil and prevents self-pollination (Kuijt 1969; Meijer and Veldkamp 1993).

Apodanthaceae (Cucurbitales, formerly regarded as Rafflesiales or Malvales) represents some of the smallest stem parasites. The exophyte consists only of the flowers, which are 2–3 mm wide (Fig. 1.4b). The three perennial closely related genera have a highly disjunctive distribution on five continents assumed to originate from the disintegration of Gondwanaland (Kuijt et al. 1985; Blarer et al. 2004; Filipowicz and Renner 2010).

Cytinaceae (Malvales) are perennial root parasites in two genera occurring in Central America, Europe, and South Africa. The endophyte is composed of rows of parenchyma cells that grow through host pericyclic derivatives and reach both phloem and xylem (De Vega et al. 2007). The exophyte (Fig. 1.4c) consists only of the inflorescence of male and female flowers (Kuijt 1969; Nickrent 2007).

Rafflesiaceae (Malpighiales) is Southeast Asian. They are perennial and mainly root parasites (Fig. 1.4d), but a few species of *Rafflesia* may occasionally occur as stem parasites (Heide-Jørgensen 2008). All vegetative parts are embedded in the host, and flower capacity is transferred to the endophyte (Kuijt 1969). *Rafflesia arnoldii* produces the largest flower in the plant kingdom with a diameter of almost 1 m (Meijer 1984; Bänziger 1991; Wurdack and Davis 2009).

Hydnoraceae (Piperales) is African and South American. The perennial plant body consists of highly modified succulent rhizomes producing short exogenous outgrowths having the potential to develop lateral haustoria or new branches (Figs. 1.2d and 1.4f). The endophyte contains both xylem and phloem (Tennakoon et al. 2007). Flowers (Fig. 1.4e) and fruits are partly or completely subterranean (Musselman and Visser 1989; Tennakoon et al. 2007).

Balanophoraceae (Balanophorales but Santalales has been suggested) occurs throughout the more humid tropical–subtropical regions. It includes 17 genera of root parasites (Hansen 1972; Nickrent and Franchina 1990), which produce only a terminal (primary) haustorium, which develops into a tuber (occasionally up to 60 cm) that partly consists of host tissue. The inflorescence arises from the tuber, and specialized conductive cells connect the vascular system of the inflorescence with the endophyte (Gedalovich-Shedletzky and Kuijt 1990; Hsiao et al. 1995). The flowers are highly reduced and several species were earlier considered to be fungi (Fig. 1.4g).

1.8 Parasite Look-Alikes

Some plants, like epiphytic orchids and bromeliads, may look as if they are parasites (Heide-Jørgensen 2008). But these plants only attach to other plants for support and do not acquire any water or nutrients from their supporting plant. Other flowering plants, the myco-heterotrophs, lost all or nearly all chlorophyll and are involved in a three-part relationship with a mycotrophic fungus that indirectly connects them to an autotrophic vascular plant from which they obtain nutrients (Leake 1994, 2004; Imhof 2010).

References

Amico GC, Aizen MA (2000) Mistletoe seed dispersal by a marsupial. Nature 408:929–930

Bänziger H (1991) Stench and fragrance: unique pollination lure of Thailand's largest flower, *Rafflesia kerrii* Meijer. Nat Hist Bull Siam Soc 39:19–52

Bennett JR, Mathews S (2006) Phylogeny of the parasitic plant family Orobanchaceae inferred from phytochrome A. Am J Bot 93:1039–1051

Beyer C, Forstreuter W, Weber HC (1989) Anatomical studies of haustorium ontogeny and the remarkable mode of penetration of the haustorium in *Nuytsia floribunda* (Labill.) R. Br. Bot Acta 102:229–235

Bhandari NN, Mukerji KG (1993) The haustorium. Wiley, New York

Blarer A, Nickrent DL, Endress PK (2004) Comparative floral structure and systematics in Apodanthaceae (Rafflesiales). Plant Syst Evol 245:119–142

Calder M, Bernhardt P (eds) (1983) The biology of mistletoes. Academic, Sydney

Calladine A, Pate JS (2000) Haustorial structure and functioning of the root hemiparasitic tree *Nuytsia floribunda* (Labill.) R. Br. and water relationships with its hosts. Ann Bot 85:723–731

Calvin CL (1967) Anatomy of the endophytic system of the mistletoe *Phoradendron flavescens*. Bot Gaz 128:117–137

Calvin CL, Wilson CA (2006) Comparative morphology of epicortical roots in Old and New World Loranthaceae with reference to root types, origin, patterns of longitudinal extension and potential for clonal growth. Flora 201:51–64

Calvin CL, Wilson CA (2009) Epiparasitism in *Phoradendron durangense* and *P. falcatum* (Viscaceae). Aliso 27:1–12

Carlón L, Gómez Casares G, Laínz M, Moreno Moral G, Sánchez Pedraja Ó, Schneeweiss GM (2008) Más, a propósito de algunas *Phelipanche* Pomel, *Boulardia* F. W. Schultz y *Orobanche* L. (Orobanchaceae) del oeste del Paleártico. Documentos Jard Bot Atlántico (Gijón) 6:1–128

Coetzee JF, Fineran BA (1987) The apoplastic continuum, nutrient absorption and plasmatubules in the dwarf mistletoe *Korthalsella lindsayi* (Viscaceae). Protoplasma 136:145–153

Coleman E (1934) Notes on *Exocarpus*. Victorian Nat 51:132–139

De Candolle AP (1813) Théorie élémentaire de la botanique. Déterville, Paris

De Vega C, Ortiz PL, Arista M, Talavera S (2007) The endophytic system of mediterranean *Cytinus* (Cytinaceae) developing on five host Cistaceae species. Ann Bot 100:1209–1217

dePamphilis CW, Palmer JD (1990) Loss of photosynthetic and chlororespiratory genes from the plastid genome of a parasitic flowering plant. Nature 348:337–339

Der JP, Nickrent DL (2008) A molecular phylogeny of Santalaceae (Santalales). Syst Bot 33:107–116

Dobbins DR, Kuijt J (1973) Studies of the haustorium of *Castilleja* (Scrophulariaceae). II. The endophyte. Can J Bot 51:923–931

Dobbins DR, Kuijt J (1974) Anatomy and fine structure of the mistletoe haustorium (*Phthirusa pyrifolia*). I. Development of the young haustorium. Am J Bot 61:535–543
Dörr I (1972) Der Anschluss der *Cuscuta*-hyphen an die Siebröhren ihrer Wirtspflanzen. Protoplasma 75:167–184
Dörr I (1997) How *Striga* parasitizes its host: a TEM and SEM study. Ann Bot 79:463–472
Dörr I, Kollmann R (1995) Symplastic sieve element continuity between *Orobanche* and host. Bot Acta 108:47–55
Feild TS, Brodribb TJ (2005) A unique mode of parasitism in the conifer coral tree *Parasitaxus ustus* (Podocarpaceae). Plant Cell Environ 28:1316–1325
Filipowicz N, Renner SS (2010) The worldwide holoparasitic Apodanthaceae confidently placed in the Cucurbitales by nuclear and mitochondrial gene trees. BMC Evol Biol 10:219
Fineran BA (1963a) Studies of the root parasitism of *Exocarpus bidwillii* Hook f. II. External morphology, distribution and arrangement of haustoria. Phytomorphology 13:30–41
Fineran BA (1963b) Parasitism in Exocarpus bidwillii Hook. f. Trans R Soc N Z Bot 2:109–119
Fineran BA (1985) Graniferous tracheary elements in haustoria of root parasites. Bot Rev 51:389–441
Fineran BA (1991) Root hemiparasitism in the Santalales. Bot Jahrb Syst 113:277–308
Fineran BA, Calvin CL (2000) Transfer cells and flange cells in sinkers of the mistletoe *Phoradendron macrophyllum* (Viscaceae), and their novel combination. Protoplasma 211:76–93
Fineran BA, Hocking PJ (1983) Features of parasitism, morphology and haustorial anatomy in Loranthaceous root parasites. In: Calder M, Bernhardt P (eds) The biology of mistletoes. Academic, Sidney, pp 205–227
Fineran BA, Ingerfeld M, Patterson WD (1987) Inclusions of graniferous tracheary elements in the root hemiparasite *Olax phyllanthi* (Olacaceae). Protoplasma 136:16–28
Gedalovich-Shedletzky E, Kuijt J (1990) An ultrastructural study of the tuber strands of *Balanophora* (Balanophoraceae). Can J Bot 68:1271–1279
Geils B, Tovar JC, Moody B (eds) (2002) Mistletoes of North American conifers. UT, Ogden
Gibson CC, Watkinson AR (1989) The host range and selectivity of a parasitic plant: *Rhinanthus minor* L. Oecologia 78:401–406
Govier RN, Nelson MD, Pate JS (1967) Hemiparasitic nutrition in Angiosperms. I. The transfer of organic compounds from host to *Odontites verna* (Bell.) Dum. (Scrophulariaceae). New Phytol 66:285–297
Hansen B (1972) The genus *Balanophora* J. R. & G. Forster. A taxonomic monograph. Dansk Bot Ark 28:1–188
Heide-Jørgensen HS (1989) Development and ultrastructure of the haustorium of *Viscum minimum* Harvey. I. The adhesive disk. Can J Bot 67:1161–1173
Heide-Jørgensen HS (1991) Anatomy and ultrastructure of the haustorium of *Cassytha pubescens* R. Br. – I. The adhesive disk. Bot Gaz 152:321–334
Heide-Jørgensen HS (2008) Parasitic flowering plants. Brill, Leiden
Heide-Jørgensen HS (2011) Parasitic plants. In: Simberloff D, Reymánek M (eds) Encyclopedia of biological invasions. University of California Press, Berkeley, pp 504–510
Heide-Jørgensen HS, Kuijt J (1993) Epidermal derivatives as xylem elements and transfer cells: a study of the host-parasite interface in two species of *Triphysaria* (Scrophulariaceae). Protoplasma 174:173–183
Heide-Jørgensen HS, Kuijt J (1995) The haustorium of the root parasite *Triphysaria* (Scrophulariaceae), with special reference to xylem bridge ultrastructure. Am J Bot 82:782–797
Hibberd JM, Bungard RA, Press MC, Jeschke WD, Scholes JD, Quick WP (1998) Localization of photosynthetic metabolism in the parasitic angiosperm *Cuscuta reflexa*. Planta 205:506–513
Hiepko P (1979) A revision of Opiliaceae I. Genera of the eastern Old World, excluding *Opilia*. Wildenowia 9:13–56
Hiepko P (1982) A revision of Opiliaceae II. *Opilia* Roxb. Wildenowia 12:161–182

Hinds TE, Hawksworth FH, McGinnies WJ (1963) Seed discharge in *Arceuthobium*: a photographic study. Science 140:1236–1238
Hooker JD (1847) The botany of the Antarctic voyage of H. M. Discovery ships Erebus and Terror in the years 1839–1843. I. Flora Antarctica. Part II, XXII. Loranthaceae, pp 289–302
Hsiao S-C, Mauseth JD, Peng C-I (1995) Composite bundles, the host/parasite interface in the holoparasitic angiosperms *Langsdorffia* and *Balanophora*. Am J Bot 82:81–91
Imhof S (2010) Are monocots particularly suited to develop mycoheterotrophy? In: Seberg O, Petersen G, Barfod AS, Davis JI (eds) Diversity, phylogeny, and evolution in the monocotyledons. Aarhus University Press, Denmark
Joel DM, Hershenhorn J, Eizenberg H, Aly R, Ejeta G, Rich P, Ransom J, Sauerborn J, Rubiales D (2007) Biology and management of weedy root parasites. Hortic Rev 33:207–349
Kirkup D (1998) Pollination mechanisms in African Loranthaceae. In: Polhill R, Wiens D (eds) Mistletoes of Africa. The Royal Botanic Gardens, Kew, pp 37–60
Kuijt J (1966) Parasitism in *Pholisma* (Lennoaceae). I. External morphology of subterranean organs. Am J Bot 53:82–86
Kuijt J (1969) The biology of parasitic flowering plants. University of California Press, Berkeley
Kuijt J (1977) Haustoria of phanerogamic parasites. Ann Rev Phytopathol 17:91–118
Kuijt J (1979) Host selection by parasitic Angiosperms. Symb Bot Upsal 22:194–199
Kuijt J (1988) Monograph of the Eremolepidaceae. Syst Bot Monogr 18:1–60
Kuijt J (2003) Monograph of *Phoradendron* (Viscaceae). Syst Bot Monogr 66:1–643
Kuijt J (2009) Monograph of *Psittacanthus* (Loranthaceae). Syst Bot Monogr 86:1–361
Kuijt J, Weber HC, Visser JH (1978) Morphological observations on leaf haustoria and related organs of the South African genus *Hyobanche* (Scrophulariaceae). Can J Bot 56:2981–2986
Kuijt J, Bray D, Olson AR (1985) Anatomy and ultrastructure of the endophytic system of *Pilostyles thurberi* (Rafflesiaceae). Can J Bot 63:1231–1240
Ladley JJ, Kelly D (1995) Explosive New Zealand mistletoe. Nature 378:766
Lam HJ (1945) Fragmenta Papuana. Sargentia 5:1–196
Lanfranco GG (1960) Cynomorium coccineum Linn., A Maltese historical plant. J Malta Hist Soc 3:53–70
Leake JR (1994) The biology of myco-heterotrophic ('saprophytic') plants. New Phytol 127:171–216
Leake JR (2004) Myco-heterotroph/epiparasitic plant interactions with ectomycorrhizal and arbuscular mycorrhizal fungi. Curr Opin Plant Biol 7:422–428
Lye D (2006) Charting the isophasic endophyte of dwarf mistletoe *Arceuthobium douglasii* (Viscaceae) in host apical buts. Ann Bot 97:953–963
Malécot V, Nickrent DL (2008) Molecular phylogenetic relationships of Olacaceae and related Santalales. Syst Bot 33:97–106
Mathiasen RL, Nickrent DL, Shaw DC, Watson DM (2008) Mistletoes: pathology, systematics, ecology, and management. Plant Dis 92:988–1006
Meijer W (1984) New species of *Rafflesia* (Rafflesiaceae). Blumea 30:209–215
Meijer W, Veldkamp JF (1993) A revision of *Mitrastema* (Rafflesiaceae). Blumea 38:221–229
Molau U (1995) Reproductive ecology and biology. In: Press MC, Graves JD (eds) Parasitic plants. Chapman & Hall, London, pp 141–176
Musselman LJ, Visser JH (1989) Taxonomy and natural history of *Hydnora* (Hydnoraceae). Aliso 12:317–321
Narasimha VL, Rabindranath V (1964) A further contribution to the host range of *Dendrophthoe falcata* (L.f.) Ettingsh. Bull Bot Surv India 6:103
Nickrent DL (2007) Cytinaceae are sister to Muntingiaceae (Malvales). Taxon 56:1129–1135
Nickrent DL (2008) Phylogenetic relationships of parasitic flowering plants. Retrieved 20 Aug 2012 from http://www.science.siu.edu/parasitic-plants/index.html
Nickrent DL (2010) The parasitic plant connection. Retrieved 20 Oct 2010 from http://www.science.siu.edu/parasitic-plants/index.html
Nickrent DL, Franchina CR (1990) Phylogenetic relationships of the Santalales and relatives. J Mol Evol 31:294–301

Nickrent DL, Malécot V (2001) A molecular phylogeny of Santalales. In: Fer A, Thalouarn P, Joel DM, Musselman LJ, Parker C, Verkleij JAC (eds) Proceedings of the 7th international parasitic weed symposium. Faculté des Sciences, Université de Nantes, Nantes, France, pp 69–74

Nickrent DL, Malécot V, Vidal-Russell R, Der JP (2010) A revised classification of Santalales. Taxon 59:538–558

Nilsson CH, Svensson BM (1997) Host affiliation in two subarctic hemiparasitic plants: *Bartsia alpina* and *Pedicularis lapponica*. Ecoscience 4:80–85

Olmstead RG, dePamphilis CW, Wolfe AD, Young ND, Elisons WJ, Reeves PA (2001) Disintegration of the Scrophularaceae. Am J Bot 88:348–361

Parker C, Riches CR (1993) Parasitic weeds of the world: biology and control. CAB, Wallingford

Pate JS, Kuo J, Davidson NJ (1990) Morphology and anatomy of the root hemiparasite *Olax phyllanthi* (Olacaceae), with special reference to the haustorial interface. Ann Bot 65:425–436

Polhill R, Wiens D (1998) Mistletoes of Africa. The Royal Botanic Gardens, Kew

Press MC, Graves JD (eds) (1995) Parasitic plants. Chapman and Hall, London

Revill MJW, Stanley S, Hibberd JM (2005) Plastid genome structure and loss of photosynthetic ability in the parasitic genus *Cuscuta*. J Exp Bot 56:2477–2486

Runyon JB, Mescher MC, De Moraes CM (2006) Volatile chemical cues guide host location and host selection by parasitic plants. Science 313:1964–1967

Sahni B (1933) Explosive fruits in *Viscum japonicum*, Thunb. J Indian Bot Soc 12:96–101

Satovic Z, Joel DM, Rubiales D, Cubero JI, Román B (2009) Population genetics in weedy species of *Orobanche*. Australas Plant Pathol 38:228–234

Simpson BB (1989) Krameriaceae. Flora Neotropica Monograph 49:1–109

Spisar K (1910) Beiträge zur Physiologie der *Cuscuta gronovii*. Ber Deutsch Bot Ges 28:329–334

Tennakoon KU, Cameron DD (2006) The anatomy of *Santalum album* (Sandalwood) haustoria. Can J Bot 84:1608–1616

Tennakoon KU, Bolin JF, Musselman LJ, Maass E (2007) Structural attributes of the hypogeous holoparasite *Hydnora triceps* Drège & Meyer (Hydnoraceae). Am J Bot 94:1439–1449

Teryokhin ES, Shibakina GV, Serafimovitch NB, Kravtsova TI (1993) Determination of broomrapes in the USSR flora (in Russian). Nauka, Sankt-Peterburg

Thomson LAJ (2006) *Santalum austrocaledonicum* and *S. yasi* (sandalwood ver. 2.1. In: Elevitch CR (ed) Species profiles for Pacific Island agroforestry. Permanent Agriculture Resources (PAR), Holualoa, pp 1–21 (http://www.traditionaltree.org)

Thorne RF (2002) How many species of seed plants are there? Taxon 51:511–512

Toth R, Kuijt J (1976) Anatomy and ultrastructure of the young haustorial gland in *Comandra* (Santalaceae). Can J Bot 54:2315–2327

Toth R, Kuijt J (1977) Anatomy and ultrastructure of the haustorium in *Comandra* (Santalaceae). Can J Bot 55:455–469

Trattinick, In: Schlechtendal DFL (1828) Nachtrag zu der *Ichthyosoma wehdemanni*. Linnaea 3:194–198

Van der Kooij TAW, Krause K, Dörr I (2000) Molecular, functional and ultrastructural characterisation of plastids from six species of the parasitic flowering plant genus *Cuscuta*. Planta 210:701–707

Veenendaal EM, Abebrese IK, Walsh MF, Swaine MD (1996) Root hemiparasitism in a West African rainforest tree *Okoubaka aubrevillei* (Santalaceae). New Phytol 134:487–493

Vidal-Russell R, Nickrent DN (2007) A molecular phylogeny of the Feathery Mistletoe *Misodendrum*. Syst Bot 32:560–568

von Tubeuf K (1923) Monographie der Mistel. R Oldenbourg, München

Watson DM (2001) Mistletoe – a keystone resource in forests and woodlands worldwide. Ann Rev Ecol Syst 32:219–249

Weber HC (1976) Über Wirtspflanzen und Parasitismus einiger mitteleuropäischer Rhinanthoideae (Scrophulariaceae). Plant Syst Evol 125:97–107

Weber HC (1980) Untersuchungen an australischen und neuseeländischen Loranthaceae/ Viscaceae. 1. Zur Morphologie und Anatomie der unterirdischen Organe von *Nuytsia floribunda* (Labill.) R. Br. Beitr Biol Pflanzen 55:77–99

Weber JZ (1981) A taxonomic revision of *Cassytha* (Lauraceae) in Australia. J Adelaide Bot Gard 3:187–262
Weber HC (1993) Parasitismus von Blütenpflanzen. Wissenschaftliche Buchgesselschaft, Darmstadt
Werth CR, Baird WV, Musselman LJ (1979) Root parasitism in *Schoepfia schreberi*. (Olacaceae). Biotropica 11:140–143
Wolswinkel P (1974) Complete inhibition of setting and growth of fruits of *Vicia faba* L., resulting from the draining of the phloem system by *Cuscuta* species. Acta Bot Neerl 23:48–60
Wurdack KJ, Davis CC (2009) Malpighiales phylogenetics: gaining ground on one of the most recalcitrant clades in the angiosperm tree of life. Am J Bot 96:1551–1570
Young ND, Steiner KE, dePamphilis CW (1999) The evolution of parasitism in Scrophulariaceae/Orobanchaceae: plastid gene sequences refute an evolutionary transition series. Ann Mo Bot Gard 86:876–893
Yuncker TG (1932) The genus *Cuscuta*. Mem Torrey Bot Club 18:109–331

Part I
The Orobanchaceae and Their Parasitic Mechanisms

Chapter 2
The Haustorium and the Life Cycles of Parasitic Orobanchaceae

Daniel M. Joel

2.1 How Do We Define the Haustorium in the Orobanchaceae?

The haustorium is the special organ of all parasitic plants, which connects them to a living host plant and provides them with the ability to extract water and nutrient from their hosts. The term haustorium was originally defined by de Candolle (1813), who took it from the Latin word *haurire* meaning "to draw up" or "to drink," which points to an active role. While Kuijt (1969) suggested that the haustorium serves as the *bridge* between host and parasite, Fahn's opinion (1982) that the haustorium is a "specialized plant organ that *draws* nutrients from other organs or tissues" resembles the original meaning of the term haustorium, assuming that this organ is an *active* organ and not a *passive* bridge. Interestingly, most definitions of parasite haustoria do not mention their invasive nature, which should be included in the definition of haustoria, as put forward by Riopel and Timko (1995), who used this term for all developmental stages of this infective structure, from initiation through invasion and until the establishment of full vascular connections.

In this book, the haustorium is defined as the special organ of parasitic plants, which invades host tissues and serves as the structural and physiological bridge that allows the parasites to withdraw water and nutrients from the conductive systems of living host plants.

D.M. Joel (✉)
Department of Weed Research, Agricultural Research Organization, Newe-Ya'ar Research Center, Ramat Yishay 30095, Israel
e-mail: dmjoel@agri.gov.il

D.M. Joel et al. (eds.), *Parasitic Orobanchaceae*,
DOI 10.1007/978-3-642-38146-1_2, © Springer-Verlag Berlin Heidelberg 2013

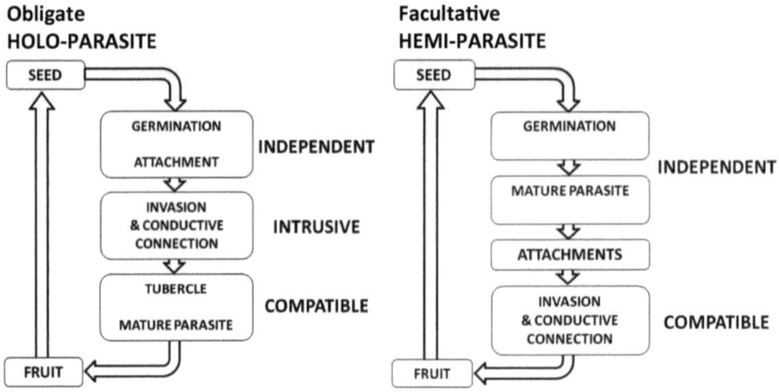

Fig. 2.1 Life cycles of obligate holoparasitic and facultative hemiparasitic Orobanchaceae

2.2 Life Cycles of Facultative and Obligate Orobanchaceae

The life cycle of the majority of the holoparasitic Orobanchaceae includes several key developmental phases (Fig. 2.1). The seedling develops independently for a short while up to the stage when it attaches to a host. This is the independent phase of the parasite development (see Chap. 9). Then comes the intrusive developmental phase, which includes (a) development of a terminal haustorium at the tip of the radicle, (b) invasion of the haustorium into host tissues (see Chap. 5), and (c) development of primary conductive connections with the host (see Sect. 3.9). Finally is the compatible phase in which the parasite development is coordinated with that of the host (see Chap. 6). The development of the haustorium depends on its ability to overcome host resistance mechanisms (see Chap. 7) and to compete with host organs on available host resources (see Chap. 6). The holoparasites can then develop roots carrying lateral haustoria (see Sect. 3.3).

The facultative hemiparasitic Orobanchaceae develop only lateral haustoria (see Sect. 3.3 and Chap. 4) after they have already established autotrophically (Fig. 2.1). Therefore, parasite–host relations in these plants differ from those of holoparasites, and the lateral haustoria face a different set of developmental and physiological challenges. However, the obligate hemiparasites, e.g., *Striga* and *Alectra*, resemble holoparasites in their short independent phase and in the development of a terminal haustorium; they also develop lateral haustoria and are autotrophic in later stages of their compatible phase.

Parasitism in plants cannot be understood until the architecture and mode of operation of the haustorium are known (Kuijt 1969). The next chapters are accordingly dedicated to the structural and developmental aspects of haustoria in the Orobanchaceae. Chapter 3 focuses on the diversity and anatomical structure of mature haustoria, with reference to the possible function of each haustorial tissue; Chap. 4 describes the stimulation and initiation of haustoria, and Chap. 5 deals with the invasion of the developing haustorium into host tissues.

References

De Candolle AP (1813) Théorie élémentaire de la botanique. Déterville, Paris
Fahn A (1982) Plant anatomy, 3rd edn. Pergamon, Oxford
Kuijt J (1969) The biology of parasitic flowering plants. University of California Press, Berkeley
Riopel JL, Timko MP (1995) Haustorial initiation and differentiation. In: Press MC, Graves JD (eds) Parasitic plants. Chapman & Hall, London, pp 39–79

Chapter 3
Functional Structure of the Mature Haustorium

Daniel M. Joel

3.1 Introduction

The mature haustorium connects the parasitic plant to its host. As the parasite and the host are two distinct entities, usually belonging to different plant families, this connection needs to address the physiological differences between them while facilitating and possibly also regulating the movement of water, nutrients and various macromolecules between the two organisms (see Chap. 6).

Surprisingly, so far there has been no common agreement regarding the boundaries of the haustorium and the function of its various parts. This chapter focuses on the structure and diversity of mature haustoria in the Orobanchaceae. It describes the various parts of the organ that has successfully invaded a host root. The initiation of haustoria and their invasion into host tissues are separately dealt with in Chaps. 4 and 5.

The structure of the haustorium is rather complex in many parasites. It is composed of several structural regions. One part of the mature haustorium is located inside host tissues and is regarded as the endophyte, whereas another part is located between the host and the main body of the parasite and is often regarded as the exophyte, upper haustorium, haustorial bridge or haustorial neck. In addition, there is an attachment organ allowing the parasite to anchor to host root surface or even to grasp the host root before and during invasion into host tissues. The haustorium connects to its parent root at the haustorial base. In this book, the term 'haustorium' includes all these regions and is used for all developmental stages of this invasive structure, from initiation, through attachment and invasion, until the establishment of full vascular connection and full maturation.

D.M. Joel (✉)
Department of Weed Research, Agricultural Research Organization, Newe-Ya'ar Research Center, Ramat Yishay 30095, Israel
e-mail: dmjoel@agri.gov.il

D.M. Joel et al. (eds.), *Parasitic Orobanchaceae*,
DOI 10.1007/978-3-642-38146-1_3, © Springer-Verlag Berlin Heidelberg 2013

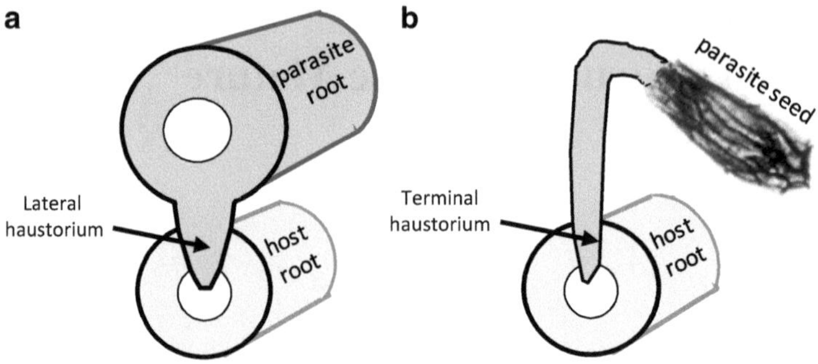

Fig. 3.1 The two types of Orobanchaceae haustoria. (**a**) Lateral haustorium, arising from a parasite root. (**b**) Terminal haustorium, arising from the radicle tip of the parasite seedling

3.2 Haustorium Diversity

Several independent transitions from hemi- to holoparasitism were identified within the Orobanchaceae by molecular phylogenetic analysis (see Chap. 14). This is reflected in the high diversity of haustoria within this family. Unlike the holoparasites, which fully depend on their hosts for all nutritional requirements, some hemiparasites may survive independently (Kuijt 1969; see Chap. 1). The haustoria of holoparasites are complex structures that facilitate massive translocation of water and minerals and in particular organic nutrients from the host. The haustorial structures in many hemiparasites are much simpler, containing xylem but not phloem, and basically mainly allow the transfer of water and mineral nutrients from the host (see Chap. 6). There are many intermediates between these two extremes. *Alectra* and *Striga* species are interesting intermediates, which develop underground as holoparasites for a while, and only then emerge and become hemiparasites that obtain at least some of their carbon requirements by photosynthesis.

3.3 Lateral and Terminal Haustoria

Orobanchaceae have two kinds of haustoria: **lateral haustoria** (also known as 'secondary haustoria'), which develop as lateral extensions of parasite roots (Fig. 3.1a), and **terminal haustoria** (also known as 'primary haustoria'), which develop at the apex of the radicle soon after germination (Fig. 3.1b).

The misleading terms 'primary haustorium' and 'secondary haustorium' (Koch 1887) should be avoided because the majority of Orobanchaceae develop only lateral haustoria. These archaic terms are also meaningless because both morphological and molecular phylogenetic analyses clearly indicated that in the Orobanchaceae the evolution of lateral haustoria preceded the evolution of terminal haustoria (Weber 1987b; Westwood et al. 2010; see Chap. 15). Sometimes haustoria may develop directly from the apical meristem of a mature parasite root (Weber 1987b); these are the **apical haustoria**.

Hemiparasitic plants develop lateral haustoria that are usually smaller than terminal haustoria and often vary in their degree of differentiation (Heide-Jørgensen and Kuijt 1995; Neumann et al. 1999; Rümer et al. 2007). The number of lateral haustoria on parasite roots depends on the ability of the parasite to sense the adjacent presence of a host (see Chap. 4). It also depends on the soil environment, and in particular on the density of neighbouring host roots (Kuijt 1977). Lateral haustoria may be erroneously thought to be terminal when the distal parts of the host root aborts and drops off (Piehl 1963).

The terminal haustorium is a characteristic feature of the holoparasitic *Orobanche* clade of the Orobanchaceae (see Chap. 16) and of a few hemiparasitic genera, including *Alectra* and *Striga* (but not *Striga angustifolia*; Krause 1990). The majority of these parasites produce numerous seeds, increasing their chance to find a host, but this comes at the expense of seed size (see Chap. 9). Their tiny seeds have only small amounts of nutrients and cannot start their life cycle without an immediate connection to a compatible host, for nutrient supply. This is achieved by the development of the terminal haustorium, which guarantees the supply of nutrients soon after germination. The ability of these parasites to produce numerous tiny seeds, which increases their dispersal potential but limits the independent survival of the seedlings, evolved following the evolution of the terminal haustorium.

Some holoparasites develop roots that may produce lateral haustoria after the development of the terminal haustorium. Both young and mature holoparasites fully rely on their immediate connection to the host, but *Striga* species and most other small-seeded hemiparasites with terminal haustoria, which develop green shoots above ground, seem to depend on a host mainly during early development. All other hemiparasitic Orobanchaceae establish a normal root system prior to attachment to a host and carry only lateral haustoria. In these plants the nutrients acquired through haustoria come in addition to the primary sources of nutrition obtained by the parasite roots and leaves. Therefore, unlike the situation with the terminal haustorium, impairment of the development of a lateral haustorium does not affect the survival of the parasite (Gurney et al. 2003). Some parasites can regenerate from lateral haustoria after the death of the terminal haustorium (Teryokhin 1997).

3.4 Morphological Features of Terminal Haustoria

3.4.1 The Tubercle

A tubercle develops following the establishment of initial vascular connection with the host by terminal haustoria of some parasites, e.g. species of *Orobanche*, *Harveya*, *Cistanche* and *Alectra*. The tubercle is a swelling in the young parasite

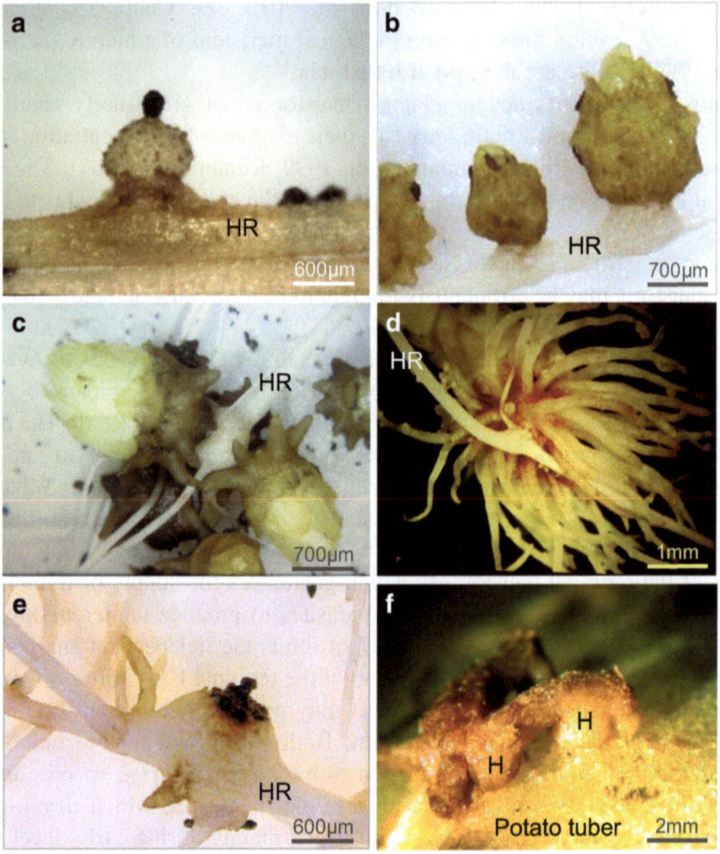

Fig. 3.2 Tubercles and adventitious roots of terminal haustoria. (**a**) Young tubercle of *Phelipanche aegyptiaca*; note the seed coat on top of the tubercle; *HR*—host root. (**b**) Slightly older tubercles of *Orobanche cumana*; note the developing apical shoot bud and the remains of the seed coat on top of the tubercles. (**c**) *O. cumana* tubercles with prominent apical shoot bud and the short adventitious roots at the base of the tubercle. (**d**) Mature *P. aegyptiaca* tubercle with numerous adventitious roots. (**e**) Tomato root thickened at the site of infestation by *O. cernua*; the parasite tubercle developed inside the thickened portion of the root. (**f**) Adventitious roots of *P. aegyptiaca* with lateral haustoria that are connected to the surface of a potato tuber; *H*—lateral haustoria

seedling just above the surface of the host root (Fig. 3.2a). It eventually develops an apical shoot meristem (or meristems) (Figs. 3.2b, c and 3.10b) and may also develop a crown of lateral, non-geotropic **adventitious roots** ('crown roots' or 'haustorial roots', but see Sect. 3.4.2) (Fig. 3.2c, d). When the seed germinates in very close contact with the host root, the tubercle may start developing within the seed coat. In rare cases the tubercle may even develop within host root cortex (Fig. 3.2e). The tubercle size may reach a few millimetres or up to few centimetres, depending on the parasite species and on the 'quality' of the host. Tubercles grow to larger sizes only in extreme cases, as in *Striga gesnerioides*, where it can reach up to 2–3 cm in diameter (Okonkwo and Nwoke 1978). A tubercle-like swelling may also develop on lateral haustoria of some parasites (see Sect. 3.12.1), but these are usually smaller than those of terminal haustoria.

The tubercle in holoparasitic Orobanchaceae accumulates starch (Fer et al. 1987; Joel and Losner-Goshen 1994; Chen et al. 2011; see Chap. 6), which is utilized during later growth stages, mainly during flower and seed development. Tubercle development depends on nutrient supply from the host (Aber et al. 1983). Its development may take a few weeks to several months in some *Orobanche* species. In *Tozzia* it lasts for several years and in *Lathraea* up to 10 years (Kuijt 1969). An important aspect of these storage organs is that their food reserves allow flowering and seed production even after the death of the host. Severing host stems or weeding root parasites do not necessarily stop seed production and do not avoid seed dispersal by the harvested flowering stems of the parasite (see Chap. 22).

Perennial holoparasites may retain functional terminal haustorium for several years (Teryokhin 1997). Cambium activity and secondary growth within the haustorium is essential for keeping it functional. In this manner the tubercles of *Boschniakia hookeri* and *Conopholis americana* develop into large tubers (Kuijt 1969; Olsen and Olsen 1981; Heide-Jørgensen 2008). Nonetheless, most hemiparasites develop additional lateral haustoria instead of adding secondary conductive tissues by cambial activity in a single haustorium.

3.4.2 Crown Roots

Adventitious roots may form a 'root crown' on the tubercle. They differ in length in the various species, 5 cm in *Phelipanche aegyptiaca* (Fig. 3.2d), 10 cm in *Orobanche crenata* or even longer in some other parasites, while in *O. cumana* they are very short (Fig. 3.2c). Root length may also differ within local populations of the same species and when parasitizing different hosts. The adventitious roots, which are often fragile, have only a poorly developed stele and rudimentary root cap and are without any root hairs.

It seems that this unique adventitious root system does not absorb water or nutrients from the soil. In some holoparasites like *Epifagus virginiana* and *O. cumana*, where the terminal haustorium is the only connection of the parasite to a host, the crown roots seem to serve only in plant stability during flowering (Kuijt 1969). However, in most other species, whenever crown roots come in close proximity with host roots, they develop functional lateral haustoria.

When the adventitious root population of a single tubercle is dense, both self-parasitism (connection of neighbouring roots of a single parasite through haustoria) and self-grafting often lead to the development of a mass of merged parasite tissue (Attawi and Weber 1980), which encloses roots of host plants to which they are connected by numerous additional lateral haustoria. In *O. grayana* the short adventitious roots thicken and develop into a complex mass of a fleshy coral-like root system (Kuijt 1969).

3.5 Roots of Hemiparasites

Unlike the tubercle-forming parasites, which develop adventitious roots on the tubercle or along underground parts of the shoot (Ba 1984), the majority of the hemiparasitic Orobanchaceae develop immediately after germination a normal root system, which resembles that of the non-parasitic plants. This root system has the ability to develop numerous lateral haustoria when they meet a host root or when a host root comes nearby (Musselman and Dickison 1975; see Chap. 4). A single *S. hermonthica* plant, for example, may develop 80 or more haustoria (Ba 1984), depending on the availability of host roots in its rhizosphere. The root system is often less developed when a host is not available (Weber 1987a).

The roots of some species develop parallel to host roots or coil around them, which allows the development of a series of consecutive haustoria along the boundary between them (Piehl 1963). It is not clear whether this coordinated root growth is thigmotropic as suggested by Musselman and Dickison (1975) or chemotropic. Self-parasitism may also occur between neighbouring parasite roots and may lead, especially in perennial species of *Aureolaria*, to the development of "a complex anastomosing mass of parasite roots" (Musselman and Dickison 1975). Self-parasitism occurs in many hemiparasitic species only when a true host root is also present nearby (Weber 1987a), possibly thanks to chemical stimuli derived from the host.

3.6 Morphological Features of Lateral Haustoria

Lateral haustoria usually develop soon after the emergence of the parasite roots, depending on the nearby availability of potential host roots (Figs. 3.3a, b; 4.1a–c; 4.3d and 5.1b). These lateral haustoria are annual in many cases, and therefore, the majority of the perennial hemiparasites develop new lateral haustoria concomitantly with the development of new roots. Old haustoria may be sealed by the development of tyloses in their vascular elements, though some persist for more than one year.

The haustorium may have various shapes (Fig. 3.3c). The simplest and most common in lateral haustoria of holoparasites is cylindrical or conical. Hemiparasite species may initially adhere to host root by haustorial hairs (see Chap. 4). In some genera the haustorium develops a mantle composed of two lobes that clasp the host organ on both sides and may encompass it in some cases, especially when the host root is young and thin, or when the host resists penetration (Piehl 1963; Kuijt 1977). A mantle may also develop by fusion of haustorial hairs (Weber 1976c). Lateral haustoria are commonly very small (ca. 0.1 mm), but in some species, they may be

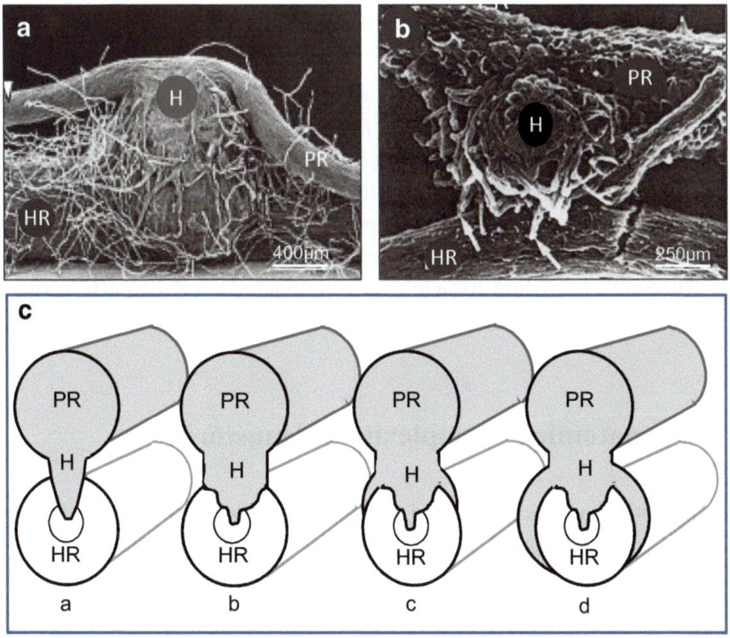

Fig. 3.3 Lateral haustoria. (**a**) Lateral haustorium of *Rhinanthus minor* on root of *Phleum* sp.; note the haustorial hairs on host surface (from Rümer et al. 2007 with permission). (**b**) Lateral haustorium of *Rhamphicarpa fistulosa*; some haustorial hairs touching the host root (from Ouédraogo et al. 1999 with permission). (**c**) Various shapes of lateral haustoria, some with lateral lobes that clasp the host root. *H*—haustorium; *PR*—parasite root; *HR*—host root

1–1.5 mm or even larger. The lateral haustoria of *Aureolaria flava* may even exceed 1 cm in extreme cases (Musselman and Dickison 1975; see Sect. 3.15 for exceptional haustoria). However, the size of haustoria often varies even on an individual parasite.

3.6.1 Lateral Haustoria Parasitizing Shoots

In spite of being regarded as 'root parasitic plants', some parasitic Orobanchaceae also connect to various other subterranean parts of host plants, such as rhizomes and tubers, in addition to their connection to host roots. In this way stems and rhizomes of plants belonging to the Cyperaceae and Poaceae are often parasitized by species of *Tozzia*, *Bartsia*, *Euphrasia* and *Odontites*, and also by *Rhinanthus*, *Melampyrum* and *Pedicularis* (Piehl 1963; Weber 1976a). Lateral haustoria of *Phelipanche aegyptiaca* can parasitize potato tubers (Joel 2007; Fig. 3.2f). The extent of stem parasitism is, however, difficult to follow because parasite roots are often very fragile and easily break when the root system is dug out of the soil. Digging up host underground organs together with the parasite is therefore essential in any field study.

3.6.2 Shoot-Borne and Leaf-Borne Haustoria

The ability of the Orobanchaceae to develop lateral haustoria is not always restricted to their roots. In *Hyobanche* spp., for example, the subterranean scale leaves develop haustoria when meeting host roots (Kuijt et al. 1978; Weber and Visser 1980; Weber 1980). Similar haustoria were also found in *Orobanche teucrii* (Weber 1980). *Agalinis linifolia* develops haustoria near the apex of its rhizomes (Musselman and Dickison 1975). *Aeginetia indica* develops lateral haustoria on its tuberous rhizomes; these haustoria allow disconnected rhizome parts to survive and flower independently (Kuijt 1969).

3.7 The Anatomical Complexity of Haustoria

Mature haustoria have been studied in only few Orobanchaceae, due to various difficulties. One major problem is that they are often connected to delicate host roots by tiny haustoria that develop on delicate or even fragile parasite roots. Another key difficulty in the study of haustoria is the complex anatomical 'mingling' of parasite cells within the host root, which divide and differentiate concomitantly with neighbouring host cells (Tate 1925). This is particularly noticeable in terminal haustoria. In addition, parasites may induce cell division or increased cambial activity in the proximal parts of the parasitized root. The study of these complex organs is difficult not only because the distinction between host and parasite cells is problematic but also because the arrangement of the tissues is highly irregular.

The vascular system is the central and most prominent element common to all functional haustoria, and its typical structure within haustoria is not commonly found in other plant organs. Early studies easily identified xylem connections in the haustorium because xylem is readily seen at low magnification (Figs. 3.4 and 3.5a), but some other individual tissues or cell types are unique to parasitic haustoria, e.g. the graniferous tracheary elements (see Sect. 3.13.1) and the hyaline tissue (see Sect. 3.12.1). However, the extent of vascular and surrounding tissues' development differs significantly among species. The simplest structures are found in lateral haustoria of certain hemiparasites, while the most complex structures develop in terminal haustoria of holoparasites (Kuijt 1969, 1977; Musselman and Dickison 1975; Weber 1976a; Musselman 1980; Sallé 1987).

The structural description of haustoria is clearer if based on its median longitudinal sections in planes that coincide either with the long axis of the parasite root, or with the long axis of host root, or with both (Fig. 3.6). Using this method the conductive connection between host and parasite can be followed, viewing the haustorium as a largely longitudinal structure. Cross sections of the haustorium along its axis are also needed to further analyse the internal structure of haustorial and relevant host tissues, but this has been described only for very few parasites (e.g. *Triphysaria* spp. by Heide-Jørgensen and Kuijt 1995 and *Lathraea clandestina* by Renaudin 1974).

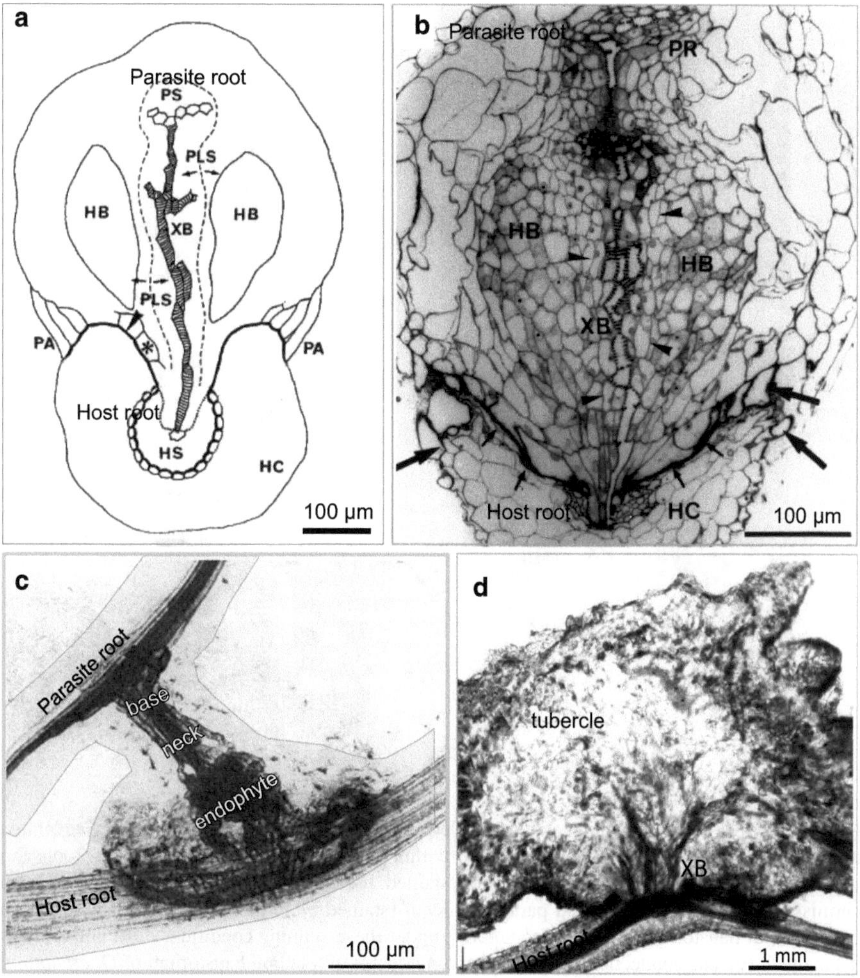

Fig. 3.4 The vascular connection. (**a**) Schematic longitudinal section of a mature lateral haustorium of *Buchnera hispida;* the parasite root and the host root are presented in cross sections. *HB*—hyaline tissue; asterisk—parasite interfacial cells; *HC*—host root cortex; *HS*—host stele; *PLS*—procambium-like strand, which produces cells towards the centre and the periphery of the haustorium (*small arrows*); *PA*—parasite papillae anchoring it to host root surface; *PS*—parasite stele; *XB*—xylem bridge (**b**) Longitudinal section through the lateral haustorium of *B. hispida*, showing the procambium-like strand (*arrowheads*), the dark staining at the lateral interface between the haustorium and the host root cortex (*thin arrows*) and the papillae that adhere to the host root surface (*thick arrows*). (**c**) Cleared haustorium of *Rhamphicarpa fistulosa*; note the differential structure of the xylem in the various parts of the haustorium. (**d**) Longitudinal section through the tubercle and the haustorium of *Phelipanche aegyptiaca* showing the xylem threads that connect the parasite to the xylem system of the host root. (**a** and **b** from Neumann et al. 1999, **c** from Neumann et al. 1998, with permission)

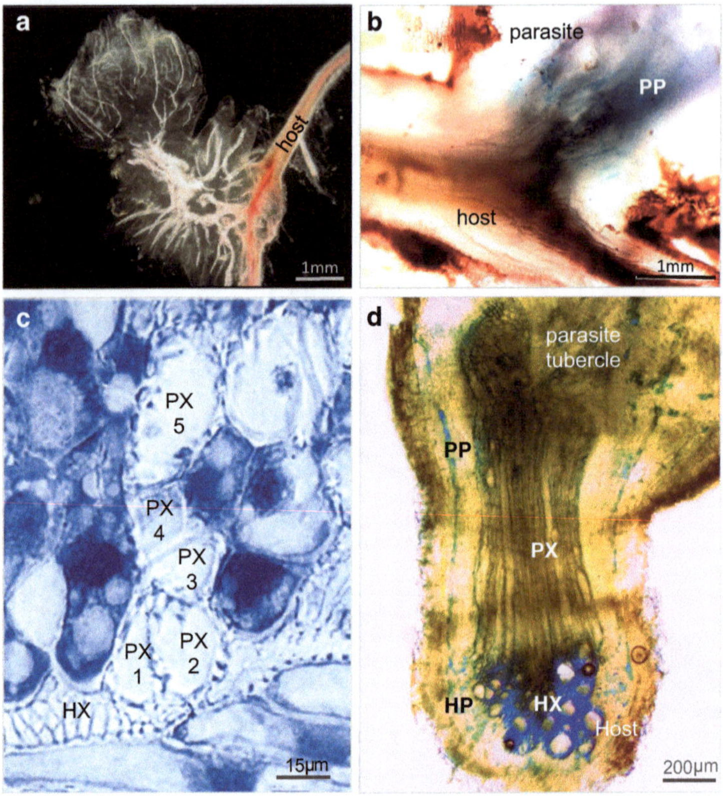

Fig. 3.5 Xylem and phloem connections. (**a**) Xylem connection between *Phelipanche aegyptiaca* tubercle and tomato root, and its arrangement within the parasite tubercle, adventitious roots and young shoot; the material was cleared and stained for lignin (from Zhou et al. 2004, with permission). (**b**) Spirally arranged parasite phloem (stained *blue*, PP) in a cleared thick section of a terminal haustorium of *Orobanche minor*; under these staining conditions, host phloem was not stained. (**c**) The connection site of the vascular centre of a lateral haustorium of *O. crenata* in the root tissues of *Vicia faba*; note the series of young tracheary elements (PX1–5) that originate from a single haustorial cell; the lower two elements and neighbouring parenchyma cells abut xylem elements of the host (from Dörr and Kollmann 1976, with permission). (**d**) Longitudinal section of a terminal haustorium of *O. cumana* on sunflower root; note the extended development of phloem (pale *blue*) in the host root, the phloem strands along the haustorium and those within the tubercle. *HX*—host xylem; *PX*—parasite xylem; *HP*—host phloem; *PP* parasite phloem

It is commonly thought that all tissues are symmetrically arranged along the haustorium, but this is not always the case. Therefore, tissues must be examined using various methods, not only in thin longitudinal and cross serial sections but particularly in cleared whole mounts and also in sections at other planes (Figs. 3.4c and 3.10c). In this way Weber (1976c) and Dobbins and Kuijt (1973b) found that xylem elements at the haustorial base of *Rhinanthus*, which were previously described as having an 'irregular' shape or 'randomly scattered', are actually xylem cells with regular shape that are arranged in a spirally oriented or undulating

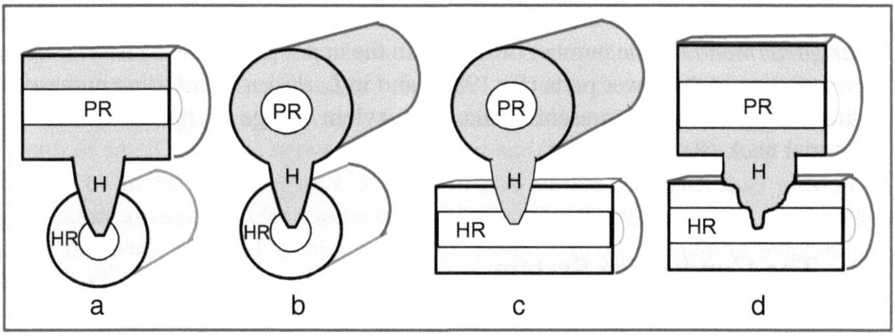

Fig. 3.6 The four section directions of lateral haustoria in relation to host and parasite roots axes. *H*—haustorium; *HR*—host root; *PR*—parasite root

strands. Similarly the phloem in the haustorium of *Phelipanche* is often arranged spirally and thus difficult to follow unless the section is thick enough to accommodate significant parts of the tissue, as demonstrated in Fig. 3.5b. Electron microscopy and confocal microscopy are essential tools that should be used for higher-resolution studies of the haustorium, in particular for the distinction between host and parasite cells and for identification of the phloem (see Sect. 3.9.3).

3.8 Tissue Organization Within the Mature Haustorium

Understanding the internal structure of mature haustoria is easier when separately considering the three main regions of lateral haustoria: the endophyte, the haustorial neck and the base of the haustorium (Fig. 3.4c; see Sect. 3.1). In terminal haustoria the two latter regions are integrated within the parasite hypocotyl, forming the tubercle (Fig. 3.4d). These regions will be dealt with below, but first we will see the structure of the conductive system, which is continuous along the haustorium.

One of the simplest haustoria is the small (0.2–0.5 mm) keel-shaped lateral haustorium of *Triphysaria* spp. (Heide-Jørgensen and Kuijt 1995). This haustorium is enclosed in parenchymatic cortex that is covered by an exodermal cell layer, both being continuous with the corresponding tissues in the parent root. The **xylem bridge**, which is composed of 1–5 xylem strands in the centre of the *Triphysaria* haustorium, is surrounded by parenchymatic cells containing dense cytoplasm.

The connection to the parent root at the base of the lateral haustorium is characterized by a mass of smaller xylem elements, oriented parallel to the long axis of the parent root. The other pole of the xylem bridge is in contact with the host xylem. This basic structural framework (Fig. 3.4a, b) is typical for all haustoria, but the extent of expression of the various tissues may vary. For example, in *Rhamphicarpa fistulosa* the number of xylem strands in the haustorial neck is

3.10 Developmental Aspects of the Vascular System

The development of the vascular system is crucial for the maturation of the haustorium and determines the eventual absorptive ability of this organ. The mode of vascular differentiation seems to be determined by hormonal relations between host and parasite, and different parasite-host combinations may therefore develop the conductive connection to different extents. Intrusive cells differentiate into tracheary elements only after they approach the central cylinder of the host root, i.e. crossed the root endodermis, and only then a whole xylem strand differentiates and directly connects to the main vascular system of the parasite (Musselman and Dickison 1975; White-Pennypacker et al. 1979; Hood et al. 1998). Tissue polarity, which determines the route of auxin flow along the haustorium, may determine the orientation and the precise location of xylem differentiation within the haustorium (Bar-Nun et al. 2008). In general, one side of a regenerating plant organ should act as an auxin source, i.e. take the role of a shoot, while the other side should act as an auxin sink, i.e. take the role of a root (Sachs 1991). This root system polarity is not obvious when a haustorium bridges between a parasite root and a host root. Then the polarity could develop either way, depending on the nature of the two partners.

Infestation by terminal haustorium of *P. aegyptiaca* is inhibited when an antagonist of IAA is applied to the roots, suggesting that auxin originating in the host shoot moves to the developing parasite through the young haustorium and that in this case the parasite acts as a root, i.e. as a sink for IAA flow, leading to the development of a continuous vascular system along the haustorium towards the parasite base (Bar-Nun et al. 2008). Since the development of a continuous vascular system requires connection to a host, haustoria failing to develop such a connection do not develop a continuous vascular system (see Sect. 3.15). Similarly, discontinuous aggregations of vessel elements develop in parasite callii that are incompatible with the host to which they are attached, whereas a continuous system develops in compatible callii (Zhou et al. 2004).

The xylem of young terminal haustoria of *Phelipanche* connects to host xylem strands that are directed towards the base of the host root, not towards the distal part of the host root, which is consistent with auxin flow from the host shoot to the developing parasite via the developing haustorium (Bar-Nun et al. 2008). This polarity changes when the parasite develops a tubercle and a shoot. Then the parasite xylem is continuous also towards host root apex (Zhou et al. 2004). Similar bidirectional connections are also found in terminal haustoria of *S. hermonthica* (Krause 1990). Intermediate forms of connection, with many xylem strands towards host root base and only few in the direction of the root apex, are also common in *Epifagus* (Schrenk 1894) and *O. cumana* (Krenner 1958).

A parasite may stimulate intensive tissue proliferation and/or cambial activity in the host root, so that host roots connected to parasites are often typically thicker (Fig. 3.2c; see also Baird and Riopel 1986), and the site of contact with the parasite may be significantly swollen (Schmucker 1959; Dörr and Kollmann 1974). This is particularly common with terminal haustoria but may also develop with lateral haustoria, as described for *P. ramosa*. In this latter case, the roots of certain host plants do not develop root hairs at the contact area (Attawi and Weber 1980). In some host species, the haustorium also induces the development of lateral host roots at the proximal neighbourhood of the tubercle (Kuijt 1969; Weber 1975). All these phenomena seem to result from excessive auxin supply by the parasite. In this way the perennial holoparasite *Conopholis americana* stimulates host tissue proliferation at the site of infestation, leading to the formation of a 'gall-like' mass of host and parasite tissues (Musselman and Mann 1978). An extreme example is found in *Alectra*, where following the invasion by its (lateral or terminal) haustorium into *Vigna* roots, their tissues interlock and divisions of host cells give rise to lateral host roots. These '**haustorial host roots**', which emerge from within the tubercle rather than nearby, are not to be confused with the adventitious roots of parasite tubercles (Visser et al. 1977; Dörr et al. 1977; Nwoke 1982; Nwoke and Okonkwo 1978).

The sequence of developmental events during vascular tissue differentiation also seems to relate to hormonal balance. Whereas xylem in *S. hermonthica* differentiates simultaneously along the haustorium, in *Rhamphicarpa fistulosa* and *Cordylanthus orcuttianus*, the differentiation may also proceed from the haustorial base (Neumann 1999; Niranjana 1994; Chuang and Heckard 1971).

The factors controlling the development of parasite-to-host phloem and plasmodesmata connections are not known, but these are also likely to depend on hormonal relations between the haustorium and the host.

3.11 The Mature Endophyte

The endophyte is the part of the haustorium that is located within the host and comes in direct contact with host tissues. The structure of the mature endophyte is complex in some terminal haustoria and rather simple in lateral haustoria of some hemiparasites.

All endophytes are intimately connected with tissues within the central cylinder of the host. Many parasites develop an intimate connection with all host tissues across the interface, in both the cortex and the stele (Figs. 3.9a and 3.10b; Baird and Riopel 1986), while in other cases, crushed cells, cell wall lignification and sometimes also a secreted substance occur in the interface of host cortex with the endophyte (Figs. 3.10a and 3.4b), restricting the intimate connection only to the host vascular cylinder. This latter case typically occurs in the interaction of host roots with many hemiparasites, e.g. *Striga asiatica*, *Rhinanthus minor* and *Buchnera hispida* (Stephens 1912; Neumann et al. 1999; Cameron and Seel 2007).

Xylem cells in the endophytes are usually accompanied with parenchymatic cells, and both come in intimate contact with host cells. Parenchyma cells often form the main connection with the host stele and in *S. hermonthica* may even extend further into the proximal part of host xylem and phloem (Mayer et al. 1997).

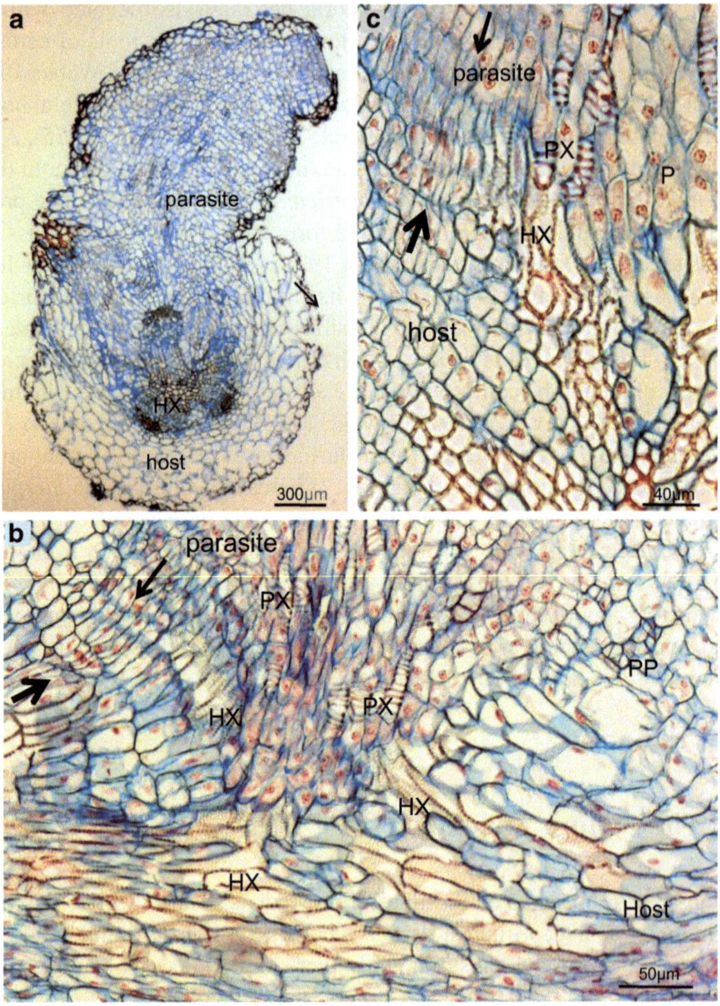

Fig. 3.9 *Orobanche crenata* connection with *Pisum sativum* root. (**a**) The intimate connection of the haustorium (in longitudinal section) with host root tissues (in cross section), including the cortex and the stele. (**b–c**) Details of xylem connection and cambium continuity. (**b**) Portion of a longitudinal section of both the haustorium and the host root; the host tracheary elements (HX) are pitted whereas those of the parasite (PX) have ring thickenings; note the association of well-developed parenchyma cells with dense cytoplasm and large nuclei that directly reach the host xylem; the *arrows* point on the aligned host and parasite cambia. (**c**) Longitudinal section of the haustorium combined with cross section of the host root; note the alignment of host and parasite xylem and cambia (*arrows*) (courtesy of A. Perez de Luque)

Parasites not only invade host roots and connect to existing host conductive elements; they can also induce an extensive development of conductive tissues within the host root, which are directed towards the corresponding conductive elements of the endophyte. This is pronounced to various degrees in different

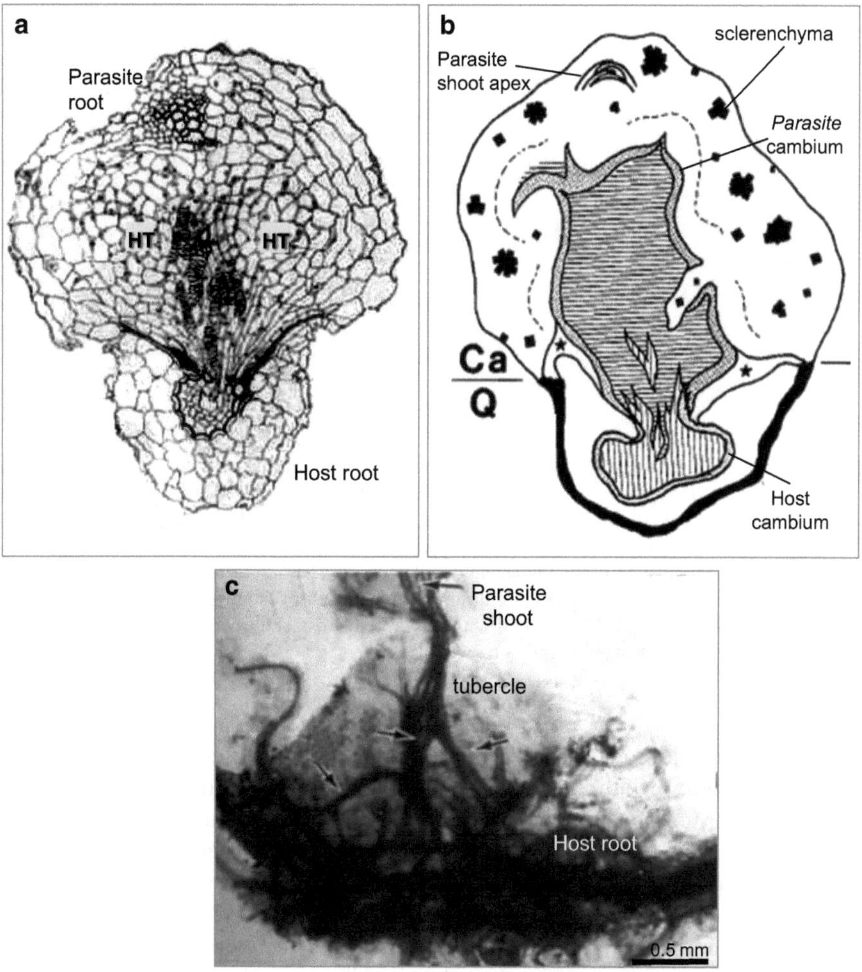

Fig. 3.10 Vascular connection. (**a**) Transverse section of lateral haustorium of *Striga asiatica* (syn. *S. lutea*) on maize root; note the direct connection only with the host central cylinder; the haustorial contact with the host cortex is sealed; the hyaline tissue (HT) is prominent on both sides of the central vascular system (from Stephens 1912, with permission). (**b**) Schematic transverse section of *Conopholis americana* tubercle and its connection with an oak root; the centre of both the haustorium and the host root is occupied by the vascular system (mainly secondary); an intermediate parasite parenchymatic tissue with large nuclei (*) adhered to the host root cortex (from Baird and Riopel 1986, with permission). (**c**) Cleared terminal haustorium of *Alectra vogelii* on cowpea root showing xylem connections with proliferation of xylem strands in both the parasite and the host, and their connection (from Nwoke and Okonkwo 1978, with permission)

parasites even during early stages of parasite development in the host. The degree of induced host tissue differentiation at the infestation zone is especially high in terminal haustoria. In *Alectra vogelii*, for example, there is an “intimate blending” of host and parasite tissues at the infestation site (Dörr et al. 1979), and young host

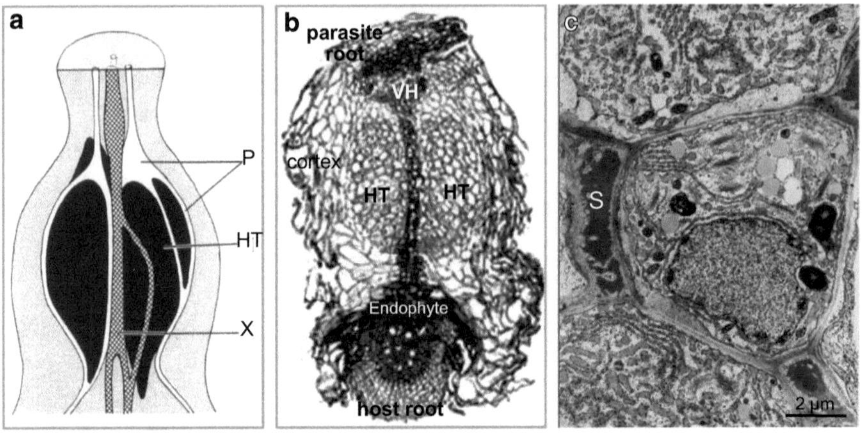

Fig. 3.11 The hyaline tissue. (**a**) The neck region of the terminal haustorium of *Alectra vogelii*; the hyaline tissue (HT) is seen on both sides of the central vascular strand and is traversed by both xylem (X) and phloem (P). (**b**) Longitudinal section of the lateral haustorium of *Rhinanthus minor* on a compatible host root, showing the vascular head (VH) at the connection with the parent parasite root and the hyaline tissue in the haustorial neck. (**c**) Hyaline cells of *A. vogelii* with dense cytoplasm and large nucleus, rough endoplasmic reticulum, dictyosomes and mitochondria; note that the intercellular spaces (S) are filled with a secreted substance (**a**,**c** from Visser et al. 1984; **b** from Rümer et al. 2007, with permission)

cells, which result from tissue proliferation at the contact zone, are induced to differentiate into sieve tubes that have direct contact with intruding parasite cells.

The haustorium of some species develops an active cambium that is aligned in the endophyte with host cambium, allowing direct vascular connection between the resulting secondary vascular elements of the host and those of the parasite (Figs. 3.9b, c and 3.10b; Percival 1931; Olsen and Olsen 1981; Baird and Riopel 1986). Host cambium in the infestation site, which is continuous with the normal root cambium, is in some cases oriented perpendicular to the main axis of the host root, i.e. in line with the cambium of the haustorium. The identity of the xylem elements at the connection zone can easily be realized in cases when parasite and host tracheary elements differ in their pattern of wall thickenings (Fig. 3.9b, c).

3.12 The Haustorial Neck

The haustorial neck is prominent in lateral haustoria, where it stands between host and parasite roots (Figs. 3.4b and 3.11a, b). In terminal haustoria of some hemiparasites (e.g. *S. gesnerioides* and *S. hermonthica;* Okonkwo and Nwoke 1978; Neumann et al. 1999), the haustorial neck is also prominent due to the presence of the hyaline tissue, which contributes to neck thickening (see Sect. 3.12.1), but in most holoparasites, it is difficult to define because their endophyte is continuous with the parasite shoot or tubercle (Fig. 3.4d).

The haustorial neck contains a rhizodermis and a cortex, which resemble and are continuous with the corresponding tissues in the parent plant. The cortex of the majority of the genera is composed solely of vacuolated parenchyma cells that often accumulate starch. In *Rhamphicarpa fistulosa* both the root cortex and the cortex of the haustoria are composed of aerenchyma (Neumann et al. 1998), a tissue with large air cavities that allows exchange of gases in the flooded habitats where this parasite is found. Idioblastic sclereids and sclereid nests may also be present in the cortex of some species, e.g. *Dasistoma* sp. and *Conopholis americana* (Musselman and Dickison 1975; Baird and Riopel 1986).

The core tissues in the centre of the neck is primarily composed of parenchyma (see below), which is traversed by the vascular bridge. Whereas both xylem and phloem are well developed in the neck of terminal haustoria of certain holoparasitic plants, like species of *Phelipanche* and *Orobanche* (Figs. 3.5b, d and 3.12a–c), only xylem is usually present in the haustorial neck of lateral haustoria (Fig. 3.4) though phloem can occasionally also be found in lateral haustoria of *P. aegyptiaca* (Fig. 3.12d).

Three different types of the vascular system were identified in the centre of lateral haustoria (Musselman and Dickison 1975). The *Striga* type consists only of few vessel elements with only limited development of parenchymatic core around; the *Aureolaria* type consists of a solid mass of xylem elements, and the *Siphonostegia* type consists of a hollow pear-shaped structure that encloses the parenchyma core, borne on an elongate neck. However, the xylem bridge may develop to various degrees even on the same plant. For example, the xylem bridge of some lateral haustoria of *Striga forbesii* was composed of only one non-branched strand, and in neighbouring haustoria, it branched and interconnected several times (Krause 1990). These differences seem to depend on the local signalling interaction between each individual haustorium and the adjacent host tissues, as discussed in Sect. 3.10.

3.12.1 Vascular Parenchyma and the Hyaline Tissue

The extent of parenchyma development in the haustorial neck and its location differs significantly in the various genera. Whereas in some genera, e.g. *Rhamphicarpa* and *Triphysaria*, only few inter-tracheidal parenchyma cells are associated with the vascular strands (Neumann et al. 1998; Heide-Jørgensen and Kuijt 1995), in some other parasites, a secondary parenchymatic tissue also develops in the haustorial neck, leading to significant neck thickening.

When the parasite root is stimulated to develop a haustorium, there is an expansion of its internal cortical cells, which—together with rhizodermis cells—establish the young haustorium (see Sect. 4.2). While the highly vacuolated **cortex** of the haustorium develops from external cells of the root cortex, internal cortical

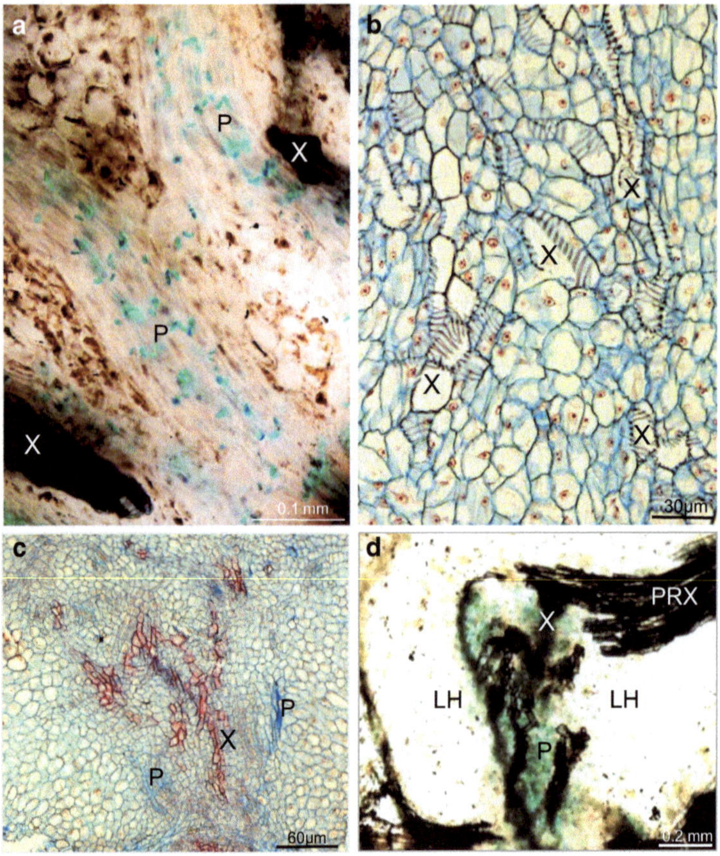

Fig. 3.12 Phloem and xylem. (**a**) Phloem strands (P; callose stained *blue*) and xylem strands (X) in the neck of the terminal haustorium of *Orobanche cumana*. (**b**) Anastomosing xylem strands in the haustorium of *O. crenata*. (**c**) Arrangement of xylem and phloem in the terminal haustorium of *O. crenata*. (**d**) Longitudinal section of a lateral haustorium (LH) of *Phelipanche aegyptiaca* stained for phloem localization (P, *blue*); *PRX*—parasite root xylem; *X*—haustorial xylem (**a**,**d** prepared by Hammam Ziadne; **b**,**c** courtesy of A. Perez de Luque)

cells develop dense cytoplasm and comprise the central **parenchymatic core**[1] of the developing haustorium. Once contact is established with the host, some core cells divide and differentiate either into **procambium** or directly into xylem elements. This formation of the central **vascular strand**(s) constitutes the initial vascular connection between the parasite and the host. At the same time, peripheral core cells may also divide, either forming **secondary parenchymatic tissues** (see below) with additional vascular strands or—as seen in *Cordylanthus* spp.—forming a **vascular cambium**. In the mature haustorium of *Cordylanthus*, a massive central

[1] Confusingly named *nucleus* in some early publications.

parenchyma tissue develops, surrounding the primary vascular strand and surrounded by well-developed **secondary xylem** (Chuang and Heckard 1971). Cell groups within the parenchymatic core in lateral haustoria of *P. ramosa* retain meristematic abilities throughout the development of the haustorium. These cells are responsible for the gradual formation of additional vascular strands (Attawi and Weber 1980). In lateral haustoria of *Castilleja*, thick-walled collenchyma cells comprise the core, at least at early developmental stages (Dobbins and Kuijt 1973a).

A massive secondary parenchymatic tissue, known as the **hyaline tissue**,[2] is prominent in some genera, e.g. *Alectra, Striga, Buchnera* and *Lathraea*, and is often seen in anatomical sections as two massive lobes around the centre of the haustorial neck (Figs. 3.4a, b and 3.11a, b; Renaudin 1974; Visser et al. 1984; Visser and Dörr 1987; Neumann et al. 1999). Its cells have large nuclei (Fig. 3.11c) and often also dense cytoplasm and can thus easily be distinguished from the neighbouring highly vacuolated cortex cells. This tissue is named '*hyaline*' because in some species it has special optical characteristics (Stephens 1912; Rogers and Nelson 1962; Visser et al. 1984), due in part to lack of air in the intercellular spaces, dense cytoplasm, lack of starch grains and lack of large vacuoles.

The hyaline tissue in lateral haustoria of the perennial holoparasite *Lathraea clandestina* also serves as a storage tissue. It stores protein during summer, the active season of its deciduous host tree, and consumes it during winter when it develops flowering stalks and rhizomes while its host is dormant (Ziegler 1955; Renaudin 1974).

Typically, the intercellular spaces of the hyaline tissue, and also of some other parenchymatic tissues that are associated with the vascular system, are filled with extracellular deposits composed of carbohydrates and proteins (Fig. 3.11c; Visser et al. 1984; Ba 1988; Heide-Jørgensen and Kuijt 1995; Neumann et al. 1998). In *Lathraea* the intercellular spaces are filled with secreted material when loading host nutrients and are empty when the hyaline tissue becomes a storage tissue (Renaudin 1974).

The hyaline tissue and presumably also the other parenchymatic tissues that are associated with the vascular system in the haustorium are believed to be involved in metabolizing host nutrients, in their transient storage, and in further regulating their supply to the developing parasite. The occurrence of intercellular secretion in *Lathraea* only when loading host nutrients and the presence of large nuclei and numerous mitochondria in these cells are consistent with this function. *Striga* has apoplastic continuity that extends from host xylem to both the intercellular spaces and the cell walls in the hyaline tissue (Neumann 1999), and ATPase activity was found in the periphery of *Striga* hyaline cells (Ba and Kahlem 1979), which is indicative of high membrane transport activity as well as active apoplastic transfer in these parenchymatic tissues.

[2] Formerly also mentioned as the *hyaline body* or *parenchyma core* and incorrectly as the *haustorial nucleus*.

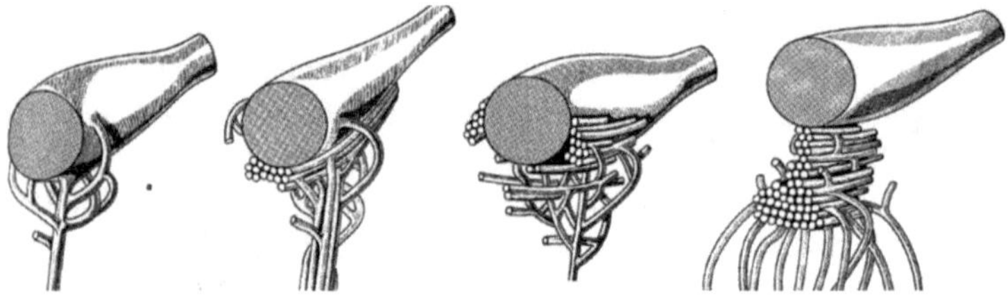

Fig. 3.13 Types of xylem arrangement at the base of lateral haustoria (from Weber 1993, with permission)

Unloading and transport of nutrients from the hyaline tissue to the developing parasite organs is probably via the phloem, which reaches the haustorial neck and comes in close vicinity with cells within the hyaline tissue of many species (Fig. 3.11a; Visser et al. 1984).

3.13 The Base of Lateral Haustoria

The base of the lateral haustorium is the site where its conductive tissues connect to those of the parent root. Unlike the endophyte, which copes with host compatibility, the base of the haustorium is involved in the regulation of physiological continuity with the parasite root. Whereas the phloem connection at this region still needs to be explored, the xylem connection, which is easily seen by light microscopy, was described for many species.

At this transition zone, xylem cells of many species form a massive tissue, the **vascular head**[3] (Fineran 1963), which is mainly composed of small tracheids that are arranged with their long axis parallel to the axis of the parent root (Figs. 3.4a, b, 3.11b and 3.13). These tracheids are often irregularly cuboid in the centre and more elongated towards the proximal and the distal sides of the parent root. The structural difference between the xylem in the vascular head and the xylem in the vascular bridge reflects their different ontogenetic origins. Whereas the xylem bridge in the centre of the haustorium originates from procambium, the vascular head develops by re-differentiation of cells in the pericycle, endodermis and vascular parenchyma of the parent root (Neumann et al. 1998), keeping their original orientation in parallel with the longitudinal axis of the parent root. Xylem tissue with similar characteristics occurs in the base of lateral roots of certain cereals (McCully and Mallett 1993) and was suggested to have a role in inhibiting pathogen movement from lateral to main roots. Would this function also apply to the base of parasite haustoria? A hint in this direction is found in *Triphysaria*. Fungal hyphae from host origin penetrated xylem elements in *Triphysaria* haustoria but did not cross the haustorial base (Heide-Jørgensen and Kuijt 1995).

[3] Formerly also named *plate xylem* and *vascular core*.

The xylem head can extend into the upper part of the haustorial neck in some genera (Fig. 3.4b), e.g. *Striga* and *Buchnera* (Krause 1989; Neumann 1999). This region, which originates from upper haustorial core cells, contains a mass of irregularly shaped xylem elements with various orientations. These cells form the extended upper part of the central xylem strand of the haustorial neck and are sheathed in parenchyma cells having dense cytoplasm and a large nucleus with several nucleoli (Weber 1976c) that are probably also involved in nutrient transport.

Several types of vascular connections to the parent root have been described (Fig. 3.13). The simplest occurs in some annual haustoria, where single central strands connect to the parent root through several ventral and lateral xylem elements. More complex vascular heads develop, e.g. in perennial parasites, where cambial activity occurs at the parent root and in some species also in the haustorium itself (Weber 1976c).

Xylem cells at the vascular head in some parasites may contain starch grains and other particles. These cells are described in the following section.

3.13.1 Graniferous Tracheary Elements

Peculiar xylem cells, the **graniferous tracheary elements**,[4] which contain starch grains (Fig. 3.14a, b) and sometimes also other contents, occur in lateral haustoria and are usually confined to the vascular head at the base of the haustorium. This unique type of tracheary element occurs in various different genera, e.g. *Lathraea, Pedicularis, Castilleja, Euphrasia* and *Triphysaria* (Heinricher 1896; Renaudin 1974; Musselman and Dickison 1975; Fineran 1985; Heide-Jørgensen and Kuijt 1995), and probably exists in some other members of the Orobanchaceae, but not in all (e.g. not in *Rhamphicarpa fistulosa*; Neumann et al. 1998).

Graniferous tracheary elements seem to have a physiological role in the regulation of parasite-host interaction, because they occur only in haustoria—not only in members of the Orobanchaceae but also in parasites belonging to other plant families. They are likely to have a mechanical role in adding resistance to the flow of the xylem sap, thus regulating the pressure difference between the host and the parasite. This may be of particular advantage when the same parasite is simultaneously connected to different roots, or to roots of various plants with different xylem characteristics (Fineran 1985).

They may also prevent cavitation (filling xylem elements with air) in the host roots, which can block water supply to the parasite. Especially root xylem is vulnerable to cavitation (Choat et al. 2005), and cavitation could develop within the host root as the result of its exposure to negative hydrostatic pressures in two opposite direction—towards the host and at the same time towards the parasite. The

[4] Formerly named *phloeotracheids*.

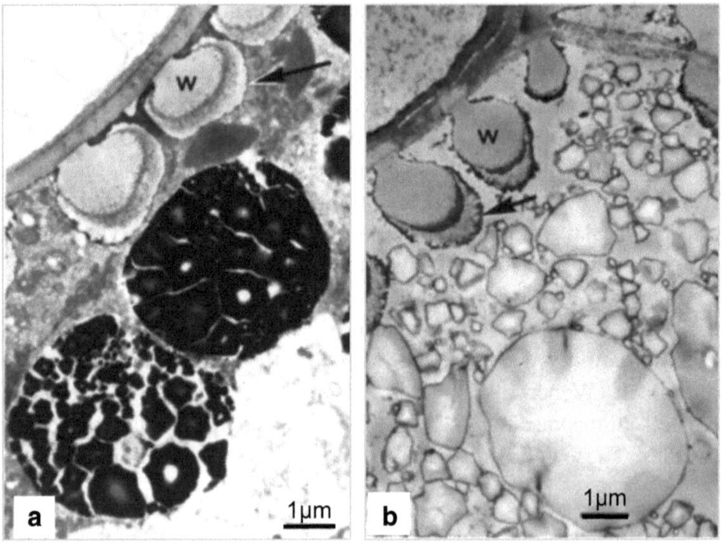

Fig. 3.14 Graniferous tracheary elements. Transmission electron micrographs of graniferous tracheary elements in the haustorium of *Lathraea squamaria*. (**a**) Nearly mature graniferous tracheary element; note the differentiated secondary wall thickenings (W), the protoplast that started disintegrating and the large amyloplasts, one of them releasing its starch grains; also note the additional layer of wall material on the wall thickening (*arrow*). (**b**) Mature graniferous tracheary element; the cell contains numerous free starch grains of different size, and no living protoplasm (from Fineran 1985, with permission)

ability to limit cavitation by regulation of the haustorial hydrostatic pressure at the graniferous tracheary tissue would therefore be of advantage to the parasite.

In addition to regulating the hydrostatic pressure difference, the parasite may also need to prevent backflow in the event that a stronger sink develops in host organs. Could the starch grains within the graniferous tracheary elements address both cavitation and backflow challenges by freely moving within the tracheids and blocking pit membranes in the relevant direction in a manner resembling the occlusion of sieve plate pores when phloem is injured? Starch grains are already known to have a physical role in plants, enabling gravity perception in root cap cells (Blancaflor et al. 1999; Morita 2010). So far the study of graniferous tracheary elements has been based only on structural examination of this tissue, but the newly developed means for in vivo observation of fluid flow within plant organs under real-time confocal microscopy and with the aid of fluorescent tracers should also facilitate a functional study of this system. These methods, which are successfully used for the study of phloem function (e.g. Knoblauch and van Bel 1998; Knoblauch et al. 2001), should help in studying the flow of xylem sap between host and parasite and allow following the behaviour of the starch grains within these cells under different host-parasite interactions.

The development of these peculiar xylem cells starts during the maturation of the vessel elements at the base of the haustorium. Numerous starch grains are then released from amyloplasts to the lumen (Fig. 3.14a, b) and seem to be freely floating within the developing xylem sap in the mature cells (Fineran 1985; Heide-Jørgensen and Kuijt 1995). While they are easily washed out during traditional anatomical examination because they are not bound to any cytoplasmic matrix, they can be preserved in the cells and detected under the light microscope in thick sections that include undamaged xylem cells from which the contents was not washed way, or in embedded material that retains all cell components. They can be identified especially after staining with iodine or with the periodic-acid-Schiff reagent or by using polarized light. The starch grains are presently best seen using transmission and scanning electron microscopy (Renaudin 1974; Fineran 1985; Heide-Jørgensen and Kuijt 1995).

3.14 The Base of Terminal Haustoria

The base of young terminal haustoria significantly differs from that of lateral haustoria because it directly connects the haustorium to the parasite shoot rather than to its root. In this respect it is homologous to plant hypocotyls (Schmucker 1959). The vascular system in this part of the parasite is in some species a transition from the protostele-like arrangement in the endophyte to arrangement in vascular bundles in the parasite shoot (Figs. 3.4d and 3.5a). This has clearly been shown in *Conopholis americana*, where both endarch vascular bundles, which are typical in dicotyledonous shoots, and exarch bundles, which are typical of roots, occur in the same cross section of the mature tubercle (Percival 1931). In *Striga* and similar hemiparasites, a hypocotyl is already present in the embryo within the seed, and after germination and attachment to a host, it becomes the base of the terminal haustorium. In holoparasite genera like *Orobanche* and *Conopholis*, the embryo is undifferentiated and does not include cotyledons (see Chap. 9), but after connecting to a host, the part of the seedling that remains outside the host root develops into a tubercle that carries shoot apices. Its location between the endophyte and the shoot apex is another indication that the tubercle is homologous to hypocotyl. Further evidence for the hypocotyl homology of the tubercle is found in many holoparasitic Orobanchaceae, in which the tubercle bears lateral adventitious roots similar to those on the hypocotyl of hemiparasitic members of this family (Kuijt 1969; Krause 1990; Weber 1993).

Additional conductive tissues develop during the maturation of the terminal haustorium, often gradually overshadowing the primary vascular system. These vascular tissues may either arise by division and re-differentiation of host and parasite cortex cells, which leads to addition of vascular bundles that bridge between the two partners (Fig. 3.10c), or by the development of a continuous cambium that produces a continuous secondary xylem around the primary vascular system (see Sect. 3.9.1).

3.15 Exceptional Haustoria

Haustoria may develop abnormally due to unsuccessful coordination with host tissues. Up to 40 % of the lateral haustoria of the annual hemiparasite *Triphysaria* have no xylem bridge at all (Heide-Jørgensen and Kuijt 1993, 1995). These 'wart haustoria' are tightly adhered to a host root but remain very small and are apparently non-functioning (Kuijt 1969; Musselman and Dickison 1975; Weber 1976b).

In other cases, the parasite may develop haustoria that do not have an intimate connection with the host, due to death or malfunctioning of the host root when the parasite began developing a haustorium. These 'meta-haustoria' (or 'abortive haustoria') often have a massively enlarged vascular core and are sometimes rather large and may have suberized surface in perennial parasites (Weber 1976b; Attawi and Weber 1980). While their function is still not known, Piehl (1963) suggested that they may function as a water absorption organ.

Members of the *Cymbaria-Siphonostegia* clade of the Orobanchaceae (see Sect. 14.2.2) carry haustoria with an exceptionally different structure, partly resembling Santalaceae haustoria (Musselman and Dickison 1975; Weber and Mickler 1986). These haustoria possess an endodermis that surrounds a long neck with a hollow vascular core and have a layer of sclereids between the vascular core and the vascular head (Musselman and Dickison 1975).

3.16 Are Haustoria Homologous to Roots?

The homology of haustoria to other plant organs has been controversial. It is tempting to believe that haustoria of the Orobanchaceae are homologous to roots, because terminal haustoria develop at the tip of the embryonic radicle, and lateral haustoria develop as lateral extensions of parasite roots. The vascular connection at the base of mature lateral haustoria also resembles that of lateral roots (see Sect. 3.13), but the initiation of lateral haustoria and the manner of tissue organization within haustoria are often considerably different from that of roots.

The sequence of events leading to the development of lateral haustoria and the internal structure of haustoria seem to have changed during the evolution of the haustorium from roots in the various groups within the Orobanchaceae. While in some Orobanchaceae, lateral haustoria development initiates in the root pericycle, similar to lateral roots in non-parasites (e.g. *Melampyrum*, Weber 1976c; *Parentucellia*, Alexander and Weber 1985), in other parasites cortical rather than pericycle cells are the first to change in the process of lateral haustorium initiation (e.g. *Agalinis purpurea*; Riopel and Musselman 1979; Riopel and Timko 1995; *Striga* spp., Krause 1990; see Chap. 4). Similarly, while in some species the vascular system within the endophyte is organized in the form of protostele (e.g. *Triphysaria*; Heide-Jørgensen and Kuijt 1995), in other parasites, and particularly in holoparasites, the structure of the vascular system is more complex and can no more be regarded as typical protostele.

Interestingly, a typical endodermis, which occurs around the vascular cylinder in roots, does not occur in the haustoria of the majority of the Orobanchaceae parasites, with the exception of only few primitive Orobanchaceae genera belonging to the *Cymbaria-Siphonostegia* clade (Musselman and Dickison 1975). Its absence in the haustoria is consistent with the fact that water and nutrient supply to the haustorium come from host tissues rather than directly from the surrounding soil.

3.17 Concluding Remarks

Being the key organ that allows a plant to become parasitic, the haustorium deserves detailed research, which should further explore not only its structure but also the function of each tissue during the various phases of its activity. Most structural studies on haustoria were conducted by light microscopy, yet their fine structure, which can be detected only under higher resolution, is not known. Surprisingly, electron microscopy, which is one of the most powerful instruments in the study of cells and tissues, has only rarely been used for the study of haustoria. Moreover, confocal microscopy has been a key tool in biological research for more than a decade, yet very little use of this technology has so far been employed in functional studies of the various haustorium components. Laser-capture microdissection, which facilitates the isolation of single cells from microscopical sections and handling them for genomic and chemical analyses, should also be employed in search for better understanding of the function of each element within the haustorium.

One of the important issues that need to be addressed in the study of haustoria is the intercellular communication between parasite cells and those of the compatible host at the connection zone. Electron microscopy together with some histochemical methods added some knowledge on tissue and cell characteristics. In this way Dörr and Kollmann (1995) used ultrastructural differences, and Neumann et al. (1998) used differences in cell wall staining to distinguish between cells of the host and those of the parasite. But in most other studies, only some general aspects of the cells were used, usually considering cytoplasmic density as a character that may differentiate between the two. However, one should not exclude the possibility that host cells and cell walls may resemble those of the parasite, particularly when affected by the invasion. Clearly, more objective methods are needed to avoid misinterpreting the identity of cells. One possible way to achieve this goal is the use of DNA-specific probes. Reporter proteins, like GFP and YFP, can also be used as differential markers for parasite *vs* host cells. This is achievable due to the recently developed transformation protocols for *Triphysaria* (Tomilov et al. 2007), *Phtheirospermum* (Ishida et al. 2011) and *Phelipanche* (Fernández-Aparicio et al. 2011) and the better knowledge of parasite genomes (Westwood et al. 2010), but care should be taken to exclude possible movement of markers from cells of one plant to another (see Aly et al. 2011).

References

Aber M, Fer A, Sallé G (1983) Étude du transfert des substances organiques de l'hôte (*Vicia faba*) vers le parasite (*Orobanche crenata* Forsk.). Z Pflanzenphysiol 112:297–308

Alexander T, Weber HC (1985) Zur parasitischen Lebensweise von *Parentucellia latifolia* (L.) Caruel (Scrophulariaceae). Beitr Biol Pflanz 60:23–34

Aloni R, Griffith M (1991) Functional xylem anatomy in root-shoot junctions of six cereal species. Planta 184:123–129

Aly R, Hamamouch N, Abu-Nassar J, Wolf S, Joel DM, Eizenberg H, Kaisler E, Cramer C, Gal-On A, Westwood JH (2011) Movement of protein and macromolecule between host plant and the parasitic weed *Phelipanche aegyptiaca* Pers. Plant Cell Rep 30:2233–2241

Attawi FAJ, Weber HC (1980) Zum Parasitismus und zum morphologisch-anatomischen Struktur der Sekundärhaustorien von *Orobanche*-Arten (*Orobanchaceae*). Flora 169:55–83

Ba AT (1984) Morphology, anatomy and ultrastructure of some parasites species of the genus *Striga* (Scrophulariaceae). In: Ayensu ES, Doggett H, Keynes RD, Musselman LJ, Parker C, Pickering A (eds) *Striga* biology and control. JCSU Press, Paris, pp 47–61

Ba AT (1988) Structure et ultrastructure de l'haustorium de *Striga hermonthica*, une scrophulariace!e parasite du mil (*Pennisetum typhoides*). Can J Bot 66:2111–2117

Ba AT, Kahlem G (1979) Mise en évidence d'activités enzymatiques au niveau de I'haustorium d'une phanérogame parasite: *Striga hermonthica* (Scrophulariaceae). Can J Bot 57:2564–2571

Baird WV, Riopel JL (1986) The developmental anatomy of *Conopholis americana* (Orobanchaceae) seedlings and tubercles. Can J Bot 64:710–717

Bar-Nun N, Sachs T, Mayer AM (2008) A role for IAA in the infection of *Arabidopsis thaliana* by *Orobanche aegyptiaca*. Ann Bot 101:261–265

Blancaflor EB, Fasano JM, Gilroy S (1999) Laser ablation of root cap cells: implications for models of graviperception. Adv Space Res 24:731–738

Cameron DD, Seel WE (2007) Functional anatomy of haustoria formed by *Rhinanthus minor*: linking evidence from histology and isotope tracing. New Phytol 174:412–419

Chen Q-L, Jia Y-M, Wang Z-F, Shan C-G, Zhu J-B, Guo Y-H (2011) Postembryonic development of *Cistanche tubulosa* (Schrenk) Wight. Pak J Bot 43:1823–1830

Choat B, Lahr E, Melcher PJ, Zwieniecki MA, Holbrook NM (2005) The spatial pattern of air seeding thresholds in mature sugar maple trees. Plant Cell Environ 28:1082–1089

Chuang TI, Heckard LR (1971) Observations on root parasitism in *Cordylanthus* (Scrophulariaceae). Am J Bot 58:218–228

Dobbins DR, Kuijt J (1973a) Studies on the haustorium of *Castilleja* (Scrophulariaceae). I. The upper haustorium. Can J Bot 51:917–922

Dobbins DR, Kuijt J (1973b) Studies on the haustorium of *Castilleja* (Scrophulariaceae). II. The endophyte. Can J Bot 51:923–931

Dörr I (1997) How *Striga* parasitizes its host: a TEM and SEM study. Ann Bot 79:463–472

Dörr I, Kollmann R (1974) Strukturelle Grundlage des Parasitismus bei *Orobanche*. I. Wachstum der Haustorialzellen im Wirtsgewebe. Protoplasma 80:245–259

Dörr I, Kollmann R (1975) Strukturelle Grundlagen des Parasitismus bei *Orobanche*. II. Die Differenzierung der Assimilat-Leitungsbahn im Haustorialgewebe. Protoplasma 83:185–199

Dörr I, Kollmann R (1976) Strukturelle Grundlagen des Parasitismus bei *Orobanche*. III. Die Differenzierung des Xylemanschlusses bei *O. crenata*. Protoplasma 89:235–249

Dörr I, Kollmann R (1995) Symplasmic sieve element continuity between *Orobanche* and its host. Bot Acta 108:47–55

Dörr I, Visser JH, Albers F (1977) On the parasitism of *Alectra vogelii* Benth. (Scrophulariaceae). II. Origin of lateral roots in the contact area of the haustorium. Z Pflanzenphysiol 85:349–359

Dörr I, Visser JH, Kollmann R (1979) On the parasitism of *Alectra vogelii* Benth. (Scrophulariaceae). III. The occurrence of phloem between host and parasite. Z Pflanzenphysiol 94:427–439

Fer A, De Bock F, Renaudin S, Rey L, Thalouarn P (1987) Relations trophiques entre les angiospermes parasites et leurs hôtes respectifs, II-Voies de transport et mécanismes impliqués dans le transfert des substances trophiques à l'interface hôte-parasite. Bull Soc Bot Fr 134:109–120

Fernández-Aparicio M, Rubiales D, Bandaranayake PGC, Yoder JI, Westwood JH (2011) Transformation and regeneration of the holoparasitic plant *Phelipanche aegyptiaca*. Plant Methods 7:36

Fineran BA (1963) Studies on the root parasitism of *Exocarpus bidwillii* Hook. f. III. Primary structure of the haustorium. Phytomorphology 13:42–54

Fineran BA (1985) Graniferous tracheary elements in haustoria of root parasitic angiosperms. Bot Rev 51:389–441

Gal-On A, Naglis A, Leibman D, Kathiravan K, Lapidot M, Ziadna H, Aly R (2009) Broomrape can acquire viruses from its hosts. Phytopathology 99:1321–1329

Gurney AL, Grimanelli D, Kanampiu F, Hoisington D, Scholes JD, Press MC (2003) Novel sources of resistance to *Striga hermonthica* in *Tripsacum dactyloides*, a wild relative of maize. New Phytol 160:557–568

Heide-Jørgensen HS (2008) Parasitic flowering plants. Brill, Leiden

Heide-Jørgensen HS, Kuijt J (1993) Epidermal derivatives as xylem elements and transfer cells: a study of the host–parasite interface in two species of *Triphysaria* (Scrophulariaceae). Protoplasma 174:173–183

Heide-Jørgensen HS, Kuijt J (1995) The haustorium of the root parasite *Triphysaria* (Scrophulariaceae), with special reference to xylem bridge ultrastructure. Am J Bot 82:782–797

Heinricher E (1896) Anatomischer Bau und Leistung der Saugorgane der Schuppenwurz Arten (*Lathraea clandestina* Lam. und *L. squamaria* L.). Beitr Biol Pflanzen 7:315–406

Hood ME, Condom JM, Timko MP, Riopel JL (1998) Primary haustorial development of *Striga asiatica* on host and nonhost species. Phytopathology 88:70–75

Ishida JK, Yoshida S, Ito M, Namba S, Shirasu K (2011) *Agrobacterium rhizogenes*-mediated transformation of the parasitic plant *Phtheirospermum japonicum*. PLoS One 6:e25802

Joel DM (2007) Direct infection of potato tubers by the root parasite *Orobanche aegyptiaca*. Weed Res 47:276–279

Joel DM, Losner-Goshen D (1994) The attachment organ of the parasitic angiosperms *Orobanche cumana* and *O. aegyptiaca* and its development. Can J Bot 72:564–574

Knoblauch M, van Bel AJE (1998) Sieve tubes in action. Plant Cell 10:35–50

Knoblauch M, Peters WS, Ehlers K, van Bel AJE (2001) Reversible calcium-regulated stopcocks in legume sieve tubes. Plant Cell 13:1221–1230

Koch L (1887) Die Entwicklungsgeschichte der Orobanchen, mit besonderer Berucksichtigung ihrer Beziehungen zu den Kulturpflanzen. C. Winter's Universitëtsbuchhandlung, Heidelberg

Krause D (1989) *Striga* – Biologie, Schaden, Kontrolle. Mikrokosmos 78:289–294

Krause D (1990) Vergleichende morphologisch/anatomische Untersuchungen an *Striga* Arten (Scrophulariaceae). Dissertation, Philipps-University, Marburg

Krenner JA (1958) The natural history of the sunflower broomrape (*Orobanche cumana* Wallr.). Acta Bot Acad Sci Hungar 4:113–144

Kuijt J (1969) The Biology of parasitic flowering plants. University of California Press, Berkeley

Kuijt J (1977) Haustoria of phanerogamic parasites. Ann Rev Phytopathol 17:91–118

Kuijt J (1991) The haustorial interface: what does it tell us? In: Ransom J, Musselman LJ, Warsham AD, Parker C (eds) Proceedings of the 5th international symposium on parasitic weeds. CIMMYT, Nairobi

Kuijt J, Dobbin DR (1971) Phloem in the haustorium of *Castilleja* (Scrophulariaceae). Can J Bot 49:1735–1737

Kuijt J, Toth R (1985) Structure of the host-parasite interface of *Boschniakia hookeri* (Orobanchaceae). Acta Bot Neerl 34:257–270

Kuijt J, Visser JH, Weber HC (1978) Morphological observations on leaf haustoria and related organs of the South African genus *Hyobanche* (Scrophulariaceae). Can J Bot 56:2981–2986

Labrousse P, Delmail D, Arnaud MC, Thalouarn P (2010) Mineral nutrient concentration influences sunflower infection by broomrape (*Orobanche cumana*). Botany 88:839–849

Mayer MJ, Steel J, Child DV, Hargreaves JA, Bailey JA (1997) Early stages of infection of maize (*Zea mays*) and *Pennisetum setosum* roots by the parasitic plant *Striga hermonthica*. Eur J Plant Pathol 103:815–827

McCully ME, Mallett JE (1993) The branch root of *Zea*. 3. vascular connections and bridges for nutrient recycling. Ann Bot 71:327–341

Morita MT (2010) Directional gravity sensing in gravitropism. Ann Rev Plant Biol 61:705–720

Musselman LJ (1980) The biology of *Striga*, *Orobanche*, and other root-parasitic weeds. Ann Rev Phytopathol 18:463–489

Musselman LJ, Dickison WC (1975) The structure and development of the haustorium in parasitic Scrophulariaceae. Bot J Linn Soc 70:183–212

Musselman LJ, Mann WF (1978) Root parasites of southern forests. Forest Service, USDA. General Technical Report SO-20

Ndambi B, Cadisch G, Elzein A, Heller A (2011) Colonization and control of *Striga hermonthica* by *Fusarium oxysporum* f. sp. *strigae*, a mycoherbicide component: an anatomical study. Biol Cont 58:149–159

Neumann U (1999) Etude ontogénique, structural et immunocytochimique des suçoirs de trios Scrophulariacées parasites africaines. Thése Doctorat de l'université Pierre et Marie Curie, Paris VI

Neumann U, Sallé G, Weber HC (1998) Development and structure of the haustorium of the parasite *Rhamphicarpa fistulosa* (Scrophulariaceae). Bot Acta 111:354–365

Neumann U, Vian B, Weber HC, Sallé G (1999) Interface between haustoria of parasitic members of the Scrophulariaceae and their hosts: a histochemical and immunocytochemical approach. Protoplasma 207:84–97

Niranjana R (1994) Morphological Studies in some root parasitic angiosperms. Thesis, University of Mysore, India

Nwoke FIO (1982) Structure and development of the mature secondary haustorium in *Alectra vogelii* Benth. Ann Bot 49:677–684

Nwoke FIO, Okonkwo SNC (1978) Structure and development of the primary haustorium in *Alectra vogelii* Benth. (Scrophulariaceae). Ann Bot 42:447–454

Okonkwo SNC, Nwoke FIO (1978) Initiation, development and structure of the primary haustorium in *Striga gesnerioides* (Scrophulariaceae). Ann Bot 42:455–463

Olsen S, Olsen ID (1981) Germination and development of the soma in *Boschniakia hookeri* (Orobanchaceae). Nord J Bot 1:246–259

Ouédraogo O, Neumann U, Raynal-Roques A, Sallé G, Tuquet C, Dembélé B (1999) New insights concerning the ecology and the biology of *Rhamphicarpa fistulosa* (Scrophulariaceae). Weed Res 39:159–169

Percival WC (1931) The parasitism of *Conopholis americana* on *Quercus borealis*. Am J Bot 10:817–837

Piehl MA (1963) Mode of attachment, haustorium structure, and hosts of *Pedicularis canadensis*. Am J Bot 50:978–985

Rajanna L, Shivamurthy GR, Niranjana R, Vijay CR (2005) Occurrence of phloem in the haustorium of *Aeginetia pedunculata* Wall. – a root holoparasite of Orobanchaceae. Taiwania 50:109–116

Renaudin S (1974) Contribution a l'étude de la biologie des phanérogames parasites: Recherches sur *Lathraea clandestina* L. (Scrophulariaceae). Thése, U.E.R. des Sciences de la Nature de l'Université de Nantes, France

Riopel J, Musselman L (1979) Experimental initiation of haustoria in *Agalinis purpurea*. Am J Bot 66:570–575

Riopel JL, Timko MP (1995) Haustorial initiation and differentiation. In: Press MC, Graves JD (eds) Parasitic plants. Chapman & Hall, London, pp 39–79
Rogers WE, Nelson RR (1962) Penetration and nutrition of *Striga asiatica*. Phytopathology 52:1064–1070
Rümer S, Cameron DD, Wacker R, Hartung W, Jiang F (2007) An anatomical study of the haustoria of *Rhinanthus minor* attached to roots of different hosts. Flora 202:194–200
Sachs T (1991) Pattern formation in plant tissues. Cambridge University Press, Cambridge
Sallé G (1987) Originsalités des tissus conducteurs des phanérogames parasites. Bull Soc Bot Fr Actual Bot 134(3/4):81–95
Schmucker T (1959) Höhere Parasiten. In: Ruhland W (ed) Handbuch der Pflanzenphysiologie, Springer Verlag, Berlin, vol 11. 480–529
Schrenk H (1894) Parasitism of *Epiphegus virginiana*. In: Proceedings of the American microscopical society, 16th annual meeting, vol 15(2), pp 91–127
Stephens (1912) The structure and development of the haustorium of *Striga lutea*. Ann Bot 26:1067–1076
Tate P (1925) On the anatomy of *Orobanche hederae* Duby, and its attachment to the host. New Phytol 24:284–293
Teryokhin ES (1997) Weed broomrapes. Aufsteig Verlag, Germany
Tomilov A, Tomilova N, Yoder JI (2007) *Agrobacterium tumefaciens* and *Agrobacterium rhizogenes* transformed roots of the parasitic plant *Triphysaria versicolor* retain parasitic competence. Planta 225:1059–1071
Visser J, Dörr I (1987) The haustorium. In: Musselman LJ (ed) Parasitic weeds in agriculture, vol I, *Striga*. CRC, Boca Raton, FL, pp 91–106
Visser JH, Dörr I, Kollmann R (1977) On the parasitism of *Alectra vogelii* Benth. (Scrophulariaceae). I. Early development of the primary haustorium and initiation of the stem. Z Pflanzenphysiol 84:213–222
Visser J, Dörr I, Kollmann R (1979) On the parasitism of *Alectra vogelii* Benth. (Scrophulariaceae). III. The occurrence of phloem between host and parasite. Z Pflanzenphysiol 94:427–439
Visser JH, Dörr I, Kollmann R (1984) The "hyaline body" of the root parasite *Alectra orobanchoides* Benth. (Scrophulariaceae), its anatomy, ultrastructure and histochemistry. Protoplasma 121:146–156
Weber HC (1975) Der Parasitismus mitteleuropäischer Rhinanthoideae unter Berücksichtigung vergleichender morphologisch und anatomischer Untersuchungen an den Haustorien. Dissertation, Justus Liebig University, Giessen
Weber HC (1976a) Über Wirtspflanzen und Parasitismus einiger mitteleuropäischer Rhinanthoideae (Scrophulariaceae). Plant Syst Evol 125:97–107
Weber HC (1976b) Studies on new types of haustoria in some central European Rhinanthoideae (Scrophulariaceae). Plant Syst Evol 125:223–232
Weber HC (1976c) Anatomische Studien an den Haustorien einiger parasitischer Scrophulariaceen Mitteleuropas. Ber Dtsch Bot Ges 89:57–84
Weber HC (1980) Zur Evolution des Parasitismus bei den Scrophulariaceae und Orobanchaceae. Plant Syst Evol 136:217–232
Weber HC (1987a) Untersuchungen an parasitischen Scrophulariaceen (Rhinanthoideen) in Kultur. II. Interaktionen zwischen Parasit und Wirt. Flora 179:35–44
Weber HC (1987b) Evolution of the secondary haustoria to a primary haustorium in the parasitic Scrophulariaceae/Orobanchaceae. Plant Syst Evol 156:127–131
Weber HC (1993) Parasitism von Blütenpflanzen. Wissenschaftliche Buchgesellschaft, Darmstadt
Weber HC, Mickler KD (1986) Parasitism in the Scrophulariaceae *Lesquereuxia syriaca* (Synonym *Siphonostegia syriaca*). Ber Deuts Bot Ges 99:123–131
Weber HC, Visser JH (1980) Zur anatomischen Struktur der aus Schuppenblättern hervorgegangen Kontaktorgane von *Hyobanche* (Scrophulariaceae). Flora 169:476–497

Westwood JH, Yoder JI, Timko MP, dePamphilis CW (2010) The evolution of parasitism in plants. Trends Plant Sci 15:227–235

White-Pennypacker B, Nelson PE, Wilhelm S (1979) Anatomic changes resulting from the parasitism of tomato by *Orobanche ramosa*. Phytopathology 69:741–748

Zhou WJ, Yoneyama K, Takeuchi Y, Rungmekarat S, Chae SH, Sato D, Joel DM (2004) In vitro infection of host roots by differentiated calli of the parasitic plant *Orobanche*. J Exp Bot 55:899–907

Ziegler H (1955) *Lathraea*: ein Blutungssaftschmarotzer. Ber Deuts Bot Ges 68:311–318

Chapter 4
Haustorium Initiation and Early Development

Pradeepa C.G. Bandaranayake and John I. Yoder

4.1 Introduction

The ability to develop invasive haustoria is the key feature of parasitic angiosperms. The haustorium attaches the parasite to the host, penetrates the host while keeping its own tissues intact, develops a vascular continuity between the host and parasite and ultimately provides the conduit through which host and parasite materials flow. The ability to make haustoria distinguishes parasitic from non-parasitic plants; indeed, 'the haustorium embodies the very idea of parasitism' (Kuijt 1969).

This chapter reviews the initiation and pre-attachment development of terminal and lateral haustoria in parasitic Orobanchaceae. Haustoria have been described for many genera of Orobanchaceae, but their initiation and development has been investigated in a relatively small number of species. Most of these studies have investigated the development of terminal haustoria in the weedy species *Striga asiatica* (Saunders 1933; William 1961; Nickrent et al. 1979; Keyes et al. 2007), *S. hermonthica* (Okonkwo 1966; Olivier et al. 1991), *S. gesnerioides* (Okonkwo and Nwoke 1978), *Phelipanche aegyptiaca* (syn. *Orobanche aegyptiaca*) and *O. cumana* (Joel and Losner-Goshen 1994; Zhou et al. 2004) and *Alectra vogelii* (Nwoke and Okonkwo 1978; Visser et al. 1990). Studies in the development of lateral haustoria have primarily focussed on three facultative species: *Agalinis purpurea* (Riopel and Musselman 1979; Baird and Riopel 1984), *Castilleja exserta* (previously known as *Orthocarpus purpurascens*) (Atsatt et al. 1978) and *Triphysaria versicolor* (Jamison and Yoder 2001; Bandaranayake et al. 2010).

P.C.G. Bandaranayake
Department of Plant Sciences, University of California-Davis, Davis, CA, USA

Department of Crop Science, University of Peradeniya, Peradeniya, Sri Lanka
e-mail: pgunathilake@ucdavis.edu

J.I. Yoder (✉)
Department of Plant Sciences, University of California-Davis, Davis, CA, USA
e-mail: jiyoder@ucdavis.edu

D.M. Joel et al. (eds.), *Parasitic Orobanchaceae*,
DOI 10.1007/978-3-642-38146-1_4, © Springer-Verlag Berlin Heidelberg 2013

Chemical signalling mechanisms and early haustorium ontogeny are similar in terminal and lateral haustoria; the primary developmental difference between them is that terminal haustoria result in the end point differentiation of the root tip meristems while lateral haustoria do not terminate root growth.

4.2 Early Haustorium Development

Haustorium development in most Orobanchaceae is initiated by chemical and physical stimuli provided by contact with a host root (Saunders 1933; William 1961; Musselman and Dickison 1975; Atsatt et al. 1978; Okonkwo and Nwoke 1978; Attawi and Weber 1980; Baird and Riopel 1984; Riopel and Timko 1995). The notable exceptions are *Orobanche* species that do not require host factors for haustorium development and whose terminal haustoria are not as swollen or morphologically well defined as those of *Striga* (Joel and Losner-Goshen 1994). Lateral and terminal haustorium development can be monitored in vitro by positioning the roots of facultative parasites or the radicle tips of obligate parasites next to those of host plants on agar plates. Alternatively, host root exudates or purified inducing factors can be applied to parasite roots or radicles. While the most responsive cells are near or at the root apical meristem, cortical cells throughout the parasite root are capable of developing lateral haustoria (Riopel and Musselman 1979).

Following exposure to host root exudates, there is an almost immediate cessation in parasite tip growth (Fig. 4.1a) (Baird and Riopel 1984; Riopel and Baird 1987). In *Striga*, the incorporation of H^3-thymidine into newly replicated DNA drops dramatically drops within an hour of exposure to 2,6-dimethoxybenzoquinone (DMBQ), an active haustoria-inducing quinone, and remains low for about 24 h or until the DMBQ is removed (Keyes et al. 2000). Several cell cycle associated genes that are transcriptionally regulated in *Triphysaria* roots during this period are possibly related to this interruption in replication (Torres et al. 2005).

Within 4–8 h of host exposure, the radicle tip or the region just behind the root tip becomes swollen (Fig. 4.1a) as a result of the rounding and isodiametric expansion of cortical cells associated with vacuole enlargement and coalescence. Cell enlargement begins with cells in the inner cortical layers and then spreads to the outer layers (Fig. 4.1b, c; see also Sect. 3.16). The involvement of auxin in modifying cell shape was implicated by dissecting *Triphysaria* root tips before and after exposure to host factors and by transforming *Triphysaria* roots with an auxin-responsive reporter (Tomilov et al. 2005; see also Sect. 3.10). Several genes encoding cell wall-modifying enzymes, including expansins, are transcriptionally regulated during this period (O'Malley and Lynn 2000; Wrobel and Yoder 2001; Torres et al. 2005). These findings show that early stages in haustorium development employ existing plant mechanisms for altering cell shape, size and number.

Epidermal cells at the top of the developing haustorium begin to divide about 10 h after induction, forming a group of densely cytoplasmic cells at the haustorium

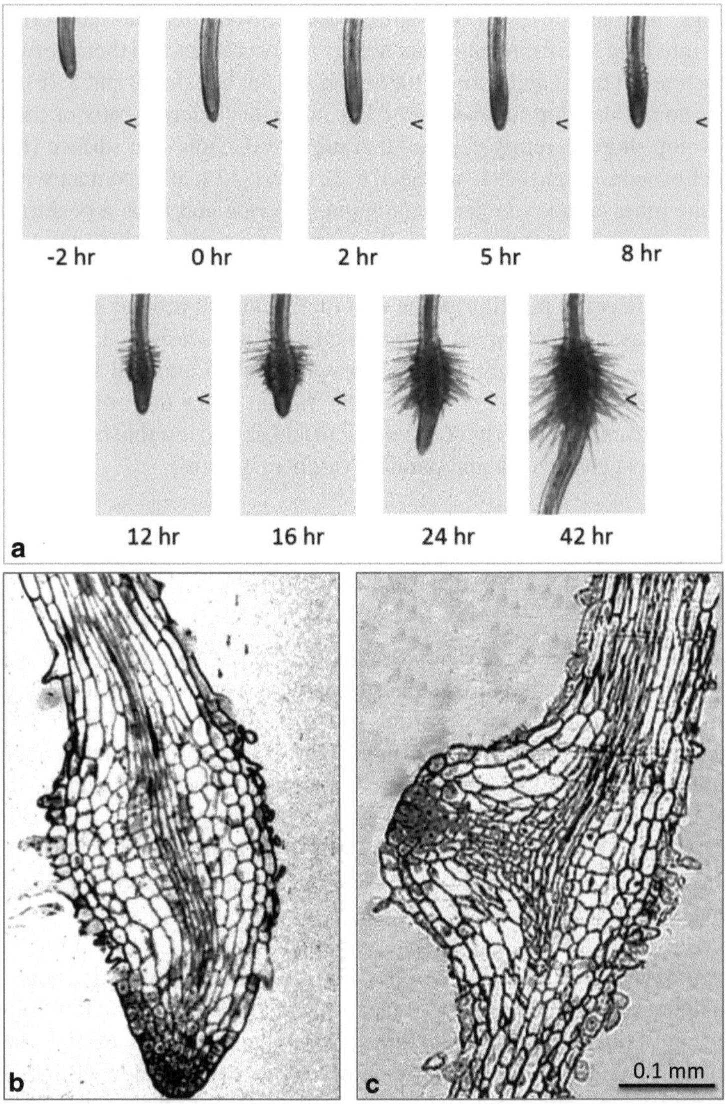

Fig. 4.1 Lateral haustorium development in *Triphysaria*. (**a**) A *Triphysaria* root growing on the surface of an agar plate, as photographed before and after exposure to DMBQ; notice the difference in root tip growth before and after application at time zero. Haustorial hairs can be observed about 5 h later and cortical swelling—by 12 h; at about 24 h the root reverts back to typical growth pattern (from Matvienko et al. 2001a, http://www.plantphysiol.org, Copyright American Society of Plant Biologists). (**b**) Longisection of *Triphysaria* root, following 48 h exposure to *Arabidopsis* root exudates; notice the rounding and isodiametric swelling of the cortical cells in the developing haustorium (from Estabrook and Yoder 1998, http://www.plantphysiol.org, Copyright American Society of Plant Biologists). (**c**) Similar to **b**, but *Triphysaria* roots were exposed to *Arabidopsis* roots for 72 h and then sectioned; this time point was after host attachment but before penetration. Animation of haustorium ontogeny can be seen at http://www.youtube.com/watch?v=9Hv1-aYNxNE

apex (Figs. 4.1c and 5.1d, e). Epidermal cells surrounding the haustorium apex elongate into long haustorial hairs that adhere to host tissues and thereby function in host attachment (Baird and Riopel 1985) (Figs. 3.3a, b, 4.3c, d, and 5.1b). Haustorial hairs do not develop in *Orobanche* but rather the external cells of the haustorium develop short secretory papillae that provide the adhesion surface (Fig. 5.1c; Joel and Losner-Goshen 1994; see Sect. 5.2). About 12 h after contact with a host, cells in the inner cortex and pericycle begin to divide and form a penetration peg that invades through the host tissues. Without host attachment the radial swelling, cell divisions and haustorial hair proliferation continue for about 24 h. If the haustorium is laterally positioned, the root meristem will resume its normal apical growth and haustoria will appear as 'beads on a string' (see Sect. 3.15). In contrast, there is no resumption of normal root growth distal to terminal haustoria unless functional host connections are established. Within a few days of host contact, a successful haustorium will have invaded the host and established a functional connection between the host and parasite vascular systems.

4.3 Haustorium Initiation Factors

The ability to monitor haustorium development in vitro provided an assay for identifying host factors that induce haustoria; these have been termed xenognosins or haustorium-inducing factors (Steffens et al. 1982; Riopel and Timko 1995). The first xenognosins identified were the flavonoids xenognosin A and xenognosin B that were isolated by fractionating a commercially available exudate from the legume *Astragalus* and assaying the fractions for haustorium-inducing activity in *Agalinis* (Lynn et al. 1981; Steffens et al. 1982). Subsequently, the flavonoid peonidin (Fig. 4.2a) was identified as a haustorium inducer for *Triphysaria* (Albrecht et al. 1999).

The only haustorium-inducing compound isolated from host roots is 2,6-dimethoxybenzoquinone (DMBQ) (Fig. 4.2b; Chang and Lynn 1986). Benzoquinones are widely present in plants and synthesized through the shikimate pathway, by oxidative decarboxylation of phenolic acids and by the enzymatic degradation of polyphenols by peroxidases and laccases (Caldwell and Steelink 1969; Krisnangkura and Gold 1979; Conn 1986). DMBQ was isolated from sorghum roots only after they were physically abraded or co-incubated with *Striga* cultures, processes that lead to the release of DMBQ through peroxidase-mediated oxidation of sorghum cellular components (Fig 4.2b) (Chang and Lynn 1986; Lynn and Chang 1990). Hydrogen peroxide generated in *Striga* radicles provides the rate-limiting substrate for host peroxidases that catalyse the conversion of their own cell wall components into haustoria-inducing benzoquinones (Keyes et al. 2000). In this model, *Striga* radicles enzymatically extract xenognosins from the surface of host roots, thereby ensuring close proximity of the parasite and host before committing to haustorium development (Keyes et al. 2007). David Lynn and colleagues propose that this may be a generalized mechanism by which even non-parasitic roots can establish subterranean spatial relationships (Palmer et al. 2009).

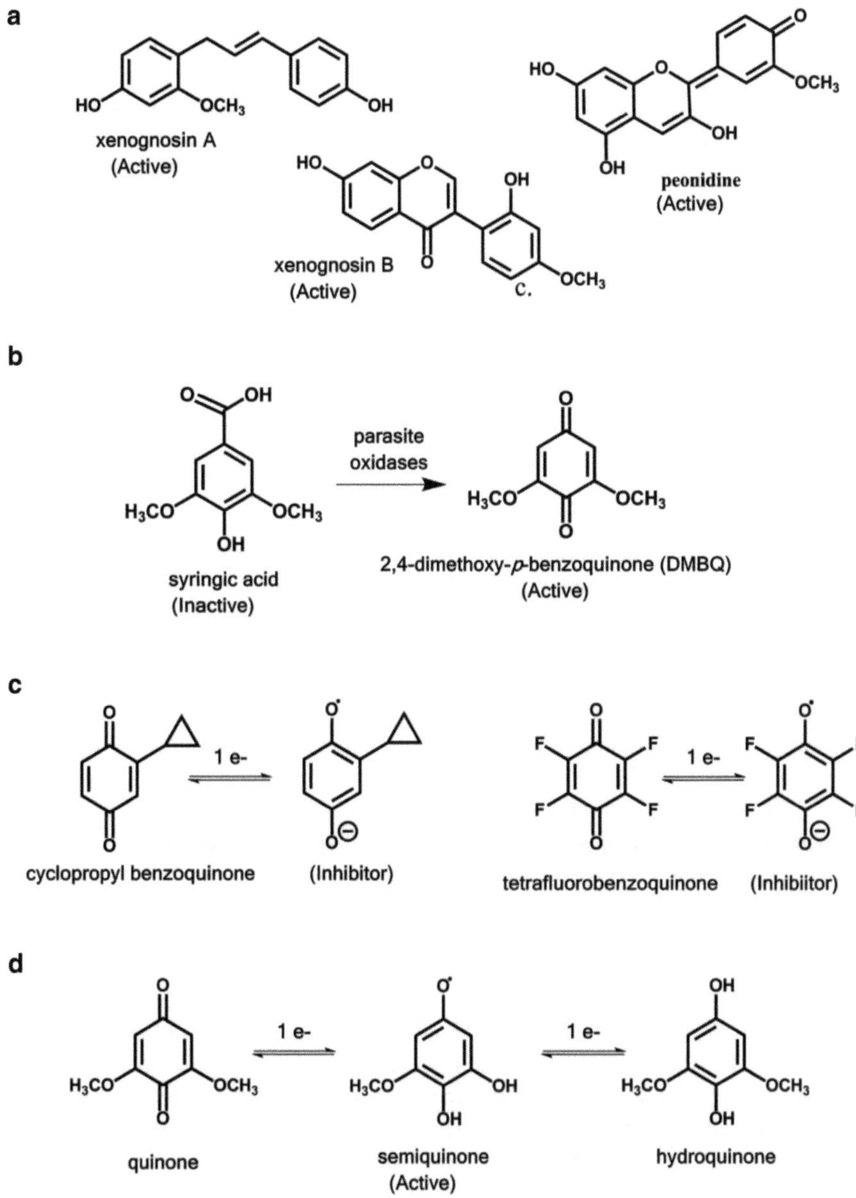

Fig. 4.2 Molecules and reactions associated with haustorium induction. (**a**) Three haustoria-inducing flavonoids, xenognosin A, xenognosin B and peonidin (Lynn et al. 1981; Steffens et al. 1982; Albrecht et al. 1999). (**b**) The enzymatic oxidation of syringic acid to the haustorium inducer DMBQ (Lynn and Chang 1990). (**c**) Single-electron reductions of cyclopropyl benzoquinone and tetrafluorohydroquinone generate inhibitors of haustorium development (Zeng et al. 1996; Keyes et al. 2000). (**d**) The single-electron reduction of DMBQ to the active semiquinone (Keyes et al. 2000); this step is catalysed by the enzyme TvQR1 in *Triphysaria* (Bandaranayake et al. 2010); further reduction of the semiquinone results in the relatively stable, inactive hydroquinone

The phenols, flavonoids and quinones that initiate haustorium development in vitro (Fig. 4.2) are structurally distinct from the strigolactones that are largely responsible for parasite seed germination (see Sect. 10.2). Not all xenognosins are equally active and different concentrations or times of exposure are needed for optimal haustorium development. For example, haustorium initiation in response to syringic acid requires several more hours of exposure than DMBQ because the phenolic acid needs to be enzymatically oxidized in order to be active (Fig. 4.2b) (Lynn and Chang 1990).

Many of the molecules associated with haustorium development are widely and commonly distributed among plants, and it is likely that host exudates contain multiple xenognosins. Therefore it might be expected that exudates from many plants should be active in inducing haustoria in parasites and this is by and large the case. Facultative parasites tend to have a broad, generalist host range (Werth and Riopel 1979; Gibson and Watkinson 1989). *Triphysaria*, for example, associates in the wild with at least 27 families of plant hosts and in pot cultures will parasitize maize, rice, legumes and *Arabidopsis* and hence will make haustoria in response to exudates from monocots and dicots (Atsatt and Strong 1970; Goldwasser et al. 2002). Obligate parasites tend to be more host specific; *Striga* species are either monocot- or dicot-specific and *Orobanche* species recognize specific dicots as hosts (see Chap. 18 and Sect. 14.4.2). However, host specificity in parasitic plants is generally not associated with haustorium initiation but rather with the ability of haustoria to functionally establish after invading the host (Nickrent et al. 1979; Hood et al. 1998; Li and Timko 2009). One known exception is the extensive variability among wild sorghum accessions in haustorium-inducing activity (Rich et al. 2004; see Sect. 21.2.1). Some sorghum lines with low levels of xenognosin activity also did not stimulate *Striga* seed germination, suggesting either that the biosynthesis of xenognosin and germination stimulants are co-regulated in sorghum or that they are inhibited by the same host factors.

The most striking examples of plants that do not induce haustoria in Orobanchaceae roots are those from closely related parasites. Self-parasitism or auto parasitism is frequently observed in mistletoes, *Cassytha* and *Cuscuta* (Sect. 1.5; Heide-Jørgensen 2008), but more rarely in Orobanchaceae (Riopel 1983, but see Sects. 3.4.2 and 3.5). In *Triphysaria* self-recognition is observed at the species level, and haustoria develop less frequently in associations between two *Triphysaria versicolor* plants than between *T. versicolor* and *Triphysaria eriantha* and much less frequently that between *T. versicolor* and *Arabidopsis* (Yoder 1997). The ability of parasites to distinguish self from non-self must have evolved soon after the origin of invasive haustoria because a plant would receive little benefit by parasitizing its own roots or those of a sibling. While the rationale for self-recognition seems obvious, the responsible mechanisms are unknown. Generalist parasites parasitize a broad range of host plants and so xenognosin signals are likely conserved among different plants. However these same signalling molecules are somehow missing, ignored, inhibited or not activated when two parasite roots come in contact. Identifying the molecular basis of this vegetative self-recognition system may suggest novel approaches for engineering resistance against parasitic weeds.

4.4 Haustorium Signal Transduction

Significant insights into the mechanism of xenognosin recognition came from the correlations between the redox potential of quinones and their haustorial inducing activity (Smith et al. 1996). The redox potential of quinones that induced *Striga* haustoria falls within a narrow redox window while inactive molecules fall outside that window. This suggests the propensity of a molecule to be oxidized or reduced is an important characteristic of xenognosins. The involvement of radical molecules in haustorium signalling was substantiated by two chemical spin traps, cyclopropyl-*p*-benzoquinone and tetrafluoro-benzoquinone, both of which inhibit haustorium formation when reduced to their semiquinone states (Fig. 4.2c) (Zeng et al. 1996; Smith et al. 1996). These experiments led to a redox model for xenognosin signalling in which semiquinone intermediates formed during transitions between quinone and phenolic states activate a redox-sensitive signal transduction pathway (Fig. 4.2d) (Keyes et al. 2001).

Redox cycling is catalysed by quinone oxidoreductases (E.C.1.6.5) that reduce quinones by either one- or two-electron transfer mechanisms (Testa 1995). Single-electron reductions generate highly reactive radical semiquinones that in the presence of oxygen form reactive oxygen species (ROS) that can be cytotoxic. Two-electron reductions do not generate reactive intermediates, and the enzymes that catalyse these reactions are considered detoxification enzymes that defend cells against electrophilic quinones (Ross et al. 2004).

Genes encoding each type of quinone oxidoreductase were isolated on the basis of their being transcriptionally regulated in *Triphysaria* roots by xenognosins (Matvienko et al. 2001b). Based on sequence comparisons and biochemical assays of protein activity, the gene *TvQR1* encodes a zeta-crystallin-like, NADPH-dependent quinone oxidoreductase (ZcQR) (EC 1.6.5.5) that catalyses single-electron quinone reductions (Bandaranayake et al. 2010) (Rao et al. 1992; Mano et al. 2000). *TvQR2* encodes a flavin binding quinone oxidoreductase (EC 1.6.5.2) that reduces quinones via the simultaneous transfer of two electrons that circumvents semiquinone intermediates (Sparla et al. 1996; Wrobel et al. 2002). While both *TvQR1* and *TvQR2* are transcriptionally up-regulated in parasite roots in response to xenognosins, only *TvQR1* is up-regulated by contact with a host root (Matvienko et al. 2001b; Bandaranayake et al. 2010). *TvQR1* regulation is also strongly correlated with haustorium development in *Triphysaria* species that form haustoria at different rates (Jamison and Yoder 2001). *Triphysaria* roots were transformed with inhibitory RNAi constructions targeting *TvQR1* or *TvQR2* for gene silencing, and roots silenced for *TvQR1* formed less haustoria than control transformants while those silenced for *TvQR2* formed haustoria at control levels (Bandaranayake et al. 2010). This work identified TvQR1 as one of the first parasite genes on the haustorium signalling pathway. Because TvQR1 generates the semiquinones that induce haustoria while TvQR2 encodes a detoxifying enzyme that eliminates them, the relative activities of these enzymes in parasite roots likely play a significant role in the parasite's commitment to haustorium development (Fig. 4.3).

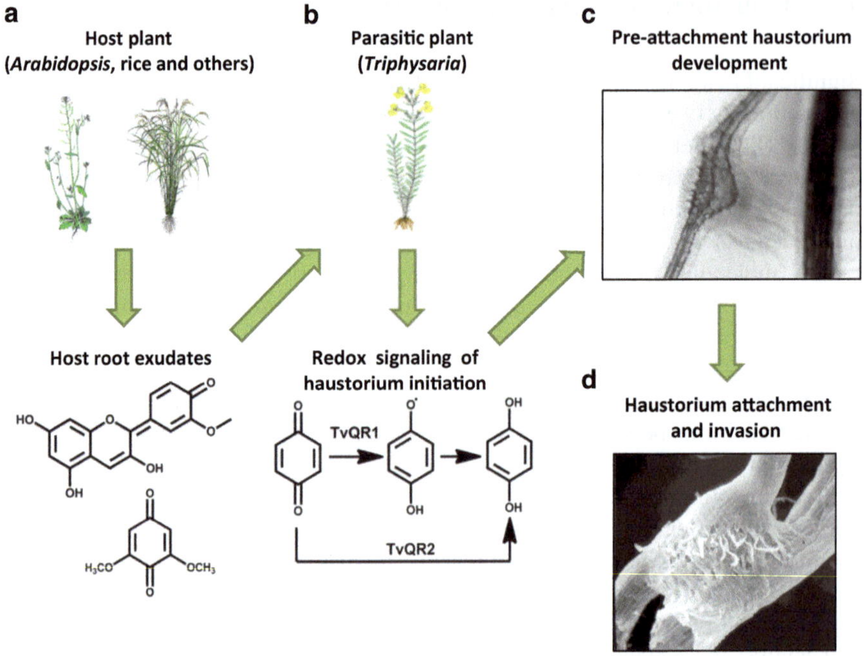

Fig. 4.3 Summary of lateral haustorium development. (**a**) Exudates released by plant roots contain molecules that induce haustorium development in parasite roots. The flavonoid peonidin and DMBQ are shown as potential haustorial inducers, though their significance as natural inducers in undisturbed root exudates is not known. (**b**) The enzyme TvQR1 catalyses the single-electron reduction of xenognosins producing the radical semiquinones that initiate haustorium signal transduction (Bandaranayake et al. 2010). (**c**) The cortical swelling and development of haustorial hairs in *Triphysaria* is a chemotropic response to xenognosins released from the lupine host root on the *right*. (**d**) Haustorial hairs attach the parasite root to the host root on the *right* (**c** and **d** from Yoder 1999, with permission from Elsevier)

Redox signal transduction pathways are common in both plant and animal systems (Ahmad et al. 2008; Foyer and Noctor 2009). Higher plants can sense, transduce and translate redox signals into appropriate cellular responses, influencing the expression of a number of genes and signal transduction pathways. For example, *NPR1*, a master regulator of defence gene expression in plants, contains cysteine residues that are targets of oxidoreduction reactions that cause conformational changes to the protein (Mou et al. 2003). Under normal conditions intermolecular disulfide bonds form structures that are sequestered to the cytoplasm. These disulfide bonds are broken under reduced conditions, and a monomerized NPR1 protein is translocated to the nucleus where it regulates a set of disease resistance genes. A similar system may act to transduce xenognosin signals from the semiquinones to the nucleus.

The generation of ROS may also be directly involved in eliciting the morphological changes associated with haustorium development. ROS have been detected using electron paramagnetic resonance in rapidly growing cells of maize roots, cucumber and *Arabidopsis* seedlings and are involved in cell wall loosening reactions associated with root growth (Liszkay et al. 2004). ROS molecules also control root hair elongation by activating plasma membrane calcium ion channels that generate a calcium gradient needed for tip growth (Foreman et al. 2003; Carol and Dolan 2006). ROS accumulation catalysed by TvQR1 may be directly responsible for the cortical cell expansion and haustorial hair elongation that are characteristic of early haustorium development.

The transcription factor-associated protein TvPirin is also necessary for effective haustorium development. *TvPirin* transcription is rapidly up-regulated in *Triphysaria* roots exposed to host root factors, and haustorium development is reduced when *TvPirin* transcripts are silenced by inhibitory RNAi (Matvienko et al. 2001a; Bandaranayake et al. 2012). The steady-state levels of several xenognosin-regulated transcripts were reduced in *Triphysaria* roots silenced for *TvPirin*, but their regulation by xenognosin exposure was not affected (Bandaranayake et al. 2012). This is consistent with *TvPirin* encoding a generalized transcription factor associated with the expression of several genes, some of which may be involved in haustorium development and others not.

While critical, chemical xenognosins are not the only stimuli that induce haustoria. Haustoria can form on inanimate surfaces such as rocks or Petri dishes, suggesting that tactile stimuli facilitate haustorium development (Kuijt 1969; Atsatt et al. 1978; Riopel and Timko 1995). Physical stimuli appear to play a larger role in the development of lateral haustoria than terminal because *Striga* seedlings develop terminal haustoria in liquid culture (Riopel and Baird 1987), but *Striga* and *Triphysaria* roots need contact with a solid support for effective lateral haustorium development (Wolf and Timko 1991; Matvienko et al. 2001a).

4.5 Evolutionary Origins

The competence to develop haustoria has originated at least 11 times during angiosperm evolution (Westwood et al. 2010; see Chaps. 1 and 14). There are two general hypotheses for the evolutionary origin of genes encoding parasitic plant functions: (1) parasitic genes were introduced into an autotrophic progenitor by horizontal gene transfer, or (2) parasite genes originated through neofunctionalization of plant genes encoding non-parasitic functions (see also Chap. 15).

Based on the morphological similarity of haustoria to nodules and crown galls, Atsatt proposed that haustoria evolved from endophytic microorganisms colonizing the plant roots (Atsatt 1973). A similar hypothesis was suggested by Kuijt (1969) who proposed that haustoria arose from mycorrhizal fungi that bridged the roots of different plants. Horizontal gene transfer between microbes and plants is well established, the most notable being the transfer of T DNA from bacteria to plants as a consequence of *Agrobacterium* infection (Nester et al. 2005). The phylogenetic placement of the *Rafflesia* mitochondrial gene *nad1B-C* into a clade associated with its host *Tetrastigma* suggests a horizontal gene transfer event over an evolutionary timescale (Davis and Wurdack 2004). Similarly, multiple *Plantago* species contain an *atp1* pseudogene that is phylogenetically related to the *atp1* homolog in *Cuscuta*, a parasite that infects *Plantago* (Mower et al. 2004). More recently, EST analysis provides evidence for the movement of a sorghum nuclear gene into the genome of *S. hermonthica* (Yoshida et al. 2010). Therefore the horizontal transfer of genes between plants and other organisms is well documented.

The alternative hypothesis is that parasite-specific functions are encoded by genes present in autotrophic plants where they perform functions unrelated to parasitism. Novel functions can arise in genes that have amplified following the duplication of genes or genomes or by ectopic expression of genes that have modified promoters (Hegarty and Hiscock 2008; Flagel and Wendel 2009). For example, many of the genes associated with floral development have homologues in non-flowering plants, indicating that these fulfil different functions in flowering and non-flowering plants (Floyd and Bowman 2007). Another example is DM13, a calmodulin-dependent protein kinase required for nodulation in legumes exposed to *Rhizobia* (Geurts et al. 2005). Because *DM13* homologues are also detected in rice and tobacco, DM13 clearly fulfils different functions in leguminous and non-leguminous plants (Mitra et al. 2004).

While there are precedents for both models, current evidence best supports the endogenous model for the origin of haustorial genes. EST databases of parasite transcripts expressed in roots during haustorium development do not contain sequences of obvious microbial origin (Torres et al. 2005). Additionally, genes that have been identified as functioning in haustorium development also function in autotrophic plants. For example, there are homologous genes to both *TvQR1* and *TvPirin* in non-parasitic plants, and the catalytic activities of the ZcQR1 enzymes are similar in *Triphysaria* and *Arabidopsis*. However, the expression profiles of these genes are different in parasitic and non-parasitic plants. It is possible that promoter mutations in the homologues of these genes in progenitors of parasitic plants altered the expression of these genes so that they were induced in parasite roots after contact with a host, thereby providing on demand of some of the components of a redox-sensitive signalling pathway. This model suggests that the evolutionary origin of plant parasitism is associated with changes in the regulation of plant genes that typically fulfil non parasitic functions (for further discussion of parasite genome evolution, see Chap. 15).

4.6 Conclusions

Phenols and quinones are common in the rhizosphere where they are known to function as signal molecules acting between plant roots and other organisms, including roots of other plants. The biological activity of these molecules is often associated with their redox state, and in some cases bioactivity is a function of the oxidoreduction cycle itself (O'Brien 1991; Appel 1993). Plants and other organisms have evolved detoxification systems that limit the cytotoxicity of radical molecules generated during redox cycling, and these mechanisms function in parasitic plants as well. However, parasitic plants have further evolved to use the redox-active molecules as signals to initiate haustorium development. In this way parasitic plants recruit biologically active and generally toxic molecules to signal the transition to a heterotrophic lifestyle.

References

Ahmad P, Sarwat M, Sharma S (2008) Reactive oxygen species, antioxidants and signaling in plants. J Plant Biol 51:167–173

Albrecht H, Yoder JI, Phillips DA (1999) Flavonoids promote haustoria formation in the root parasite *Triphysaria*. Plant Physiol 119:585–591

Appel HM (1993) Phenolics in ecological interactions – the importance of oxidation. J Chem Ecol 19:1521–1552

Atsatt PR (1973) Parasitic flowering plants: how did they evolve? Am Nat 107:502–510

Atsatt P, Strong D (1970) The population biology of annual grassland hemiparasites: I. The host environment. Evolution 24:278–291

Atsatt PR, Hearn TF, Nelson RL, Heineman RT (1978) Chemical induction and repression of haustoria in *Orthocarpus purpurascens* (Scophulariaceae). Ann Bot 42:1177–1184

Attawi FAJ, Weber HC (1980) Zum Parasitismus und zum morphologisch-anatomischen Struktur der Sekundärhaustorien von *Orobanche*-Arten (Orobanchaceae). Flora 169:55–83

Baird WV, Riopel JL (1984) Experimental studies of haustorium initiation and early development in *Agalinis purpurea* (L.) Raf. (Scrophulariaceae). Am J Bot 71:803–814

Baird WV, Riopel JL (1985) Surface characteristics of root haustorial hairs of parasitic Scrophulariaceae. Bot Gaz 146:63–69

Bandaranayake PCG, Filappova T, Tomilov A, Tomilova NB, Jamison-McClung D, Ngo Q, Inoue K, Yoder JI (2010) A single-electron reducing quinone oxidoreductase is necessary to induce haustorium development in the root parasitic plant *Triphysaria*. Plant Cell 22:1404–1419

Bandaranayake PC, Tomilov A, Tomilova NB, Ngo QA, Wickett N, dePamphilis CW, Yoder JI (2012) The TvPirin gene is necessary for haustorium development in the parasitic plant *Triphysaria versicolor*. Plant Physiol 158:1046–1053

Caldwell ES, Steelink C (1969) Phenoxy radical intermediate in the enzymatic degradation of lignin model compounds. Biochim Biophys Acta 184:420–431

Carol RJ, Dolan L (2006) The role of reactive oxygen species in cell growth: lessons from root hairs. J Exp Bot 57:1829–1834

Chang M, Lynn DG (1986) The haustorium and the chemistry of host recognition in parasitic angiosperms. J Chem Ecol 12:561–579

Conn EE (ed) (1986) Recent advances in phytochemistry: the shikimic acid pathway, vol 20. Plenum, New York

Davis CC, Wurdack KJ (2004) Host-to-parasite gene transfer in flowering plants: phylogenetic evidence from Malpighiales. Science 305:676–678

Estabrook EM, Yoder JI (1998) Plant-plant communications: rhizosphere signalling between parasitic angiosperms and their hosts. Plant Physiol 116:1–7

Flagel LE, Wendel JF (2009) Gene duplication and evolutionary novelty in plants. New Phytol 183:557–564

Floyd SK, Bowman JL (2007) The ancestral developmental tool kit of land plants. Int J Plant Sci 168:1–35

Foreman J, Demidchik V, Bothwell JH, Mylona P, Miedema H, Torres MA, Linstead P, Costa S, Brownlee C, Jones JD, Davies JM, Dolan L (2003) Reactive oxygen species produced by NADPH oxidase regulate plant cell growth. Nature 422:442–446

Foyer CH, Noctor G (2009) Redox regulation in photosynthetic organisms: signaling, acclimation, and practical implications. Antioxid Redox Signal 11:861–905

Geurts R, Fedorova E, Bisseling T (2005) Nod factor signaling genes and their function in the early stages of *Rhizobium* infection. Curr Opin Plant Biol 8:346–352

Gibson CC, Watkinson AR (1989) The host range and selectivity of a parasitic plant: *Rhinanthus minor* L. Oecologia 78:401–406

Goldwasser Y, Westwood JH, Yoder JI (2002) The use of *Arabidopsis* to study interactions between parasitic angiosperms and their plant hosts. In: Somerville C, Meyerowitz E (eds) The *Arabidopsis* book. American Society Plant Biologists, Rockville, MD, pp 1–17

Hegarty MJ, Hiscock SJ (2008) Genomic clues to the evolutionary success of polyploid plants. Curr Biol 18:R435–R444

Heide-Jørgensen HS (2008) Parasitic flowering plants. Brill Academic, Leiden, NL

Hood ME, Condon JM, Timko MP, Riopel JL (1998) Primary haustorial development of *Striga asiatica* on host and nonhost species. Phytopathology 88:70–75

Jamison DS, Yoder JI (2001) Heritable variation in quinone-induced haustorium development in the parasitic plant *Triphysaria*. Plant Physiol 125:1870–1879

Joel DM, Losner-Goshen D (1994) The attachment organ of the parasitic angiosperms *Orobanche cumana* and *O. aegyptiaca* and its development. Can J Bot 72:564–574

Keyes WJ, O'Malley RC, Kim D, Lynn DG (2000) Signaling organogenesis in parasitic angiosperms: xenognosin generation, perception, and response. J Plant Growth Regul 19:217–231

Keyes WJ, Taylor JV, Apkarian RP, Lynn DG (2001) Dancing together. Social controls in parasitic plant development. Plant Physiol 127:1508–1512

Keyes WJ, Palmer AG, Erbil WK, Taylor JV, Apkarian RP, Weeks ER, Lynn DG (2007) Semagenesis and the parasitic angiosperm *Striga asiaticae*. Plant J 51:707–716

Krisnangkura K, Gold MH (1979) Peroxidase catalysed oxidative decarboxylation of vanillic acid to methoxy-p-hydroquinone. Phytochemistry 18:2019–2021

Kuijt J (1969) The biology of parasitic flowering plants. University of California Press, Berkeley, CA

Li JX, Timko MP (2009) Gene-for-gene resistance in *Striga*-Cowpea associations. Science 325:1094

Liszkay A, van der Zalm E, Schopfer P (2004) Production of reactive oxygen intermediates $O_2^{\cdot -}$, H_2O_2, and $^{\cdot}OH$, by maize roots and their role in wall loosening and elongation growth. Plant Physiol 136:3114–3123

Lynn DG, Chang M (1990) Phenolic signals in cohabitation: implications for plant development. Ann Rev Plant Physiol Plant Mol Biol 41:497–526

Lynn DG, Steffens JC, Kamat VS, Graden DW, Shabanowitz J, Riopel JL (1981) Isolation and characterization of the first host recognition substance for parasitic angiosperms. J Am Chem Soc 103:1868–1870

Mano J, Babiychuk E, Belles-Boix E, Hiratake J, Kimura A, Inzea D, Kushnir S, Asada K (2000) A novel NADPH : diamide oxidoreductase activity in *Arabidopsis thaliana* P1 zeta-crystallin. Eur J Biochem 267:3661–3671

Matvienko M, Torres MJ, Yoder JI (2001a) Transcriptional responses in the hemiparasitic plant *Triphysaria versicolor* to host plant signals. Plant Physiol 127:272–282

Matvienko M, Wojtowicz A, Wrobel R, Jamison D, Goldwasser Y, Yoder JI (2001b) Quinone oxidoreductase message levels are differentially regulated in parasitic and non-parasitic plants exposed to allelopathic quinones. Plant J 25:375–387

Mitra RM, Gleason CA, Edwards A, Hadfield J, Downie JA, Oldroyd GED, Long SR (2004) A Ca 2+ /calmodulin-dependent protein kinase required for symbiotic nodule development: gene identification by transcript-based cloning. Proc Natl Acad Sci USA 101:4701–4705

Mou Z, Fan WH, Dong XN (2003) Inducers of plant systemic acquired resistance regulate NPR1 function through redox changes. Cell 113:935–944

Mower JP, Stefanovic S, Young GJ, Palmer JD (2004) Plant genetics: gene transfer from parasitic to host plants. Nature 432:165–166

Musselman LJ, Dickison WC (1975) The structure and development of the haustorium in parasitic Scrophulariaceae. Bot J Linn Soc 70:183–212

Nester EW, Gordon MP, Kerr A (2005) *Agrobacterium tumefaciens*: from plant pathology to biotechnology. APS Press, St. Paul, MN

Nickrent DL, Musselman LJ, Riopel JL, Eplee RE (1979) Haustorial initiation and non-host penetration in witchweed (*Striga asiatica*). Ann Bot 43:233–236

Nwoke F, Okonkwo SNC (1978) Structure and development of the primary haustorium in *Alectra vogelii* Benth. Ann Bot 42:447–454

O'Brien P (1991) Molecular mechanisms of quinone toxicity. Chem Biol Interact 80:1–41

O'Malley RC, Lynn DG (2000) Expansin message regulation in parasitic angiosperms: marking time in development. Plant Cell 12:1455–1465

Okonkwo SNC (1966) Studies of *Striga senegalensis* Benth. I. Mode of host-parasite union and haustorial structure. Phytomorphology 16:453–463

Okonkwo SNC, Nwoke FIO (1978) Initiation, development and structure of the primary haustorium in *Striga gesnerioides* (Scrophulariaceae). Ann Bot 42:455–463

Olivier A, Benhamou N, Leroux GD (1991) Cell surface interactions between sorghum roots and the parasitic weed *Striga hermonthica*: cytochemical aspects of cellulose distribution in resistant and susceptible host tissues. Can J Bot 69:1679–1690

Palmer AG, Chen MC, Kinger NP, Lynn DG (2009) Parasitic angiosperms, semagenesis and general strategies for plant-plant signaling in the rhizosphere. Pest Manag Sci 65:512–519

Rao PV, Krishna CM, Zigler JS (1992) Identification and characterization of the enzymatic activity of zeta-crystallin from guinea pig lens – a novel NADPH-quinone oxidoreductase. J Biol Chem 267:96–102

Rich PJ, Grenier C, Ejeta G (2004) *Striga* resistance in the wild relatives of sorghum. Crop Sci 44:2221–2229

Riopel JL (1983) The biology of parasitic plants: physiological aspects. In: Moore R (ed) Vegetative compatibility. Academic, New York, pp 13–34

Riopel JL, Baird WV (1987) Morphogenesis of the early development of primary haustoria in *Striga asiatica*. In: Musselman LJ (ed) Parasitic weeds in agriculture. CRC, Boca Raton, FL, pp 107–125

Riopel J, Musselman L (1979) Experimental initiation of haustoria in *Agalinis purpurea*. Am J Bot 66:570–575

Riopel JL, Timko MP (1995) Haustorial initiation and differentiation. In: Press MC, Graves JD (eds) Parasitic plants. Chapman and Hall, London, pp 39–79

Ross D, Siegel D, Helmut S, Lester P (2004) NAD(P)H:quinone oxidoreductase 1 (NQO1, DT-diaphorase), functions and pharmacogenetics. In: Sies H, Packer L (eds) Methods in enzymology. Academic, London, pp 115–144

Saunders AR (1933) Studies in phanerogamic parasitism with particular reference to *Striga lutea*. Sci Bull Union S Afr Dep Agric 128:5–56

Smith CE, Ruttledge T, Zeng Z, O'Malley RC, Lynn DG (1996) A mechanism for inducing plant development – the genesis of a specific inhibitor. Proc Natl Acad Sci USA 93:6986–6991

Sparla F, Tedeschi G, Trost P (1996) NAD(P)H-(quinone-acceptor) oxidoreductase of tobacco leaves is a flavin mononucleotide-containing flavoenzyme. Plant Physiol 112:249–258

Steffens JC, Lynn DG, Kamat VS, Riopel JL (1982) Molecular specificity of haustorial induction in *Agalinis purpurea* (L.) Raf. (Scrophulariaceae). Ann Bot 50:1–7

Testa B (1995) The metabolism of drugs and other xenobiotics: biochemistry of redox reactions. Academic, New York

Tomilov AA, Tomilova NB, Abdallah I, Yoder JI (2005) Localized hormone fluxes and early haustorium development in the hemiparasitic plant *Triphysaria versicolor*. Plant Physiol 138:1469–1480

Torres MJ, Tomilov AA, Tomilova N, Reagan RL, Yoder JI (2005) Pscroph, a parasitic plant EST database enriched for parasite associated transcripts. BMC Plant Biol 5:24

Visser JH, Dörr I, Kollman D (1990) On the parasitism of *Alectra vogelii* Benth (Scrophulariaceae) 1. Early development of the primary haustorium and initiation of the stem. Z Pflanzenphysiol 84:213–222

Werth C, Riopel JL (1979) A study of the host range of *Aureolaria pedicularia* (L.) Raf. (Scrophulariaceae). Am Midl Nat 102:300–306

Westwood JH, Yoder JI, Timko MP, dePamphilis CW (2010) The evolution of parasitism in plants. Trends Plant Sci 15:227–235

William CN (1961) Growth and morphogenesis of *Striga* seedlings. Nature 189:378–381

Wolf SJ, Timko MP (1991) In vitro root culture – a novel approach to study the obligate parasite *Striga asiatica* (L) Kuntze. Plant Sci 73:233–242

Wrobel RL, Yoder JI (2001) Differential RNA expression of alpha-expansin gene family members in the parasitic angiosperm *Triphysaria versicolor* (Scrophulariaceae). Gene 266:85–93

Wrobel RL, Matvienko M, Yoder JI (2002) Heterologous expression and biochemical characterization of an NAD(P)H : quinone oxidoreductase from the hemiparasitic plant *Triphysaria versicolor*. Plant Physiol Biochem 40:265–272

Yoder JI (1997) A species-specific recognition system directs haustorium development in the parasitic plant *Triphysaria* (Scrophulariaceae). Planta 202:407–413

Yoder JI (1999) Parasitic plant responses to host plant signals: a model for subterranean plant–plant interactions. Curr Opin Plant Biol 2:65–70

Yoshida S, Maruyama S, Nozaki H, Shirasu K (2010) Horizontal gene transfer by the parasitic plant *Striga hermonthica*. Science 328:1128

Zeng ZX, Cartwright CH, Lynn DG (1996) Cyclopropyl-p-benzoquinone – a specific organogenesis inhibitor in plants. J Am Chem Soc 118:1233–1234

Zhou WJ, Yoneyama K, Takeuchi Y, Iso S, Rungmekarat S, Chae SH, Sato D, Joel DM (2004) In vitro infection of host roots by differentiated calli of the parasitic plant *Orobanche*. J Exp Bot 55:899–907

Chapter 5
Haustorium Invasion into Host Tissues

Alejandro Pérez-de-Luque

5.1 Introduction

There are several definitions for the word 'invasion', but almost all involve entrance of something troublesome or harmful. Usually, the first definition corresponds to a military action and the second to pathogens and parasites, so it is tempting to find analogies between them. In fact, when a pathogen tries to invade a host, a 'fierce war' develops between them, and success increases the chance of survival for the 'winner'. Nevertheless, the parasite is not acting knowingly and deliberately as an individual attacking another organism. It acts following a natural behaviour resulting from an evolutionary process. Though breeders and agronomists regard the parasite as an enemy and actively construct barriers against it (see Chap. 17), in natural ecosystems the coexistence of the parasite and the host is possible and sometimes even necessary (Rowntree et al. 2011; see Sect. 16.3.3).

Invasion of host tissues is a key step for the parasite because natural barriers to penetration exist even in susceptible hosts. The pathogen must display a wide array of tools to overcome the intrinsic resistance present even in a compatible host. This chapter shows how parasitic Orobanchaceae prepare the 'machinery' for the assault against the host realm, and discusses the parasite tactics.

A. Pérez-de-Luque (✉)
IFAPA, Centro Alameda del Obispo, Área de Mejora y Biotecnología, Avda. Menédez Pidal s/n, 14004 Córdoba, Spain
e-mail: alejandro.perez.luque@juntadeandalucia.es

D.M. Joel et al. (eds.), *Parasitic Orobanchaceae*,
DOI 10.1007/978-3-642-38146-1_5,

5.2 Preparing for Penetration

Following the initiation of a haustorium (see Chap. 4), the first step in the direct attack is the development of the **attachment organ**, which facilitates anchoring to the host root surface. The development at the right place and time is crucial for the invasion. The attachment organ is a device that firmly adheres to the host root. A common feature of attachment organs of many parasites is the development of **haustorial hairs** at the periphery of the attachment site. These hairs are specialized root hairs of various lengths that serve as an additional anchoring device. Long attachment hairs, ca 150–300 μm, are formed by *Agalinis purpurea* prior to contact with the host root. Their surface is covered by a secretion, which shapes upon contact with the host surface and establish a structural bond (Fig. 5.1a). The adhesive substance of *Agalinis* was assumed to be hemicelluloses (Baird and Riopel 1983). Similar haustorial hairs develop around the radicle tip of *Striga* seedlings in contact with the host (Fig. 5.1b) (Hood et al. 1998; Reiss and Bailey 1998). Also in this latter case, a thin film is present between the hair and host root surface at the contact area. The chemical nature of the *Striga* film is also unclear, but it stains with safranin and gives a negative result in histochemical staining for carbohydrates (Musselman and Dickinson 1975). In *Orobanche cumana* and *Phelipanche aegyptiaca*, the external cells of the radicle tip differentiate into a layer of papillae, which are very short cell extensions, and these form the adhesion surface (Fig. 5.1c). The surface of these papillae is coated with a substance that stains for carbohydrates (Joel and Losner-Goshen 1994b; Pérez-de-Luque et al. 2005; Krenner 1955). Attachment hairs of *Triphysaria* spp. secrete a pectinaceous mucilage-like material (Heide-Jørgensen and Kuijt 1995).

However, not all root parasitic plants rely on haustorial hairs for adhering to host roots. For example, only few hairs of *Rhamphicarpa fistulosa* stick to the host root surface, in spite of having many haustorial hairs. In the soil penetration is obviously also supported by soil impaction (Neumann et al. 1998).

While the anchoring device develops, other cells at the apex of the initial haustorium prepare for penetration. In *Striga*, cells that are in contact with the root surface and cells in the next inner layer show a dense cytoplasm with central nucleus and numerous small vacuoles (Hood et al. 1998; Fig. 5.1d; see Sect. 4.2). These **intrusive cells** (Kuijt 1969; or digitate cells, Reiss and Bailey 1998) are usually bigger than the surrounding cells (Musselman and Dickinson 1975). In lateral haustoria of *Triphysaria*, some epidermal cells elongate tangential and perpendicular to the axis of the parasite root, forming the lateral faces of the haustorium (Heide-Jørgensen and Kuijt 1995). In *Rhamphicarpa fistulosa*, the cells that later penetrate the host derive from hypodermal cells (Fig. 5.1e) and have dense cytoplasm, a prominent nucleus with an enlarged nucleolus, numerous mitochondria, small vacuoles, rough endoplasmic reticulum, dictyosomes and lipid droplets (Neumann et al. 1998).

Once all these changes take place and the parasite is firmly attached to the host root surface, the haustorium disrupts host tissues and penetrates the root towards the vascular tissues.

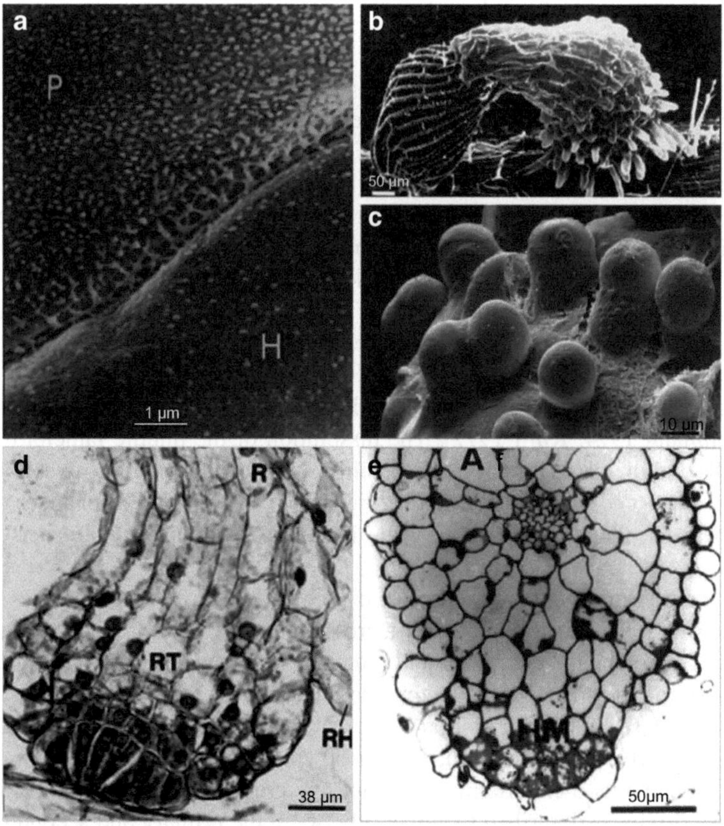

Fig. 5.1 Parasite attachment to host root. (**a**) Interface between an attachment hair of *Agalinis purpurea* (P) and host root surface (H) showing droplets of secreted substance that coalesce and bridge the gap at the contact region. (**b**) Haustorial hairs of *Striga asiatica* seedling attached to sorghum root surface. (**c**) Papillae of the root tip of *Orobanche cumana*. (**d**) Radicle cells of *Striga gesnerioides* in contact with the surface of a host root; they have dense cytoplasm; the nuclei moved into the centre of the cells; *R* parasite radicle; *RH* attachment hair; *RT* radicle tip. (**e**) *Rhamphicarpa fistulosa* root bearing a pre-haustorium; the haustorial meristem (HM) is restricted to cells of hypodermal origin; *A* root aerenchyma (**a** from Baird and Riopel 1983, **b** from Hood et al. 1998, **c** from Joel and Losner-Goshen 1994b, **d** from Reiss and Bailey 1998, **e** from Neumann et al. 1998, with permission)

5.3 Penetration

Thinking about putative similarities between parasitic plants and phytopathogenic fungi is not difficult. Such a comparison is helpful and was reviewed by Mayer (2006). Following on with the siege analogy, fungal hyphae could be compared with a small group of individual soldiers trying to find a gap in the defences. On the contrary, whereas fungal hyphae can penetrate through tiny natural openings, such as stomata or lenticels, or through lesions, a parasitic plant haustorium must open a bigger 'gate' in host tissues to reach the vascular system of the host.

5.3.1 Enzymatic Activity

Early reports about the penetration process, based on low-resolution light microscopy, claimed that *O. cumana* dissolved cell walls and cell contents of host cortical parenchyma (Krenner 1955) by means of a secretion from the intrusive cells and that the parasite is able to feed on the dissolved substances. Nowadays, despite great gaps in knowledge, we know that parasitic Orobanchaceae do not dissolve the cells found in their way, even though some parasites (e.g. *Agalinis aphylla*) may break through host cells (Musselman and Dickinson 1975). Histological observations clearly showed that only a combination of mechanical and enzymatic mechanisms, exerted by the parasite to separate host cells, allows penetration (Joel and Losner-Goshen 1994a; Neumann 1999). The intrusive cells of *Orobanche* spp. penetrate by pushing their way between host cells with the help of enzymatic processes (Dörr and Kollmann 1974; Kuijt 1977). The dissolution of the middle lamella between host cells and the concomitant mechanical pressure by the penetrating cells pushes portions of host cell walls aside so that the shape of the host cells changes and the space between them is occupied by the intrusive cells (Joel and Losner-Goshen 1994a).

The endodermis, with its cutinized or suberized Casparian strips, is another obstacle which the haustorium needs to cross on its way to host conductive tissues. Indeed, a combined anatomical and immunocytochemical study revealed that penetration of the *Phelipanche aegyptiaca* haustorium takes place between host endodermal cells by the dissolution of the cutin of the Casparian strips (Joel et al. 1998). Similarly, penetration through host endodermis by *Striga hermonthica* caused no damage to endodermal cells nor any crushing effect. In this latter case the haustorium was described to advance between the primary and secondary wall of the endodermis (Neumann 1999). A subtle difference therefore seems to exist between the parasites in their mode of penetration, but this issue needs further research.

Which enzymes are involved in the penetration process? Renaudin (1977) detected cellulolytic and proteolytic activity at the site of penetration of *Lathraea clandestina* by using tissue impressions on photographic and cellophane films. A few works presented in vitro evidence about pectolytic, cellulolytic and proteolytic enzymes being secreted by seedlings of *Phelipanche aegyptiaca* before penetration (Shomer-Ilan 1992, 1993, 1999). Singh and Singh (1993) also studied the presence of cell wall-degrading enzymes such as cellulase, polygalacturonase, xylanase and protease in tissues of the tubercle of *P. aegyptiaca* and inferred that they could also be involved in establishing haustorial connection with the host. However, neither of these works presented conclusive results as to the actual enzymes that are active in situ within host roots.

The first proof of direct involvement of enzymatic activity during the invasion process came from the work by Losner-Goshen et al. (1998). The authors showed the presence of pectin methylesterase at the penetration site using cytochemical and immunocytochemical methods with specific antibodies. In addition, the presence

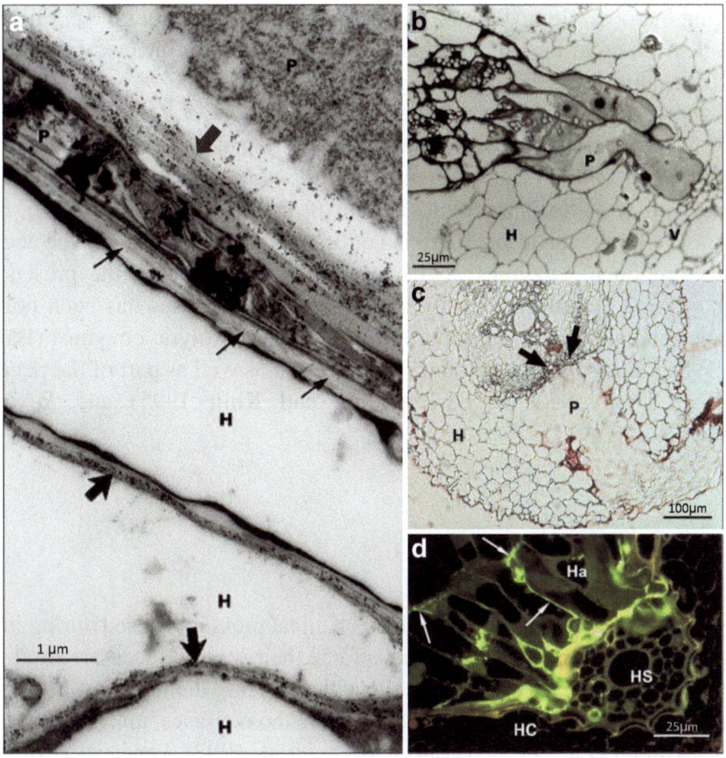

Fig. 5.2 The penetration mechanism. (**a**) Disappearance of pectins in outer cortex cell walls of sunflower root, adjacent to *Orobanche cumana* haustorium, seen after double gold labelling with JIM 5 and JIM 7 antibodies against low- and high-esterified pectins; internal host and parasite cell walls distant from the interface are labelled (*large arrows*), whereas host cell walls that touched the neighbouring parasite haustorium are not (*small arrows*); *P* parasite cell; *H* host cell. (**b**) Intruding cells of *O. cumana* reaching the vascular cylinder (V) of sunflower root. (**c**) Unsuccessful penetration attempt by *O. crenata* (P) in *Vicia sativa* root; the parasite was halted at the root endodermis, which was deformed by the exerted pressure (*arrows*). (**d**) The interface between the haustorium of *Buchnera hispida* and pearl millet root; JIM 5 antibodies labelled the parasite cell walls (*arrows*) whereas host cell walls remained unlabelled; *Ha* Haustorium; *HC* host cortex; *Hs* host stele (**a** and **b** from Losner-Goshen et al. 1998, **c** from Pérez-de-Luque et al. 2005, **d** from Neumann et al. 1999, with permission)

of pectin methylesterase was associated with the appearance of de-methylated pectins (galacturonic sequences with less than 50 % esterification) in the cell walls adjacent to intrusive cells, which is in accordance with the enzyme activity (Fig. 5.2a). This enzyme was previously identified and purified from calli and germinating seeds of *Orobanche* (Ben-Hod et al. 1993; Bar Nun et al. 1996). Other enzymes have been identified in *Orobanche* calli, such as polygalacturonase, but its involvement in host penetration still has not been proven (Ben-Hod et al. 1997). Nevertheless, Losner-Goshen et al. (1998) found degraded cell walls, supporting the possible involvement of polygalacturonase. Putative cutinase activity was also found at the endodermis penetration point by means of immunocytochemistry by Joel et al. (1998).

Cell wall-degrading enzymes were also found in *Striga*. Penetration of sorghum roots by *S. hermonthica* involved alterations of the host cell walls at the infection point (Olivier et al. 1991), and softening and dissolution of the middle lamella was observed with *S. gesnerioides* attacking cowpea (Reiss and Bailey 1998). In other hemiparasitic Orobanchaceae, there is no direct evidence indicating the accumulation or secretion of enzymes in the penetration process, but the presence of a densely staining cytoplasm in intrusive cells of the parasites has been pointed as a putative indication of the synthesis of cell wall hydrolytic enzymes (Baird and Riopel 1984). Enzymatic breakdown is implicated as well as part of the penetration process in *Triphysaria* (Heide-Jørgensen and Kuijt 1995) and *R. fistulosa* (Neumann et al. 1998).

5.3.2 Mechanical Pressure

In addition to the enzymatic process, a mechanical pressure is used during intrusion in host tissues. *Orobanche* intrusive cells force their way by successive and gradual splitting the cell walls between host cells without lysing them (Privat 1960). This is combined with elongation of the cells inside host tissues until they reach the vascular cylinder (Fig. 5.2b) (Losner-Goshen et al. 1998). Penetration of the host cortex by *Striga asiatica* implies anticlinal and periclinal divisions in the distal most cells and acropetal vacuolation of the haustorium cells (Hood et al. 1998). The distal most cells lengthen and form a palisade arrangement when they reach the endodermis and penetrate it 6–8 days after contact (see Sect. 5.4).

The existence of a mechanical force in addition to the enzymatic activity is evidenced by the presence of compacted and compressed host cells at the interface between host and parasite (Heide-Jørgensen and Kuijt 1995; Neumann et al. 1998, 1999; Reiss and Bailey 1998). When the intrusive cells find a physical resistance to penetration, for example, at the endodermis, the parasite tissues deform, and the host endodermis and pericycle are bent by the exerted pressure (Fig. 5.2c) (Pérez-de-Luque et al. 2005).

5.3.3 Internal Anchorage

A problem arises when considering a mechanical force exerted by the pathogen on the host tissues: the Newton's third law about action–reaction—if the intrusive cells are pressing against the host cells, why is the haustorium not driven out of the root? The parasitic plant must anchor in some way to the host root tissues in order to avoid being expelled by its own pressure. Several studies have shown that the interface between host and parasite plays this important role. Neumann et al. (1999) showed by means of immunocytochemistry that pectins are implicated in sticking the parasites *Buchnera hispida*, *R. fistulosa* and *S. hermonthica* to the host within its tissues (Fig. 5.2d). Similarly, an osmiophilic material was found filling the interface between *P. aegyptiaca* and the outer cortex regions of host root (Losner-Goshen et al. 1998), and a safranin-staining substance was observed at the infection points of legumes with *O. crenata* (Pérez-de-Luque et al. 2005, 2006). Such substances also stained with ruthenium red, pointing towards pectins as a component of these secretions (Pérez-de-Luque et al. 2006), which are probably similar to those secreted during the initial attachment of the parasite to host root surface; these substances similarly act as a cement that allows internal anchoring of the parasite to the host tissues (Joel et al. 1996), facilitating further physical efforts to penetrate between the host cells.

5.3.4 Reaching Host Conductive System

Once the parasite reaches the central cylinder of the host, the invasive process is almost complete, and connections with the vascular tissues must be developed. Concomitantly with endodermal penetration, differentiation of vascular elements occurs in the *S. asiatica* haustorium (Hood et al. 1998). After breaching the endodermis, *Striga* cells penetrate into host vessel elements, sometimes with more than one intrusion from a single parasite cell and developing absorbing structures termed oscula (see Sect. 3.9.2).

Further differentiation causes these haustorial cells and the oscula to lose the protoplast and become part of the water-absorbing system of the parasite in the form of xylem elements. In other parasitic plants, such as *Orobanche*, intrusive cells differentiate into transfer cells and later into xylem vessels with open connections with the host vessels (Privat 1960; Dörr and Kollmann 1976; see Sect. 3.9.1). These open xylem connections are possible when a simultaneous differentiation of adjacent host and parasite cells is induced (Dörr 1997). Open xylem connections are also present in other parasitic species (Kuijt 1977; Heide-Jørgensen and Kuijt 1995; Neumann et al. 1998; see Sect. 3.9.1). In addition, continuity between host and parasite sieve elements has been shown in *O. crenata* parasitizing *Vicia narbonensis* (Dörr and Kollmann 1995), so it is possible that connections with host phloem elements also develop in some other parasitic plants (see Sect. 3.9.3).

5.4 Duration of Penetration

The duration of all the penetration processes has not been studied in detail in terminal and lateral haustoria of the different Orobanchaceae. In some cases, the time lapse of penetration was assumed to be a week. Hood et al. (1998) developed a detailed study of the terminal haustoria of *Striga asiatica*. They found that invasion of sorghum root cortex was completed 2–3 days after parasite attachment to the host, the host endodermis was penetrated 3–4 days after first contact with the host, and the vascular connections started to be established within 6 days. This gives an interlude of a week from contact to vascular connection of *S. asiatica* in vitro. Recent experiments with *Orobanche crenata* in pea (*Pisum sativum*) have shown that the time lapse can vary (Cifuentes and Pérez-de-Luque 2011; unpublished results). No penetration of host tissues was detected until 4 days after contact of germinated parasite seeds with the host, during which the apical meristem of the *Orobanche* radicle differentiated into a haustorium. In this case the intrusive cells reached the endodermis and the central cylinder 11 days after inoculation, establishing vascular connections in 12 days. However, Losner-Goshen et al. (1998) mentioned that penetration of *P. aegyptiaca* and *O. cumana* haustoria into tomato roots is a very rapid process. All these point out differences in the duration of the penetration process, depending on several factors related to both the parasite and the host species and probably also on the experimental setup. Special attention should be paid to this question, because it can alter and distort studies requiring accurate sampling, such as the analysis of enzyme secretion and gene expression (Losner-Goshen et al. 1998).

5.5 Avoiding Defences: Tricks of War

How is it possible that the alarm is not raised in the host during the compatible invasive process? How can the parasitic plant manage to cross the natural barriers and avoid the activation of defensive mechanisms? This is one of the key issues still unknown in parasitic plant research.

In almost every plant, some cells grow between other cells. In these cases the neighbouring cells do not identify them as alien and do not react against their 'invasion'. This is the 'intrusive growth', which is the plant analogue of dendrite and axon growth in animals (Lev-Yadun 2001). Could the parasitic plant mimic the compatible intrusive growth of pollen tubes and laticifers? The question is not easy to answer, and there is almost no research on the topic. However, Joel and Portnoy (1998) showed that a susceptible host recognizes the parasite as an alien, and it does not grow in co-ordination with the host tissues. The activation of PR proteins (Joel and Portnoy 1998) and expression of 3-hydroxy-3-methylglutaryl-CoA reductase gene (Westwood et al. 1998) are evidences pointing towards the recognition of the attack not only by resistant hosts but also by compatible hosts.

Nevertheless, most of the studies are focused on resistant host genotypes and incompatible interactions (see Chap. 7), so no clear information exists about the process by which the parasite hampers the defensive responses of a compatible host. Mayer (2006) suggested two possibilities, the first one being due to the biochemical and physiological similarities between the parasite and the host, both being higher plants, and the second possibility that the parasitic plant actively prevents activation of host defence response. Being closely related, the host would find it difficult to recognize the parasitic plant as non-self. But we have already seen that during invasion, the parasite makes a real wound in the host root tissues, disrupting and separating them, and that the host recognizes it as alien. Why is the physical damage not detected? Why does the susceptible host not react to the invasion? The logical answer could be that there is an active parasitic mechanism preventing host reactions. At this point, only speculation is possible. For example, peroxidases, secreted by seedlings of *O. cumana*, were suggested as suppressors of a specific sunflower resistance (Antonova and ter Borg 1996). Mayer (2006) pointed out that phenolic compounds from the parasite could act as deterrents against host defence reactions. The lack of intracellular reactive oxygen species (ROS), either belonging to the host or to the parasite, might be an indication of the inability of the compatible host to react against parasitic attack (Mor et al. 2008). In addition, it is known that established parasitic plants interfere with the normal flux and synthesis of host hormones, e.g. abscisic acid (ABA) (Jiang et al. 2004; see Sect. 6.4), and despite that no mutual co-ordination seems to exists with the host tissues during the invasion process (Joel and Portnoy 1998). The alteration of the plant hormonal balance at the infection site (see Sect. 3.10) could perhaps delay or nullify a defensive response. Recently, Hiraoka et al. (2009) have shown, by means of suppression subtractive hybridization (SSH), that compatible interactions between *P. aegyptiaca* and *Lotus japonicus* imply the up-regulated expression of several genes in the host related with nodulation. So a new question arises: does this parasitic plant mimic the nodulation process similar to the exploitation of the mycorrhizal recognition signals? (Akiyama et al. 2005; see Chap. 10; see also Chap. 7 for host reactions to attack by the parasite).

5.6 Conclusions

During recent years, very little has been published on the cytology of host–parasitic plant interactions, and most of the studies are focused in genomics, proteomics and metabolomics, in species of *Triphysaria*, *Orobanche* or *Striga*, and centred on resistance vs. susceptibility. Knowledge about the behaviour of the parasite tissues inside the host is still lacking, and more comparative studies involving several different hosts and parasites should be conducted. These kinds of studies are not easy because of the similar nature of the two partners in the parasite–host system. In addition, the infestation takes place underground, so special and complex experiments need to be designed.

Further research is still needed on the composition of the adhesive substances allowing anchoring, the enzymes released during penetration, the development of phloem connections, a more precise understanding of the various steps during invasion in both host and parasite as well as the development of secondary haustoria from adventitious roots and their role in pathogenesis.

In addition, research is needed for the key questions of how the parasite avoids activation of host defences and how the parasite communicates with host tissues during the invasion.

Only a few genera of parasitic Orobanchaceae have been studied among the more than the 300 known species so we cannot know if all of them have the same mechanism of invasion, although some general assumptions can be made on the basis of the current knowledge. Correlated studies ranging from more primitive hemiparasitic and facultative species towards more evolved and specialized obligate holoparasitic species should be of great help for understanding particular traits in some cases and for getting a better perspective of this unique and fascinating plant group.

References

Akiyama K, Matsuzaki K, Hayashi H (2005) Plant sesquiterpenes induce hyphal branching in arbuscular mycorrhizal fungi. Nature 435:824–827

Antonova TS, ter Borg SJ (1996) The role of peroxidase in the resistance of sunflower against *Orobanche cumana* in Russia. Weed Res 36:113–121

Baird VW, Riopel J (1983) Experimental studies of the attachment of the parasitic angiosperm *Agalinis purpurea* to a host. Protoplasma 118:206–218

Baird VW, Riopel JL (1984) Experimental studies of haustorium initiation and early development in *Agalinis purpurea* (L.) Raf. (Scrophulariaceae). Am J Bot 71:803–814

Bar Nun N, Ben-Hod G, Lavi E, Mayer AM (1996) Purification of pectin methyl esterase from *Orobanche aegyptiaca*. Phytochemistry 41:403–406

Ben-Hod G, Losner D, Joel DM, Mayer AM (1993) Pectin methyl esterase in calli and germinating seeds of *Orobanche aegyptiaca*. Phytochemistry 32:1399–1442

Ben-Hod G, Bar Nun N, Tzaban S, Mayer AM (1997) Inhibitors of polygalacturonase in calli of *Orobanche aegyptiaca*. Phytochemistry 45:1115–1121

Cifuentes Z, Pérez-de-Luque A (2011) Crenate broomrape invasion of pea root: a histological time-lapse study. In: Eizenberg H, Westwood J, Vurro M (eds) 11th World congress on parasitic plants. Program and Abstracts. International Parasitic Plant Society, Martina Franca, p 27

Dörr I (1997) How *Striga* parasitizes its host: a TEM and SEM study. Ann Bot 79:463–472

Dörr I, Kollmann R (1974) Strukturelle Grundlage des Parasitismus bei *Orobanche* I. Wachstum der Haustorialzellen im Wirtsgewebe. Protoplasma 80:245–259

Dörr I, Kollmann R (1976) Strukturelle Grundlage des Parasitismus bei *Orobanche* III. Die Differenzierung des Xylemanschlusses bei *O. crenata*. Protoplasma 89:235–249

Dörr I, Kollmann R (1995) Symplasmic sieve element continuity between *Orobanche* and its hosts. Bot Acta 108:47–55

Heide-Jørgensen HS, Kuijt J (1995) The haustorium of the root parasite *Triphysaria* (Scrophulariaceae), with special reference to xylem bridge ultrastructure. Am J Bot 82:782–797

Hiraoka Y, Ueda H, Sugimoto Y (2009) Molecular responses of *Lotus japonicus* to parasitism by the compatible species *Orobanche aegyptiaca* and the incompatible species *Striga hermonthica*. J Exp Bot 60:641–650

Hood ME, Condom JM, Timko MP, Riopel JL (1998) Primary haustorial development of *Striga asiatica* on host and nonhost species. Phytopathology 88:70–75

Jiang F, Jeschke WD, Hartung W (2004) Abscisic acid (ABA) flows from *Hordeum vulgare* to the hemiparasite *Rhinanthus minor* and the influence of infection on host and parasite abscisic acid relations. J Exp Bot 55:2323–2329

Joel DM, Losner-Goshen D (1994a) Early host-parasite interaction: models and observations of host root penetration by the haustorium of *Orobanche*. In: Pieterse AH, Verkleij JAC, ter Borg SJ (eds) Biology and management of *Orobanche*. Proceedings of the third international workshop on *Orobanche* and related *Striga* research. Royal Tropical Institute, Amsterdam, pp 237–247

Joel DM, Losner-Goshen D (1994b) The attachment organ of the parasitic angiosperms *Orobanche cumana* and *O. aegyptiaca* and its development. Can J Bot 72:564–574

Joel DM, Portnoy VH (1998) The angiospermous root parasite *Orobanche* L. (Orobanchaceae) induces expression of a pathogenesis related (PR) gene in susceptible tobacco roots. Ann Bot 81:779–781

Joel DM, Losner-Goshen D, Hershenhorn J, Goldwasser Y, Assayag M (1996) The haustorium and its development in compatible and resistant host. In: Moreno MT, Cubero JI, Berner D, Joel D, Musselman LJ, Parker C (eds) Advances in parasitic plant research. Proceedings of the 6th international symposium on parasitic weeds. Junta de Andalucía, Consejería de Agricultura y Pesca, Sevilla, pp 531–541

Joel DM, Losner-Goshen D, Goldman-Guez T, Portnoy VH (1998) The haustorium of *Orobanche*. In: Wegmann K, Musselman LJ, Joel DM (eds) Current problems in *Orobanche* research. Proceedings of the 4th international workshop on *Orobanche*. Institute for Wheat and Sunflower Dobroudja, Albena, pp 101–106

Krenner JA (1955) The natural history of the sunflower broomrape (*Orobanche cumana* Wallr.). Acta Bot IV:113–144

Kuijt J (1969) The biology of parasitic plants. University of California Press, Berkeley

Kuijt J (1977) Haustoria of phanerogamic parasites. Ann Rev Phytopathol 17:91–118

Lev-Yadun S (2001) Intrusive growth – the plant analog of dendrite and axon growth in animals. New Phytol 150:508–512

Losner-Goshen D, Portnoy VH, Mayer AM, Joel DM (1998) Pectolytic activity by the haustorium of the parasitic plant *Orobanche* L. (Orobanchaceae) in host roots. Ann Bot 81:319–326

Mayer AM (2006) Pathogenesis by fungi and by parasitic plants: similarities and differences. Phytoparasitica 34:3–16

Mor A, Mayer AM, Levine A (2008) Possible peroxidase functions in the interaction between the parasitic plant, *Orobanche aegyptiaca*, and its host, *Arabidopsis thaliana*. Weed Biol Manag 8:1–10

Musselman LJ, Dickinson WC (1975) The structure and development of the haustorium in parasitic Scrophulariaceae. Bot J Linn Soc 70:183–212

Neumann U (1999) Etude ontogénique, structurale et immunocytochimique des suçoirs de trois Scrophulariacées parasites africaines. PhD Thesis, Universite Pierre et Marie Curie, Paris

Neumann U, Sallé G, Weber HC (1998) Development and structure of the haustorium of the parasite *Rhamphicarpa fistulosa* (Scrophulariaceae). Bot Acta 111:354–365

Neumann U, Vian B, Weber HC, Sallé G (1999) Interface between haustoria of parasitic members of the Scrophulariaceae and their hosts: a histochemical and immunocytochemical approach. Protoplasma 207:84–97

Olivier A, Benhamou N, Leroux G (1991) Cell surface interactions between sorghum roots and the parasitic weed *Striga hermonthica*: cytochemical aspects of cellulose distribution in resistant and susceptible host tissue. Can J Bot 69:1679–1690

Pérez-de-Luque A, Rubiales D, Cubero JI, Press MC, Scholes J, Yoneyama K, Takeuchi Y, Plakhine D, Joel DM (2005) Interaction between *Orobanche crenata* and its host legumes: unsuccessful haustorial penetration and necrosis of the developing parasite. Ann Bot 95:935–942

Pérez-de-Luque A, González-Verdejo CI, Lozano MD, Dita MA, Cubero JI, González-Melendi P, Risueño MC, Rubiales D (2006) Protein cross-linking, peroxidase and b-1,3-endoglucanase involved in resistance of pea against *Orobanche crenata*. J Exp Bot 57:1461–1469

Privat G (1960) Recherches sur les phanérogames parasites (étude d'*Orobanche hederae* Duby). Ann Sci Nat Bot 12:721–871

Reiss GC, Bailey JA (1998) *Striga gesnerioides* parasitizing cowpea: development of infection structures and mechanisms of penetration. Ann Bot 81:431–440

Renaudin S (1977) Mise en évidence d'activités enzymatiques au niveau des suçoirs de *Lathraea clandestina* L. Bull Soc Bot Fr 124:419–425

Rowntree JK, Cameron DD, Preziosi RF (2011) Genetic variation changes the interactions between the parasitic plant-ecosystem engineer *Rhinanthus* and its hosts. Philos Trans R Soc B 366:1380–1388

Shomer-Ilan A (1992) Enzymes with pectinolytic and cellulolytic activity are excreted by the haustorium of *Orobanche aegyptiaca*. Phytoparasitica 20:343

Shomer-Ilan A (1993) Germinating seeds of the root parasite *Orobanche aegyptiaca* Pers. excrete enzymes with carbohydrase activity. Symbiosis 15:61–70

Shomer-Ilan A (1999) Proteolytic activity of germinating *Orobanche aegyptiaca* seeds controls the degrading level of its own excreted pectinase and cellulase. Phytoparasitica 27:111

Singh A, Singh M (1993) Cell-wall degrading enzymes in *Orobanche aegyptiaca* and its host *Brassica campestris*. Physiol Plant 89:177–181

Westwood JH, Yu X, Foy CL, Cramer CL (1998) Expression of a defense-related 3-hydroxy-3-methylglutaryl CoA reductase gene in response to parasitization by *Orobanche* spp. Mol Plant Microbe Interact 11:530–536

Chapter 6
The Physiology of the Established Parasite–Host Association

James H. Westwood

6.1 General Physiological Considerations

The connection between parasite and host lies at the core of parasitism. The haustorium is the feature that defines parasitism (see Chap. 2), and it forms a unique point of interaction between a parasitic plant and its environment. The haustorium shares some functional characteristics with roots in that it absorbs water and minerals, but it also acts as an organ for carbon acquisition as it absorbs sugars and other organic molecules similar to the way a minor leaf vein uploads the products of photosynthesis. Depending on the parasitic species under consideration, the materials obtained from the host may include part or all of the resources needed for parasite growth. This chapter concentrates on the physiological processes associated with the mature haustorium, the transfer of materials and the relevant metabolic capacity needed to integrate with overall parasite metabolism.

Given the central importance of haustorial function in resource acquisition, it is surprising that many details of haustorium physiology remain poorly characterized. What types of materials are translocated from host to parasite? By what mechanisms do these transfers occur? Is the parasite selective in extracting certain compounds and excluding others? Do parasites induce changes in the host that increase the flow of nutrients? And how does the parasite use the acquired material?

The answers to these questions are complicated by the fact that parasite species may have widely differing nutritional dependencies on their hosts. The Orobanchaceae is remarkable in that the full trophic spectrum of parasitism is represented among its various species (Westwood et al. 2010). Levels of host dependence range from facultative hemiparasites to obligate hemiparasites and obligate holoparasites. Each of these has specific requirements of their hosts and

J.H. Westwood (✉)
Department of Plant Pathology, Physiology, and Weed Science, Virginia Tech, Blacksburg, VA, USA
e-mail: westwood@vt.edu

D.M. Joel et al. (eds.), *Parasitic Orobanchaceae*,
DOI 10.1007/978-3-642-38146-1_6,

hence different strategies for resource acquisition. When considering parasites, the specific nature of vascular continuity (xylem and phloem connections, see Sect. 3.9) is more important than whether the parasite attaches to shoot or root, so care must be taken in generalizing trends across different parasites (Hibberd and Jeschke 2001; Irving and Cameron 2009). With this in mind, it is useful to consider key features associated with xylem and phloem feeding.

6.1.1 Xylem Feeders

All parasites have xylem connections that take up water and minerals from their hosts, so a major difference among parasites lies in the extent to which they also have phloem or other symplastic connections. Parasites that rely on xylem connections as the primary route of transfer from hosts are considered to be xylem feeders. Among xylem-feeding parasites, further distinctions are made between facultative hemiparasites that have a significant functional root system of their own (and thus have a viable alternative to haustorial connections) and obligate hemiparasites that are more dependent on the haustorium for capturing host resources (see Chap. 2). The xylem-feeding facultative parasites that have been the subjects of most research into haustorium physiology and metabolism are *Rhinanthus* and *Triphysaria* spp., while studies of obligate hemiparasites have focused almost exclusively on *Striga* spp. Anatomical studies, which are described in Sect. 2.1.5.1, demonstrate direct xylem connections (including oscula), e.g. between *Rhinanthus* and its hosts (Cameron et al. 2006), with similar structures also being observed in *Triphysaria* (Heide-Jørgensen and Kuijt 1995), some *Striga* species (Dörr 1997) and other parasites. Indeed, even obligate holoparasites appear to have luminal contact with the xylem of their hosts (Dörr 1996), so there is a universal ability among parasites to acquire resources from the host xylem.

Dissolved nutrients moving from host to parasite in the xylem stream appear to be assimilated in the parasite haustorium. High metabolic capacity in this region is suggested by the hyaline tissue anatomy (see Sect. 3.12.1) present in the haustoria of many hemiparasites such as *Striga* (Dörr 1997; Neumann et al. 1999) and *Rhinanthus* (Jiang et al. 2010). In *Triphysaria*, which lacks a distinctive hyaline tissue, the cells of the haustorium adjacent to the xylem bridge nevertheless have densely staining cytoplasm and invaginated plasma membranes characteristic of hyaline tissue cells (Heide-Jørgensen and Kuijt 1995). These cells are similar to xylem parenchyma cells of non-parasitic plants that efficiently import solutes from xylem into neighbouring tissues. The transfer of sugars from the xylem would provide metabolic energy to cells of the haustorial region, permit sugar conversion into parasite-specific compounds, and avoid potentially unfavourable accumulation of solutes in the xylem.

Parasite maintenance of a distinct metabolic profile, which generates a favourable osmotic gradient relative to the host and drives the uptake of water, is an important requirement for parasite competitiveness with respect to the host. It is therefore reasonable to hypothesize that an important role of the haustorium is the initial metabolism of host-derived resources.

6.1.2 Phloem Feeders

The phloem-feeding holoparasites that have received the most research attention among Orobanchaceae are species of the genera *Orobanche* and *Phelipanche*. These parasites have direct symplastic connections between their own cells and the sieve elements of their hosts as observed by electron microscopy (Dörr and Kollmann 1995). The discovery of such symplastic connections is supported by physiological and tracer studies that indicate direct phloem transmission of dyes, proteins (Aly et al. 2011) and even viruses (Gal-On et al. 2009). A potential addition to this category could include *Striga gesnerioides*, which has chimeric plasmodesmata bridging its cells with those of the host rather than direct phloem connections (Dörr 1996). The precise role of this latter symplastic feeding is uncertain because the plasmodesmatal connections may be short lived (Reiss and Bailey 1998), but measures of carbon gain indicate that *S. gesnerioides* obtains nearly all of its photosynthates from hosts, which is similar to known phloem feeders (Press and Graves 1991).

Symplastic continuity has important implications for host–parasite transfer of materials. Rather than relying on extraction of nutrients from the host xylem stream, symplast-feeding parasites may absorb a potentially wider range of photo-assimilates directly from the host phloem. These parasites have much lower water needs because they grow underground for a prolonged period of time before their shoots emerge from the soil, and even then they lack transpiring expanded leaves. This suggests a different type of metabolic interaction of holoparasites with their hosts as compared to hemiparasites, a difference that is reflected in anatomy (see Sect. 3.8). The haustoria of *Orobanche* and *Phelipanche* lack the hyaline tissue that is typical of some hemiparasites. This may indicate a lower rate of nutrient uptake from the xylem stream, although it is also possible that the function of the hyaline tissue is accomplished in the tubercle (see Sect. 3.14). No physiological studies have directly addressed this issue.

6.1.3 Apoplastic vs. Symplastic Phloem Loading

The parallels between haustorium feeding strategies described above and physiological mechanisms of phloem loading are informative. Phloem loading has typically been studied in relation to sugar movement from mesophyll cells into phloem cells of minor veins in leaves for subsequent translocation to other parts of the plant. However, loading is also important in the context of parasitism, because the parasite haustorium functions in acquiring carbon from the host analogously to the leaf minor veins. Furthermore, variations in phloem loading also correspond with the different parasite feeding strategies, so the general concept has potential application to all parasites.

Two broad categories of phloem loading, symplastic and apoplastic, are recognized in plants, although gradations exist between these two and some species may not fall clearly into either camp (Gamalei 1989; Rennie and Turgeon 2009). Symplastic loading takes place when sugars are transported into phloem companion cells from surrounding cells via plasmodesmata. This process is passive and based on diffusion along concentration gradients. Apoplastic loading, in contrast, involves exporting sugars from cells around the vein into the apoplastic space and then rapidly re-importing them into the phloem companion cells. This process requires active transport of sugars and carries an energetic cost. Symplastic- and apoplastic-transporting plants differ anatomically in the number of plasmodesmatal connections between companion cells and surrounding cells. Symplastic loaders have many more connections, as would be required for the transport of sugars via plasmodesmata. Apoplastic loading species have phloem that is more isolated symplastically, with fewer plasmodesmatal connections to neighbouring cells. There is also a biochemical difference, as all apoplastic loaders transport sucrose whereas symplastic loading species typically translocate raffinose series oligosaccharides (raffinose, stachyose, verbascose and galactinol). The use of these sugars enables 'polymer trapping', whereby simple sugars move from one cell to another where they are polymerized into higher-order polysaccharides that do not readily re-diffuse out of the cell, thereby creating a concentration gradient that facilitates further movement. The principles of how apoplastic and symplastic loading could be applied to the xylem- and phloem-feeding strategies of parasites are illustrated in Fig. 6.1.

The ability to load phloem via apoplastic vs. symplastic mechanisms could have been a pre-adaption in the evolution of parasitism. The loading method tends to be a characteristic feature of plant families, with most members of a family usually having the same mechanism of phloem loading. Unfortunately, a comprehensive study of loading in the Orobanchaceae has not been completed and the literature contains conflicting reports. The only study of phloem loading mechanisms in the Orobanchaceae reported that the leaves of the hemiparasites *Castilleja* and *Pedicularis* have the symplastic type, while the hemiparasites *Cymbaria* and *Melampyrum* (which belong to different clades of the Orobanchaceae; see Sect. 14.2.2 and Fig. 14.2) have variations of apoplastic loading (Gamalei 1989). Study of three non-parasitic members of the related family Scrophulariaceae concluded that they use a combination of symplastic loading with polymer trapping (Rennie and Turgeon 2009). Finding raffinose series oligosaccharides in parasites would be an indirect indicator of symplastic loading style, but the only report of stachyose or related sugars in parasitic Orobanchaceae was in *Orobanche foetida* growing on faba bean where it seems that these carbohydrates are derived from the host rather than synthesized by the parasite (Abbes et al. 2009).

An alternate hypothesis is that parasite success is more related to the phloem loading mechanism of the hosts than to that of the parasites themselves. After all,

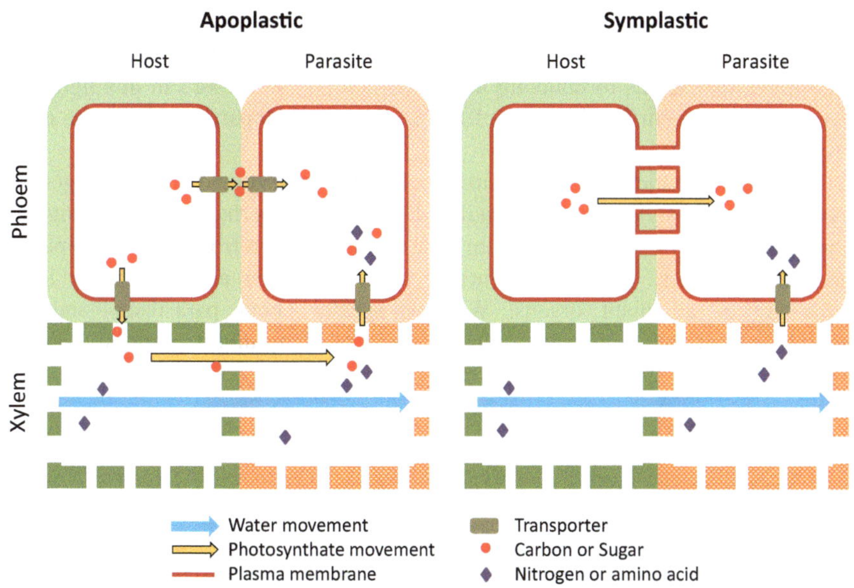

Fig. 6.1 Pathways of nutrient transfer from host to parasite. Schematic diagram of apoplastic and symplastic pathways of nutrient transfer from host to parasite. Apoplastic movement requires unloading sugars into the apoplast and then uptake into the parasite cell. Uptake may occur directly in adjacent cells or after some distance of travel in xylem. Symplastic transfer of sugars occurs through plasmodesmata or sieve pore connections between host and parasite. Uptake of water and nutrients from host xylem may still be important for phloem-feeding parasites

host resources must be available for uptake by the parasite, and in the case of xylem-feeding parasites, there would be clear advantages to attacking hosts that are genetically preprogrammed to export resources into the apoplastic space (as in Fig. 6.1). Whether by coincidence or evolution, this appears to hold true for many hosts of xylem-feeding parasites. Apoplastic loading is regarded as a more evolutionarily advanced development and is characteristic of grasses and herbaceous plants (Gamalei 1989). In barley (and apparently also in maize and other Poaceae) the loading and unloading of phloem appears to be apoplastic (Evert et al. 1996). In certain parasitic plant interactions with hosts, the host becomes enriched in levels of specific nitrogenous compounds needed by the parasite (Pageau et al. 2003), suggesting that part of the parasite effect is the reprogramming of host metabolism in a way that promotes export into the apoplast of resources needed by the parasite.

Maintaining a high concentration of osmotically active solutes in the parasite relative to the host is evolutionarily advantageous for the parasite. In this sense the use of apoplastic pathways of resource capture seems to maintain membrane separation between parasite and host. A membrane barrier is necessary for maintaining osmotic differentials to create a driving force for water and solutes into the parasite. The barrier may also help protect the parasite from risks associated with open connections with the host. Unregulated connections could result in accumulation of unwanted materials from the host, including salts, phytoalexins, and even pathogens detrimental to parasite health (see Sects. 3.9.1 and 3.13). Such separation also seems to maintain metabolic autonomy. Viewed in this light, the evolution of phloem feeding (symplastic loading) in holoparasites is all the more remarkable.

There is no evidence that parasitic Orobanchaceae evolved from a progenitor having a specific style of phloem loading. However, it is perhaps not surprising that core functionalities of the haustorium have features known from other plant physiological processes. The mechanisms of nutrient transfer are part of the basic repertoire of all plant organs. What is special about parasitism—and the haustorium—is the evolutionary reconfiguring of many diverse processes into a new organ.

6.2 Nutrient Acquisition and Transport

6.2.1 Water and Mineral Nutrients

Water and mineral nutrients move from host to parasite in what appears to be a continuous xylem pathway. Electron micrographs of the host–parasite junction reveal parasite xylem elements in open connection with their hosts (Dörr 1996) that facilitate transmission of macromolecules of at least up to 70 kDa (Aly et al. 2011). Therefore, the point of regulation for uptake of xylem contents exists not at the host–parasite boundary, but within the parasite system at the interface of all cells where there is unloading of xylem contents into the symplast. The water potential of parasite cells in the haustorium, stem and leaves (in hemiparasites) determines the flow.

Important aspects of xylem feeding by parasites include the types of solutes available in the host xylem stream, how the solutes are accumulated by the parasite and how the overall movement of water is regulated. In general, the flow of mineral nutrients into the parasite reflects the content of the host xylem (Irving and Cameron 2009). Nevertheless, parasites selectively accumulate certain cations and potassium is foremost among these, always occurring in concentrations higher than those recorded in corresponding host tissues (Table 6.1). The facultative parasite *Odontites lutea* accumulates 1.5 times more potassium than its host, while obligate parasites have four- to fivefold higher concentrations than their hosts. The obligate parasites appear to reach this same level of potassium accumulation despite different rates of xylem utilization between various species, e.g. between *Striga* and *Orobanche* (Hibberd et al. 1999). Calcium, magnesium and sodium are also among the minerals with highest concentrations in parasites (Abbes et al. 2009), but these rarely accumulate to levels higher than those found in host tissues.

Table 6.1 Concentrations of mineral nutrients in some Orobanchaceae parasites

Mineral	*Odontites lutea* leaf[a] (μmol g^{-1} DW)	*Striga hermonthica* leaf[b] (μmol g^{-1} DW)	*Orobanche cernua* xylem sap[c] (mM)	*O. fasciculate* tissue[d] (mM)
K^+	75 (1.50)	1,381 (5.35)	19.4 (4.04)	501.3 (4.45)
Ca^{2+}	146 (0.49)	45 (1.38)	1.85 (1.54)	331.8 (0.47)
Mg^{2+}	46 (0.66)	617 (2.94)	1.78 (2.44)	148.1 (0.49)
Na^+	–	–	0.41 (1.05)	69.6 (1.60)
Fe^{2+}	0.99 (0.14)	5 (0.75)	–	21.4 (0.53)
Zn^{2+}	0.54 (0.93)	1 (0.32)	–	0.3 (0.75)
Cu^{2+}	0.16 (1.53)	–	–	0.2 (0.59)
Mn^{2+}	–	3 (1.08)	–	0.8 (0.65)

The number in parentheses is the ratio of cation concentration in parasites divided by the concentration in the corresponding host tissue. Values from references b and c were converted to mM

O. fasciculate = *Aphyllon fasciculatum*

[a]Llugany et al. (2009)

[b]Stewart et al. (1984)

[c]Hibberd et al. (1999)

[d]Brotherson et al. (2005)

The driving force for water flow is the gradient of water potential between host and parasite (Ehleringer and Marshall 1995). Parasites maintain lower water potentials relative to hosts through accumulation of solutes, open stomata or a combination of these, depending on the type of parasite. Parasites lower their water potentials by accumulating high levels of osmotically active compounds such as mineral ions (e.g. potassium), sugars and sugar alcohols (Sect. 6.2.2.2). In hemiparasites water potentials are decreased by maintaining open leaf stomata even when the host is under severe water stress (Smith and Stewart 1990). For example, *Rhinanthus* and *S. hermonthica* plants that are attached to hosts keep open stomata (Jiang et al. 2003), resulting in a high rate of transpiration that preserves high flow of xylem contents into the parasite. It is notable that the stomata in free-living *Rhinanthus* are regulated in a more usual manner, being closed much of the time, but after *Rhinanthus* attaches to a host, the leaf stomata remain continuously open (Jiang et al. 2003).[1] Models of host resource acquisition by *Rhinanthus* indicate that water use by the parasite is a relatively minor fraction of the total water flux of the host (Fig. 6.2). Holoparasites have lower transpiration rates than hemiparasites because holoparasites are underground for much of their lives and even after emergence lack the expanded leaves necessary for effective transpiration.

The flow of water from host to parasite could be limited by the number of haustorial attachments (Cameron et al. 2005) and the size of the connections. The cross-sectional area for xylem transport of a lateral haustorium is small, measured

[1] Guttation through glandular hydathodes may also contribute to xylem flow, as suggested by Renaudin and Garrigues (1967, Sur l'ultrastructure des glandes en bouclier de *Lathraea clandestina* L. et leur role physiologique. C R Acad Sci Paris 264:1984–1987).

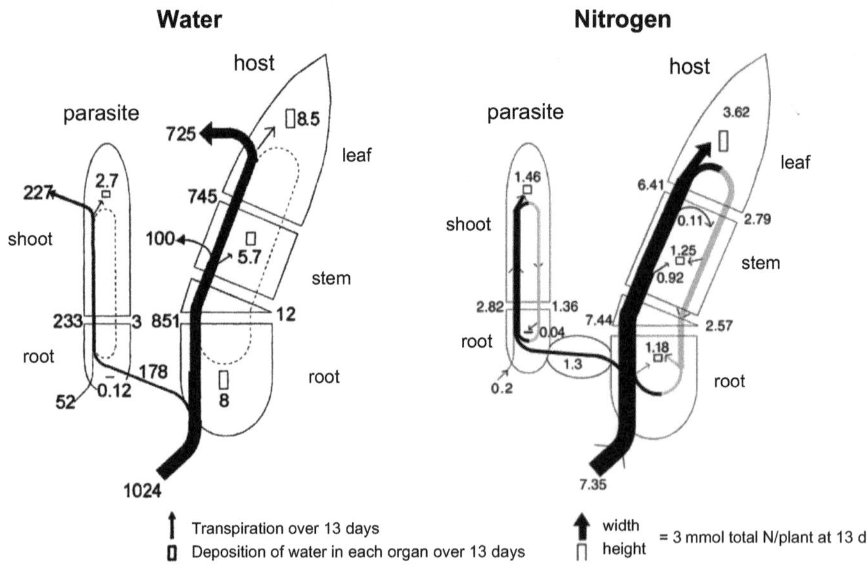

Fig. 6.2 Water and nitrogen transfer from host to parasite. Diagrams for flow of water and nitrogen from barley to *Rhinanthus minor*. Width of the arrows represents flow for xylem (*black arrows*) and phloem (*dashed/grey arrows*). The numbers indicate the water transpiration (ml) or nitrogen flux (mmol) in the respective figures (figure is compiled from Jiang et al. (2003) with permission of Oxford University Press and Jiang et al. (2004a) with permission from CSIRO Publishing)

at a tenth of the xylem area of a host stem in one case (Hibberd et al. 1999). Even a mature parasite with multiple haustoria has much less physical capacity for translocation compared to the vascular system of an autotrophic plant. The small haustorial cross section is compensated by a decrease of the hydraulic conductivity of the parasite tissues, which could be attributed in part to anatomical adaptations. *Rhinanthus* roots apparently lack Casparian strips in the endodermis and hypodermis (Jiang et al. 2003) that would restrict apoplastic water movement. Also, the high abscisic acid (ABA) concentrations typical of parasites (see Sect. 6.4.1) are correlated with increased hydraulic conductivity of cell membranes (Hose et al. 2001), which may further facilitate efficient water uptake through the haustoria.

6.2.2 *Carbon*

6.2.2.1 Carbon Assimilation

The proportion of organic carbon and nitrogen derived from hosts is directly related to the level of parasite dependence on hosts (Irving and Cameron 2009). Hemiparasites are able to fix some or all their carbon through photosynthesis so are primarily parasitic for nitrogen (inorganic or organic forms), while holoparasites require both nitrogen and carbon from their hosts. In terms of carbon, a rule of thumb is that facultative parasites obtain 10 % of their carbon from hosts, obligate hemiparasites such as *Striga* (emerged) take about 30 % and holoparasites such as *Orobanche* take 100 % (Irving and Cameron 2009). Of course, the exact proportion of carbon acquired depends on many factors including growth stage of the parasite (e.g. underground *Striga* is 100 % dependent on hosts), host species and environmental availability of resources for host and parasite. This may explain the wide ranges of host carbon assimilation that are documented in the literature.

Estimated carbon budgets for host–parasite interactions highlight differences among parasite species and the pronounced influence of host species on parasite metabolism. The heterotrophic carbon gain for the hemiparasites *Euphrasia rostkoviana* and *R. minor* was estimated at 25 and 50 %, respectively, with the greater carbon gain by *R. minor* being consistent with its greater negative impact on hosts (Těšitel et al. 2010). The estimates of photosynthesis rates for *Striga* species suggest that the parasites are not capable of fixing enough carbon themselves to meet their own respiration and growth needs, thus being dependent on their hosts for most carbon (Press et al. 1991). Specifically, the percent of carbon *S. hermonthica* gained from hosts was estimated to be around 80 % when growing on millet, 66 % on maize and 40 % on sorghum. *S. gesnerioides* parasitizing cowpea gained greater than 99 % of its carbon from cowpea, placing it at the extreme of carbon dependence for these genera (Press and Graves 1991).

Holoparasites, by definition, derive all of their carbon needs from the host, with various species taking up and accumulating different sugars. For example, it is estimated that greater than 99 % of *O. cernua* carbon is taken from tobacco phloem (Hibberd et al. 1999). The question of which form of carbon is acquired by the parasites is more difficult to determine because the parasites rapidly convert host compounds into parasite metabolites. The main host-derived compound in *O. crenata* growing on $^{14}CO_2$-fed faba beans was sucrose, followed by glucose and fructose (Aber et al. 1983). The concentrations of all sugars in parasite tubercles were between six- and eightfold higher than concentrations of the same compounds in the host roots, indicating a strong accumulation by the parasite. Sucrose is not accumulated in *P. ramosa* but rather is metabolized to other compounds such as hexoses, mannitol and starch. It was hypothesized that cleavage of sucrose into glucose and fructose by invertases and the reverse action of sucrose synthases improve the osmotic potential of the parasite. Indeed, at least one invertase gene from *P. ramosa* was expressed at high levels during seed germination and in all subsequent growth stages (Draie et al. 2011). Ultimately, sugars are converted into storage products, and holoparasites such as *O. foetida* parasitizing faba bean accumulate starch, especially in the tubercles as compared to shoots (Abbes et al. 2009). *Orobanche foetida* growing on faba bean substantially changed the carbohydrate profile. Whereas the faba bean phloem sap contained primarily raffinose and stachyose, with significant levels of fructose, glucose and sucrose, tubercles and shoots of *O. foetida* contained highest levels of glucose and fructose, followed by stachyose, mannitol and sucrose (Abbes et al. 2009).

Additional insights into parasite carbon acquisition have come from experimental manipulations of atmospheric CO_2 levels. For the facultative parasite *R. minor* growing on *Poa pratensis*, elevated CO_2 concentrations (650 ppm compared to 350 ppm) led to greater growth of the parasite, but did not significantly alter the percentage of total carbon or nitrogen in the parasite tissue, indicating that the parasite was able to correspondingly increase its level of nitrogen along with carbon (Hwangbo et al. 2003). Raising CO_2 levels in this case did not benefit the host, which decreased in biomass in response to parasitism regardless of CO_2 availability. Raising CO_2 levels from 350 to 700 ppm in the *S. hermonthica* interaction with a C_4 grass, *Eragrostis pilosa*, led to an increase in parasite photosynthesis rate and increased soluble sugar content, but did not translate into an increase in parasite biomass (Watling and Press 1998). In this case *Striga* derived 27 % of its carbon from the host under ambient CO_2, and this increased to 39 % under high CO_2, suggesting that the *Striga* benefitted more from increases in host photosynthesis than from changes in its own photosynthesis. Again, the host did not benefit from higher CO_2 in terms of growth as its biomass accumulation was reduced by half by *Striga* parasitism under either CO_2 regime. In contrast to these observations, elevated CO_2 levels (550 ppm) did not significantly affect growth of the holoparasite *O. minor* on *Trifolium repens*, but did produce greater biomass accumulation by the host, which largely alleviated the negative impact of parasitism on the host (Dale and Press 1998). In sum, the impacts of changes in CO_2 levels on host–parasite interactions will vary depending on the parasite capacity for photosynthesis. Rising atmospheric CO_2 levels are likely to benefit hemiparasites that are able to take advantage of gains in photosynthesis, while holoparasites may not benefit.

6.2.2.2 Mannitol Metabolism

The capacity to biosynthesize the polyol mannitol has long been recognized in parasitic plants and has been detected widely in Orobanchaceae genera, including *Orobanche* and *Phelipanche* (Harloff and Wegmann 1993), *Striga* (Robert et al. 1999), *Euphrasia*, *Lathraea*, *Orthocarpus*, *Odontites*, *Parentucellia*, *Pedicularis* and *Rhinanthus* (Press 1995). Mannitol has also been reported in other parasitic lineages such as *Thesium humile* (Santalaceae) (Simier et al. 1994). Other polyols such as galactitol may be accumulated in addition to mannitol or may even be the principal polyol, as is the case for *Melampyrum* (Press 1995).

While these sugar alcohols provide a strong osmoticum that drives the flow of water into the parasite (Ehleringer and Marshall 1995), polyols are not limited to parasitic plants. Mannitol is the most common sugar alcohol in nature and has been found in over 100 species of non-parasitic plants besides the Orobanchaceae, including some host plants (Stoop et al. 1996). Nevertheless, mannitol can be a useful solute for plants because of its many physiological functions. In addition to acting as a storage form for reduced carbon, it functions as a compatible solute that can be accumulated to high physiological concentrations without damaging the cell.

Mannitol has features of an osmoprotectant, reducing cellular stress by substituting for water in coating macromolecules, thereby preserving their activity and function. Finally, mannitol can scavenge reactive oxygen species. These same benefits of mannitol are thought to play an important role in the success of pathogenic fungi (Vélëz et al. 2007), so it appears that mannitol is an especially advantageous chemical for pathogens of many types, including parasitic plants.

Mannitol is commonly synthesized as a product of photosynthesis in source leaves and transported to sink tissues where it may be converted back to mannose and related sugars such as fructose (Noiraud et al. 2001; Stoop et al. 1996). The site of mannitol synthesis in parasitic plants has not been demonstrated, but *S. hermonthica* sap contained high levels of mannitol (58 % of soluble sugar) whereas no mannitol was detected in sap of the sorghum host (Press et al. 1991). Mannitol represented up to 77 % of total soluble sugars in *S. asiatica* leaves, and although *Orobanche hederae* stems contained a lower concentration of just 34 %, this is still a substantial portion of the carbon reserve. *Orobanche foetida* also synthesizes mannitol as an important component of the carbohydrate profile (Abbes et al. 2009).

The key metabolic step in mannitol synthesis in plants is the conversion of mannose-6-P to mannitol-1-P by the enzyme mannose 6-phosphate reductase (M6PR). The importance of mannitol to parasites has prompted suggestions that M6PR would be an ideal target for selectively disrupting parasite growth (Robert et al. 1999). This was tested using trans-specific gene silencing (see Sect. 6.5.1) to specifically reduce the levels of M6PR gene expression in *P. aegyptiaca* attached to transgenic tomato hosts (Aly et al. 2009). The experiment caused a decrease in M6PR transcript levels in the parasite and reduced the percentage of mannitol in the tubercles by up to 50 % in the best case. The silencing appeared to lead to increased mortality of tubercles, suggesting that mannitol accumulation is in fact a vital process for parasites. The fact that less than complete parasite death resulted from the decreased mannitol may be due to inefficiencies in the transmission of gene silencing from the host plant such that M6PR enzyme production was not completely shut off or alternatively that mannitol accumulation is not of vital importance for the parasite. It will be interesting to know how parasites would survive if mannitol synthesis was completely disrupted.

6.2.3 Nitrogen

6.2.3.1 Nitrogen Assimilation

The acquisition of fixed nitrogen from another plant is perhaps the greatest benefit of parasitism for hemiparasites. By tapping into a host, the parasite gains access to an expanded root system and a direct supply of fixed inorganic nitrogen. Considering that nitrogen is the limiting nutrient in many ecosystems, this is an important advantage in conditions of high competition and nutrient scarcity. An additional gain is that the parasite can accomplish this with relatively low investment in its own root system and assimilation machinery. Prior to emergence from the ground *S. hermonthica* depended on its host for 100 % of its carbon but only up to 59 % of its nitrogen (Aflakpui et al. 2005), with the remainder perhaps coming directly from the soil through *Striga* roots (Igbinnosa and Thalouarn 1996). In contrast, the emerged parasite was more self-sufficient for carbon, with mature plants taking only about 35 % of their carbon from the host, but the reverse was true for nitrogen. As the parasite became older, its dependency on host inorganic nitrogen increased until it was nearly 100 % in the mature parasite (Aflakpui et al. 2005).

In plants, the first two major steps in nitrogen assimilation are the reduction of nitrate to nitrite by the enzyme nitrate reductase and the conversion of nitrite to ammonium by nitrite reductase. Nitrate can be taken up by cells of roots or leaves, stored in vacuoles and transported in xylem from root to shoot. Nitrate reductase can occur in roots or shoots, depending on the species, but is always located in the cytoplasm of cells. The nitrite reductase enzyme functions in plastids but is coordinated with nitrate reductase ensuring that all phytotoxic nitrite is converted to ammonium. Nitrate reductase activity is substrate-inducible by nitrate and is co-regulated with photosynthesis (light) and carbon metabolism, thereby ensuring sufficient energy for the reduction reactions.

Parasites have less capacity for nitrate reduction than fully autotrophic plants. For example, the *S. hermonthica* nitrate reductase activity is much lower than that of host plants and the enzyme is not inducible by addition of nitrate (Press et al. 1986). *O. crenata* has no detectable nitrate reductase activity (Press et al. 1986; Stewart et al. 1984). The finding that nitrate reductase is low in hemiparasites and absent in holoparasites is consistent with parasites being able to obtain from their hosts all or a significant proportion of their nitrogen in fully reduced forms such as ammonium or amino acids.

The next step in assimilating inorganic nitrogen into organic compounds is the incorporation of ammonium into the amino acid glutamine by glutamine synthetase (GS). Glutamine synthetase generally occurs in two forms in plants, with each form localized to distinct regions of the cell and having a specialized role in metabolism (McNally et al. 1983). GS1 is located in the cytosol of cells and is expressed more prominently in roots where it functions in primary assimilation of ammonium. GS2 is located in chloroplasts, is the most abundant form in leaves and is thought to have a major role in re-assimilation of ammonium from photorespiration. Whereas the expression of GS2 is associated with photosynthetic cells of leaves and stems, GS1 expression is associated with phloem. Non-parasitic plants generally have two to four genes for GS1 and one gene for GS2, but parasites are unusual in that the GS2 form is generally missing or reduced in activity. That parasites would contain primarily GS1 is consistent with this form of glutamine synthetase being more closely associated with the phloem and with expression in non-photosynthetic tissues such as etiolated tissues or roots.

S. asiatica, *S. hermonthica* and *S. gesnerioides* contain both GS1 and GS2 forms, with GS1accounting for 80–90 % of their total glutamine synthetase activity (Press et al. 1986; Stewart et al. 1984). Parasites that entirely lack GS2 include the holoparasites *O. cernua*, *O. hederae*, *O. minor* and *P. ramosa* (McNally et al. 1983). The holoparasite *Lathraea clandestina* was initially reported to be among those species that lack GS2, but subsequent studies detected low levels of the enzyme activity, with highest activity in the scale leaves (Thalouarn et al. 1987). These authors used antibodies to identify the two forms of glutamine synthetase and found GS1 in the cytosol and a form weakly corresponding to GS2 in the stroma of amyloplasts. Further investigation of this species demonstrated the presence of several other nitrogen metabolic enzymes including nitrate reductase, nitrite reductase, glutamate synthase, glutamate oxoglutarate aminotransferase and glutamate dehydrogenase, suggesting that the parasite is equipped to deal with a variety of nitrogenous compounds coming from the host (Thalouarn et al. 1988).

Taken together, the reports of missing or reduced activity of nitrogen assimilation machinery could imply that some parasites—and especially holoparasites—have undergone an evolutionary reduction that leaves them dependent on their hosts for nitrogen that has already been converted to reduced and organic forms, such as ammonium and amino acids. The main argument against this is the ability for parasites such as *Striga* and *Phelipanche* to grow and develop on minimal media culture (Deeks et al. 1999; Zhou et al. 2004). While all basic culture media contain ammonium, the fact that parasites growing in such conditions can produce recognizable structures such as infective roots, shoots and in some cases even flowers indicates that they are able to process inorganic nitrogen with their own metabolic systems. The fact that herbicidal inhibitors of amino acid synthesis (e.g. glyphosate and acetolactate synthase inhibitors; see Sect. 23.2.1) kill attached parasites provides further evidence that they rely substantially on their own enzymes for amino acid synthesis (Eizenberg et al. 2012; Gressel 2009).

Asparagine synthase is another important enzyme in nitrogen assimilation, transferring ammonium from glutamine to asparagine for subsequent storage, transport and metabolism and also playing an important role in parasite metabolism (Pageau et al. 2003). Asparagine has the second highest N:C ratio of any amino acid, which may explain why it evolved to be a storage amino acid. The expression of asparagine synthase is suppressed by light and enhanced in darkness and is thus most abundant in etiolated tissues. Asparagine synthase of *S. hermonthica* is encoded by a small gene family of at least two genes, one of which is not suppressed in light (Simier et al. 2005). The *Triphysaria versicolor* enzyme is also encoded by a small gene family of 2–3 genes, and asparagine synthase gene expression in the roots was induced by host root exudates, suggesting that it is induced in the parasite prior to host contact (Delavault et al. 1998). This asparagine synthase was not induced by the haustorium-inducing quinone DMBQ, so its expression pattern is distinct from the haustorial induction pathway (see Sect. 4.3).

6.2.3.2 Nitrogen Uptake and Translocation

Accessing nitrogen from the vascular system of a host plant presents 'challenges' for a parasite given that the form of translocated nitrogen varies considerably depending which host species is encountered. For example, some plant species translocate nitrate from roots to shoots, while other species directly metabolize nitrate to amino acids or other organic forms in root cells prior to transport to the shoots (Pate 1973). Most plants use some combination of these approaches, so parasites drawing from the host xylem stream encounter compounds ranging from nitrate and amino acids to ammonium, amides and ureides (Miller et al. 2009; Pate 1973; Schjoerring et al. 2002; Xu et al. 2012). Parasites would be expected to have the ability to adapt to nutrient content of their hosts, and as discussed above, the xylem parenchyma cells of the haustorium are likely geared for the rapid extraction of host nitrogenous compounds. In principle, absorption of translocated nitrogen into parasite cells could take place at any point along the parasite continuum from the haustorium to the shoot tip. This would not require unusual metabolic capacity on part of the parasite because nitrate and amino acid transporters are known to occur in all parts of plants to regulate nitrogen uptake, metabolism and regeneration (Miller et al. 2009).

Facultative hemiparasites can absorb mineral nitrogen from soil using their own roots. Nevertheless, haustoria provide a significant supplement and it was estimated that *R. minor* acquired 17 % of its mineral nitrogen from the xylem of a grass host *Cynosurus cristatus* (Cameron and Seel 2007). The parasites appear to have no selective mechanism to accomplish this uptake from the host, as the proportion of mineral nitrogen taken into the parasite from xylem of barley roots was roughly the same as that of phosphate and potassium (~20 % for each) and water (Jiang et al. 2003, 2004a) (Fig. 6.2). This is consistent with an unregulated exchange between continuous xylem cells of the two species.

To answer the question of whether host transport of different nitrogen forms is a factor in parasite success, *R. minor* was grown on *Vicia faba* plants that obtained their nitrogen from either roots containing N_2-fixing nodules or roots lacking nodules but fed a complete nutrient solution. The rationale was that nodules produce organic forms of nitrogen whereas nutrient solution supplies inorganic forms. No differences in parasite growth were attributable to differences in host nitrogen composition, although the overall greater nitrogen levels in the fertilized system resulted in higher nitrogen content in both hosts and parasites (Jiang et al. 2008). It is noteworthy that although *R. minor* takes a relatively small fraction of the host mineral nitrogen, this is still a detriment to the host and the competition for nitrogen between *R. minor* and its host *Poa pratensis* appears to be more important than competition for carbon (Hwangbo et al. 2003).

Nitrogen uptake by obligate hemiparasites also occurs primarily through xylem. Rapid transfer of host root-applied ^{15}N-nitrate occurred between *Sorghum bicolor* and *Striga hermonthica* (Pageau et al. 2003). The labelled nitrogen in the *Sorghum* xylem sap was distributed about evenly among nitrate and free amino acids, indicating *Sorghum* ability to fix some nitrogen in the roots and to translocate both inorganic and organic forms within the plant. The parasite appeared to take up the nitrogen in approximately the same ratios as present in the host xylem, although nitrate and amino acid concentrations in *S. hermonthica* xylem were several times higher than those of the host (Pageau et al. 2003) (Fig. 6.3). The labelled nitrate was

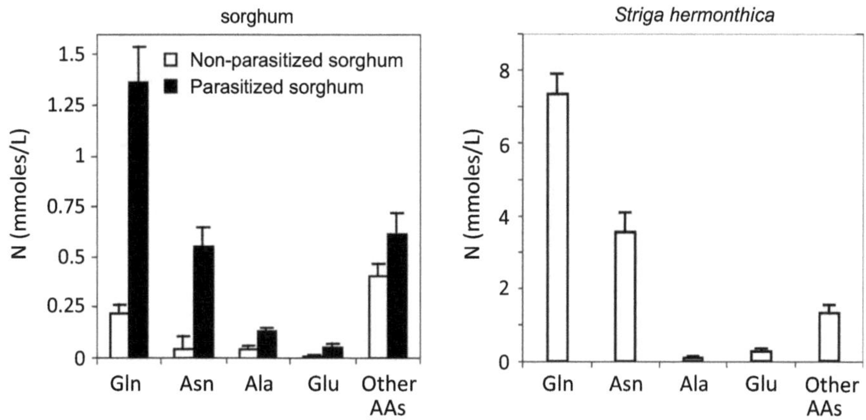

Fig. 6.3 Free amino acids in xylem sap of host and parasite. Nitrogen content of the major free amino acids (mmol N L^{-1}) in xylem sap from non-parasitized and parasitized sorghum (*left*) and from *Striga hermonthica* parasitizing sorghum (*right*) (figure is from Pageau et al. (2003) with permission of Oxford University Press)

assimilated in the parasite into predominantly glutamine and asparagine. These amino acids are the first products in the process of assimilating ammonium (more on this below) and are important transport forms of nitrogen, so it is not surprising that they are the first to appear in the parasite. This study also measured accumulation of ^{15}N-labelled compounds in different organs and suggests distinct patterns of metabolism. Specifically, the haustorium had a tenfold higher concentration of nitrate than parasite roots or shoots or even host roots (Pageau et al. 2003). In contrast, free amino acids occurred at low concentrations in the haustorium, but much higher concentrations in the shoots. A striking example is asparagine, which was measured at 0.05 μmol g^{-1} FW in haustoria but more than two orders of magnitude higher in *S. hermonthica* roots and three orders of magnitude in shoots. Such differences argue for a distinct set of metabolic functions between the haustorium and the rest of the parasite.

The profiles of amino acids are parasite species specific, reflecting different metabolic requirements and resources available from hosts. However, in general parasites tend to accumulate aspartate, asparagine, glutamate and glutamine. Other accumulated amino acids include alanine and arginine, depending on the parasite in question (Nandula et al. 2000; Press et al. 1986). The levels for selected amino acids found in *Striga*, *Orobanche* and *Phelipanche* are shown in Table 6.2. Although the data reflect various parasites, hosts and experimental conditions, the broad trends are similar and show common metabolic equilibrium points for these parasites.

Many amino acids, including aspartate and glutamate, accumulate in xylem sap of plants under stress, which would suit the needs of parasites (Nemec 1995). Parasites can thus influence the quality of resources coming from the host. *S. hermonthica* parasitism induced an increase in both nitrate and free amino acid levels in xylem sap of sorghum (Pageau et al. 2003). Furthermore, the

Table 6.2 Examples of free amino acid content reported for parasites

Amino acid	*Striga hermonthica* xylem sap[a] (nmol cm^{-3})	*S. hermonthica* shoot tissue[b] (μmol g^{-1} FW)	*Orobanche foetida* tubercle tissue[c] (mM)	*Phelipanche aegyptiaca* tubercle tissue[d] (μmol g^{-1} DW)
Aspartate		4.6	11.7	3.3[e]
Asparagine	25–50	81.4	17.1	
Glutamate	500–1,500	3.5	2.5	3.4[e]
Glutamine	50–750	2.4	0.2	
Alanine	10–20	2.0	0.4	3.9
Arginine	–	–	1.0	6.7
Glycine	Trace	–	0.1	0.3
Isoleucine	Trace	–	0.3	0.9
Leucine	10–20	–	0.4	1.1
Serine	Trace	–	0.5	2.4
Threonine	Trace	–	0.36	1.3
Valine	5–30	–	0.56	1.5

[a]Stewart et al. (1984)
[b]Pageau et al. (2003)
[c]Abbes et al. (2009)
[d]Nandula et al. (2000)
[e]Combined asp + asn and glu + gln

concentrations of xylem glutamine and asparagine increased dramatically (Fig. 6.3). Carrot leaves had similar or higher amino acid levels when parasitized by *P. aegyptiaca* compared to those of non-parasitized plants (Nandula et al. 2000). In contrast, roots of parasitized carrots had similar or lower levels than non-parasitized plants. The parasite tubercles were at the same time higher in asparagine/aspartate and glutamine/glutamate than associated host roots, and this held true for most amino acids, indicating preferential accumulation in the parasite. The differences in parasite effect on host amino acid composition between *Striga* and *Phelipanche* may reflect the different feeding styles of the parasites and hence different stresses placed on the host.

Changes in the nutritional status of the host would be expected to affect the parasite since parasites are linked directly to their hosts for their nutritional supply (see Sect. 22.3.3). Several experiments have looked at this possibility by varying nitrogen or carbon status of the host. Increasing nitrogen fertilization in a sorghum-*S. hermonthica* association led to greater nitrogen incorporation into parasite leaves compared to hosts, demonstrating the proficiency of these parasites at extracting nitrogen (Cechin and Press 1993). The photosynthesis rate of *S. hermonthica* also increased with higher nitrogen supply, but the overall effect of high nitrogen was that the parasites suffered. *S. hermonthica* grew well at lower rates of nitrogen, but its ability to gain carbon decreased from 27 % to just 6 % as nitrogen rates increased (Cechin and Press 1993). Another study failed to find this nitrogen effect in the interaction between *S. hermonthica* and maize, but the concentrations of N used were lower and may not have caused the same level of stress in the parasite (Aflakpui et al. 2005).

The fact that these parasites combine high capacity for nitrogen uptake with a reduced capacity for nitrogen metabolism may explain why they generally do not grow well under conditions of high soil nitrogen fertility (Igbinnosa and Thalouarn 1996). For example, high rates of ammonium nitrate applied to sorghum decreased biomass accumulation of attached *S. hermonthica* (Cechin and Press 1993). Toxicity in this case may be due to elevated uptake of nitrogen from the host and accumulation in parasite leaves, leading to an overload of the parasite's metabolic system. These parasites are especially well adapted to conditions of low nitrogen and seem to have limited capacity to deal with excess nitrate and ammonium. This may be particularly true of certain obligate parasites, which are especially susceptible in the seedling stage when they have limited biomass and metabolic reserves. This could amount to a counter-adaptation for dealing with the metabolic demands of transporting and metabolizing unusually high levels of nitrogen (Britto and Kronzucker 2002). The concept of fertilizing to reduce parasitic weeds has been demonstrated repeatedly, although many conflicting reports also exist (e.g. Kamara et al. 2007) and various factors, such as soil types, nitrogen forms and weather, likely confound the effect. It is also possible that soil fertility affects parasite success through its impact on germination signalling more than on later metabolic interactions with the host (Fernández-Aparicio et al. 2011). Nevertheless, adequate fertility can be part of an integrated approach to control of certain parasitic weeds (Hearne 2009; Tesso and Ejeta 2011) (see Sect. 22.3.3 for discussion of agronomic aspects).

6.3 Direction of Movement

Most of the literature on parasite nutrition assumes that movement occurs as a rule from host to parasite. Indeed, the strong translocation of materials from host to parasite is well documented in the literature, but it is not clear whether lack of evidence for parasite-to-host movement is due to actual negative results or because the question is rarely addressed. Considering the physiology of vascular connections, the most likely cases of parasite-to-host translocation should be observed in holoparasites where phloem connections may allow for bidirectional flow of materials. In fact, some movement from parasite to host has been demonstrated for both xylem and phloem feeders. One example of 'reverse flow' comes from an experiment in which radiolabelled CO_2, urea and sulphur were used to trace translocation between sorghum and *S. senegalensis* (Okonkwo 1966). Low amounts of radiolabelled material applied to the parasite also moved into the host, with above-background levels of photo-assimilate detected in host tissues and in other non-labelled parasites attached to the same host. A similar study in which ^{14}C-urea or $^{14}CO_2$ were applied to the hemiparasite *Odontites verna* detected low levels of radiolabelled products in the host (Govier et al. 1967).

The above results require that xylem flow be reversed in the host–parasite interaction, at least temporarily. This could be possible if the osmotic pull of the parasite were overcome by the host or otherwise interrupted (see Chap. 3 for structural aspects). An experiment using dye loaded into the cut root tip of *P. aegyptiaca* showed rapid movement of the dye into the host, ultimately appearing in the host leaves (Aly et al. 2011). The experiment required cutting the parasite root, which disrupted the integrity of the parasite system, but nevertheless it demonstrated that open flow is possible between host and parasite and suggested that the direction of flow can be reversed by physical damage to the parasite.

RNAs appear to move bidirectionally between parasites and hosts (see Sect. 6.5.1). The evidence to date strongly suggests that RNAs move both into and out of *Triphysaria*. *Triphysaria* (Tomilov et al. 2008) can bridge two different host plants and transmit RNAi signals across a span of parasite tissue (similar to the Convolvulaceae parasite *Cuscuta*; Birschwilks et al. 2006), demonstrating both entry and exit from the parasite. Only host-to-parasite movement of viruses has been indicated for *Orobanche* (Gal-On et al. 2009), but the possibility of transmission in the reverse direction has not been tested.

Taken together, these data support a model of parasite physiology in which the haustorial connections allow bidirectional flow of materials, but the dominant flow occurs towards the parasite. A reversal of flow could occur as part of the tension between host and parasite or any environmental or physiological conditions that temporarily weaken the osmotic advantage of the parasite.

It is also possible that parasites have a mechanism that allows flow of compounds such as toxins or effectors into the host that facilitate parasitism. It has long been speculated that parasites such as *Striga* secrete substances into their hosts that cause stunting or otherwise reconfigure host metabolism (Musselman 1980), and backflow could be one mechanism to accomplish that goal.

6.4 Hormone Interactions

6.4.1 Abscisic Acid and Cytokinin

A general feature of the parasitic Orobanchaceae is the accumulation of ABA and cytokinin in the parasites following attachment to hosts. Cytokinin levels are low in unattached parasites but increase nearly 100-fold once the parasite has attached to the host (Lechowski and Bialczyk 1996). ABA levels follow a similar trend, increasing dramatically after attachment to the host. For example, the hemiparasite *R. minor* has about 35 times more ABA than its barley host even while growing autotrophically, but after attachment the parasite ABA concentration increases to 53 times higher than its host (Fig. 6.4) (Jiang et al. 2003). Xylem sap of the hemiparasite *Melampyrum arvense* has about a third of the ABA level of its *Capsella* host when growing autotrophically, but the ABA concentration of attached parasites exceeds that of the host by 50 % (Lechowski 1996). Pre- and

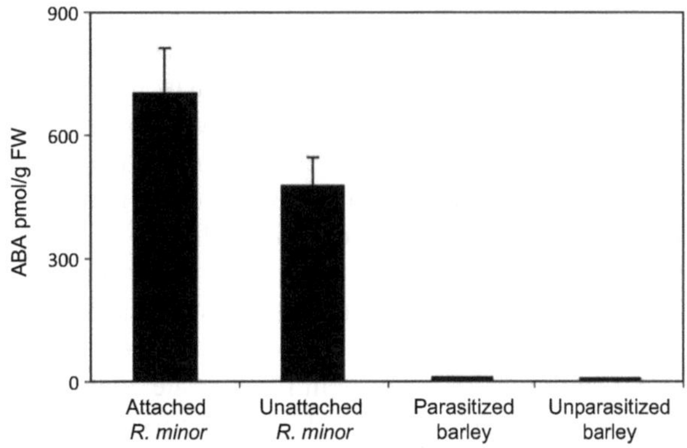

Fig. 6.4 ABA status in host–parasite association. ABA concentrations in leaves of *Rhinanthus minor* and barley living in parasitic association and independently (figure is from Jiang et al. (2003) with permission of Oxford University Press)

post-attachment ABA levels are not documented for obligate parasites, but the approximately tenfold higher ABA levels in *S. hermonthica* leaves as compared to its maize host suggest that elevated ABA levels in parasites are common (Taylor et al. 1996).

The reports of high ABA levels have led to much speculation about the mechanisms of ABA accumulation and its role in the life of the parasite. The high ABA concentration is primarily due to increased ABA synthesis in the parasites, although there is certainly a contribution from host xylem. In fact, many parasites stimulate their hosts to increase synthesis of ABA, which is then available for uptake into the parasite or may directly affect host physiology to the benefit of the parasite (Lechowski 1996). The ABA levels in sorghum leaf tissue and xylem sap were approximately doubled following parasitism by *S. hermonthica* (Frost et al. 1997). ABA levels were also higher in maize tissues parasitized by *Striga* (Taylor et al. 1996). This does not seem to hold true for barley (*Hordeum vulgare*), which maintained ABA levels following parasitism by *R. minor* (Jiang et al. 2004b). It is possible that low water potentials in the parasite stimulate the increase in ABA concentrations due to the role of ABA as a drought-associated hormone, but no clear explanation for high ABA levels in the host is currently available (Jiang et al. 2010).

The functions of elevated ABA concentrations in the host–parasite association are not certain, although effects in host and parasite are likely different. In host plants, high ABA may have a role in reducing host stomatal apertures and lead to reduced photosynthesis and plant growth (Frost et al. 1997). While detrimental to host growth, this can lead to a shift in water usage from the host to the parasite, thereby ensuring sufficient water flow into the parasite. ABA also suppresses the salicylic acid-mediated defence response in plants, and increasing ABA levels in host tissues—through either direct synthesis by the pathogen or induction of host biosynthesis—is a common theme in fungal and bacterial pathogenesis (Cao et al. 2011). The increase in host ABA concentrations could contribute to the observed lack of salicylic acid associated defence responses in parasitized host roots (Griffitts et al. 2004; Vieira Dos Santos et al. 2003).

In most plants ABA acts to induce closing of stomata, but parasite stomata remain open almost constantly despite exceptionally high ABA concentrations. Parasites are less sensitive to ABA than non-parasites as evidenced by higher concentrations of ABA needed to induce closing of *R. minor* stomata compared with host stomata (Jiang et al. 2003). Another possible explanation for open stomata in parasites is high cytokinin levels, which may antagonize the ABA effect and force open the stomata. Stomata of *Melampyrum arvense* are closed in darkness when this hemiparasite is growing without a host but open wide and become insensitive to cytokinins after the parasite attaches to a host (Lechowski 1997).

Root hydraulic conductivity is also regulated by ABA in plants (Markhart et al. 1979) and high ABA levels in the parasite could contribute to greater water flow into the parasite (Jiang et al. 2004b). The parasite–host connections are not limited by physical barriers such as the Casparian strips (see Sect. 3.16) and the effect of ABA on increasing membrane permeability would be advantageous to the parasite.

6.4.2 Auxin

The role of IAA in established parasites has received less attention than ABA or cytokinin. The central role of auxin in plant growth and development suggests that this hormone must also be important to parasites, but no role has been documented in the mature haustorium. Auxin is important in the development of vascular tissue in the haustorium and may be important in establishing the directionality of xylem differentiation between host and parasite (Bar-Nun et al. 2008). Auxin is also an important component for regulation of haustorial formation in *Triphysaria* (Tomilov et al. 2005). *Triphysaria* forms haustoria more readily when exposed to IAA (Tomilov et al. 2004), and inhibitors of auxin and ethylene reduced the number of haustoria in *Triphysaria* (Tomilov et al. 2005).

6.5 Macromolecules

6.5.1 Proteins and RNA

In addition to translocating small molecules from host to parasite, the haustorium may also transmit large molecules such as proteins and nucleic acids. As discussed above for small molecules, the type of molecules transmitted may depend on the anatomy of the specific parasite and the metabolic needs of that parasite. Many reports of macromolecular trafficking between hosts and parasites come from *Cuscuta*, which unlike the Orobanchaceae appears to have exceptionally open connections to host vascular tissues. *Cuscuta* takes up many substances from the host phloem, including dye tracers (Birschwilks et al. 2006), soluble proteins

(Haupt et al. 2001), mRNA (Roney et al. 2007) and viruses (Birschwilks et al. 2006). It appears that these same compounds are also mobile into at least certain holoparasitic members of the Orobanchaceae.

The phloem-localized dye carboxyfluorescein and the green fluorescent protein (GFP) are both readily mobile into *P. aegyptiaca* (Aly et al. 2011). Importantly, the GFP had to be expressed in host phloem cells and in a form soluble in the cytoplasm in order to move to the parasite. Parallel host transformants containing GFP with an endoplasmic reticulum targeting signal did not move. The GFP was localized to the parasite phloem, suggesting that it was translocated through phloem connections.

Single-stranded RNA and DNA viruses can also translocate from hosts to *P. aegyptiaca* (Gal-On et al. 2009). The question of whether mRNAs are mobile between hosts and Orobanchaceae has not been satisfactorily addressed, but there is evidence for movement of small RNAs associated with gene silencing. The best example to date is from *T. versicolor* plants that were transformed to constitutively express the *GUS* (β-glucuronidase) reporter gene, the expression of which was shut down following parasitism of lettuce that expressed a silencing construct for *GUS* (Tomilov et al. 2008). Both the *GUS* mRNA levels and the activity of the GUS enzyme were suppressed in the parasite near the point of contact and in young root tissues that developed after parasitizing the host, indicative of host-to-parasite transport of the silencing signal. Although the GUS silencing was strongest immediately around the point of haustorial connection to the host, silencing also occurred away from the haustorium, indicating that the signal was transmitted some distance through the parasite tissue.

A second example of this trans-specific gene silencing process comes from a study aimed at suppressing mannose 6-phosphate reductase (M6PR) expression in *P. aegyptiaca* (Aly 2007). In this case, tomato was transformed with a silencing construct targeting the parasite version of M6PR, which was discussed above for its role in mannitol metabolism and osmotic regulation in the parasite (Sect. 6.2.2.2). Parasite M6PR transcript levels were reduced in the tubercles of plants parasitizing the transgenic tomato, but no information is available on the spatial distribution of the silencing effect in the parasite.

In contrast to the situation in *Triphysaria* and *Phelipanche*, the trans-specific gene silencing approach has not yet succeeded in silencing *Striga asiatica* (L. Kuntze) *genes*. RNA interference constructs were generated for five essential *S. asiatica* genes and were transformed into maize (*Zea mays*), but parasites grown on these hosts showed no measurable effect (de Framond et al. 2007). Although this report was from a work in progress and several experimental factors could have contributed to the results, it suggests that caution is required in generalizing about uptake dynamics of different parasites. It is reasonable to hypothesize that *Triphysaria* and *Striga* are equivalent in their mechanisms of xylem feeding, but the details of RNAi signal transmission and function in host–parasite interactions need further study to understand the apparent differences among parasite species.

Beyond the demonstration of macromolecule movement between hosts and parasites, the biological function of such exchange is unknown. The *Triphysaria* and *Phelipanche* experiments with gene silencing provide a strong indication that gene silencing signals not only move but may function in the parasites. Post-transcriptional gene silencing in plants takes many forms but involves the generation of 21–24 bp double-stranded RNA by a dicer-like enzyme, leading in many cases to a systemic signal that can propagate throughout a plant. At the core of this signal is a single-stranded RNA about 21 nucleotides long (Kehr and Buhtz 2008). In order for this to have practical significance in altering parasite gene expression, the host must express a silencing RNA that has high homology to a corresponding section of parasite RNA (see Sect. 24.4.2). The extent to which this occurs will become clearer as more parasite gene sequences become available.

6.5.2 DNA

Horizontal gene transfer between parasitic plants and their hosts is another indicator of nucleic acid transfer. In these cases, horizontal gene transfer is usually discovered as part of phylogenetic studies in which some genes align with greater homology to counterparts in distantly related species rather than to close relative species as judged by traditional phylogenetic arrangements (see Sect. 15.5). This indicates the evolutionary transfer of a gene through a means other than typical vertical inheritance from parent to progeny. The process of nucleic acid movement between species required for horizontal gene transfer is facilitated by direct haustorial connections between donor and recipient plants, so it is not surprising that parasitic plants are well represented among cases of horizontal gene transfer.

Orobanchaceae are involved in several examples of horizontal gene transfer that involve genes from mitochondrial (Mower et al. 2004), plastid (Park et al. 2007) and nuclear genomes (Yoshida et al. 2010). At this time little is known about the mechanisms of horizontal gene transfer or movement of RNAs and protein in plants, so it is only possible to speculate about the mechanisms involved in haustorial transfer of these materials. The movement of large sections of DNA between plants has been proposed as one mechanism for horizontal gene transfer (Mower et al. 2010), although the movement of a gene into *Striga* carries evidence of missing introns and a polyadenylated tail sequence that suggests the transfer was mediated by an mRNA molecule. To the extent that the haustorium functions as a continuation element of host and parasite vascular systems, it should be expected that any macromolecules capable of moving systemically in the host are equally capable of moving into the parasite. Messenger RNAs are known to move systemically in phloem (Lough and Lucas 2006), but no such evidence exists for movement of DNA. Large sections of DNA may transfer short distances in the area of a graft junction (Stegemann and Bock 2009) and haustoria share some features with grafts (Kuijt 1983), including inter-specific symplastic connections (Sect. 2.1.5.1), so it would be interesting to know whether DNA exchange can take place in the haustorial region. Stable incorporation of foreign DNA into a plant under this scenario would require that cells of the haustorial junction give rise to adventitious shoots that could flower and transmit the new gene(s). So far parasite regeneration from a haustorium was not described for the Orobanchaceae. Considering the current interest in horizontal gene transfer, parasitic plants are sure to receive additional attention.

6.6 Conclusions

The mature haustorium is the central feature in the interactions between host and parasite, yet many aspects of haustorium physiology and function remain unknown. Questions surround the precise mechanisms of uptake of the diverse range of materials transferred from host to parasite, as well as the metabolic fate of host molecules within the parasite. Complicating the subject are the significant biological differences represented in the range of host dependences encompassed by species of the Orobanchaceae. The haustoria of holoparasites that make both xylem and phloem connections to the host represent a different evolutionary direction than haustoria of hemiparasites that connect to just the xylem. These basic differences in function are compounded by other variations among parasitic species and the fact that parasite interactions may be influenced by specific hosts or environmental conditions, resulting in a diverse body of literature on the physiological interactions of various parasites and hosts.

Despite differences in parasite haustoria, some common themes span all parasites and point to the core functions of haustoria. Both hemi- and holoparasites have evolved mechanisms of nutrient uptake and transport to enable their growth. Foremost, haustoria are anatomical connections that enable exchange of resources between hosts and parasites. Regardless of whether parasites form direct connection with host symplast, all appear to have some ability to absorb reduced carbon compounds from hosts. Furthermore, haustoria appear to be the first line of parasite metabolism, actively converting host resources into chemicals used by the parasite. These include osmotic compounds that contribute to the process of drawing water and dissolved solutes into the parasite. The accumulation of osmotic solutes is important even in hemiparasite species in which water potential differentials are aided by transpiring leaves, as all Orobanchaceae appear to accumulate ions (primarily potassium) and sugar alcohols (mannitol), and have elevated synthesis and accumulation of ABA. Nitrogen (in either mineral or reduced forms) is an absolute requirement of parasites, and reliance on hosts for this nutrient is reflected in their diminished capacity to incorporate nitrogen via nitrate reductase and glutamine synthetase. The haustorium appears to be active in converting nitrogenous compounds to preferred amino acids (glutamine and asparagine) and in processing host-derived sugars into preferred simple sugars, sugar alcohols and starch.

Much of the research on physiological interactions between parasites and hosts has focused on large organs such as stems, leaves and roots. This is understandable because the haustorium is small and embedded in host tissues underground, so it presents a challenge to study. Nevertheless, this structure holds key insights into understanding parasitism and the tools of modern biology will allow detailed investigation into the metabolism and gene expression of tissues and cells within the haustorium to understand how it regulates interactions between host and parasite. Most transport and metabolic functions of the haustorium have parallels in other tissues or stages of plant development but have likely been recruited through evolution to function in the haustorium. It will be interesting to learn more about how these functions have been combined and coordinated in this remarkable structure.

References

Abbes Z, Kharrat M, Delavault P, Chaïbi W, Simier P (2009) Nitrogen and carbon relationships between the parasitic weed *Orobanche foetida* and susceptible and tolerant faba bean lines. Plant Physiol Biochem 47:153–159

Aber M, Fer A, Sallé G (1983) Etude du transfert des substances organiques de l'hôte (*Vicia faba*) vers le parasite (*Orobanche crenata* Forsk.). Z Pflanzenphysiol 112:297–308

Aflakpui GKS, Gregory PJ, Froud-Williams RJ (2005) Carbon (^{13}C) and nitrogen (^{15}N) translocation in maize-*Striga hermonthica* association. Exp Agric 41:321–333

Aly R (2007) Conventional and biotechnological approaches for control of parasitic weeds. In Vitro Cell Dev Biol Plant 43:304–317

Aly R, Cholakh H, Joel DM, Leibman D, Steinitz B, Zelcer A, Naglis A, Yarden O, Gal-On A (2009) Gene silencing of mannose 6-phosphate reductase in the parasitic weed *Orobanche aegyptiaca* through the production of homologous dsRNA sequences in the host plant. Plant Biotechnol J 7:487–498

Aly R, Hamamouch N, Abu-Nassar J, Wolf S, Joel D, Eizenberg H, Kaisler E, Cramer C, Gal-On A, Westwood J (2011) Movement of protein and macromolecules between host plants and the parasitic weed *Phelipanche aegyptiaca* Pers. Plant Cell Rep 30:2233–2241

Bar-Nun N, Sachs T, Mayer AM (2008) A Role for IAA in the Infection of *Arabidopsis thaliana* by *Orobanche aegyptiaca*. Ann Bot 101:261–265

Birschwilks M, Haupt S, Hofius D, Neumann S (2006) Transfer of phloem-mobile substances from the host plants to the holoparasite *Cuscuta* sp. J Exp Bot 57:911–921

Britto DT, Kronzucker HJ (2002) NH_4^+ toxicity in higher plants: a critical review. J Plant Physiol 159:567–584

Brotherson JD, Simmons BT, Ball T, Anderson WR (2005) Nutrient relationships between *Orobanche fasciculata* Nutt and its host *Artemisia pygmaea* Gray in the Uinta Basin of Utah, USA. West N Am Nat 65:242–247

Cameron DD, Seel WE (2007) Functional anatomy of haustoria formed by *Rhinanthus minor*: linking evidence from histology and isotope tracing. New Phytol 174:412–419

Cameron DD, Hwangbo JK, Keith AM, Geniez JM, Kraushaar D, Rowntree J, Seel WE (2005) Interactions between the hemiparasitic angiosperm *Rhinanthus minor* and its hosts: from the cell to the ecosystem. Folia Geobot 40:217–229

Cameron DD, Coats AM, Seel WE (2006) Differential resistance among host and non-host species underlies the variable success of the hemiparasitic plant *Rhinanthus minor*. Ann Bot 98:1289–1299

Cao F, Yoshioka K, Desveaux D (2011) The roles of ABA in plant-pathogen interactions. J Plant Res 124:489–499

Cechin I, Press MC (1993) Nitrogen relations of the sorghum-*Striga hermonthica* host-parasite association: growth and photosynthesis. Plant Cell Environ 16:237–247

Dale H, Press MC (1998) Elevated atmospheric CO_2 influences the interaction between the parasitic angiosperm *Orobanche minor* and its host *Trifolium repens*. New Phytol 140:65–73
de Framond A, Rich P, McMillan J, Ejeta G (2007) Effects on *Striga* parasitism of transgenic maize armed with RNAi constructs targeting essential *S. asiatica* genes. In: Ejeta G, Gressel J (eds) Integrating new technologies for *Striga* control: towards ending the witch-hunt. World Scientific, Hackensack, NJ, pp 185–196
Deeks SJ, Shamoun SF, Punja ZK (1999) Tissue culture of parasitic flowering plants: methods and applications in agriculture and forestry. In Vitro Cell Dev Biol Plant 35:369–381
Delavault P, Estabrook E, Albrecht H, Russell W, Yoder JI (1998) Host-root exudates increase gene expression of asparagine synthetase in the roots of a hemiparasitic plant *Triphysaria versicolor* (Scrophulariaceae). Gene 222:155–162
Dörr I (1996) New results on interspecific bridges between parasites and their hosts. In: Moreno MT, Cubero JI, Berner D, Joel D, Musselman LJ (eds) Advances in parasitic plant research. Junta de Andalucia, Cordoba, Spain, pp 195–201
Dörr I (1997) How *Striga* parasitizes its host: a TEM and SEM study. Ann Bot 79:463–472
Dörr I, Kollmann R (1995) Symplasmic sieve element continuity between *Orobanche* and its host. Bot Acta 108:47–55
Draie R, Péron T, Pouvreau J-B, Véronési C, Jégou S, Delavault P, Thoiron S, Simier P (2011) Invertases involved in the development of the parasitic plant *Phelipanche ramosa*: characterization of the dominant soluble acid isoform, PrSAI1. Mol Plant Pathol 12:638–652
Ehleringer J, Marshall J (1995) Water relations. In: Press MC, Graves JD (eds) Parasitic plants. Chapman & Hall, London, pp 125–140
Eizenberg H, Aly R, Cohen Y (2012) Technologies for smart chemical control of broomrape (*Orobanche* spp. and *Phelipanche* spp.). Weed Sci 60:316–323
Evert RF, Russin WA, Botha CEJ (1996) Distribution and frequency of plasmodesmata in relation to photoassimilate pathways and phloem loading in the barley leaf. Planta 198:572–579
Fernández-Aparicio M, Westwood JH, Rubiales D (2011) Agronomic, breeding, and biotechnological approaches to parasitic plant management through manipulation of germination stimulant levels in agricultural soils. Botany 89:813–826
Frost DL, Gurney AL, Press MC, Scholes JD (1997) *Striga hermonthica* reduces photosynthesis in sorghum: the importance of stomatal limitations and a potential role for ABA? Plant Cell Environ 20:483–492
Gal-On A, Naglis A, Leibman D, Ziadna H, Kathiravan K, Papayiannis L, Holdengreber V, Guenoune-Gelbert D, Lapidot M, Aly R (2009) Broomrape can acquire viruses from Its hosts. Phytopathology 99:1321–1329
Gamalei Y (1989) Structure and function of leaf minor veins in trees and herbs. Trees 3:96–110
Govier RN, Nelson MD, Pate JS (1967) Hemiparasitic nutrition in angiosperms. New Phytol 66:285–297
Gressel J (2009) Crops with target-site herbicide resistance for *Orobanche* and *Striga* control. Pest Manag Sci 65:560–565
Griffitts AA, Cramer CL, Westwood JH (2004) Host gene expression in response to Egyptian broomrape (*Orobanche aegyptiaca*). Weed Sci 52:697–703
Harloff HJ, Wegmann D (1993) Evidence for a mannitol cycle in *Orobanche ramosa* and *Orobanche crenata*. J Plant Physiol 141:513–520
Haupt S, Oparka KJ, Sauer N, Neumann S (2001) Macromolecular trafficking between *Nicotiana tabacum* and the holoparasite *Cuscuta reflexa*. J Exp Bot 52:173–177
Hearne SJ (2009) Control – the *Striga* conundrum. Pest Manag Sci 65:603–614
Heide-Jørgensen HS, Kuijt J (1995) The haustorium of the root parasite *Triphysaria* (Scrophulariaceae), with special reference to xylem bridge ultrastructure. Am J Bot 82:782–797
Hibberd JM, Jeschke WD (2001) Solute flux into parasitic plants. J Exp Bot 52:2043–2049

Hibberd JM, Quick WP, Press MC, Scholes JD, Jeschke WD (1999) Solute fluxes from tobacco to the parasitic angiosperm *Orobanche cernua* and the influence of infection on host carbon and nitrogen relations. Plant Cell Environ 22:937–947
Hose E, Clarkson DT, Steudle E, Schreiber L, Hartung W (2001) The exodermis: a variable apoplastic barrier. J Exp Bot 52:2245–2264
Hwangbo J-K, Seel WE, Woodin SJ (2003) Short-term exposure to elevated atmospheric CO_2 benefits the growth of a facultative annual root hemiparasite, *Rhinanthus minor* (L.), more than that of its host, *Poa pratensis* (L.). J Exp Bot 54:1951–1955
Igbinnosa I, Thalouarn PA (1996) Nitrogen assimilation enzyme activities in witchweed (*Striga*) in hosts presence and absence. Weed Sci 44:224–232
Irving LJ, Cameron DD (2009) You are what you eat: interactions between root parasitic plants and their hosts. Adv Bot Res 50:87–138
Jiang F, Jeschke WD, Hartung W (2003) Water flows in the parasitic association *Rhinanthus minor*/*Hordeum vulgare*. J Exp Bot 54:1985–1993
Jiang F, Jeschke WD, Hartung W (2004a) Solute flows from *Hordeum vulgare* to the hemiparasite *Rhinanthus minor* and the influence of infection on host and parasite nutrient relations. Funct Plant Biol 31:633–643
Jiang F, Jeschke WD, Hartung W (2004b) Abscisic acid (ABA) flows from *Hordeum vulgare* to the hemiparasite *Rhinanthus minor* and the influence of infection on host and parasite abscisic acid relations. J Exp Bot 55:2323–2329
Jiang F, Jeschke WD, Hartung W, Cameron DD (2008) Does legume nitrogen fixation underpin host quality for the hemiparasitic plant *Rhinanthus minor*? J Exp Bot 59:917–925
Jiang F, Jeschke W, Hartung W, Cameron D (2010) Interactions between *Rhinanthus minor* and its hosts: a review of water, mineral nutrient and hormone flows and exchanges in the hemiparasitic association. Folia Geobot 45:369–385
Kamara AY, Menkir A, Chikoye D, Omoigui LO, Ekeleme F (2007) Cultivar and nitrogen fertilization effects on *Striga* infestation and grain yield of early maturing tropical maize. Maydica 52:415–423
Kehr J, Buhtz A (2008) Long distance transport and movement of RNA through the phloem. J Exp Bot 59:85–92
Kuijt J (1983) Tissue compatibility and the haustoria of parasitic angiosperms. In: Moore R (ed) Vegetative compatibility responses in plants. Baylor University, Waco, TX, pp 1–12
Lechowski Z (1996) Abscisic acid content in the root hemiparasite *Melampyrum arvense* L. before and after attachment to the host plant. Biol Plant 38:489–494
Lechowski Z (1997) Stomatal response to exogenous cytokinin treatment of the hemiparasite *Melampyrum arvense* L. before and after attachment to the host. Biol Plant 39:13–21
Lechowski Z, Bialczyk J (1996) Cytokinins in the hemiparasite *Melampyrum arvense* L before and after attachment to the host. Biol Plant 38:481–488
Llugany M, Lombini A, Dinelli E, Poschenrieder C, Barceló J (2009) Transfer of selected mineral nutrients and trace elements in the host–hemiparasite association, *Cistus*–*Odontites lutea*, growing on and off metal-polluted sites. Plant Biol 11:170–178
Lough T, Lucas W (2006) Integrative plant biology: role of phloem long-distance macromolecular trafficking. Annu Rev Plant Biol 57:203–232
Markhart AH, Fiscus EL, Naylor AW, Kramer PJ (1979) Effect of abscisic-acid on root hydraulic conductivity. Plant Physiol 64:611–614
McNally SF, Hirel B, Gadal P, Mann AF, Stewart GR (1983) Glutamine synthetases of higher plants. Plant Physiol 72:22–25
Miller AJ, Shen Q, Xu G (2009) Freeways in the plant: transporters for N, P and S and their regulation. Curr Opin Plant Biol 12:284–290
Mower J, Stefanovic S, Young G, Palmer J (2004) Plant Genetics: gene transfer from parasitic to host plants. Nature 432:165–166

Mower J, Stefanovic S, Hao W, Gummow J, Jain K, Ahmed D, Palmer J (2010) Horizontal acquisition of multiple mitochondrial genes from a parasitic plant followed by gene conversion with host mitochondrial genes. BMC Biol 8:150

Musselman LJ (1980) The biology of *Striga*, *Orobanche*, and other root parasitic weeds. Ann Rev Phytopathol 18:463–489

Nandula VK, Foster JG, Foy CL (2000) Impact of Egyptian broomrape (*Orobanche aegyptiaca* (Pers.) parasitism on amino acid composition of carrot (*Daucus carota* L.). J Agric Food Chem 48:3930–3934

Nemec S (1995) Stress-related compounds in xylem fluid of blight-diseased citrus containing *Fusarium solani* naphthazarin toxins and their effects on the host. Can J Microbiol 41:515–524

Neumann U, Vian B, Weber HC, Sallé G (1999) Interface between haustoria of parasitic members of the Scrophulariaceae and their hosts: a histochemical and immunocytochemical approach. Protoplasma 207:84–97

Noiraud N, Maurousset L, Lemoine R (2001) Transport of polyols in higher plants. Plant Physiol Biochem 39:717–728

Okonkwo SNC (1966) Studies on *Striga senegalensis*. II. Translocation of C^{14}-labelled photosynthate, urea-C^{14} and sulphur-35 between host and parasite. Am J Bot 53:142–148

Pageau K, Simier P, Le Bizec B, Robins RJ, Fer A (2003) Characterization of nitrogen relationships between *Sorghum bicolor* and the root-hemiparasitic angiosperm *Striga hermonthica* (Del.) Benth. using $K^{15}NO_3$ as isotopic tracer. J Exp Bot 54:789–799

Park J-M, Manen J-F, Schneeweiss GM (2007) Horizontal gene transfer of a plastid gene in the non-photosynthetic flowering plants *Orobanche* and *Phelipanche* (Orobanchaceae). Mol Phylogenet Evol 43:974–985

Pate JS (1973) Uptake, assimilation and transport of nitrogen compounds by plants. Soil Biol Biochem 5:109–119

Press MC (1995) Carbon and nitrogen relations. In: Press MC, Graves JD (eds) Parasitic plants. Chapman & Hall, London, pp 103–124

Press MC, Graves JD (1991) Carbon relations of angiosperm parasites and their hosts. In: Wegmann K, Musselman LJ (eds) Progress in *Orobanche* research. Eberhard-Karls-Univrersitat, Tübingen, pp 55–65

Press MC, Shah N, Stewart GR (1986) The parasitic habit: trends in metabolic reductionism. In: ter Borg SJ (ed) Biology and control of *Orobanche*. LH/VPO, Wageningen, pp 96–106

Press MC, Smith S, Stewart GR (1991) Carbon acquisition and assimilation in parasitic plants. Funct Ecol 5:278–283

Reiss GC, Bailey JA (1998) *Striga gesnerioides* parasitising cowpea: development of infection structures and mechanisms of penetration. Ann Bot 81:431–440

Rennie EA, Turgeon R (2009) A comprehensive picture of phloem loading strategies. Proc Natl Acad Sci USA 106:14162–14167

Robert S, Simier P, Fer A (1999) Purification and characterization of mannose 6-phosphate reductase, a potential target for the control of *Striga hermonthica* and *Orobanche ramosa*. Aust J Plant Physiol 26:233–237

Roney JK, Khatibi PA, Westwood JH (2007) Cross-species translocation of mRNA from host plants into the parasitic plant dodder. Plant Physiol 143:1037–1043

Schjoerring JK, Husted S, Mäck G, Mattsson M (2002) The regulation of ammonium translocation in plants. J Exp Bot 53:883–890

Simier P, Renaudin S, Fer A (1994) Characteristics of the mannitol pathway in a root hemiparasitic species, *Thesium humile* Vahl. (Santalaceae). J Plant Physiol 143:33–38

Simier P, Delavault P, Demarsy E, Pouvreau J-B, Pageau K, Le Bizec B, Fer A, Thalouarn P (2005) Characterization of an unusually regulated gene encoding asparagine synthetase in the parasitic plant *Striga hermonthica* (Scrophulariaceae). Physiol Plant 123:9–20

Smith S, Stewart GR (1990) Effect of potassium levels on the stomatal behavior of the hemiparasite *Striga hermonthica*. Plant Physiol 94:1472–1476

Stegemann S, Bock R (2009) Exchange of genetic material between cells in plant tissue grafts. Science 324:649–651

Stewart G, Nour J, MacQueen M, Shah N (1984) Aspects of the biochemistry of *Striga*. In: Proceedings of the international workshop on *Striga* biology and control. ICSU, pp 161–178

Stoop JMH, Williamson JD, Mason Pharr D (1996) Mannitol metabolism in plants: a method for coping with stress. Trends Plant Sci 1:139–144

Taylor A, Martin J, Seel WE (1996) Physiology of the parasitic association between maize and witchweed (*Striga hermonthica*): is ABA involved? J Exp Bot 47:1057–1065

Těšitel J, Plavcová L, Cameron D (2010) Heterotrophic carbon gain by the root hemiparasites, *Rhinanthus minor* and *Euphrasia rostkoviana* (Orobanchaceae). Planta 231:1137–1144

Tesso TT, Ejeta G (2011) Integrating multiple control options enhances management and sorghum yield on heavily infested soils. Agron J 103:1464–1471

Thalouarn P, Rey L, Hirel B, Renaudin S, Fer A (1987) Activity and immunocytochemical localization of glutamine synthetase in *Lathraea clandestina* L. Protoplasma 141:95–100

Thalouarn P, Philouze V, Renaudin S (1988) Nitrogen metabolism key enzyme activities in a Scrophulariaceae holoparasite *Lathraea clandestina* L. J Plant Physiol 132:63–66

Tomilov A, Tomilova N, Yoder JI (2004) *In vitro* haustorium development in roots and root cultures on the hemiparasitic plant *Triphysaria versicolor*. Plant Cell Tissue Organ Cult 77:257–265

Tomilov AA, Tomilova NB, Abdallah I, Yoder JI (2005) Localized hormone fluxes and early haustorium development in the hemiparasitic plant *Triphysaria versicolor*. Plant Physiol 138:1469–1480

Tomilov AA, Tomilova NB, Wroblewski T, Michelmore R, Yoder JI (2008) Trans-specific gene silencing between host and parasitic plants. Plant J 56:389–397

Vélëz H, Glassbrook NJ, Daub ME (2007) Mannitol metabolism in the phytopathogenic fungus *Alternaria alternata*. Fungal Genet Biol 44:258–268

Vieira Dos Santos C, Letousey P, Delavault P, Thalouarn P (2003) Defense gene expression analysis of *Arabidopsis thaliana* parasitized by *Orobanche ramosa*. Phytopathology 93:451–457

Watling JR, Press MC (1998) How does the C_4 grass *Eragrostis pilosa* respond to elevated carbon dioxide and infection with the parasitic angiosperm *Striga hermonthica*? New Phytol 140:667–675

Westwood JH, Yoder JI, Timko MP, dePamphilis CW (2010) The evolution of parasitism in plants. Trends Plant Sci 15:227–235

Xu G, Fan X, Miller AJ (2012) Plant nitrogen assimilation and use efficiency. Ann Rev Plant Biol 63:153–182

Yoshida S, Maruyama S, Nozaki H, Shirasu K (2010) Horizontal gene transfer by the parasitic plant *Striga hermonthica*. Science 328:1128

Zhou WJ, Yoneyama K, Takeuchi Y, Iso S, Rungmekarat S, Chae SH, Sato D, Joel DM (2004) *In vitro* infection of host roots by differentiated calli of the parasitic plant *Orobanche*. J Exp Bot 55:899–907

Chapter 7
Host Reaction to Attack by Root Parasitic Plants

Michael P. Timko and Julie D. Scholes

7.1 Introduction

Plants are constantly challenged by a wide variety of pathogenic and parasitic organisms, including bacteria, fungi and oomycetes, viruses, insects, nematodes and parasitic plants. While our understanding of plant–plant pathogen interactions has improved dramatically in the past 25 years, the cellular, molecular and genetic factors that govern the interaction of parasitic plants with host and non-host species are still poorly understood. With improved genetic and molecular tools, including the ability to rapidly genotype individuals and track alterations in gene expression throughout development, information contributing to our understanding of mechanisms operating during the interaction of root parasitic angiosperms with potential host and non-host plant species is emerging rapidly. As a consequence researchers are beginning to better understand the similarities and often subtle differences that exist in host–parasite interactions and the processes that drive host selection and parasite virulence, and alternatively the processes of host resistance and susceptibility.

As described in Chap. 1, root parasitic angiosperms can be facultative (attaching to a suitable host if available, but capable of completing their life cycle without host contact), hemiparasitic (deriving part of their nutrition from a host and requiring host association to complete their life cycle) or holoparasitic (deriving all of their nutrition from a host species and, therefore, completely dependent upon the host to complete their life cycle). The root parasitic angiosperms successfully attach to the roots of a potential host species using a unique organ known as a haustorium (see Sect. 2.1 and Chap. 3; Kuijt 1969). Differentiation of the haustorium into a

M.P. Timko (✉)
Department of Biology, University of Virginia, Charlottesville, VA 22903, USA
e-mail: mpt9g@virginia.edu

J.D. Scholes
Department of Animal and Plant Sciences, University of Sheffield, Sheffield S10 2TN, UK

D.M. Joel et al. (eds.), *Parasitic Orobanchaceae*,
DOI 10.1007/978-3-642-38146-1_7,

penetration peg, transversing the host root cortex, breaching the endodermal barrier and formation of vascular connections with the host vascular system are essential developmental events that ensure parasite survival (see Chap. 5). For the purposes of our discussion, we refer to those plants capable of supporting parasite growth to maturity and to which the parasite has adapted its life cycle as hosts. In contrast, plants not normally found in association with the parasite and incapable of supporting parasite development are referred to as non-host (Press and Graves 1995). The delineation between host and non-host is artificial at best, since under strong selective pressure to survive, a parasite will challenge both in an attempt to find a suitable source of nutrition necessary to complete its life cycle.

It is also evident that within the host species range of a given parasite, individual host genotypes may display varied levels of resistance, tolerance or susceptibility to parasite attack, and the mechanism of host resistance to attempted parasitism can vary. We define **resistance** as the ability of the host to withstand parasite attack in a manner that prevents parasite establishment and growth, whereas **tolerance** connotes the ability to withstand damages inflicted by the parasite or by host defence. Few host species demonstrate complete resistance to parasite infection, and resistance is less frequent in the cultivated (domesticated) germplasm of most agronomically important plants, whereas wild relatives of crop species show a greater tendency to be fully or partially resistant, or tolerant to parasitism (Scholes and Press 2008; Hearne 2009). Host resistance can be multidimensional, involving both general and specific defence mechanisms constitutively deployed or activated to specifically interfere with critical steps throughout the parasite life cycle. Similarly, plants that cannot serve as a host for parasitic weeds may exhibit complex, multicomponent forms of resistance, including both constitutive and inducible defences.

One can broadly categorize observed mechanisms of resistance against parasitic weeds as being either pre-attachment or post-attachment resistance (Cameron et al. 2006; Scholes and Press 2008; Rodenburg et al. 2010). The former group includes all mechanisms that allow a potential host to avoid or prevent parasite attachment, including the absence or reduced production of germination stimulant, germination inhibition, inhibition or reduction of haustorium formation, partial inhibition of haustorium development and formation of mechanical barriers to infection such as thickened host root cell walls. Post-attachment resistance occurs once the haustorium has formed and the parasite attempts to penetrate the host root tissues and connect to the vascular system. During these developmental stages, different constitutive or induced incompatibility or host resistance mechanisms can be activated. These include the synthesis and release of cytotoxic compounds (e.g. phenolic acids, phytoalexins) by the challenged host root cells, a process generally referred to as abiosis, the formation of physical barriers to prevent possible pathogen ingress and growth (e.g. lignification and suberization of cell walls), programmed cell death (PCD) in the form of a hypersensitive response (HR) at the point of parasite attachment to limit parasite development and retard its penetration and prevention of the parasite establishing the essential functional vascular continuity (i.e. xylem-to-xylem and/or phloem-to-phloem connections) with the host. In this chapter, we primarily focus on post-attachment resistance responses. Most early descriptions of parasitic plant–host interactions examined how a single host species or set of closely related hosts responded to a particular parasite, leading many investigators to conclude that resistance responses were either absent or fairly uniform. As our understanding of parasite diversity increased (e.g. the existence of geographic variants and functional pathotypes or races), and as investigators began to better appreciate the subtle differences between the response of true hosts and non-hosts, a greater diversity in resistance responses were uncovered that represent a range of post-attachment resistance responses in host and non-host interactions. What is now clear is that host and non-host plants may differ significantly in their defence responses to parasites and the underlying molecular genetic mechanisms governing the response similarly differ, or host and non-host resistance can share components of the signalling pathways in common and exhibit similar resistance phenotypes (Thordal-Christensen 2003).

7.2 General Mechanisms of Host Resistance

Among the most widely accepted models of how plants respond to pathogen challenge is the 'zigzag' model, which suggests that plants respond to pathogen challenge using a two-level innate immune response system (Jones and Dangl 2006). Although derived largely from examinations of pathogenic microbe–plant interactions, it can easily be applied to discussions of resistance to root parasitic angiosperms. The first level responds to slowly evolving molecules (e.g. flagellin, lipopolysaccharides and elongation factor Tu found in Gram-negative bacteria, chitin and β-glucan present in fungi) collectively referred to as pathogen-associated molecular patterns (PAMPs) or microbe-associated molecular patterns (MAMPs) (Boller and He 2009). Various receptor-like kinases, termed pattern recognition receptors (PRRs), serve as host sensors that recognize MAMPS and PAMPs and activate host defence responses referred to as pathogen-triggered immunity (PTI) (Boller and Felix 2009; Ronald and Beutler 2010).

To evade this host surveillance mechanism, phytopathogens have evolved specific effectors and virulence factors that are capable of entering the host cell and suppressing PTI. A variety of effector proteins capable of suppressing PTI have now been characterized from various phytopathogenic bacteria, fungi, oomycetes and nematodes (Abramovitch et al. 2006; Bent and Mackey 2007; Torto-Alalibo et al. 2009). How these effectors enter into plant cells varies depending on their microbial origin; generally they enter plant cells via discrete secretion machinery (Göhre and Robatzek 2008; Tyler 2009). Interestingly, despite the general similarity of strategies shared among pathogens in overcoming host defence responses, little structural similarity is shared among effectors from diverse pathogens (McDowell and Simon 2008; Tyler 2009).

In an effort to further limit pathogen growth, plants have developed a second level of defence that perceives and attempts to disable such effectors, called effector-triggered immunity (ETI). ETI involves a second class of receptor proteins, typically containing a nucleotide binding site (NBS) and a leucine-rich repeat (LRR) domain (Takken et al. 2006; Tameling and Joosten 2007; Caplan et al. 2008). NBS-LRR proteins, commonly referred to as resistance (R) proteins, are generally encoded by genes identified by plant breeders as the major R genes that protect against specific strains (or races) of a pathogen (Ellis et al. 1997, 2007b). It is now clear that ETI corresponds to what was initially described by Flor (1971) as gene-for-gene resistance. Historically, pathogen genes encoding effectors detected by plant R genes have been called avirulence (Avr) genes, because they prevent infection of host plants containing those R genes.

How R-proteins function to bring about protection is still under active investigation, but it appears that they act mainly, but not exclusively, intracellularly and confer protection against biotrophic pathogens (viruses, bacteria, fungi, oomycetes, nematodes and parasitic plants) that need living host tissues for their proliferation and survival (Boller and He 2009; Ronald and Beutler 2010). From this one can infer that when Avr proteins or effectors enter host plant cells, they are detected either directly or indirectly by the R-proteins (Ellis et al. 2007a, b; Greenshields and Jones 2008; van der Hoorn and Kamoun 2008). When recognition occurs, a complex series of signalling events is activated leading to defence responses that limit pathogen or parasite growth.

7.3 Histological Characteristics of the Host Resistance Responses

Successful penetration of the host root cell layers and establishment of host–parasite vascular continuity are essential for parasite survival of *Orobanche* and *Striga*. From germination, through initial host contact and until the embedded haustorial penetration peg successfully establishes vascular connections and/or differentiates cells capable of nutrient transfer, the parasite is still an independent organism, utilizing its own resources for growth (Kuijt 1969; Riopel and Timko 1995). At each stage of parasite ingress, the challenged host has an opportunity to mount a resistance response that could effectively block parasite ingress or development. We often think of susceptibility (compatible host–parasite interactions) as a null situation where the apparent lack of host response belies their inability to recognize the invading haustorial penetration peg and activate the cascade of signalling events required to turn on a defence response. In fact, parasitic plants do not behave as a fully compatible partner even in susceptible interactions. The rate of parasite establishment, tubercle development and the final number of emerged shoots more properly reflect the ability of the parasite to overcome the variety of different resistance mechanisms activated to block parasitism. We know that the parasite is active in its attempt to enter the host. Studies of germinating *Orobanche* seedlings in the presence of various substrates indicate that the parasite secretes pectin methylesterase (PME), polygalacturonase and endocellulase, but not exocellulase, β-glucosidase or xylanase (Losner-Goshen et al. 1998; Shomer-Ilan 1993; Ben-Hod et al. 1993), to help facilitate entry into the host. These observations are supported by in situ immunocytochemical studies demonstrating that PME is present in *Orobanche* haustorial cells and detectable in the apoplast of adjacent host tissue (Losner-Goshen et al. 1998). There was also a measurable decrease in the amount of methylated pectins in the host middle lamella near the parasite intrusive cells. Similar observations were in a study examining the interactions of *Rhamphicarpa fistulosa* with its host (Neumann et al. 1999). As will be discussed later, the lack of a visible resistance response may also be the result of active suppression of host defences by the parasite through the release of a range of avirulence factors and effectors that specifically disrupt host defence pathways.

Resistance responses resulting from incompatible interactions with a parasite can occur at discrete times during parasite ingress and are either rapid or delayed. Hosts, as well as non-hosts, use a variety of mechanisms to offset parasite challenge including the rapid accumulation of electron-dense (poly)phenolic compounds at the host–parasite interface as seen in *S. hermonthica*–sorghum associations (Olivier et al. 1991a) and *S. asiatica* on a range of non-host plants (Hood et al. 1998). A rapid accumulation of phytoalexins has been reported to occur at the interface of *O. cumana* (syn. *O. cernua* ssp. *cumana*) and sunflower (Jorrín et al. 1996), and an accumulation and excretion of phytoalexins into the apoplast outside of cortical cells in sunflower and the central cylinder of *Medicago truncatula* infected by *Orobanche* has been observed (Echevarría-Zomeño et al. 2006; Lozano-Baena et al. 2007). In resistant vetch (*Vicia* spp.) genotypes, response to *Phelipanche aegyptiaca* (syn. *O. aegyptiaca*) parasitism is associated with the appearance of a reddish substance in the apoplastic space between host and parasite cells, and there is a measurable increase in the concentrations of phenolics and lignin and greater peroxidase activity than in susceptible species (Goldwasser et al. 1999, 2000).

The accumulation of polyphenolics and phytoalexins in response to attempted parasitism is often accompanied by the formation of physical barriers within the host root cortex. Cell wall thickening resulting from lignification, suberization and callose deposition was reported at the point of parasite penetration in cowpea (*Vigna unguiculata*) infected with *S. gesnerioides* (Lane et al. 1996; Botanga and Timko 2005), sunflower infected with *O. cumana* (Labrousse et al. 2001; Shergini et al. 2001; Echevarría-Zomeño et al. 2006; Letousey et al. 2007) and various legumes infected with *O. crenata* (Pérez-de-Luque et al. 2005a, b, 2007, 2008) (Fig. 7.1a, b). In resistant sunflower/*O. cumana* interaction, cell wall reinforcement was preceded by increased levels of *HaGLS1* gene transcripts encoding callose synthase (Letousey et al. 2007). Roots of pea plants (*Pisum sativum*) resistant to *O. crenata* have increased levels of peroxidase transcripts, an enzyme known to be involved in cell wall cross-linking; in situ hybridization studies have shown that a peroxidase and a β-glucanase are differentially expressed during the resistance response (Pérez-de-Luque et al. 2006a). Resistance to *Rhinanthus* parasitism was also shown to be controlled at the cellular level by either lignification or host cell death at the host–parasite interface (Irving and Cameron 2009). For example, when

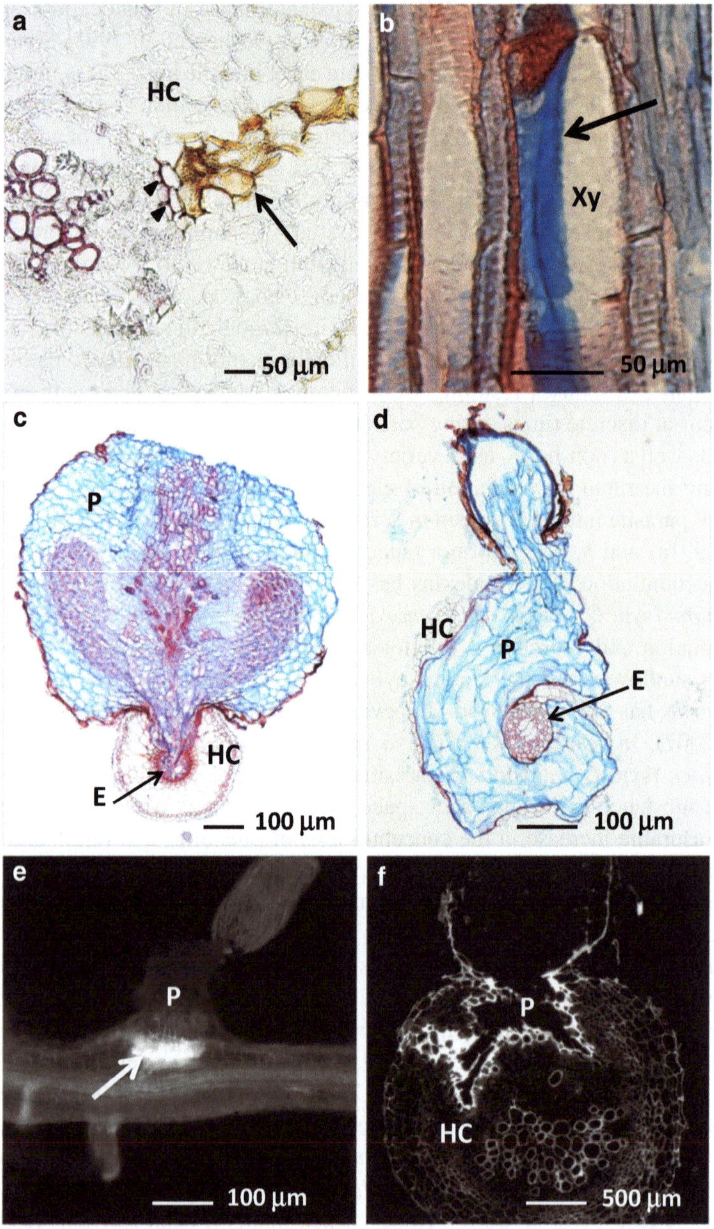

Fig. 7.1 Different types of resistance reactions to *Orobanche* and *Striga* species. (**a**) An unsuccessful *O. crenata* penetration in root of a resistant vetch cultivar 20 days after inoculation, showing lignification of host cells (stained *red* after phloroglucinol-HCl staining, *arrowheads*); *arrow* indicates accumulation of a *brown* secreted material. (**b**) Longitudinal section of a resistant vetch root showing a xylem vessel filled with mucilage 30 days after inoculation with *O. crenata*; *arrow* indicates accumulation of Alcian blue staining of carbohydrates in host vessel elements.

R. minor was grown on the non-grass forbs *Leucanthemum vulgare* and *Plantago lanceolata*, two natural host species, the invading parasite was encapsulated by lignin preventing access to the host's vascular system. In contrast, when *Cynosurus cristatus*, a grass, served as host, no resistance response was observed (Fig. 16.1; Cameron and Seel 2007). These investigators also noted that the mechanism of resistance differed between *L. vulgare* and *P. lanceolata*, with *L. vulgare* displaying lignifications of the host cell walls at the host–parasite interface and *P. lanceolata* host cells showing lignification as well as necrosis (see Sect. 16.2.1 and Fig. 16.1).

A HR-like rapid browning and localized death of host cells at the site of parasite attachment that appears similar to the HR observed in other incompatible plant–plant pathogen interactions have been described in some *Striga*–host and *Orobanche*–host associations. Lane et al. (1993) reported browning of host cortical tissues and rapid necrosis of the attached parasite in associations of resistant cowpea cultivars with *S. gesnerioides*. Mohamed et al. (2003) reported that several cultivated varieties of sorghum and the wild sorghum accession P47121 similarly exhibit an HR-like response upon *S. asiatica* infection. Gurney et al. (2006) described an HR-like response in the rice cultivar Nipponbare following infection by *S. hermonthica*. In both sorghum and rice, necrosis of host root tissues first became visible at the site of parasite attachment 3 days postinfection (dpi) of the roots and reached a maximum at 7–12 dpi. This timescale is longer than that observed for many fungal or bacterial pathogens that elicit an HR response, and it is not clear whether the underlying molecular and biochemical changes are similar.

The appearance of an HR blocking haustorial penetration of *P. aegyptiaca* was described in *Vicia atropurpurea* (Goldwasser et al. 1997), and the failure of attached *Orobanche* seedlings to thrive and eventual necrosis of the parasite has been observed in interactions of the parasite with a number of resistant hosts including sunflower (Labrousse et al. 2001), various commercial legumes (e.g. *Cicer arietinum, Vicia faba, Lathyrus cicera and Pisum sativum*) (ter Borg 1999; Rubiales et al. 2003, 2005; Sillero et al. 2005; Pérez-de-Luque et al. 2009; Fernández-Aparicio and Rubiales 2010) and the model legume *Medicago truncatula* (Rodriguez-Conde et al. 2004; Fernández-Aparicio et al. 2008). Interestingly, browning and death of attached *Orobanche* seedlings without the presence of an HR or necrosis of host cells near the site of parasite attachment has also been broadly reported in *Orobanche*–legume interactions (Rubiales 2001; Rubiales et al.

Fig. 7.1 (continued) (**c**, **d**) Cross section through a susceptible (**c**, IAC 165) and resistant (**d**, Nipponbare) rice root 9 days after infestation with *S. hermonthica*; in the susceptible interaction, the haustorium is well developed with xylem continuity between parasite and host; in the resistant interaction, the haustorium invades the host root cortex but is not able to penetrate the endodermis to establish host–parasite xylem connectivity. (**e**, **f**) Non-host resistance reaction in *Arabidopsis thaliana* roots 7 days after inoculation with *S. hermonthica*, viewed under UV light; lignification of host tissue around the invading parasite is visible as bright fluorescence; (**e**) non-sectioned material cleared in chloral hydrate; (**f**) cross section through a root; the haustorium penetrates into the cortex but does not form connections with the host xylem. *P* parasite, *HC* host cortex, *Xy* xylem vessel (**a**–**b** by A. Pérez-de-Luque; **c**–**f** by J. Scholes)

2006; Fernández-Aparicio and Rubiales 2010). A correlation between the presence of an HR and the failure of *R. minor* to produce functional haustoria was reported by Rümer et al. (2007) who suggested that this might be indicative of direct action by phytotoxins and reactive oxygen species (ROS) excreted by *P. lanceolata*. The endodermis is generally considered a substantial barrier to vascular penetration by root parasitic weeds and has been reported as the site of resistance expression in several crop species, including sorghum to *S. asiatica* (Maiti et al. 1984), rice to *S. hermonthica* (Gurney et al. 2006), and, the wild maize *Zea diploperennis* inbred line ZD05 to *S. hermonthica* (Amusan et al. 2008). The sorghum cultivar N-13 exhibits what has been referred to as 'mechanical' resistance at the root endodermal barrier (Haussmann et al. 2004), that is, a failure to be able to breach this cellular layer.

Unlike the susceptible rice cultivar IAC 165 (Fig. 7.1c), the rice cultivar Nipponbare exhibits strong resistance to some ecotypes of *S. hermonthica*. In this case the parasite invades the cortex but the haustorium grows around the vascular cylinder suggestive of a lesion in the signalling pathway(s) involved in penetrating between endodermal cells (Gurney et al. 2006; Fig. 7.1d). This phenotype has subsequently been observed at different frequencies in other rice cultivars (Yoshida and Shirasu 2009). The endodermis is also reported to be the major barrier to infection of sunflower by *O. cumana* and vetch and faba bean by *O. crenata* (Dörr et al. 1994; Labrousse et al. 2001; Pérez-de-Luque et al. 2005a, 2007, 2008) where lignification of the endodermal cells is often observed (Pérez-de-Luque et al. 2005a).

Resistance can also occur following successful connection of the parasite to the host vascular system and can take a number of forms. Occlusion or sealing of host vessels by gel- or gum-like substances, peroxidase-related lignification, deposition of mucilage (mainly composed of non-esterified pectins and other complex carbohydrates) and haustorium disorganization have all been implicated as the cause of necrosis and death of developing *Orobanche* tubercles before their emergence (Labrousse et al. 2001; Pérez-de-Luque et al. 2006a, b, 2008; Fernández-Aparicio et al. 2008, 2009). In each case, the blockage effectively disrupts the flux of water and nutrients from host to parasite preventing the parasite's survival.

Another potential mechanism is abiosis, the transfer of toxic compounds to the host vascular system for translocation into the parasite through the haustorium. For example, when *Tripsacum dactyloides* is parasitized by *S. hermonthica*, the parasite makes connections to the host xylem and, although there is no blockage of the vessels, the parasite dies (Gurney et al. 2003). When *S. hermonthica* individuals attached to susceptible maize roots were manipulated so that secondary haustoria were attached to the roots of *T. dactyloides*, thus bridging the two hosts, the secondary haustoria on *T. dactyloides* failed to differentiate properly, and subsequent secondary haustoria formation on the susceptible maize host was also impaired. These observations suggest that an inhibitory compound or toxin is being translocated from the host into the parasite and mobile within the parasite haustorial root system (Gurney et al. 2003). It has also been reported that both the number and size of *S. hermonthica* attached to the resistant rice cultivar Kasalath are smaller than that observed on the highly susceptible cultivar IAC 165 suggesting that some level of abiosis may be involved in the interaction (Gurney et al. 2006). Toxin movement has also been reported to occur in *O. crenata* tubercles on chickpea (*Cicer arietinum*; Pérez-de-Luque et al. 2006c) and *M. truncatula* (Lozano-Baena et al. 2007).

S. gesnerioides exhibits significant host specialization, and as a consequence when it encounters a host to which it is not adapted, it elicits a resistance response exemplified by a failure of the parasite to thrive after attachment, a phenomenon referred to as tubercle arrest (Botanga and Timko 2005). This is readily observed when *S. gesnerioides* adapted to cowpea attempts to parasitize wild legumes or when *S. gesnerioides* isolates adapted to wild legumes attempt to attack cowpea, including cultivars susceptible to cowpea-adapted isolates. Histological examination found little response by the host root tissue but revealed internal disorganization within the parasite tubercle typified by a lack of vascular differentiation. Recently, Yoshida and Shirasu (2009) investigated the interactions of *S. hermonthica* on various non-host eudicots and identified at least four types of incompatible interactions that can be distinguished by the host root cell layer at which invasion stops: layer I, after vascular connection; layer II, at endodermal cell layers; layer III, at cortex cell layers; and layer IV, before *S. hermonthica* attachment. It is worth noting that these investigators also found that the incompatibility phenotypes varied among different isolates of the parasite.

7.4 Genetic Basis of Resistance

Resistance is the ability of a plant, whether host or non-host, to evade parasite attack or following attack to prevent establishment and growth of the parasite. Two forms of resistance are recognized as operating in plant–parasite interactions, non-host resistance and host resistance. Plant species not parasitized by a particular genus of parasite under natural conditions are considered non-hosts. Non-host resistance is the most common form of resistance found in plants and appears to involve the activation of multiple defence response cascades under complex genetic control. Because of the general nature of the response, plants expressing non-host resistance to a particular parasite are unlikely to become part of that parasite host range over time. At the present time, little information beyond histological characterizations is available on non-host resistance to parasitic weeds, and what genes and gene products are involved remains unknown. In contrast, the co-evolution of root parasites with particular host species has led to adaptation and specialization of the parasites and to the development of specific avoidance, tolerance and resistance mechanisms to parasite attack. Among host species, responses to parasitism can vary from susceptibility (no evidence of a resistance response) to partial resistance and complete resistance, depending on the particular plant–parasitic weed interaction under consideration. For example, most plants that do not belong to the Poaceae (i.e. non-graminaceous plants) are not suitable hosts for *S. asiatica*,

S. aspera and *S. hermonthica*, and *S. gesnerioides* rarely parasitizes plants outside of a select group of dicots. Within the dicots, *S. gesnerioides* will show host preference and specialization, and within a particular host species, genotypes can be found that are either susceptible or resistant. The identification of genetic variants within host germplasm that differentially resist parasite attack has allowed researchers to study the genetic heritability of resistance responses and ultimately to identify particular genes or groups of genes that confer the resistance or susceptibility phenotype.

7.4.1 Monogenic and Quantitative Resistance to Striga

Reports describing cultivars and landraces with variable levels of resistance to *Striga* have appeared in the literature for over half a century, although the genetic basis for the observed resistance was often unclear (Ramaiah 1987; Ejeta 2007). Segregation analysis of the resistance phenotype in F1, F2 and advanced populations created by crossing "resistant" and "susceptible" phenotypes indicates that in most cases, resistance to *Striga spp.* in the grasses appears polygenic involving both major and minor genes with a large genotype X environment interaction (Scholes and Press 2008). Moreover, resistance appears to involve several mechanisms, is often weak and tends to break down in the presence of either new geographic or physiologically specialized forms of the parasite. In contrast, resistance to *S. gesnerioides* in its dicot hosts appears to be mainly monogenic in character (Timko et al. 2007).

In sorghum, resistance resulting from a low germination stimulant production phenotype has been found to be controlled by a single nuclear recessive gene (*lgs*) with largely additive gene action (Volgler et al. 1996). In contrast, resistance associated with low production of haustorial initiation factor appears to be controlled by a single dominant gene (Ejeta 2007). Few cultivated and wild sorghums exhibit post-attachment resistance to *Striga*, and the character is not always fully penetrant (Olivier et al. 1991b; Rich et al. 2004; Haussmann et al. 2001; Grenier et al. 2007; Ejeta 2007). Volgler et al. (1996) reported that resistance to *S. hermonthica* in sorghum populations derived from the resistant line SRN-39 was controlled by a single nuclear recessive gene. In contrast, Haussmann et al. (2001) found that resistance in sorghum was controlled by a major recessive gene and several minor genes. These same researchers were able to identify 9–11 QTLs that explained 77–82 % of the phenotypic variation (Haussmann et al. 2004). Among these QTLs, the most significant was the one associated with the *lgs* locus identified previously. A cross between the wild sorghum species *S. arundinaceum*, which exhibits an HR-like resistance response, with two cultivated sorghum species revealed that the resistance trait was controlled by two nuclear genes HR1 and HR2 which were associated with two different markers on the genetic linkage map (Mohamed et al. 2003, 2010).

There are no reports of resistance to *Striga* spp. in cultivated maize, although resistance has been observed in some of its wild progenitors. For example, *Z. diploperennis* harbours some resistance to *S. hermonthica* (Lane et al. 1997a) and *T. dactyloides* is unable to support post-attachment growth of the parasite (Gurney et al. 2003). Parker and Riches (1993) have suggested that this is likely due to the fact that maize is a New World crop and likely did not encounter *Striga* during its domestication and, therefore, would not have maintained the resistance genes present in wild progenitors or been placed under selective pressures to develop new resistance genes to the parasite.

Two cultivated rice species (*Oryza sativa* and *Oryza glaberrima*) show a range of responses to different species of *Striga* (*S. hermonthica*, *S. asiatica* and *S. aspera*), but the number of known resistant and tolerant cultivars is still limited. In general, cultivars of the African rice species (*O. glaberrima*) more frequently show *Striga* resistance than *O. sativa* genotypes (Rodenburg et al. 2010). The genotypes resistant to *S. hermonthica* within *O. sativa* are from ssp, *indica* and *japonica*, with the *O. sativa* ssp. *japonica* variety Nipponbare being very resistant (Gurney et al. 2006). The resistance phenotype in Nipponbare includes host cell necrosis at the site of attachment and an inability to penetrate the endodermis (Gurney et al. 2006; Fig. 7.1d). QTL analysis using a mapping population of backcross inbred lines (BILs) identified six QTL derived from Nipponbare alleles that account for 23 % of the phenotypic variance associated with the resistance trait. Although resistance in Nipponbare is polygenic, allelic substitutions at each QTL altered the phenotype by at least 0.5 of a phenotypic standard deviation relative to the parental lines suggesting that resistance may be conferred by a few genes of major effect (Gurney et al. 2006).

Significantly less is known about the heritability of resistance to *Striga* among other cultivated grasses. In wild pearl millet (*P. glaucum* subsp. *monodii*), resistance too appears to be polygenic in nature (Wilson et al. 2000, 2004).

The first mention of cowpea resistance came from Burkina Faso where cultivars 58-57 and Suvita-2 (Gorom Local) were observed to support no or low levels of parasite growth in the field (Aggarwal et al. 1984, 1986). Unfortunately these cultivars showed no resistance when grown in other countries, suggesting that there were some geographic-based differences in parasite virulence. Athena Lane and her colleagues (Lane et al. 1993, 1994, 1996) were the first to systematically document the differential resistance of cowpea genotypes to *Striga* isolates from different geographic origins leading to the suggestion that distinct races of *S. gesnerioides* exist in West Africa (Lane et al. 1997b). Based on extensive studies of the differential response of various cowpea cultivars, landraces and breeding lines to attempted parasitism by isolates of *S. gesnerioides* collected widely throughout West Africa, and careful genotyping of these *S. gesnerioides* isolates, Botanga and Timko (2006) determined that there are at least seven distinct races of *S. gesnerioides* parasitic on cowpea designated: SG1 (Burkina Faso), SG2 (Mali), SG3 (Nigeria and Niger), SG4 (Benin), SG4z (localized to the Zakpota region of Benin), SG5 (Cameroon) and SG6 (Sénégal). SG1 and SG5 are the most closely related, while SG4 and SG3 are the most diverged. SG4 is notable since it appears that under strong host selection to have mutated giving rise to a hypervirulent form in the Zakpota region (designated SG4z). While SG4 and SG4z are virtually indistinguishable based on their molecular genetic profiles (Botanga and Timko 2006), they elicit dramatically different responses on B301, a cowpea cultivar from Botswana initially identified by Parker and Polniaszek (1990) that appeared to be resistant throughout West Africa (Singh and Emechebe 1990; Parker and Polniaszek 1990; Atokple et al. 1995). However, in the mid-1990s resistance in B301 began to break down in the Zakpota region of Benin but remains resistant elsewhere in the country suggesting that a new virulence had arisen (Lane et al. 1994).

Studies of the heritability of cowpea resistance to *S. gesnerioides* indicate that race-specific resistance is conferred by single dominant genes (Timko et al. 2007; Timko and Singh 2008) or a combination of dominant and recessive loci (Touré et al. 1997). Using independently derived F_2 and recombinant inbred populations generated from crosses of resistant and susceptible cowpea genotypes and various molecular marker techniques (AFLPs, RAPDs, SSRs), we have mapped the various race-specific resistance genes within the cowpea genome (Ouédraogo et al. 2001, 2002a, b; Timko et al. 2007). Resistance to *S. gesnerioides* races SG1, SG2, SG3 and SG5 (present in the resistant cowpea lines B301, Tvu14676, IT82D-849 and Tvu14676, respectively) maps to Linkage Group 1 (LG1), whereas resistance to SG1 and SG4z (found in Suvita-2 and IT81D-994) maps to LG6 (Timko et al. 2007).

Using a positional cloning approach, Li and Timko (2009) recently reported the isolation and characterization of a full-length gene (designated *RSG3-301*) required for resistance of cowpea to *S. gesnerioides* race 3(SG3). *RSG3-301* encodes an R-protein (RSG3-301) with a coiled-coil (CC) protein–protein interaction domain at the N-terminus, a nucleotide binding site (NBS), and a leucine-rich repeat domain at the C-terminus. Structurally, RSG3-301 is similar to other R-proteins identified in plants that recognize various disease agents and pests (Tameling and Joosten 2007; Caplan et al. 2008) and is most similar (~54 % identity) to the soybean R-protein Rpg1-b that confers resistance to *Pseudomonas syringae* pv. *glycinea* containing the Avr factor *avrB*. When RSG3-301 was silenced in the resistant cultivar B301 [using virus-induced gene silencing (VIGS)], compatibility to *S. gesnerioides* race 3 was restored, but the cultivar remained resistant to other races of the parasite (SG2 and SG5) supporting the hypothesis of a gene-for-gene interaction (Li and Timko 2009).

7.4.2 Monogenic and Quantitative Resistance to Orobanche *and* Phelipanche

Vrânceanu et al. (1980) were the first to suggest that resistance in sunflower to *O. cumana* races was monogenically inherited and governed by a gene-for-gene mechanism. Five single dominant genes, designated *Or1–Or5*, were identified that were capable of conferring resistance to five pathogenic races of *O. cumana* (designated A to E) found in Romania (Molinero-Ruiz et al. 2006; Letousey et al. 2007). Races B and C were subsequently found in Spain along with other *O. cumana* variants that exhibited higher levels of virulence than reported for the parasites in Romania (Alonso et al. 1996; Melero-Vara et al. 2000). One of these variants, race F, overcomes resistance conferred by *Or5* (Alonso et al. 1996). Race F also appears to have multiple population variants with different intrinsic pathogenicities that may be environmentally controlled (Molinero-Ruiz et al. 2008, 2009). Whereas resistance to races A to E is controlled by single dominant genes, resistance to race F obtained from cultivated sunflower was initially reported to be controlled by recessive alleles at two loci (Rodríguez-Ojeda et al. 2001; Akhtouch et al. 2002) and subsequently by up to six QTLs some of which were not race specific (Pérez-Vich et al. 2004a, b; Velasco et al. 2006).

The identity of *Or1–Or5* is presently unknown, but *Or5* has been mapped to a telomeric region of LG3 in the sunflower genetic map (Tang et al. 2003; Pérez-Vich et al. 2004a) that contains multiple NBS-LRR type R-gene homologues. This region of the genome is derived from *H. tuberosus*, a common source of *Orobanche* resistance genes (Fernández-Martínez et al. 2009), leading to the suggestion that *Or5* might be part of a cluster R gene (Radwan et al. 2008). In addition to *Or5,* four QTLs with minor effects, some of which are not race E specific, have also been reported to influence infection (Pérez-Vich et al. 2004a). That a quantitative component is also involved in resistance to race E in sunflower is consistent with reports indicating that several resistance mechanisms appear to be operating that can be distinguished histologically and by differential patterns of defence gene expression (Labrousse et al. 2001; Letousey et al. 2007; de Zélicourt et al. 2007). Cytological and cytochemical observations of compatible and incompatible interactions similarly suggest that several different mechanisms may be active against race F (Echevarría-Zomeño et al. 2006).

Resistance to *Orobanche* and *Phelipanche* spp. in legumes is polygenic, with low heritability and highly influenced by the environment (Valderrama et al. 2004; Rubiales et al. 2006). There does not appear to be any race structure within the parasites (Radwan et al. 1988), and the level of parasite infestation appears to affect both the penetrance of the resistance phenotype and degree of epistatic effects. In faba bean, susceptibility is usually dominant over resistance, whereas in common vetch, the reverse is observed (Gil et al. 1987; Cubero and Moreno 1999).

Three QTLs linked to resistance, designated *Oc1*, *Oc2* and *Oc3*, were identified in faba bean using an F_2 population segregating for resistance to *O. crenata* (Román et al. 2002). These QTLs explained 74 % of the phenotypic variation, with *Oc1* alone accounting for 37 % of the character and *Oc2* and *Oc3* explaining 11 % and 25 % of the variation observed, respectively. Surprisingly, when a recombinant inbred line (RIL) population derived from this same cross was evaluated over two growing seasons in two locations, *Oc1* was not a stable QTL and could not be located (Díaz et al. 2005). In contrast, *Oc2* and *Oc3* were still detected in the RIL population in addition to a new QTL (Díaz et al. 2005). In subsequent studies,

Diaz-Ruiz et al. (2010) used an F6 RIL population derived from the same F2 population initially evaluated by Román et al. (2002) to detect additional QTLs when resistance was evaluated in several environments. Similar to earlier studies, *Oc1* was not significant in the advanced lines, whereas four QTLs (*Oc2*–*Oc5*) could be mapped. *Oc2* and *Oc3* were found to be associated with *O. crenata* resistance in at least two of the three environments tested, while *Oc4* and *Oc5* appear only in selected environments.

In pea, four QTLs conferring resistance to *O. crenata* were identified using RILs derived from a cross between a *P. sativum* ssp. *syriacum*-resistant line and a susceptible pea and assaying different phases of the parasite cycle to enable identification of genes governing specific mechanisms of resistance (Fondevilla et al. 2010). These QTLs explained 38–59 % of the variation depending on the trait.

7.5 Cell Signalling and Gene Expression in Host Defence Responses

Prior to the advent of microarray technologies and low-cost high-throughput sequencing (transcriptomic analysis), examinations of changes in gene expression during parasitic plant–host interactions were performed using candidate gene approaches in which gene targets were selected for investigation based upon their observed involvement in other plant–pathogen interactions and general plant stress response mechanisms. As a result, only a limited number of genes were evaluated for their potential involvement in plant defence against parasitic plant attack. In some cases, subtractive hybridization strategies were applied to identify genes differentially regulated during susceptible (compatible) and resistant (incompatible) host–parasite interactions. In other instances, analyses took the form of transgene overexpression and knockout studies.

Among the earliest studies using the candidate gene approach was the work of Westwood et al. (1996, 1998) who expressed a *GUS* gene encoding β-glucuronidase reporter gene under the control of a tomato *hmg2* promoter in tobacco plants and demonstrated localized GUS expression in roots around the site of *P. aegyptiaca* attachment as early as 1 day after parasite attachment to the root. The *hmg2* gene encodes 3-hydroxy-3-methylglutaryl CoA reductase, an enzyme known to be involved in defence-related isoprenoid biosynthesis associated with phytoalexin and sesquiterpene production. In another set of studies, two well-known defence-related metabolic pathways (i.e. isoprenoid and phenylpropanoid biosynthesis) were found to be activated in tobacco roots in response to *P. aegyptiaca* infection (Joel and Portnoy 1998; Griffitts et al. 2004). Promoter fusion experiments showed that the pathogenesis-related (PR) gene *PR-1b*, but not *PR-1a*, was expressed in tobacco roots following *P. aegyptiaca* infection (Griffitts et al. 2004). Expression of the PR1 family of genes is linked to the systemic acquired resistance (SAR) in plants (Eyal and Fluhr 1991).

Vieira Dos Santos et al. (2003a) examined changes in the expression pattern of 20 candidate signal transduction and resistance response genes activated in other plant–pathogen interactions during infection of the susceptible host *Arabidopsis thaliana* by *P. ramosa* (syn. *O. ramosa*). These investigators found that most of the general response signalling pathways regulated by jasmonic acid (JA) and ethylene (ETH) were induced in a transient manner even before parasite attachment to the *A. thaliana* roots. In contrast, no effect was observed on the expression of genes involved in salicylic acid (SA)-mediated defence responses. Vieira Dos Santos et al. (2003b) investigated gene expression in *A. thaliana* roots inoculated with *P. ramosa* and reported that among the differentially expressed cDNAs isolated using a suppression subtractive hybridization (SSH) strategy, 12 showed a transient induction in host roots occurring as early as 1–2 h after infection. A majority of the genes identified encoded proteins previously implicated in the response of *A. thaliana* to attack by other pathogens, detoxification of reactive oxygen species (ROS), cell wall reinforcement and JA and ETH signalling. Similarly, upregulation of a cell wall-associated receptor kinase was shown to occur at an early stage of parasitism early in tomato roots in response to interaction with germinated *P. ramosa* seeds (Lejeune et al. 2006).

The most comprehensive studies of transcriptomic changes during host resistance responses to *Orobanche* parasitism have used SSH to evaluate *O. cumana*–sunflower and *O. crenata–M. truncatula* associations. Three cDNAs isolated by an SSH were shown to be strongly induced in a resistant sunflower genotype (LR1) 8 days postinfection (dpi) when the first *O. cumana* attachments occurred (Letousey et al. 2007). These cDNAs putatively encode a methionine synthase, a glutathione *S*-transferase and a quinone oxidoreductase, components of ROS detoxification, implicating the possibility of an oxidative burst during this incompatible interaction. A targeted approach was adopted to compare the expression of 11 defence-related genes in susceptible and resistant sunflower genotypes before (early response) and after connection of the parasite to host vascular tissues (late response) (Letousey et al. 2007). The results suggest that the resistant genotype exhibited a stronger overall defence response (early and late) against *O. cumana* than the susceptible one, involving some marker genes of the JA, SA and phenylpropanoid pathways, but not of the ETH pathway. Among the genes studied, the SA-responsive gene *Hadefl*, encoding a defensin, exhibited a strong induction a few days after attachment and before necrosis of the parasite tubercle, whereas all other studied genes were severely repressed. The authors suggested that the strong *Hadefl* overexpression was determinant in sunflower resistance and that the putative encoded peptide, a defensin, was involved in the necrosis of the attached parasite. This was confirmed when the corresponding recombinant peptide Ha-DEF1, produced in *E. coli*, was assayed for its possible toxic effect on *Orobanche* seedlings (de Zélicourt et al. 2007). Indeed, while Ha-DEF1 toxicity towards fungi is low, compared with other plant defensins, it acts at much lower concentration on *Orobanche* seedlings by inducing cell death at the radical apex. Interestingly, defensin was found to be ineffective on *S. hermonthica* and on *A. thaliana* seedlings. Thus, *O. cumana* and *P. ramosa* respond specifically to defensin, through a signalling pathway that remains to be described. In somewhat related work, Mabrouk et al. (2007a, b) found that pea roots inoculated with *Rhizobium leguminosarum* strain 248, an *nodc* mutant with altered production of chitin oligosaccharide backbones, showed a 74 % reduction in *O. crenata* infections. The reduced parasite load was also associated with activation of lipoxygenase and phenylpropanoid/isoflavonoid pathways and accumulation of derived toxins like phenolics and pisatin (pea phytoalexin).

Studies of the interaction of *O. crenata* with the non-host *M. truncatula* found that resistance in the accession SA4087 develops late and was related to necrosis of well-developed parasite tubercles (Die et al. 2007). An SSH cDNA library representing upregulated transcripts was made, comprising some 288 cDNA fragments with good BLAST homology to 81 genes. Most of the isolated cDNAs belong to defence genes involved in well-known defence processes against biotic stresses, like JA, phenylpropanoid and phytoalexin pathways, and cell wall modifications. Among the most highly expressed transcripts was a cDNA encoding a dehydrin-like protein with high homology to an EST isolated from drought-stressed *M. truncatula* plantlets. Interestingly, the *Hadef1* gene, triggered in resistant sunflower roots upon *O. cumana* infection, has been shown to be upregulated by abscisic acid. Moreover, in comparison with a susceptible cultivar, a resistant pea accession was shown by proteomic studies to exhibit a higher proportion of an ABA-responsive protein (Castillejo et al. 2004, 2009).

To date only limited information is available on transcription changes in hosts in response to *Striga* parasitism. Gowda et al. (1999) used a differential display strategy to identify 23 genes whose expressions are upregulated in the roots of *Tagetes erecta* during invasion by the incompatible *S. asiatica*. One of these upregulated genes, (non-host resistance to *S. asiatica*) encodes a protein that is highly homologous to the disease-resistance (R) proteins identified in several plants. NRSA-1 was systemically induced, and under the control of the JA response cascade.

Examination of the differential expression of genes encoding candidate disease-resistance and signal transduction components during the interactions of *S. gesnerioides* with various cowpea cultivars revealed that PR-5 transcript levels were dramatically elevated in the roots of cowpea genotypes resistant to *Striga* compared to uninfected roots and roots of cowpea challenged with a race of *Striga* to which it was susceptible or adapted to another host species (Li et al. 2009). In contrast, transcript levels of *COI1* and *EDS1* increased during susceptible and non-host interactions but were unchanged during resistance response. The *COI1* gene product plays a pivotal role in the activation of JA-mediated response cascades and in some cases serves as an inhibitor of SA signalling. *EDS1* encodes lipase-like protein that controls defence activation and programmed cell death in plants. Induction of *COI1* gene expression in compatible *Striga*–cowpea interaction suggests that *COI1* may downregulate or suppress the resistance response and block SA signalling pathway in cowpea plants, thus allowing *Striga* attachment and further development.

Hiraoka and Sugimoto (2008) identified 30 genes that were upregulated in response to *S. hermonthica* parasitism. *Striga* parasitism was found to induce JA-responsive genes and suppress SA-responsive genes in the roots of highly susceptible cultivars, suggesting that susceptible hosts recognize *Striga* parasitism as wounding stress rather than microbial stress. In contrast, the less-susceptible sorghum cultivar appears to recognize *Striga* parasitism as not only wounding stress but also as microbial stress because *Striga* parasitism was observed to induce SA- and JA-responsive genes in the roots. The gene expression in the less-susceptible cultivar and the inhibition of *Striga* development in the roots of the sorghum cultivars following treatment with SA suggest that SA-responsive genes are involved in host resistance against *Striga*. In a subsequent study, Hiraoka et al. (2009) used an SSH strategy to examine the interaction of *Lotus japonicus* with *P. aegyptiaca* and *S. hermonthica*, compatible and incompatible interaction, respectively. They found little or no overlap among the *Phelipanche*- and *Striga*-induced transcripts suggesting that *L. japonicus* roots are able to distinguish the compatible parasite from the incompatible one. Among the genes specifically induced by *P. aegyptiaca* were those encoding components of JA biosynthesis, whereas *S. hermonthica* parasitism induced genes in phytoalexin biosynthesis.

Using microarray technology, Swarbrick et al. (2008) have characterized the global patterns of gene expression in Nipponbare, an *S. hermonthica*-resistant rice cultivar, and IAC 165, a *Striga*-susceptible cultivar. The levels of a large number of transcripts in rice roots were found to be either positively (upregulated) or negatively (downregulated) affected by parasitism. Among the genes upregulated in the *Striga*-resistant cultivar Nipponbare were genes encoding HR protein homologues, PR-proteins associated with microbial pathogenesis including endochitinases (PR-3), glucanases (PR-2) and thaumatin-like proteins (PR-5); pleiotropic drug-resistance ABC transporters; and enzymes in phenylpropanoid metabolism. In addition, transcripts encoding several WRKY transcription factors (TFs) (e.g. OsWRKY45, OsWRKY62 and OsWRKY76) were more abundant in parasitized roots. OsWRKY45 and OsWRKY62 have been previously implicated in other SA-dependent resistance responses (Ryu et al. 2006). Large-scale downregulation of gene expression was observed in the susceptible cultivar IAC 165, particularly transcripts whose encoded products annotate to Gene Ontology (GO) functional categories of plant growth regulator signalling and metabolism, biogenesis of cellular components, and cell division. Interestingly, a majority of the genes downregulated in both IAC 165 and Nipponbare roots following attempted *Striga* parasitism encode products that annotate as proteins of unknown function.

The mechanics of the resistance response and the nature of the gene expression changes occurring during compatible and incompatible *Striga*–cowpea associations has been recently investigated by oligonucleotide microarrays capable of interrogating ~43,253 cowpea unigenes (Timko et al. 2008; Huang et al. 2012). In these studies, global changes in gene expression were monitored in the roots of B301 cowpea plants challenged with both host-adapted S. *gesnerioides* races to which it was resistant (e.g. SG3) and susceptible (SG4z), and isolates adapted to another host (SG6i; an *Indigofera*-adapted isolate). Among the most highly induced genes in the early resistance response (6 dpi) of B301 to SG3 are genes involved in response to biotic stimuli, abiotic stimuli, wounding, oxidative stress and HR, and components of the JA and ETH signalling pathways. Among those genes most highly induced in the late resistance response (13 dpi) were genes involved in cell differentiation, pattern formation, morphogenesis/multicellularity and vascularization consistent with the induction and establishment of a physical barrier and turning on of PCD. In contrast, among the most highly suppressed genes in B301 during a susceptible interaction with SG4z were genes that annotated to phenylpropanoid and lignin biosynthesis, primary and secondary cell wall biosynthesis and cell wall thickening and modification, and components of the SA and JA signal transduction pathways.

There is little published information on the changes in gene expression in non-host species challenged with *Striga*. Recently, we have examined the transcriptome of *A. thaliana* roots following inoculation with pre-germinated *S. hermonthica* seeds using whole-genome oligonucleotide arrays (Affymetrix, UK Ltd) (Scholes JD, Vasey R, and Press MC, unpublished results). Following inoculation, the parasite haustorium penetrated the host root cortex but failed to develop parasite–host xylem continuity essential for the transfer of solutes from host to parasite. This non-host interaction was characterized by the deposition of lignin (Fig. 7.1e, f), pectin-like substances and the accumulation of phenolic compounds at the site of attachment. Analysis of changes in host gene expression revealed that approximately equal numbers of genes (~1,500) were significantly up- and downregulated in infected compared to control roots. The greatest number of genes was most strongly upregulated 24 h postinoculation, consistent with an active host resistance response. Functional analysis revealed that large numbers of genes involved in cell wall synthesis, defence signalling, regulation of transcription and protein synthesis, oxidative stress and primary and secondary metabolism were upregulated. One of the most striking results was the upregulation of many genes (*EDS1*, *EDS5*, *PAD3*, *NPR1*, *NIMIN1*, *PR2*) involved in the SA signalling pathway together with the upregulation of a key WRKY transcription factor (AtWRKY70) that regulates the expression of genes involved in the SA signalling pathway and is thought to have a role in determining the balance between SA and JA signalling. In addition, there was evidence for the activation of genes involved in the JA and ethylene biosynthetic pathways.

7.6 Conclusions and Perspective

The coevolution of parasitic plants with their hosts has enforced selective pressures resulting in great diversity and specialization among the parasites. In response host species have tried to keep pace by the establishment of multilevel resistance to fend off these parasites. During post-attachment stages of host–parasite interactions, different constitutive or induced incompatibility/host resistance mechanisms can be activated. A synthesis of the current molecular work on *Striga* and *Orobanche*/

Phelipanche–host interactions indicates that the host plant responds through elicitation of a set of classical biotic stress-related defence pathways, which in some cases share common mechanisms/pathways to other pest and pathogen-activated pathways and to abiotic stress pathways such as drought responses.

Similar suites of genes are observed to be upregulated during resistance responses to *Orobanche* and *Striga*, and it is becoming increasingly clear that the SA signalling pathway and to a lesser extent the JA signalling pathway play important roles in the activation of resistance to parasitic plants. In other plant–plant pathogen interactions, the SA and JA defence pathways can interact either antagonistically or synergistically (Mur et al. 2006). The SA pathway is often activated in response to biotrophic fungal pathogens leading to the expression of suites of PR genes, whereas the JA pathway is often important in resistance to necrotrophic pathogens and insect pests. A number of studies have evaluated the effectiveness of SA applications to hosts to prevent parasitism by *Orobanche* spp. SA consistently promoted resistance, in the case of clover roots infected with *O. minor*, by the activation of defence responses leading to lignification of the endodermis (Kusumoto et al. 2007). To date SA-inducing chemicals have not been widely applied to field-grown crops to test their effectiveness as part of a control strategy. The induction of genes involved in JA biosynthesis has been observed in compatible interactions between *Orobanche* species and their hosts (Hiraoka et al. 2009), but the involvement of JA in resistance is less clear (Kusumoto et al. 2007). In a recent study of resistance in tomato to the shoot parasite *Cuscuta pentagona*, a role for both JA and SA was also proposed (Runyon et al. 2010).

Relative to other host–pathogen interaction, we are just beginning to unravel the dynamic aspects of the interaction of parasitic plants with their host. Understanding the differences and similarities in host and non-host responses to parasitism, and the aspects of parasite structure, composition and physiology that elicit dynamic responses from potential host will be important for designing strategies to breed for or genetically engineer durable resistance. Of the aspects of plant–parasitic plant interactions where we know the least, the determinants of virulence and host specification are among the least well understood. Clearly, future research needs to be directed at understanding the nature of the avirulence gene products and other effectors present in *Striga, Orobanche* and other root parasitic species that are recognized directly or indirectly by the host surveillance system (R-proteins and other host sensor proteins) leading to activation of host defence responses since these molecules have evolved to alter host cell structure or function in order to suppress defence responses and allow parasite ingress.

Acknowledgements The authors thank the present and former members of our laboratories for their research contributions over the years, especially Kan Huang, Karolina Mellor and Monica Fernández-Aparicio for their critical reading and comments. This work was supported by grants from the National Science Foundation (DBI-0701748 and IBN-0322420) and Kirkhouse Trust to MPT and the Biotechnology and Biological Sciences Research Council (BBSRC) to JDS.

References

Abramovitch RB, Anderson JC, Martin GB (2006) Bacterial elicitation and evasion of plant innate immunity. Nat Rev Mol Cell Biol 7:601–611

Aggarwal VD, Muleba N, Drabo I, Souma J, Mbewe M (1984) Inheritance of *Striga gesnerioides* resistance in cowpea. In: Parker C, Musselman LJ, Polhill RM, Wilson AK (eds) Proceedings of the 3rd international symposium on parasitic weeds. ICARDA-IPSPRG, Aleppo, pp 143–147

Aggarwal VD, Haley SD, Broockman FE (1986) Present status of breeding cowpea for resistance to *Striga* at IITA. In: ter Borg SJ (ed) Proceeding of a workshop on biology and control of *Orobanche*. LH/VPO, Wageningen, pp 176–180

Akhtouch B, Muñoz-Ruz J, Melero-Vara JM, Fernández-Martínez JM, Domínguez J (2002) Inheritance of resistance to race F of broomrape (*Orobanche cumana* Wallr.) in sunflower lines of different origin. Plant Breed 121:266–269

Alonso LC, Fernández-Escobar J, López G, Rodríguez-Ojeda M, Sallago F (1996) New highly virulent sunflower broomrape (*Orobanche cernua* Loefl.) pathotypes in Spain. In: Moreno MT, Cubero JI, Berner D, Joel D, Musselman LJ, Parker C (eds) Advances in parasitic plant research. Proceedings of the 6th international symposium on parasitic weeds. Dirección General de Investigación Agraria, Consejería de Agricultura y Pesca, Sevilla, Spain, pp 639–644

Amusan IO, Rich PJ, Menkir A, Housley T, Ejeta G (2008) Resistance to *Striga hermonthica* in a maize inbred line derived from *Zea diploperennis*. New Phytol 178:157–166

Atokple IDK, Singh BB, Emechebe AM (1995) Genetics of resistance to *Striga* and *Alectra* in cowpea. J Hered 86:45–49

Ben-Hod G, Losner D, Joel DM, Mayer M (1993) Pectin methylesterase in calli and germinating seeds of *Orobanche aegyptiaca*. Phytochemistry 32:1399–1402

Bent AF, Mackey D (2007) Elicitors, effectors, and R genes: the new paradigm and a lifetime supply of questions. Annu Rev Phytopathol 45:399–436

Boller T, Felix G (2009) A renaissance of elicitors: perception of microbe-associated molecular patterns and danger signals by pattern-recognition receptors. Annu Rev Plant Biol 60:379–406

Boller T, He S-Y (2009) Innate immunity in plants: an arms race between pattern recognition receptors and effectors in microbial pathogens. Science 324:742–744

Botanga CJ, Timko MP (2005) Genetic structure and analysis of host and nonhost interactions of *Striga gesnerioides* (Witchweed) from Central Florida. Phytopathology 95:1166–1173

Botanga CJ, Timko MP (2006) Phenetic relationships among different races of *Striga gesnerioides* (Willd.) Vatke from West Africa. Genome 49:1351–1365

Cameron DD, Seel WE (2007) Functional anatomy of haustoria formed by *Rhinanthus minor*: linking evidence from histology and isotope tracing. New Phytol 174:412–419

Cameron DC, Coats AM, Seel WE (2006) Differential resistance among host and non-host species underlies the variable success of the hemiparasitic plant *Rhinanthus minor*. Ann Bot 98:1289–1299

Caplan J, Padmanabhan M, Dinesh-Kumar SP (2008) Plant NB-LRR immune receptors: from recognition to transcriptional reprogramming. Cell Host Microbe 3:126–135

Castillejo MA, Amiour N, Dumas-Gaudot E, Rubiales D, Jorrín JV (2004) A proteome approach to studying plant response to crenate broomrape (*Orobanche crenata*) in pea (*Pisum sativum*). Phytochemistry 65:1817–1828

Castillejo MA, Maldonado AM, Dumas-Gaudot E, Fernández-Aparicio M, Susin R, Rubiales D, Jorrín J (2009) Differential expression proteomics to investigate responses and resistance to *Orobanche crenata* in *Medicago truncatula*. BMC Genomics 10:294

Cubero JI, Moreno MT (1999) Studies on resistance to Orobanche crenata in Vicia faba. In: Cubero JI, Moreno MT, Rubiales D, Sillero J (eds) Resistance to Orobanche: the state of the art. Junta de Andalucía, Consejería de Agricultura y Pesca, Seville, Spain, pp 9–15 de Zélicourt A, Letousey P, Thoiron S, Campion C, Simoneau P, Elmorjani K, Marion D, Simier P, Delavault P (2007) *Ha-DEF1*, a sunflower defensin, induces cell death in *Orobanche* parasitic plants. Planta 226:591–600

Díaz R, Šatović Z, Román B, Rubiales D, Cubero JI,Torres AM (2005) QTL analysis of broomrape resistance in faba bean (*Vicia faba* L.). In: Kovacevic V, Jovanovac S (eds) Proceedings of the 40th Croatian symposium on agriculture. Faculty of Agriculture, Zagreb, Croatia, pp 181–182

Diaz-Ruiz R, Torres AM, Šatović Z, Gutierrez MV, Cubero JI, Román B (2010) Validation of QTLs for *Orobanche crenata* resistance in faba bean (*Vicia faba* L.) across environments and generations. Theor Appl Genet 120:909–919

Die JV, Dita MA, Krajinski F, Gonzalez-Verdejo CI, Rubiales D, Moreno MT, Román B (2007) Identification by suppression subtractive hybridization and expression analysis of *Medicago truncatula* putative defence genes in response to *Orobanche crenata* parasitization. Physiol Mol Plant Pathol 70:49–59

Dörr I, Staack A, Kollmann R (1994) Resistance of *Helianthus annuus* to *Orobanche* – histological and cytological studies. In: Pieterse AH, Verkleij JAC, ter Borg SJ (eds) Biology and management of Orobanche. Proceedings of 3rd international workshop on *Orobanche* and related *Striga* research. Royal Tropical Institute, Amsterdam, pp 276–289

Echevarría-Zomeño S, Pérez-de-Luque A, Jorrín J, Maldonado AM (2006) Pre-haustorial resistance to broomrape (*Orobanche cumana*) in sunflower (*Helianthus annuus*): cytochemical studies. J Exp Bot 57:4189–4200

Ejeta G (2007) Breeding for *Striga* resistance in sorghum: exploitation of an intricate host–parasite biology. Crop Sci 47:S216–S227

Ellis J, Lawrence G, Ayliffe M, Anderson P, Collins N, Finnegan J, Frost D, Luck J, Pryor T (1997) Advances in the molecular genetic analysis of the flax–flax rust interaction. Annu Rev Phytopathol 35:271–291

Ellis JG, Dodds PN, Lawrence GJ (2007a) Flax rust resistance gene specificity is based on direct resistance-avirulence protein interactions. Annu Rev Phytopathol 45:289–306

Ellis JG, Lawrence GJ, Dodds PN (2007b) Further analysis of gene-for-gene disease resistance specificity in flax. Mol Plant Pathol 8:103–109

Eyal Y, Fluhr R (1991) Cellular and molecular biology of pathogenesis related proteins. Oxf Surv Plant Mol Cell Biol 7:223–254

Fernández-Aparicio M, Rubiales D (2010) Characterisation of resistance to crenate broomrape (*Orobanche crenata* Forsk.) in *Lathyrus cicera* L. Euphytica 173:77–84

Fernández-Aparicio M, Pérez-de-Luque A, Prats E, Rubiales D (2008) Variability of interactions between barrel medic (*Medicago truncatula*) genotypes and *Orobanche* species. Ann Appl Biol 153:117–126

Fernández-Aparicio M, Sillero JC, Rubiales D (2009) Resistance to broomrape species (*Orobanche* spp.) in common vetch (*Vicia sativa* L.). Crop Prot 28:7–12

Fernández-Martínez JM, Domínguez J, Pérez-Vich B, Velasco L (2009) Current research strategies for sunflower broomrape control in Spain. Helia 32:47–55

Flor HH (1971) Current status of the gene-for-gene concept. Annu Rev Phytopathol 9:275–296

Fondevilla S, Fernández-Aparicio M, Šatović Z, Emeran AA, Torres M, Moreno MT, Rubiales D (2010) Identification of quantitative trait loci for specific mechanisms of resistance to *Orobanche crenata* in pea. Mol Breed 25:259–272

Gil J, Martín LM, Cubero JI (1987) Genetic of resistance in *Vicia sativa* L. to *Orobanche crenata* Forks. Plant Breed 99:134–143

Göhre V, Robatzek S (2008) Breaking the barriers: microbial effector molecules subvert plant immunity. Annu Rev Phytopathol 46:189–215

Goldwasser Y, Kleifeld Y, Plakhine D, Rubin D (1997) Variation in vetch (*Vicia* spp.) response to *Orobanche aegyptiaca*. Weed Sci 45:756–762

Goldwasser Y, Hershenhorn J, Plakhine D, Kleifeld Y, Rubin B (1999) Biochemical factors involved in vetch resistance to *Orobanche aegyptiaca*. Physiol Mol Plant Pathol 54:87–96

Goldwasser Y, Plakhine D, Kleifeld Y, Zamski E, Rubin B (2000) The differential susceptibility of vetch (*Vicia* spp.) to *Orobanche aegyptiaca*: anatomical studies. Ann Bot 85:257–262

Gowda BS, Riopel JL, Timko MP (1999) *NRS1*: a resistance gene homolog expressed in roots of nonhost plants following parasitism by *Striga asiatica* (witchweed). Plant J 20:217–230

Greenshields DL, Jones JDG (2008) Plant pathogen effectors: getting mixed messages. Curr Biol 18:R128–R130

Grenier C, Ibrahim Y, Haussmann BIG, Kiambi D, Ejeta G (2007) Marker-assisted selection for *Striga* resistance in sorghum. In: Ejeta G, Gressell J (eds) Integrating new technologies for *Striga* control: towards ending the witch-hunt. World Scientific, Singapore, pp 159–172

Griffitts AA, Cramer CL, Westwood JH (2004) Host gene expression in response to Egyptian broomrape (*Orobanche aegyptiaca*). Weed Sci 52:697–703

Gurney AL, Grimanelli D, Kanampiu F, Hoisington D, Scholes JD, Press MC (2003) Novel sources of resistance to *Striga hermonthica* in *Tripsacum dactyloides*, a wild relative of maize. New Phytol 160:557–568

Gurney AL, Slate J, Press MC, Scholes JD (2006) A novel form of resistance in rice to the angiosperm parasite *Striga hermonthica*. New Phytol 169:199–208

Haussmann BIG, Hess DE, Omanya O, Reddy BVS, Welz HG, Geiger HH (2001) Major and minor genes for stimulation of *Striga hermonthica* seed germination in sorghum, and interaction with different populations. Crop Sci 41:1507–1512

Haussmann BIG, Hess DE, Omanya GO, Folkertsma RT, Reddy BVS, Kayentao M, Welz HG, Geiger HH (2004) Genomic regions influencing resistance to the parasitic weed *Striga hermonthica* in two recombinant inbred populations of sorghum. Theor Appl Genet 109:1005–1016

Hearne SJ (2009) Control – the *Striga* conundrum. Pest Manag Sci 65:603–614

Hiraoka Y, Sugimoto Y (2008) Molecular responses of sorghum to purple witchweed (*Striga hermonthica*) parasitism. Weed Sci 56:356–363

Hiraoka Y, Ueda H, Sugimoto Y (2009) Molecular responses of *Lotus japonicus* to parasitism by the compatible species *Orobanche aegyptiaca* and the incompatible species *Striga hermonthica*. J Exp Bot 60:641–650

Hood ME, Condon JM, Timko MP, Riopel JL (1998) Primary haustorial development of *Striga asiatica* on host and nonhost species. Phytopathology 88:70–75

Huang K, Mellor KE, Paul SN, Lawson MJ, Mackey AJ, Timko MP (2012) Global changes in gene expression during compatible and incompatible interactions of cowpea (*Vigna unguiculata* L.) with the root parasitic angiosperm *Striga gesnerioides*. BMC Genomics 13:402

Irving LJ, Cameron DD (2009) You are what you eat: interactions between root parasitic plants and their hosts. Adv Bot Res 50:87–138

Joel DM, Portnoy VH (1998) The angiospermous root parasite *Orobanche* L. (Orobanchaceae) induces expression of a pathogenesis related (PR) gene in susceptible tobacco roots. Ann Bot 81:779–781

Jones JDG, Dangl JL (2006) The plant immune system. Nature 444:323–329

Jorrín J, De Ruck E, Serghini K et al (1996) Biochemical aspects of the parasitism of sunflower by *Orobanche*. In: Moreno MT, Cubero JI, Berner D, Joel D, Musselman LJ, Parker C (eds) Proceedings of the 6th international symposium on parasitic weed, advances in parasitic plant research. Dirección General de Investigación Agraria, Servicio de Publicaciones y Divulgación, Cordoba, Spain, pp 552–558

Kuijt J (1969) The biology of parasitic flowering plants. University of California Press, Berkeley, CA

Kusumoto D, Goldwasser Y, Xie X, Yoneyama K, Takeuchi Y (2007) Resistance of red clover (*Trifolium pratense*) to the root parasitic plant *Orobanche minor* is activated by salicylate but not by jasmonate. Ann Bot 100:537–544

Labrousse P, Arnaud M-C, Serieys H, Bervillé A, Thalouarn P (2001) Several mechanisms are involved in resistance of *Helianthus* to *Orobanche cumana* Wallr. Ann Bot 88:859–868

Lane JA, Bailey JA, Butler RC, Terry PJ (1993) Resistance of cowpea [*Vigna unguiculata* (L.) Walp.] to *Striga gesnerioides* (Willd.) Vatke, a parasitic angiosperm. New Phytol 125:405–412

Lane JA, Moore THM, Child DV, Cardwell KF, Singh BB, Bailey JA (1994) Virulence characteristics of a new race of the parasitic angiosperm *Striga gesnerioides* from southern Benin on cowpea (*Vigna unguiculata*). Euphytica 72:183–188

Lane JA, Moore THM, Child DV, Cardwell KF (1996) Characterization of virulence and geographic distribution of *Striga gesnerioides* on cowpea in West Africa. Plant Dis 80:299–301

Lane JA, Child DV, Reiss GC, Entcheva V, Bailey JA (1997a) Crop resistance to parasitic plants. In: Crute IR, Holub EB, Burdon JJ (eds) The gene-for-gene relationship in plant-parasite interactions. CAB International, Wallingford, Oxon, pp 81–97

Lane JA, Moore THM, Child DV, Bailey JA (1997b) Variation in virulence of *Striga gesnerioides* on cowpea: new sources of resistance. In: Singh BB, Mohan Raj DR, Dashiell KE, Jakai LEN (eds) Advances in cowpea research. International Institute of Tropical Agriculture (IITA) and Japan International Research Center for Agricultural Sciences (JIRCAS), Ibadan, Nigeria, pp 225–230

Lejeune A, Constant S, Delavault P, Simier P, Thalouarn P, Thoiron S (2006) Involvement of a putative *Lycopersicon esculentum* wall-associated kinase in the early steps of tomato–*Orobanche ramosa* interaction. Physiol Mol Plant Pathol 69:3–12

Letousey P, de Zélicourt A, Vieira Dos Santos C, Thoiron S, Monteau F, Simier P, Thalouarn P, Delavault P (2007) Molecular analysis of resistance mechanisms to *Orobanche cumana* in sunflower. Plant Pathol 56:536–546

Li J, Timko MP (2009) Gene-for-gene resistance in *Striga*-cowpea associations. Science 325:1094

Li J, Lis KE, Timko MP (2009) Molecular genetics of race-specific resistance of cowpea to *Striga gesnerioides* (Wilid.). Pest Manag Sci 65:520–527

Losner-Goshen D, Portnoy VH, Mayer AM, Joel DM (1998) Pectolytic activity by the haustorium of the parasitic plant *Orobanche* L. (Orobanchaceae) in host roots. Ann Bot 81:319–326

Lozano-Baena MD, Prats E, Moreno MT, Rubiales D, Pérez-de-Luque A (2007) *Medicago truncatula* as a model for non-host resistance in legume-parasitic plant interactions. Plant Physiol 145:437–449

Mabrouk Y, Simier P, Delavault P, Delgrange S, Sifi B, Zourgui L, Belhadj O (2007a) Molecular and biochemical mechanisms of defence induced in pea by *Rhizobium leguminosarum* against *Orobanche crenata*. Weed Res 47:452–460

Mabrouk Y, Simier P, Arfaoui A, Sifi DP, Zourgui L, Belhadj O (2007b) Induction of phenolic compounds in pea (*Pisum sativum* L.) inoculated by *Rhizobium leguminosarum* and infected with *Orobanche crenata*. J Phytopathol 155:728–734

Maiti RK, Ramaiah KV, Bisen SS, Chidley VL (1984) A comparative study of the haustorial development of *Striga asiatica* (L.) Kuntze on sorghum cultivars. Ann Bot 54:447–457

McDowell JM, Simon SA (2008) Molecular diversity at the plant-pathogen interface. Dev Comp Immunol 32:736–744

Melero-Vara JM, Domínguez J, Fernández-Martínez JM (2000) Update on sunflower broomrape situation in Spain: racial status and sunflower breeding for resistance. Helia 23:45–56

Mohamed A, Ellicott A, Housley TL, Ejeta G (2003) Hypersensitive response to *Striga* infection in sorghum. Crop Sci 43:1320–1324

Mohamed AH, Housley TL, Ejeta G (2010) Inheritance of hyper sensitive response to *Striga* parasitism in sorghum [*Sorghum bicolor* (L.) Moench]. Afr J Agric Res 19:2720–2729

Molinero-Ruiz M, Melero-Vara J, Garcia-Ruiz R, Dominguez J (2006) Pathogenic diversity within field populations of *Orobanche cumana* and different reactions on sunflower genotypes. Weed Res 46:462–469

Molinero-Ruiz ML, Pérez-Vich B, Pineda-Martos R, Melero-Vara JM (2008) Indigenous highly virulent accessions of the sunflower root parasitic weed *Orobanche cumana*. Weed Res 48:169–178

Molinero-Ruiz ML, Garcia-Ruiz R, Melero-Vara JM, Domínguez J (2009) *Orobanche cumana* race F: performance of resistant sunflower hybrids and aggressiveness of populations of the parasitic weed. Weed Res 49:469–478

Mur L, Kenton P, Atzorn R, Miersch O, Wasternack C (2006) The outcomes of concentration-specific interactions between salicylate and jasmonate signalling include synergy, antagonism, and oxidative stress leading to cell death. Plant Physiol 140:249–262

Neumann U, Vian B, Weber HC, Sallé G (1999) Interface between haustoria of parasitic members of the Scrophulariaceae and their hosts: a histochemical and immunocytochemical approach. Protoplasma 207:84–97

Olivier A, Benhamou N, Leroux GD (1991a) Cell surface interactions between sorghum roots and the parasitic weed *Striga hermonthica*: cytochemical aspects of cellulose distribution in resistant and susceptible host tissues. Can J Bot 69:1679–1690

Olivier A, Ramaiah KV, Leroux GD (1991b) Selection of sorghum (*Sorghum bicolor* Moench) varieties resistant to the parasitic weed *Striga hermonthica* (Del) Benth. Weed Res 31:219–225

Ouédraogo JT, Maheshwari V, Berner DK, St-Pierre C-A, Belzile F, Timko MP (2001) Identification of AFLP markers linked to resistance of cowpea (*Vigna unguiculata* L.) to parasitism by *Striga gesnerioides*. Theor Appl Genet 102:1029–1036

Ouédraogo JT, Olivier A, Timko MP, Belzile FJ (2002a) AFLP markers for resistance against *Striga gesnerioides* race 1 in cowpea (*Vigna unguiculata*). Genome 45:787–793

Ouédraogo JT, Gowda BS, Jean M, Close TJ, Ehlers JD, Hall AE, Gillaspie AG, Roberts PA, Ismail AM, Bruening G, Gepts P, Timko MP, Belzile FJ (2002b) An improved genetic linkage map for cowpea (Vigna unguiculata L.) combining AFLP, RFLP, RAPD and biochemical markers and biological resistance traits. Genome 45:175–188

Parker C, Polniaszek TI (1990) Parasitism of cowpea by *Striga gesnerioides*: variation in virulence and discovery of a new source of host resistance. Ann Appl Biol 116:305–311

Parker C, Riches CR (1993) Parasitic weeds of the world: biology and control. CAB International, Wallingford, 332 pp

Pérez-de-Luque A, Rubiales D, Cubero JI, Press MC, Scholes J, Yoneyama K, Takeuchi Y, Plakhine D, Joel DM (2005a) Interaction between *Orobanche crenata* and its host legumes: unsuccessful haustorial penetration and necrosis of the developing parasite. Ann Bot 95:935–942

Pérez-de-Luque A, Jorrín J, Cubero JI, Rubiales D (2005b) *Orobanche crenata* resistance and avoidance in pea (*Pisum* spp.) operate at different developmental stages of the parasite. Weed Res 45:379–387

Pérez-de-Luque A, González-Verdejo CI, Lozano MD, Dita MA, Cubero JI, González-Melendi P, Risueño MC, Rubiales D (2006a) Protein cross-linking, peroxidase and β-1, 3-endoglucanase involved in resistance of pea against *Orobanche crenata*. J Exp Bot 57:1461–1469

Pérez-de-Luque A, Lozano MD, Cubero JI, González-Melendi P, Risueño MC, Rubiales D (2006b) Mucilage production during the incompatible interaction between *Orobanche crenata* and *Vicia sativa*. J Exp Bot 57:931–942

Pérez-de-Luque A, Lozano MD, Rubiales D (2006c) Resistencia post-haustorial a *Orobanche crenata* en garbanzo. In: De Los Mozos Pascual M, Giménez Alvear MJ, Rodríguez Conde MF, Sánchez Vioque R (eds) Nuevos retos y oportunidades de las leguminosas en el sector agroalimentario español. Consejería de Agricultura de Castilla-La Mancha, Toledo, Spain, pp 361–368

Pérez-de-Luque A, Lozano MD, Moreno MT, Testillano PS, Rubiales D (2007) Resistance to broomrape (*Orobanche crenata*) in faba bean (*Vicia faba*): cell wall changes associated with pre-haustorial defensive mechanisms. Ann Appl Biol 151:89–98

Pérez-de-Luque A, Moreno MT, Rubiales D (2008) Host plant resistance against broomrapes (*Orobanche* spp.): defence reactions and mechanisms of resistance. Ann Appl Biol 152:131–141

Pérez-de-Luque A, Fondevilla S, Pérez-Vich B, Aly R, Thoiron S, Simier P, Castillejo MA, Fernández JM, Jorrín J, Rubiales D, Delavault P (2009) Understanding broomrape-host plant interaction and developing resistance. Weed Res 49:8–22

Pérez-Vich B, Akhtouch B, Knapp SJ, Leon AJ, Velasco L, Fernández-Martínez JM, Berry ST (2004a) Quantitative trait loci for broomrape (*Orobanche cumana* Wallr.) resistance in sunflower. Theor Appl Genet 109:92–102

Pérez-Vich B, Akhtouch B, Mateos A, Velasco L, Jan CC, Fernández J, Domínguez J, Fernández-Martínez JM (2004b) Dominance relationships for genes conferring resistance to sunflower broomrape (*Orobanche cumana* Wallr.). Helia 27:183–192

Press MC, Graves JD (eds) (1995) Parasitic plants. Chapman & Hall, London

Radwan MS, Abdalla MMF, Fischbeck G, Metwally AA, Darwish DS (1988) Variation in reaction of faba bean lines to different accessions of *Orobanche crenata* Forks. Plant Breed 101:208–216

Radwan O, Gandhi S, Heesacker A et al (2008) Genetic diversity and genomic distribution of homologs encoding NBS-LRR disease resistance proteins in sunflower. Mol Genet Genom 280:111–125

Ramaiah KV (1987) Control of *Striga* and *Orobanche* species – a review. In: Weber H Chr, Forstreuter W (eds) Parasitic flowering plants. Proceedings of the 4th ISPFP, Marburg, FRG, pp 637–664

Rich PJ, Grenier C, Ejeta G (2004) *Striga* resistance in the wild relatives of sorghum. Crop Sci 4:2221–2229

Riopel JL, Timko MP (1995) Haustorial initiation and differentiation. In: Press MC, Graves JD (eds) Parasitic plants. Chapman & Hall, London, pp 39–73

Rodenburg J, Riches CR, Kayeke JM (2010) Addressing current and future problems of parasitic weeds in rice. Crop Prot 29:210–221

Rodriguez-Conde MF, Moreno MT, Cubero JI, Rubiales D (2004) Characterization of the *Orobanche-Medicago truncatula* association for studying early stages of the parasite–host interaction. Weed Res 44:218–223

Rodríguez-Ojeda MI, Fernández-Escobar J, Alonso LC (2001) Sunflower inbred line (KI-374) carrying two recessive genes for resistance against a highly virulent Spanish population of *Orobanche cernua* Loefl./*O. cumana* Wallr. race "F". In: Fer A, Thalouarn P, Joel D, Musselman LJ, Parker C, Verkleij JAC (eds) Proceedings of the 7th international parasitic weed symposium. University of Nantes, France, pp 208–211

Román B, Torres AM, Rubiales D, Cubero JI, Šatović Z (2002) Mapping of quantitative trait loci controlling broomrape (*Orobanche crenata* Forks.) resistance in faba bean (*Vicia faba* L.). Genome 45:1057–1063

Ronald PC, Beutler B (2010) Plant and animal sensors of conserved microbial signatures. Science 330:1061–1064

Rubiales D (2001) Parasitic plants: an increasing threat. Grain Legumes 33:10–11

Rubiales D, Pérez de Luque A, Joel DM, Alcántara C, Sillero JC (2003) Characterization of resistance in chickpea to crenate broomrape (*Orobanche crenata*). Weed Sci 51:702–707

Rubiales D, Moreno MT, Sillero JC (2005) Search for resistance to crenate broomrape (*Orobanche crenata*) in pea germplasm. Genet Resour Crop Evol 52:853–861

Rubiales D, Pérez-de-Luque A, Fernández-Aparicio M, Sillero JC, Román B, Kharrat M, Khalil S, Joel DM, Riches C (2006) Screening techniques and sources of resistance against parasitic weeds in grain legumes. Euphytica 147:87–199

Rümer S, Cameron DD, Wacker R, Hartung W, Jiang F (2007) An anatomical study of the haustoria of *Rhinanthus minor* attached to roots of different hosts. Flora 202:194–200

Runyon J, Mescher M, Felton G, de Moraes C (2010) Parasitism by *Cuscuta pentagona* sequentially induces JA and SA defence pathways in tomato. Plant Cell Environ 33:290–303

Ryu H, Han M, Lee S, Cho J, Ryoo N, Heu S, Lee Y, Bhoo S, Wang G, Hahn T, Jeon JS (2006) A comprehensive expression analysis of the WRKY gene superfamily in rice plants during defence response. Plant Cell Rep 25:836–847

Scholes JD, Press MC (2008) *Striga* infestation of cereal crops – an unsolved problem in resource limited agriculture. Curr Opin Plant Biol 11:180–186

Shergini K, Pérez de Luque A, Castejón Muñoz M, García-Torres L, Jorrín JV (2001) Sunflower (*Helianthus annuus* L.) response to broomrape (*Orobanche cernua* Loefl.) parasitism: induced synthesis and excretion of 7-hydroxylated simple coumarins. J Exp Bot 52:2227–2234

Shomer-Ilan A (1993) Germinating seeds of the root parasite *Orobanche aegyptiaca* Pers. excrete enzymes with carbohydrase activity. Symbiosis 15:61–70

Sillero JC, Moreno MT, Rubiales D (2005) Sources of resistance to crenate broomrape among species of *Vicia*. Plant Dis 89:23–27

Singh BB, Emechebe AM (1990) Inheritance of *Striga* resistance in cowpea genotype B301. Crop Sci 30:879–881

Swarbrick PJ, Huang K, Liu G, Slate J, Press MC, Scholes JD (2008) Global patterns of gene expression in rice cultivars undergoing a susceptible or resistant interaction with the parasitic plant *Striga hermonthica*. New Phytol 179:515–529

Takken FL, Albrecht M, Tameling WI (2006) Resistance proteins: molecular switches of plant defence. Curr Opin Plant Biol 9:383–390

Tameling WIL, Joosten MHAJ (2007) The diverse roles of NB-LRR proteins in plants. Physiol Mol Plant Pathol 71:126–134

Tang S, Heesacher A, Kishore VK, Fernandez A, Sadik ES, Cole G, Knapp SJ (2003) Genetic mapping of the *Or5* gene for resistance to *Orobanche* race E in sunflower. Crop Sci 43:1021–1028

ter Borg SJ (1999) Broomrape resistance in faba bean: what do we know? In: Cubero JI, Moreno MT, Rubiales D, Sillero JC (eds) Resistance to broomrape, the state of the art. Junta de Andalucıa, Sevilla, Spain, pp 25–41

Thordal-Christensen H (2003) Fresh insights into processes of nonhost resistance. Curr Opin Plant Biol 6:351–357

Timko MP, Singh BB (2008) Cowpea, a multifunctional legume. In: Moore PH, Ming R (eds) Genomics of tropical crop plants. Springer Science + Business Media LLC, New York, pp 227–258

Timko MP, Gowda BS, Ouedraogo J, Ousmane B (2007) Molecular markers for analysis of resistance to *Striga gesnerioides* in cowpea. In: Ejeta G, Gressell J (eds) Integrating new technologies for *Striga* control: towards ending the witch-hunt. World Scientific, Singapore, pp 115–128

Timko MP, Rushton PJ, Laudeman TW, Bokowiec MT, Chipumuro E, Cheung F, Town CD, Chen X (2008) Sequencing and analysis of the gene-rich space of cowpea. BMC Genomics 9:103

Torto-Alalibo T, Collmer CW, Lineberg M, Bird D, Collmer A, Tyler BM (2009) Common and contrasting themes in host cell-targeted effectors from bacterial, fungal, oomycete and nematode plant symbionts described using the gene ontology. BMC Microbiol 9(Suppl 1):S3. doi:10.1186/1471-2180-9-S1-S3

Touré M, Olivier A, Ntare BR, Lane JA, St. Pierre C-A (1997) Inheritance of resistance to *Striga gesnerioides* biotypes from Mali and Niger in cowpea (*Vigna unguiculata* (L.) Walp.). Euphytica 94:273–278

Tyler BM (2009) Entering and breaking: virulence effector proteins of oomycete plant pathogens. Cell Microbiol 11:13–20

Valderrama MR, Román B, Satovic Z, Rubiales D, Cubero JI, Torres AM (2004) Locating quantitative trait loci associated with *Orobanche crenata* resistance in pea. Weed Res 44:1–6

van der Hoorn RAL, Kamoun S (2008) From guard to decoy: a new model for perception of plant pathogen effectors. Plant Cell 20:2009–2017

Velasco L, Pérez-Vich B, Jan CC, Fernández-Martínez JM (2006) Inheritance of resistance to broomrape (*Orobanche cumana* Wallr.) race F in a sunflower line carrying resistance genes from wild sunflower species. Plant Breed 126:67–71

Vieira Dos Santos C, Letousey P, Delavault P, Thalouarn P (2003a) Defence gene expression analysis of *Arabidopsis thaliana* parasitized by *Orobanche ramosa*. Phytopathology 93:451–457

Vieira Dos Santos C, Delavault P, Letousey P, Thalouarn P (2003b) Identification by suppression subtractive hybridization and expression analysis of *Arabidopsis thaliana* putative defence genes during *Orobanche ramosa* infection. Physiol Mol Plant Pathol 62:297–303

Volgler RK, Ejeta G, Butler LG (1996) Inheritance of low production of *Striga* germination stimulant in sorghum. Crop Sci 36:1185

Vrânceanu AV, Tudor VA, Stoenescu FM, Pirvu N (1980) Virulence groups of *Orobanche cumana* Wallr., differential hosts and resistance source genes in sunflower. In: International Sunflower Association (ed) Proceedings of the 9th international sunflower conference. LH/VPO, Torremolinos, Spain, pp 74–82

Westwood J, Yu X, Foy C, Cramer C (1996) Parasitization by *Orobanche* induces expression of a defence-related gene in tobacco. In: Moreno MT, Cubero JI, Berner D, Joel D, Musselman LJ, Parker C (eds) Proceedings of the 6th international parasitic weed symposium, advances in parasitic plant research. Dirección General de Investigación Agraria, Cordoba, Spain, pp 543–550

Westwood JH, Yu X, Foy CL, Cramer CL (1998) Expression of a defence-related 3-hydroxy-3-methylglutaryl CoA reductase gene in response to parasitization by *Orobanche* spp. Mol Plant Microbe Interact 11:530–536

Wilson JP, Hess DE, Hanna WW (2000) Resistance to *Striga hermonthica* in wild accessions of the primary gene pool of *Pennisetum glaucum*. Phytopathology 90:1169–1172

Wilson JP, Hess DE, Hanna WW, Kumar KA, Gupta SC (2004) *Pennisetum glaucum* subsp. *monodii* accessions with *Striga* resistance in West Africa. Crop Prot 23:865–870

Yoshida S, Shirasu K (2009) Multiple layers of incompatibility to the parasitic witchweed, *Striga hermonthic*a. New Phytol 183:180–189

Chapter 8
Seed Production and Dispersal in the Orobanchaceae

Daniel M. Joel

Most facultative hemiparasites of the Orobanchaceae, which develop only lateral haustoria, produce seeds that resemble those of non-parasitic plants. The germination of these facultative parasites is triggered only by seasonal cues. However, the seeds of the majority of obligate parasitic Orobanchaceae, which develop terminal haustoria following germination, are unique in their germination requirements. The following four chapters are dedicated mainly to the unique seeds of the obligate root parasitic Orobanchaceae, which require chemical stimulation from adjacent plant roots for germination. The description of their seeds and seedling is presented in Chap. 9; the stimulation of their germination is dealt with in Chaps. 10 and 12, whereas their germination ecophysiology is discussed in Chap. 11.

The seeds of the obligate parasitic Orobanchaceae are small. In some genera the seeds are dust-like, ranging between 0.2 mm (e.g. in some *Phelipanche*, *Striga* and *Orobanche* species; Figs. 8.1a and 8.2a, b), while in others up to 2 mm (e.g. in *Cistanche* and *Conopholis* species; Fig. 8.1b; Musselman and Dickison 1975; Teryokhin 1997; Baird and Riopel 1986). Only 200 cells comprise the seed of some *Orobanche* species (Joel et al. 1995). Seeds of facultative species can be much larger, e.g. up to 6 mm in *Melampyrum* (Fischer 2004) and 1.5 mm in *Triphysaria* (Fig. 8.2c).

Many obligate parasites produce numerous small seeds. Furthermore, their longevity in soil is often longer than the longevity of facultative parasites (Bekker and Kwak 2005). *Phelipanche aegyptiaca*, for example, often remains viable for several decades in highly infested agricultural fields in Israel when non-host crops (e.g. citrus trees) are grown in the field for a long while. Similarly, *Alectra vogelii* seeds persist in the soil even for 15 years until a host plant is planted in the field (Kroschel 1998). Unlike some facultative hemiparasites, e.g. *Melampyrum* and

D.M. Joel (✉)
Department of Weed Research, Agricultural Research Organization, Newe-Ya'ar Research Center, Ramat-Yishay 30095, Israel
e-mail: dmjoel@agri.gov.il

D.M. Joel et al. (eds.), *Parasitic Orobanchaceae*,
DOI 10.1007/978-3-642-38146-1_8, © Springer-Verlag Berlin Heidelberg 2013

Fig. 8.1 Open seed capsules. (**a**) *Phelipanche aegyptiaca*; (**b**) *Conopholis americana*

some *Rhinanthus* species, which produce only few seeds per capsule, a single capsule of obligate parasites like *Orobanche*, *Striga*, *Alectra*, *Phelipanche* and *Conopholis* contains around 500 seeds (Fig. 8.1a, b). A single plant can produce annually tens of capsules, sometimes also hundreds, leading to annual dispersal of 10,000–500,000 seeds per single plant (Baird and Riopel 1986; Joel et al. 1995, 2007; Rich and Ejeta 2007). The production of numerous long-living miniscule seeds (microspermy) increases the probability that at least some seeds find a host even when host plants are temporarily and spatially uncommon. This strategy allows survival of the parasites in natural habitats. In agricultural fields, where host plants are abundant, this nature of weedy parasites is a key element in their rapid propagation and consequently also in their serious economic impact, and provides the parasites with genetic adaptability to changes in host availability and host resistance (Rich and Ejeta 2007; see Sect. 22.1.2).

Another strategy that increases the reproductive potential of some obligate Orobanchaceae is apomixis, an asexual reproduction mechanism that allows the development of seeds from unfertilized ovules (Jensen 1951; Greilhuber and Weber 1975; Heckard and Chuang 1975; Teryokhin 1997; Pazy 1998; Plitmann 2002). The flower of some broomrape species has the ability to develop seeds in three different ways: by cross-pollination, by self-pollination and by apomixis. This allows the plant to produce seeds of various genetic compositions even if the flowers have not been visited by pollinators (Teryokhin 1997; see Chap. 19). Interestingly a similar phenomenon is also found in insects, where asexual reproduction occurs among parasitic insects while it is rare in predatory species (Price 1980).

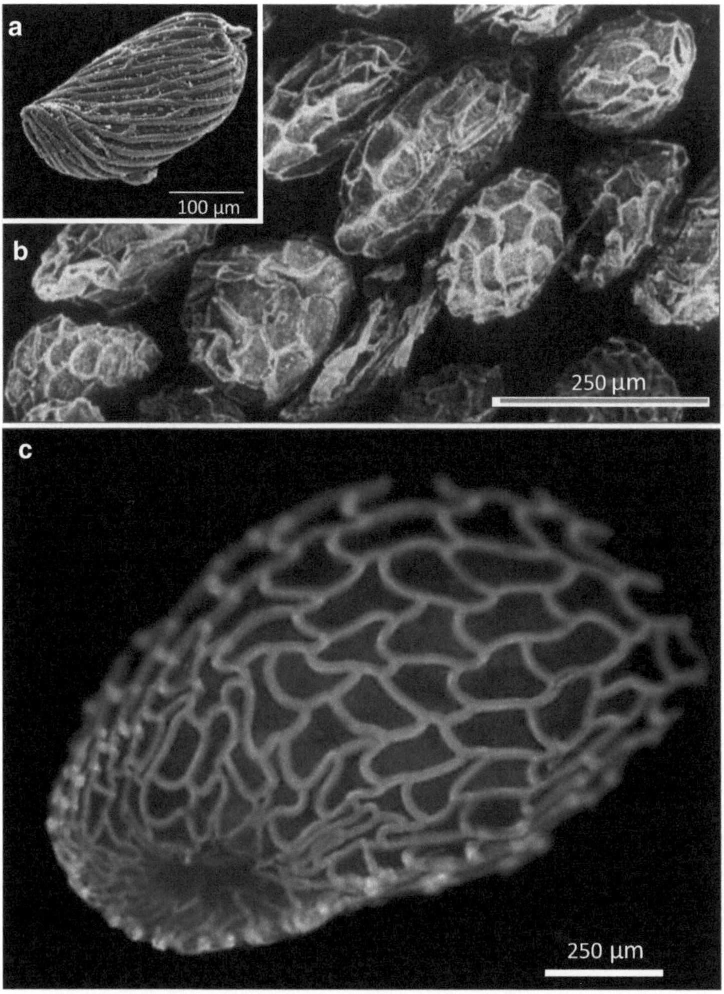

Fig. 8.2 Seed size of various parasitic Orobanchaceae. (**a**) *Striga aspera* (SEM from Krause and Weber 1990); (**b**) *P. aegyptiaca* (SEM); (**c**) *Triphysaria versicolor* (epifluorescence micrograph)

References

Baird VM, Riopel JL (1986) The developmental anatomy of *Conopholis americana* (Orobanchaceae) seedlings and tubercles. Can J Bot 64:710–717

Bekker RM, Kwak MM (2005) Life history traits as predictors of plant rarity, with particular reference to hemiparasitic Orobanchaceae. Folia Geobot 40:231–242

Fischer E (2004) Scropulariaceae. In: Kubitzki K (ed) The families and genera of vascular plants. Springer, Berlin, pp 333–432

Greilhuber J, Weber A (1975) Aneusomaty in *Orobanche gracilis*. Plant Syst Evol 124:67–77

Heckard LR, Chuang TI (1975) Chromosome numbers and polyploidy in *Orobanche* (Orobanchaceae). Brittonia 27:179–186

Jensen HW (1951) The normal and parthenogenetic forms of *Orobanche uniflora* L. La Cellule 54:135–141

Joel DM, Steffens JC, Matthews DE (1995) Germination of weedy root parasites. In: Kigel J, Galili G (eds) Seed development and germination. Marcel Dekker, New York, pp 567–598

Joel DM, Hershenhorn Y, Eizenberg H, Aly R, Ejeta G, Rich PJ, Ransom JK, Sauerborn J, Rubiales D (2007) Biology and management of weedy root parasites. In: Janick J (ed) Horticultural reviews, vol 33. Wiley, Hoboken, NJ, pp 267–349

Krause D, Weber HC (1990) SEM observations on seeds of *Striga* spp. and *Buchnera americana* (Scrophulariaceae). Plant Sys Evol 170:257–263

Kroschel J (1998) *Striga* – how will it affect African agriculture in future? An ecological perspective. Plits (Germany) 16:137–158

Musselman LJ, Dickison WC (1975) The structure and development of the haustorium in parasitic Scrophulariaceae. Bot J Linn Soc 70:183–212

Pazy B (1998) Diploidization failure and apomixis in Orobanchaceae. Bot J Linn Soc 128:99–103

Plitmann U (2002) Agamospermy is much more common than conceived: a hypothesis. Isr J Plant Sci 50:S111–S117

Price PW (1980) Evolutionary biology of parasites, vol 15, Monographs in evolution biology. Princeton University Press, Princeton NJ

Rich PJ, Ejeta G (2007) Biology of host-parasite interaction in *Striga* species. In: Ejeta G, Gressel J (eds) Integrating new technologies for *Striga* control: towards ending the witch-hunt. World Scientific, Singapore, pp 19–32

Teryokhin ES (1997) Weed broomrapes: systematic, ontogenesis, biology, evolution. Aufstieg, Landshut

Chapter 9
The Seed and the Seedling

Daniel M. Joel and Hilla Bar

9.1 Surface Structure

As detailed in Chap. 8, the seeds of the obligate parasitic Orobanchaceae are very small, ranging 0.2–2 mm (Figs. 8.1 and 8.2a, b). The micromorphology of the seed surface is highly diverse and has been described for various species, particularly of the genera that are of economic importance. Seed surface sculpturing is best seen under the scanning electron microscope (Figs. 8.2a, b and 9.1a–e; Musselman and Mann 1976; Musselman and Parker 1981; Jones and Safa 1982; Joel 1987a; Abu Sbaih and Jury 1994; Krause and Weber 1990). It can also be explored by light microscopy. The seed coat is usually dark and opaque, but bleaching seeds allows a clear view of some surface features, particularly the contours and pitting of the seed coat epidermal cells (Ungurean 1986). Seeds can also be examined by fluorescence microscopy (Fig. 8.2c) exploiting the auto fluorescence of the lignified outer cell walls. The latter method is particularly helpful in the diagnosis of soil-borne seeds that lost their original shape and pigmentation but retained their seed coat sculpturing and remained viable (Joel 1987b). All these methods provide information on seed size and on the arrangement of the external seed coat cells as well as the pattern of thickenings and pitting of the lateral and inner tangential cell walls of the seed coat (Fig. 9.1a–f).

The seed surface ornamentation is of taxonomic and diagnostic importance (Chuang and Heckard 1983; Joel 1987b; Teryokhin and Kravtsova 1987; Krause and Weber 1990; Abu Sbaih and Jury 1994; Fischer 2004; Domina and Colombo 2005). However, care should be taken when interpreting the observations, in particular those taken under the SEM, because in many genera the outer tangential wall of mature seeds is very thin (see Sect. 9.2.4) and may either disintegrate and

D.M. Joel (✉) • H. Bar
Department of Weed Research, Agricultural Research Organization, Newe-Ya'ar Research Center, Ramat Yishay 30095, Israel
e-mail: dmjoel@agri.gov.il

D.M. Joel et al. (eds.), *Parasitic Orobanchaceae*,
DOI 10.1007/978-3-642-38146-1_9, © Springer-Verlag Berlin Heidelberg 2013

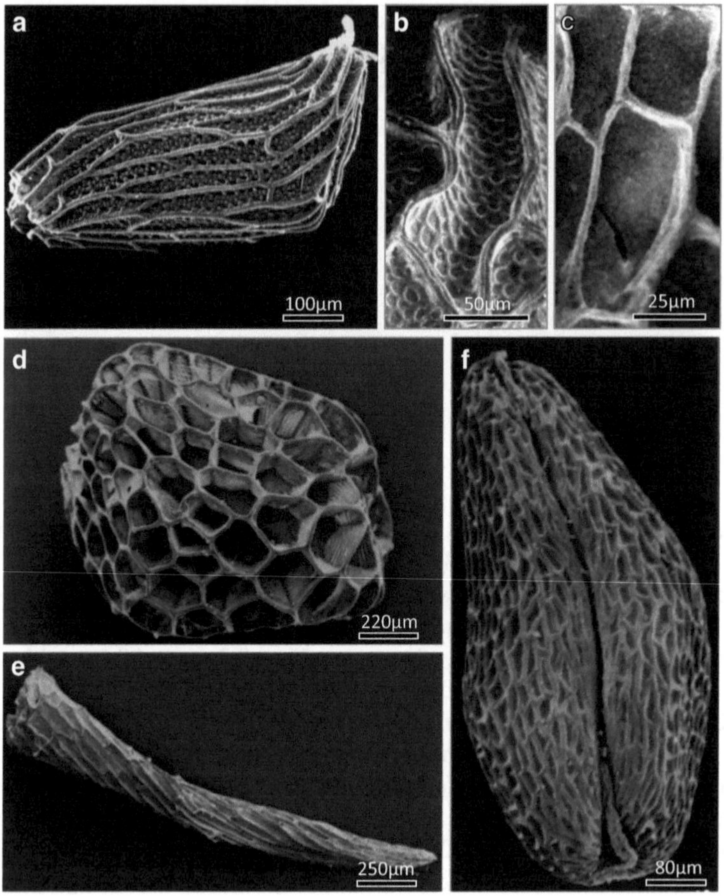

Fig. 9.1 Diversity of Orobanchaceae seeds. (**a**) *Buchnera americana* (from Krause and Weber 1990). (**b**) Portion of *Phelipanche* seed with reticulate thickenings of inner tangential wall of epidermal cells (SEM). (**c**) Portion of *Phelipanche* seed showing start of outer wall breakdown in one epidermal cell; (**d**) *Cistanche* sp.; (**e**) *Alectra fruticosa*; (**f**) *Pedicularis palustris* (**d–f** from Fischer 2004)

disappear or collapse and cling to the inner walls (Fig. 9.1b, c), respectively exposing or covering the typical ornamentation of the inner walls of these cells (Joel 1987a). The extent of external wall modifications reflects seed age and dryness and is also affected by storage conditions. Freshly harvested seeds often have an intact outer wall while dry seeds often lose it.

Seed coat sculpturing can be helpful in the analysis of soil seed banks, but due to the variations in seed shape, size and ornamentation even within a single capsule, unequivocal morphological diagnosis should rely on a set of seeds rather than on a single one (Joel 1987b; Domina and Colombo 2005). A more accurate diagnostic identification of single seeds, including soil-borne seeds, can be achieved by DNA analysis (Joel et al. 1996, 1998; Portnoy et al. 1997; see Chap. 20).

The Orobanchaceae exhibit a wide variety of different shapes and surface characteristics (Figs. 8.2 and 9.1). These morphological characteristics, and in particular seed sculpturing, are key features in determining the manner of their dispersal. Depending on their vector the seeds can be divided into different ecological types: water dispersed (*Epifagus*), animal fur dispersed (*Phacellanthus*), passed through animal digestive tracts (*Conopholis*), wind dispersed (*Aeginetia*) and wind and water dispersed (*Boschniakia*) (Kuijt 1969; Teryokhin 1997). Since these seed characteristics do not directly relate to the parasitic habit, the seed surface is not further discussed in this chapter.

9.2 Anatomy

Whereas the ovule in Orobanchaceae flowers is composed of three main parts—integument, nucellus and embryo sac (Fig. 9.2a)—the seeds which develop from these ovules are composed of four main parts, each with a different genetic origin (Figs. 9.2b–d and 9.8a). Following the double fertilization within the embryo sac, the diploid embryo develops by cell proliferation of the zygote after the fusion of the first sperm cell of the pollen tube with the egg cell, while the triploid endosperm develops following fusion of the second sperm cell with the two polar nuclei of the embryo sac (Michell 1915; Teryokhin 1997; Chen and Hsiao 2011). The maternal tissues of the ovule give rise to the perisperm and the seed coat (testa); the perisperm develops by differentiation of the unilayered diploid nucellus (Plakhine et al. 2012), while the seed coat (testa) develops by differentiation of the diploid integument (Joel et al. 2012). Typically, an endosperm haustorium is formed during early stages of seed development, which transfers nutrients to the developing embryo (the embryology of the Orobanchaceae is reviewed by Teryokhin 1997 and Fischer 2004). These developmental stages seem to be common to all Orobanchaceae. Further development of the endosperm and the perisperm differs, however, in the various genera. The perisperm is particularly conspicuous in seeds of obligate parasites (Fig. 9.2b, c, see Sect. 9.2.3), while the endosperm, which occupies the main volume around the embryo in facultative hemiparasites like *Triphysaria* (Fig. 9.2d), is poorly developed in some holoparasitic species (Figs. 9.2b and 9.3c) and almost absent in *Aeginetia* (Fig. 9.2b).

The seed anatomy was so far studied mainly under the light microscope, and therefore details on seed fine structure and on the function of the various seed parts are missing for the great majority of Orobanchaceae. A comprehensive analysis of seed ultrastructure was only done on *Phelipanche aegyptiaca* (Joel et al. 2012) though some ultrastructural information is also provided for species of *Orobanche* and *Striga* (Aber and Sallé 1983; Kravtzova and Teryokhin 1987; Sauerborn et al. 1996). In general, the seed structure of the majority of the obligate parasitic Orobanchaceae has a similar pattern, and therefore the description below should be relevant to most species. The seeds of *P. aegyptiaca* contain only a reduced embryo. Lipids compose their main seed storage material; the major fatty acids in

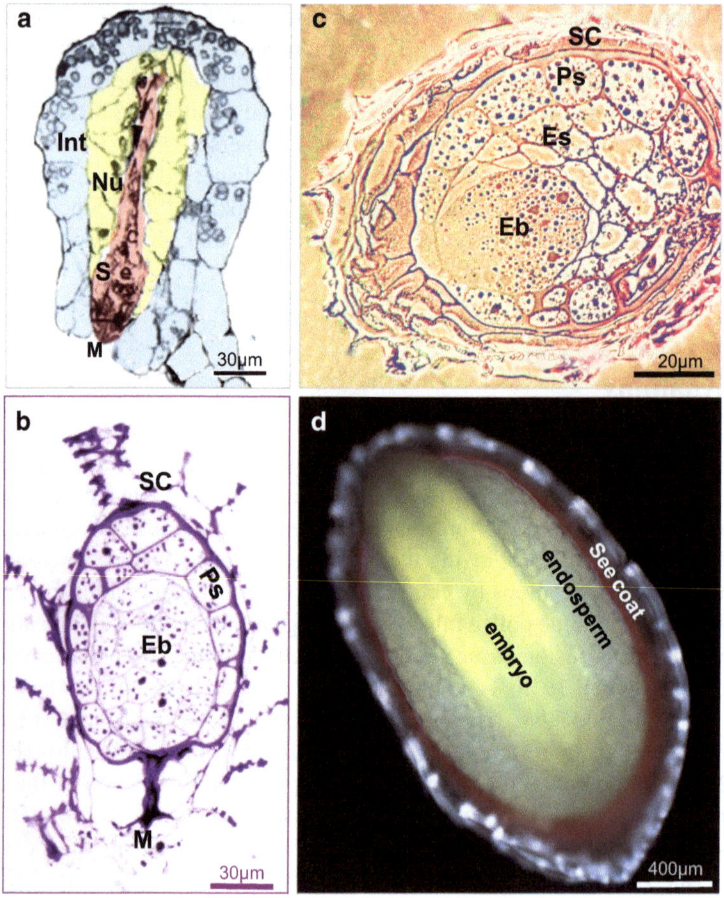

Fig. 9.2 The ovule and the seed. (**a**) Mature ovule of *Aeginetia indica*. (**b**) *A. indica* seed (**a** and **b** from Chen and Hsiao 2011, with permission); (**c**) median section of *P. aegyptiaca* seed; (**d**) median section of *Triphysaria versicolor* seed. *Eb* embryo, *Es* endosperm, *Int* integument, *M* micropyle, *Nu* nucellus, *Ps* perisperm, *S* embryo sac, *SC* seed coat

the seeds of *Orobanche* and *Phelipanche* spp. are oleic and linoleic acids, but their respective occurrence varies in the different species (Sauerborn et al. 1996; Velasco et al. 2000; Bar-Nun and Mayer 2002).

9.2.1 The Embryo

Obligate hemiparasitic Orobanchaceae, like *Alectra* and *Striga* species, have a weakly differentiated heart-shaped embryo that includes a small radicle, two small embryonic cotyledons, and between them—a hypocotyl with a procambium along the central axis of the embryo (Fig. 9.3a).

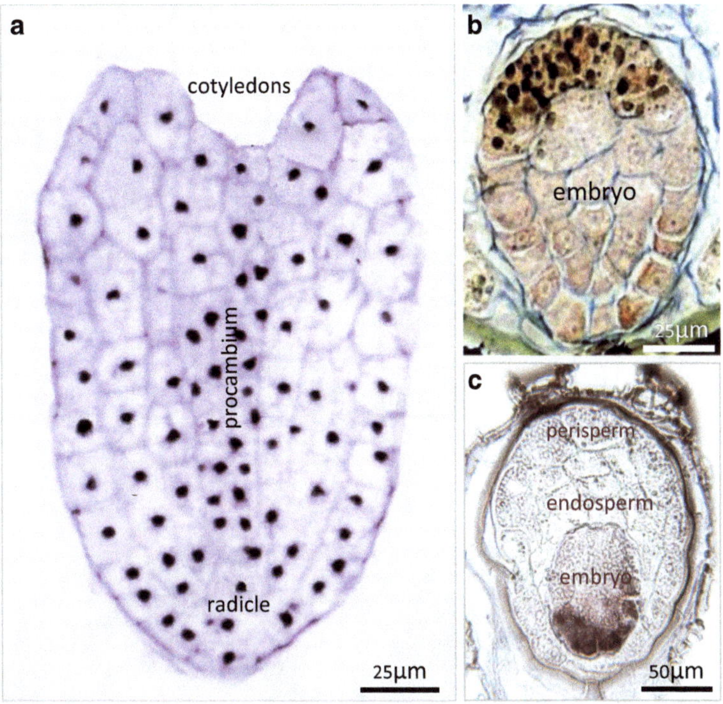

Fig. 9.3 The embryo. (**a**) *Striga* embryo, median section showing the small cotyledons at the chalazal pole, the radicle at the radicular pole and the procambium along the embryo axis. (**b**) *Phelipanche* embryo, protein bodies accumulate in cells at the chalazal pole. (**c**) Expression of the ABA catabolic gene *PrCYP707A1* in the radicular embryo pole in conditioned seed of *P. ramosa* (from Lechat et al. 2012, with permission)

The embryo of the holoparasites *Conopholis*, *Phelipanche*, *Aeginetia* and *Orobanche*, which do not develop green shoots, is a small spherical or slightly elongated body, composed of around 100 cells (Figs. 9.2b, c and 9.3b), without morphologically identified cotyledons or a shoot meristem (Kumar 1977; Baird and Riopel 1986; Joel et al. 1995; Teryokhin 1997). These highly reduced embryos are located in the central part of the seed close to the micropyle and are composed of thin-walled small cells with no intercellular spaces (Fig. 9.4a, b). The cells that give rise to the emerging radicle are usually smaller (Fig. 9.3a, b).

In addition to a large nucleus and many mitochondria, the embryo cells of the Orobanchaceae contain oil bodies that are commonly organized in concentric circles around protein bodies (Fig. 9.4a; Joel et al. 2012). In hemiparasites such as *Alectra* and *Striga*, the oil bodies are mainly located in the small cotyledons (Okonkwo and Nwoke 1978), while in the holoparasites that do not develop cotyledons, the oil and protein bodies typically fill the cells at the chalazal pole of the embryo (Figs. 9.3b and 9.4a). This embryonic region can thus be regarded as the reduced cotyledonary region of the embryo.

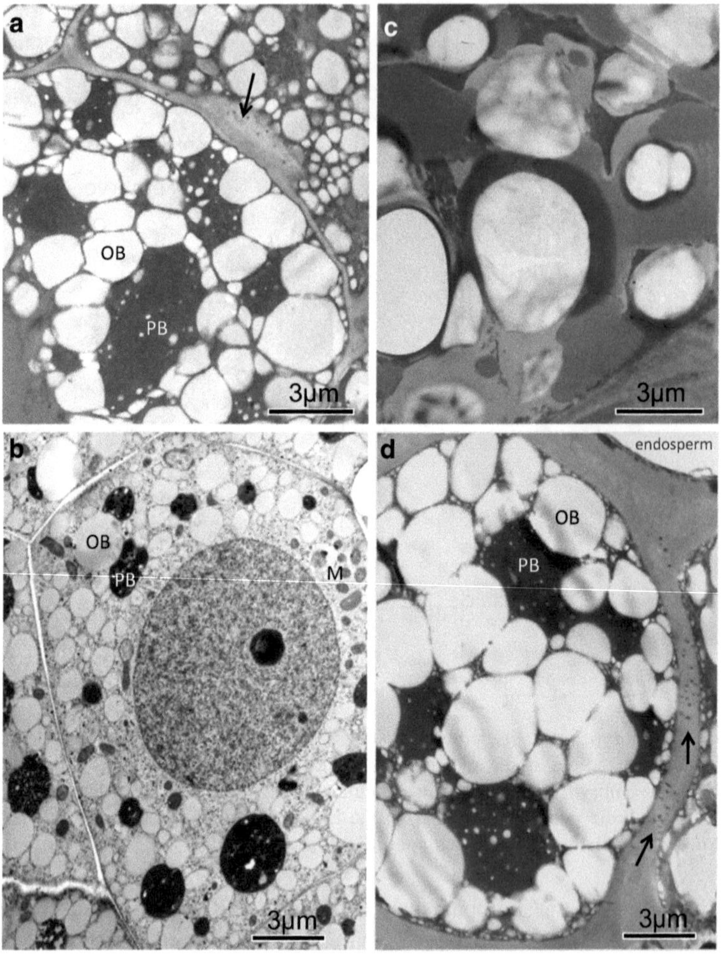

Fig. 9.4 Cell types in *Phelipanche aegyptiaca* seeds. (**a**) Portion of two embryo cells at the chalazal pole, showing plasmodesmata in the cell wall between them (*arrow*), large protein bodies (PB) and oil bodies (OB). (**b**) Embryo cell from the radicular pole, with a large nucleus, numerous mitochondria (M) and protein and oil bodies that are smaller than those at the chalazal pole. (**c**) Portion of an endosperm cell, note the oil bodies with diffuse boundaries. (**d**) Portion of a perisperm cell, with large protein bodies and oil bodies; note the plasmodesmata that connect neighbouring perisperm cells (*right*) and absence of plasmodesmata with the neighbouring endosperm cell (*top*)

Early studies often described the embryos of holoparasitic Orobanchaceae as 'undifferentiated'. This is true when referring to cell organization, though there is a clear polarization in the embryo and a gradient in cell size from the chalazal to the radicular pole. In addition, a clear physiological polarization can be seen not only when the seeds start germinating but also in dormant seeds and in imbibed seeds before perception of a germination stimulant. As seen above, oil bodies and large protein bodies typically accumulate in the chalazal pole of the embryo, while the cells at the radicular pole have a dominant large nucleus and many mitochondria (Fig. 9.4b). Furthermore, during seed conditioning the embryo cells facing the micropyle of *P. ramosa* significantly express the *PrCYP707A1* gene encoding an abscisic acid (ABA) carboxylase (Fig. 9.3c). The expression of this gene coincides with a 6.3-fold decrease in ABA levels during the first day of seed conditioning (Lechat et al. 2012), indicating a preliminary dormancy relief of the radicle in preparation for potential stimulant perception that in turn facilitates germination (see Sect. 11.4.1 on hormonal control of dormancy relief and germination).

9.2.2 The Endosperm

In the majority of Orobanchaceae, the endosperm is located in the mature seed between the embryo and the perisperm (Fig. 9.2c). It is composed of thin-walled small cells, usually containing large amounts oil bodies with diffuse boundaries (Fig. 9.4c), a phenomenon that correlates with a reduced amount of the proteins that are associated with the oil bodies. No organelles were found in the endosperm of mature *P. aegyptiaca* seeds (Joel et al. 2012), and cell compartmentation was broken. This is interpreted as being an internal space for reserve material that will be absorbed during germination by the neighbouring embryo and perisperm cells (see Sect. 9.5). In seeds of some genera, like *Aeginetia*, the endosperm is poorly developed and the main volume of the seed within the seed coat is occupied by the embryo and the perisperm (Fig. 9.2b; see Fig. 5 in Chen and Hsiao 2011).

9.2.3 The Perisperm

The Lamiales, including the Orobanchaceae, are tenuinucellate, i.e. their mature ovule contains a nucellus that is composed of only one cell layer (Fig. 9.2a; Michell 1915; Fischer 2004). In some genera, such as *Conopholis*, the nucellus seems to disappear during seed development (Fig. 9.2d), but in *Striga*, *Alectra*, *Phelipanche*, *Orobanche*, *Aeginetia* and many other genera, it remains after fertilization and further develops into a unique cell layer, the *perisperm*, which is involved in various key signalling and metabolic functions before and during germination (Joel et al. 2012, and see below).

The perisperm lies under to the seed coat and surrounds the endosperm and the embryo (Figs. 9.2b, c and 9.8a). Apart from the cells near the micropyle, the perisperm cells typically contain numerous large oil bodies and large protein bodies (Figs. 9.4d and 9.5a). This cell layer had previously been incorrectly interpreted as being the external layer of the endosperm (or the internal layer of the seed coat) and regarded as the 'aleuronic layer' of the seed (Privat 1960; Egley 1972; Okonkwo and Nwoke 1978; Teryokhin et al. 1993; Krause 1990; Teryokhin 1997). However,

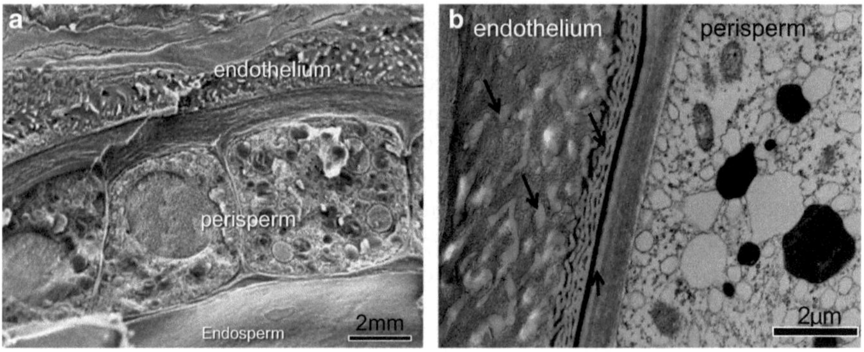

Fig. 9.5 The endothelium and the perisperm. (**a**) Portion of a freeze-fractioned seed of *Striga hermonthica*; the perisperm cells have a large nucleus and numerous oil and protein bodies; the endothelial cells contain wall labyrinth (courtesy of B. Zwanenburg). (**b**) Transmission electron micrograph of *Phelipanche aegyptiaca* seed near the micropyle, showing the thin cuticle (*short arrow*) between the endothelium and the perisperm, the perisperm cell with small oil and protein bodies, numerous vesicles and mitochondria; note the mucilage between wall protuberances (*long arrows*) of the endothelium cell

this single cell layer differs from the endosperm in cell arrangement and in cytoplasm differentiation. Also, no plasmodesmata connections exist between these two tissues, in contrast to the numerous plasmodesmata between perisperm cells (Fig. 9.4d; Joel et al. 2012).

These differences between the perisperm and the endosperm correspond to their different ontogenetic origin and to the difference in their ploidy levels. Whereas the endosperm is triploid (Teryokhin 1997; Chen and Hsiao 2011), the perisperm arises by differentiation of the ovular nucellus, is diploid and maternal (Plakhine et al. 2012).

9.2.4 The Seed Coat

The seed coat (testa) develops from the single integument of the ovule, which is composed of three cell layers. However, the mature seed coat is basically composed of only two cell layers, the outer epidermis and the endothelium (Fig. 9.6a). Cell walls of a third layer, the hypodermis, may remain in some species between these two cell layers, mainly at the chalazal zone (Fig. 9.6b; Joel et al. 2012).

The outer epidermis of the mature seed coat is composed of dead cells with thin outer walls (Fig. 9.6a), whereas the inner and side walls of the epidermis are lignified and have pitted or reticulate wall thickenings (Figs. 8.2, 9.1, and 9.6a, b). The thin outer cell wall often breaks and disintegrates, exposing the inner wall to the outside (see Sect. 9.1). The cell walls at the micropyle are thinner, which enable opening the micropyle during seed conditioning (see Sect. 9.3).

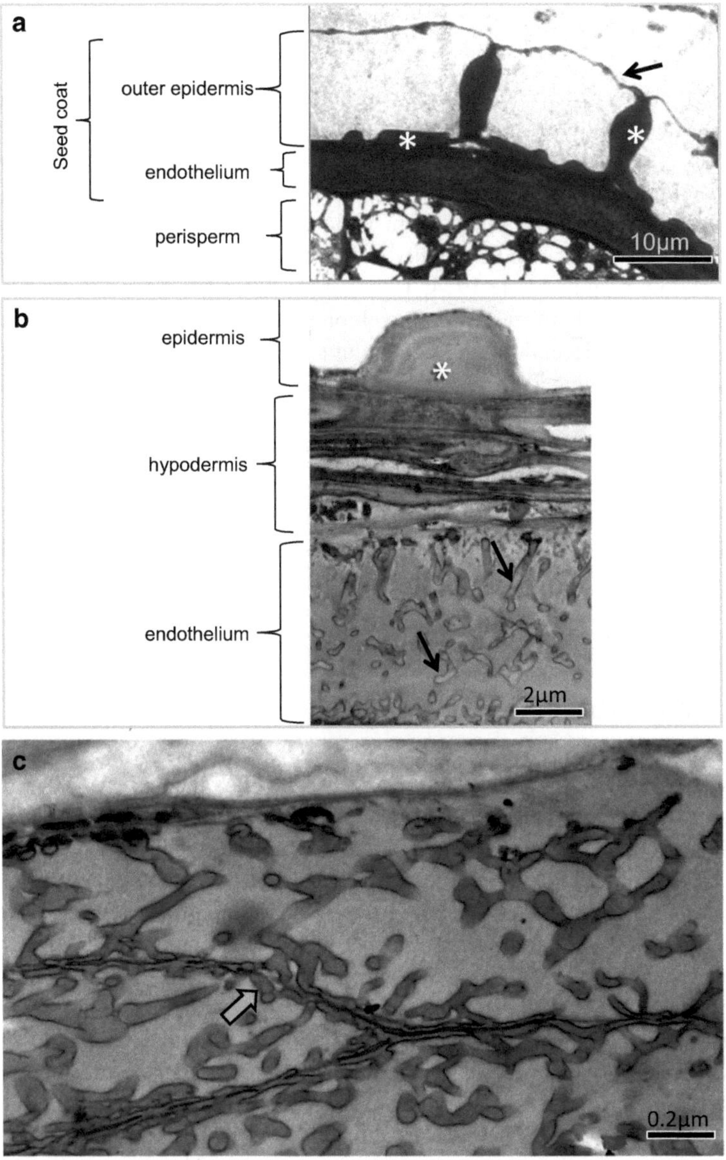

Fig. 9.6 *Phelipanche* seed coat. (**a**) The outer tissues; thin external cell wall of seed coat epidermis (*arrow*), inner epidermis wall thickenings (*asterisks*) (from Joel et al. 2012, with permission). (**b**) Seed coat structure at the chalazal pole. Note the endothelial wall protuberance (*arrows*), the wall remains of hypodermal cells and the internal thickening (*asterisk*) of the inner tangential wall of an epidermal cell. (**c**) Details of the endothelium; note the continuity of wall protuberances in neighbouring cells (*arrow*) and the mucilage that fills the space between protuberances

9.2.5 The Endothelium

The endothelium, which is the inner cell layer of the seed coat (Fig. 9.6.a), arises from the inner epidermis of the integument. The endothelial cells are unique in having an internal wall labyrinth, which is composed of an elaborate network of long branching cell wall ingrowths (Figs. 9.5a, b and 9.6b, c; Joel et al. 2012), and resemble transfer cells of plants from various tissues that are involved in short-distance cell-to-cell transport, mainly at bottlenecks for apoplast–symplast solute exchange (Gunning and Pate 1974; Talbot et al. 2002; Thompson et al. 2001). The internal wall protuberances start developing after fertilization and eventually occupy much of the endothelial cell volume (Joel et al. 2012). The endothelial cells are often elongated in parallel to the main seed axis, up to 60 μm long in *P. aegyptiaca*, and usually the walls between them have a diagonal position (Figs. 9.5a and 9.6c) so that the contact area between cells is enlarged.

The endothelial cells in the mature seed are anucleated, do not contain any cytoplasm and are filled with mucilage, unlike transfer cells. In addition, the internal wall protuberances are often continuous in contiguous cells, which provide apoplastic continuity within the endothelium (Fig. 9.6c; Joel et al. 2012).

9.3 Water Absorption

The route of water penetration into the seed is of relevance when considering the pathway of germination stimulants to the sites of stimulant perception. This has so far only been elucidated for *P. aegyptiaca*. The dry stimulant-dependent seeds are sealed by the impermeable outer layer of the seed coat. Once water is available to *Phelipanche* seeds, it is first absorbed by the endothelium through the micropyle (Joel et al. 2012). The rapid water absorption by the endothelium, often within less than 1 h, is facilitated by the presence of mucilage within the endothelium that acts as a sponge and by the continuous system of wall protuberances in neighbouring endothelial cells. In this respect the endothelium is analogous to conductive tissues.

Direct water movement from the endothelium into the perisperm is prevented by the waxy cuticular layer between them (Figs. 9.5b and 9.7a). The micropyle is also sealed in the dry seed (Fig. 9.7a). It becomes permeable and opens only when a sufficient amount of water accumulates in the seed coat. Then the swollen endothelial cells force the micropyle open (Fig. 9.7b), allowing water entry into the seed interior (Joel et al. 2012). Water then becomes available to the embryo, endosperm and perisperm after being absorbed by the endothelium.

The mucilage that fills the endothelium provides the seeds also with the ability to keep the seed tissues hydrated for days, thus securing the ability of seeds to activate metabolic processes during seed conditioning (see Sect. 11.4.1), even when water has only been available for a short while, which is an advantage particularly in dry habitats (Joel et al. 2012).

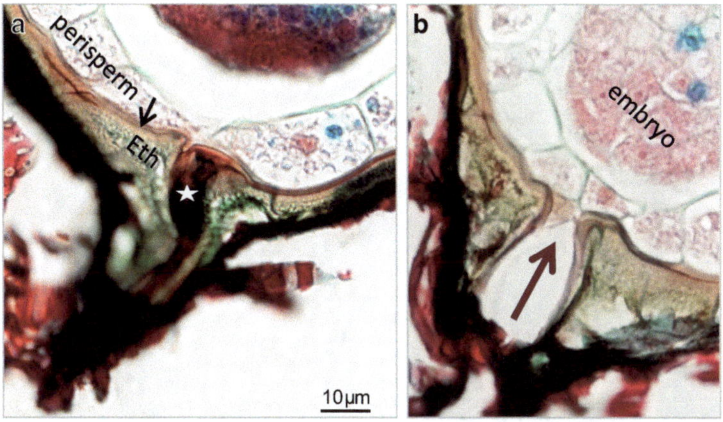

Fig. 9.7 The micropyle opening. (**a**) Longitudinal section of the sealed micropyle of a dry *Phelipanche* seed; note the stained substance that seals the micropyle (*asterisk*) and the cuticle (*arrow*) between the endothelium (Eth) and the perisperm. (**b**) Similar section, but of an imbibed seed; note the open micropyle (*arrow*) leading to the perisperm (from Joel et al. 2012, with permission)

9.4 Site of Signal Perception

As mentioned above, water reaches the inner parts of the seed through the micropyle; thus the perisperm cells, which lie immediately below the micropyle, are the first to obtain water when the micropyle opens. Accordingly, it was assumed that the route of signal movement from the rhizosphere to the living parts of the seeds would follow this direction and first meet the perisperm cells that lie immediately underneath the micropyle. An indication in this direction was obtained in a genetic study establishing that the dependence of *Orobanche* seed germination on external chemical stimuli is genetically controlled by perisperm cells (Plakhine et al. 2012). In this study spontaneous germination (germination without chemical stimulation) was expressed among F_3 seeds obtained by reciprocally crossing the closely related species *O. cumana* and *O. cernua*, while F_1 and F_2 seeds germinated only after chemical stimulation, indicating that the genetic control of stimulant-dependent germination is expressed in maternal seed tissue, i.e. in the perisperm. Accordingly the perisperm cells at the micropylar area were hypothesized to be the sensory cells that perceive germination stimuli from the rhizosphere (Plakhine et al. 2012; Fig. 9.8a). This hypothesis is further supported by data showing that a cut or puncture through the perisperm at the micropylar side of the seed, but not elsewhere, induces *Striga* seed germination (Egley 1972).

Interestingly, the ABA catabolism gene *PrCYP707A1* is expressed in the perisperm cells close to the micropyle (Fig. 9.8b) a few hours after chemical stimulation of *P. ramosa* seeds (Lechat et al. 2012). The expression of this gene and the associated decline in ABA content in the seed may be a major component not only during seed conditioning (see above in Sect. 9.2.1) but also in the seed

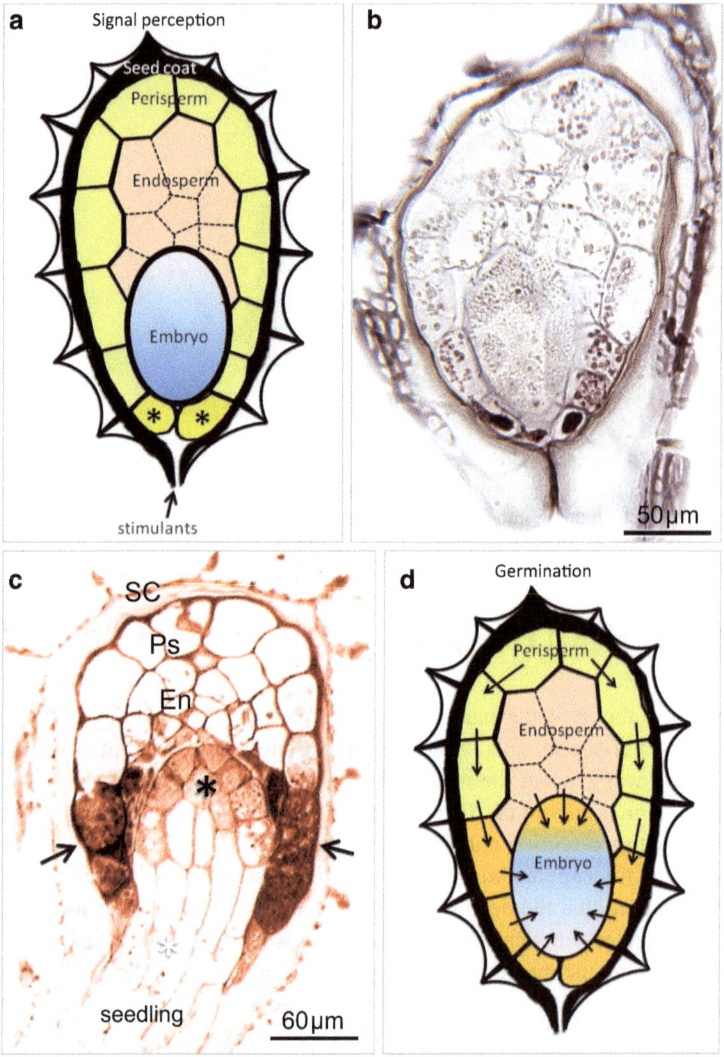

Fig. 9.8 Gene expression and nutrient transfer in the seeds. (**a**) The supposed site (*asterisk*) for germination stimulant perception (from Plakhine et al. 2012, with permission). (**b**) Expression of the ABA catabolic gene *PrCYP707A1* in perisperm cells underneath the micropyle of *P. ramosa* seed after 6h exposure to a germination stimulant (from Lechat et al. 2012, with permission). (**c**) Germinating *Orobanche* seed; two cell groups contain dense cytoplasm and are therefore assumed to be active in nutrient transfer to the developing seedling: the embryo cells at the chalazal pole (*asterisk*) and the perisperm sheath cells (*arrows*) that tightly surround the base of the emerging seedling. (**d**) The supposed routes of nutrient transport to the developing seedling of *Phelipanche* seed during germination

response to germination stimulants, which is the final stage in dormancy relief before germination. ABA is known to be involved in the control of seed dormancy of many non-parasitic and parasitic plants (see Sect. 11.4.1).

While the above results indicate the involvement of perisperm cells in dormancy control of the parasite seeds, some very recent genetic experiments indicated that the differential seed response to specific germination stimulants (strigolactones vs. sesquiterpene lactones, see Sect. 10.3.1) is determined in the embryo and not in the perisperm (Plakhine et al., in preparation), which is consistent with the assumption that the stimulant receptors are located in the embryo.

9.5 Nutrient Transfer During Germination

The stored nutrients are gradually transferred to the seedling that emerges through the open micropyle at the onset of germination. The precise physiological mechanism of this nutrient transfer has not been reported, but the anatomy and the ultrastructure of the seeds provides an insight to possible routes of nutrient transfer. The relevant structural elements are wall thickness and composition, the existence of plasmodesmata and dense cytoplasm.

Dense cytoplasm, which indicates high metabolic activity, typically develops in two seed regions in the holoparasites *Orobanche* and *Phelipanche*: in cells at the chalazal pole of the embryo and in the perisperm sheath cells that tightly surround the base of the emerging seedling at the micropyle (Fig. 9.8c; Aber and Sallé 1983; Teryokhin 1997; Joel et al. 2012). These cells, which have prominent nuclei and many small vacuoles and vesicles ('sponge vacuoles' according to Teryokhin 1997), seem to be actively involved in nutrient transport from the endosperm and from the perisperm.

Wall characteristics in the various seed tissues provide a hint on the routes of nutrient transport. Plasmodesmata are typically abundant between all perisperm cells (Fig. 9.4d) and between the embryo cells (Fig. 9.4a). This is consistent with movement of nutrients along the perisperm to the developing seedling and between the embryo cells towards the radicle at the micropylar pole of the embryo. Nutrient transfer between endosperm and perisperm is less likely because the walls at their boundary are thick and without plasmodesmata (Joel et al. 2012).

Thus, it seems that once the seed has been stimulated for germination, the perisperm becomes a key tissue involved in nutrient supply to the developing embryo, and cells at the chalazal pole of the embryo are simultaneously involved in transferring nutrients from the endosperm to the growing portions of the seedling (Fig. 9.8d).

The perisperm sheath cells remain active throughout germination, with dense cytoplasm, nuclei and numerous small vacuoles. The rest of the perisperm gradually loses cell contents during early germination. Based on these observations, Teryokhin (1997) suggested that the sheath cells regulate the growth and development of the emerging seedlings (see Sect. 9.4).

Similar perisperm sheath cells are active during germination of the hemiparasites *Striga* and *Alectra* (see Fig. 10 in Okonkwo and Raghavan 1982). However, the chalazal pole of the embryo in these hemiparasites, which bears embryonic cotyledons, does not contain dense cytoplasm during germination and does not seem to be involved in nutrient transfer from the endosperm. No ultrastructural data are available regarding the cell walls and plasmodesmatal organization within these seeds, but we may assume that similar to embryonic cotyledons of many non-parasitic plants, the cotyledons of *Striga* and *Alectra* are also coated by a thin impermeable cuticle and that the perisperm, acting like aleurone tissues in seeds of some other plants, is active in transferring nutrients from the endosperm to the developing seedling. Nonetheless, understanding the routes of nutrient transfer in seeds of hemiparasitic plants should be based at least on detailed yet unavailable knowledge of the anatomical features of the various seed components.

9.6 The Seedling

Several different names, like 'germ-tube', 'tube-like organ' and 'procaulôme', were previously given to the unique seedling type of obligate parasitic Orobanchaceae; but as discussed in Sect. 3.14, the organ emerging out of the seed coat during germination results from the elongation of two embryonic regions, the radicle and the hypocotyl. Therefore in this book the organ emerging during germination is regarded as the 'seedling'. The limited seed resources limit its growth to only a few millimetres (Fig. 9.9a–d).

9.6.1 Seedling Structure

During germination the obligate hemiparasites *Striga* and *Alectra* elongate by both cell growth and apical cell divisions. The radicle of these parasites includes an active apical meristem with a quiescent centre, as demonstrated by ^{3}H-thymidine incorporation experiments (Raghavan and Okonkwo 1982). The young seedlings of these parasites contain a central procambium that was already present in the embryo (see Sect. 9.2.1).

In the holoparasites *Cistanche*, *Orobanche*, *Conopholis* and *Phelipanche*, the emerging seedling grows by cell elongation. Some apical cells may divide in later stages of seedling development, but there is no typical root meristem—both the quiescence centre and the root cap are missing—and there is no procambium (Fig. 9.9e; Aber and Sallé 1983; Joel and Losner-Goshen 1994). The resulting seedling is a thin organ, only several cells thick, with no procambium. The main body of the seedling is composed of highly vacuolated and thin-walled elongated cells (Fig. 9.9d), while its apex is composed of several small cells with large nuclei

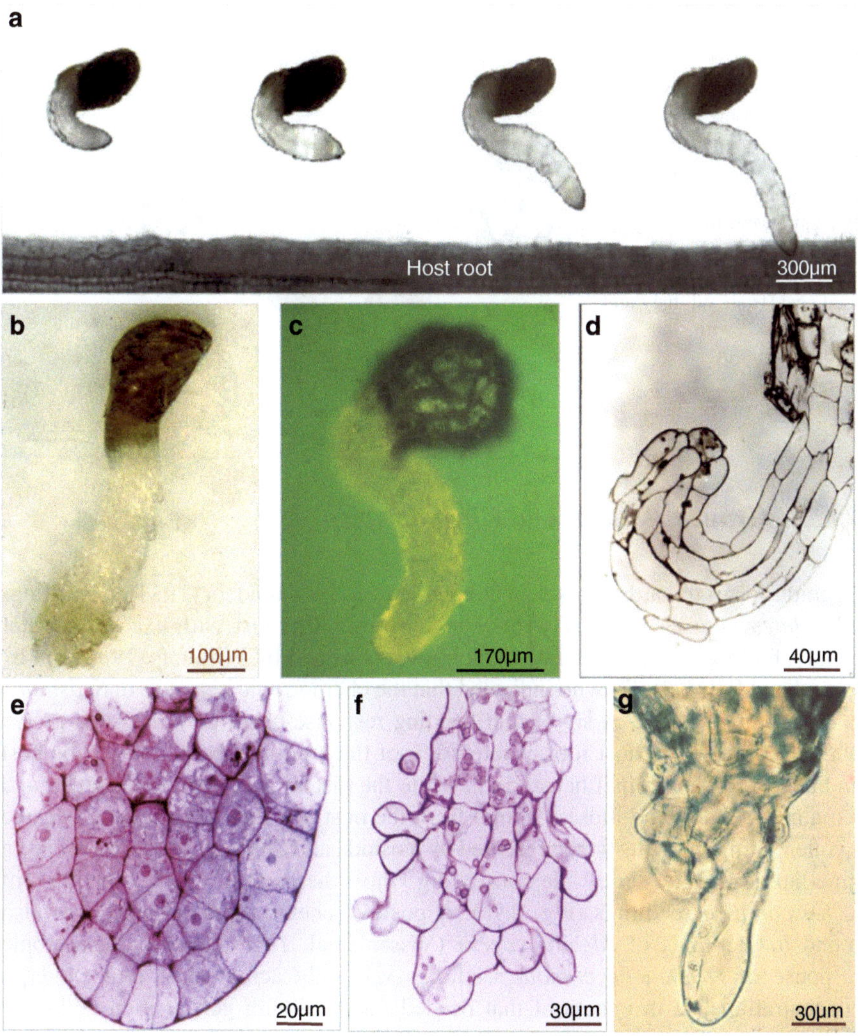

Fig. 9.9 Seedling development. (**a**) Time-lapse image series showing, at 6 h intervals, the chemotropic growth behaviour of *S. hermonthica* seedling near rice root (from Yoshida and Shirasu 2009, with permission). (**b**) *Striga hermonthica* seedling (courtesy of B. Zwanenburg). (**c**) *Phelipanche aegyptiaca*. (**d**) Longitudinal section of *Orobanche cumana* seedling—notice the elongated vacuolated cells; only the margin of the seedling apex is included in this section. (**e**) Mature seedling apex of *P. aegyptiaca*, notice the cell divisions in the various cell layers. (**f**) Papillate attachment organ at the apex of *P. aegyptiaca* seedling. (**g**) An intrusive cell elongating at the tip of an *Orobanche* seedling

and a dense cytoplasm. Once the seedling encounters a host root, these apical cells extend outwards giving rise to the attachment organ (Figs. 5.1b and 9.9f) and some of them become the intrusive cells (Figs. 5.2b and 9.9g).

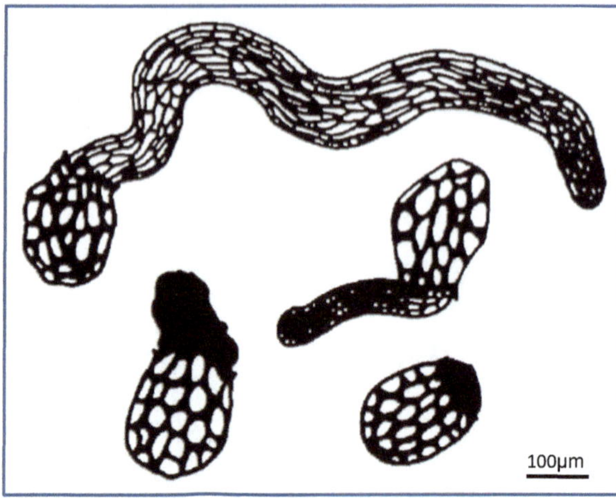

Fig. 9.10 Influence of stimulant concentration on seedling length. Low stimulant concentrations lead to the development of elongated *Orobanche crenata* seedlings, but when stimulated by high concentrations, the radicle does not elongate and thus it does not emerge out of the seed coat (from Whitney and Carsten 1981, with permission)

9.6.2 Growth Pattern and Chemotropism

Chemotropism towards hosts was first recognized by Saunders (1933) in seedlings of *S. lutea* (syn. *S. asiatica*). This chemotropism was lost with excessive water supply, hinting at signal dilution, which was confirmed by Williams (1961) and by Riopel and Baird (1987) who indicated that the chemotropic response of *S. asiatica* is distance dependent. A significant bending response occurred only when grown within 4 mm from a host root, where 76 % of the seedlings bent towards the host and developed a terminal haustorium, while the distant seedlings were longer and remained slender, and those 4–8 mm from the host lost their polarity and became swollen. Similar results were presented by Yoshida and Shirasu (2009), who made a time-lapse image series demonstrating the chemotropic root bending of *S. hermonthica* seedlings (Fig. 9.9a). A positive chemotropic response was also found in seedlings of *Alectra vogelii* (Visser et al. 1977). Strong chemotropic response of *Orobanche crenata* seedlings was only achieved at root exudates concentration just in excess of that needed for maximum germination (Whitney and Carsten 1981).

Exudate concentration is an important factor influencing cell elongation, thus seedlings length negatively correlated with distance from the root, which seems to increase the likelihood that seedlings reach host roots even from a distance (Whitney and Carsten 1981).

When conducting germination experiments, one should keep in mind that increase of stimulant concentration may lead to reduction in radicle length to the extent that the radicle does not emerge out of the seed coat during germination (Fig. 9.10; Whitney and Carsten 1981). Thus, recorded reduced germination percentages may not always indicate lack of germination and/or inactivity of the stimulant, but rather non-optimal concentrations of the germination stimulant.

9.7 Concluding Remarks

Whereas the ability to develop haustoria is unique to all parasitic plants, some seed characteristics that are described above are unique only to the obligate root parasites. The seeds of these Orobanchaceae survive in soil as miniature dormant plants much longer than the active, significantly larger, mature parasites—they often persist in soil for decades, and thus maintain the parasite ability to mature whenever a suitable host plant occasionally grows nearby. This ability, which is crucial for the continual existence of obligate parasites, is facilitated by structural features that accommodate dormancy and dormancy control mechanisms as well as mechanisms that allow perception of the nearby presence of a potential hosts.

Further study of these seeds is therefore important in any attempt to better understand and manage the dormancy and germination of obligate parasitic Orobanchaceae. The key issues that need to be addressed are the intercellular communication between the various seed compartments, including the transduction of host-derived signals from perisperm to embryo cells, and the mechanisms that facilitate and regulate the transfer of stored nutrients to the emerging seedling during early germination. Both need to be treated by interdisciplinary research, employing structural, physiological, biochemical and genomic approaches.

References

Aber M, Sallé G (1983) Graine et procaulôme d'*Orobanche crenata* Forsk: étude histocytologique et cytochimique. Can J Bot 61:3302–3313

Abu Sbaih HA, Jury SL (1994) Seed micromorphology and taxonomy in *Orobanche* (Orobanchaceae). Flora Mediterr 4:41–48

Baird VM, Riopel JL (1986) The developmental anatomy of *Conopholis americana* (Orobanchaceae) seedlings and tubercles. Can J Bot 64:710–717

Bar-Nun N, Mayer AM (2002) Composition of and changes in storage compounds in *Orobanche aegyptiaca* seeds during preconditioning. Isr J Plant Sci 50:278–279

Chen JS, Hsiao SC (2011) Study on seed morphogenesis of Orobanchaceae in Taiwan. Taiwania 56:267–278

Chuang TI, Heckard LR (1983) Systematic significance of seed-surface features in *Orthocarpus* (Scrophulariaceae subtribe Castillejinae). Am J Bot 70:877–890

Domina G, Colombo P (2005) Seed micromorphology and its variation in Sicilian *Orobanche* (*Magnoliopsida*). Flora Mediterr 15:99–107

Egley GH (1972) Influence of the seed envelope and growth regulators upon seed dormancy in witchweed (*Striga lutea* Lour.). Ann Bot 36:755–770

Fischer E (2004) Scrophulariaceae. In: Kubitzki K (ed) The families and genera of vascular plants. Springer, Berlin, pp 333–432

Gunning BE, Pate JS (1974) Transfer cells. In: Robards AW (ed) Dynamic aspects of plant ultrastructure. McGraw-Hill, London, pp 441–480

Joel DM (1987a) Identification of *Orobanche* seeds. In: Weber HCh, Forstreuter W (eds) Parasitic flowering plants. Proceedings of the 4th ISPFP. Marburg, FRG, pp 437–444

Joel DM (1987b) Detection and identification of *Orobanche* seeds using fluorescence microscopy. Seed Sci Technol 15:119–124

Joel DM, Steffens JC, Matthews DE (1995) Germination of weedy root parasites. In: Kigel J, Galili G (eds) Seed development and germination. Marcel Dekker, New York, pp 567–598

Joel DM, Losner-Goshen D (1994) The attachment organ of the parasitic angiosperms *Orobanche cumana* and *O. aegyptiaca* and its development. Can J Bot 72:564–574

Joel DM, Portnoy V, Katzir N (1996) Identification of single tiny seeds of *Orobanche* using RAPD analysis. Plant Mol Biol Rep 14:243–248

Joel DM, Portnoy V, Katzir N (1998) Use of DNA fingerprinting for soil-borne seed identification. Aspects Appl Biol 51:23–27

Joel DM, Bar H, Mayer AM, Plakhine D, Ziadne H, Westwood JH, Welbaum GF (2012) Seed ultrastructure and water absorption pathway of the root-parasitic plant *Phelipanche aegyptiaca* (Orobanchaceae). Ann Bot 109:181–195

Jones MG, Safa SB (1982) Variation of seed-coat ornamentation in *Striga hermonthica* (Scrophulariaceae). Ann Bot 50:629–634

Krause D (1990) Vergleichende morphologisch/anatomische Untersuchungen an *Striga* Arten (Scrophulariaceae). Dissertation, Philipps-University, Marburg

Krause D, Weber HC (1990) SEM observations of seeds of *Striga* spp. and *Buchnera americana* (Scrophulariaceae). Plant Syst Evol 170:257–263

Kravtzova TI, Teryokhin ES (1987) Histochemical analysis of the seed coat in some representatives of the Orobanchaceae family of the flora of the USSR. Bot Zh SSSR 72:644–651

Kuijt J (1969) The biology of parasitic flowering plants. University of California Press, Berkeley

Kumar U (1977) Morphogenetic regulation of seed germination in *Orobanche aegyptiaca* Pers. Can J Bot 55:2613–2621

Lechat MM, Pouvreau JB, Péron T, Gauthier M, Montiel G, Véronési C, Todoroki Y, Le Bizec B, Monteau F, Macherel D, Simier P, Thoiron S, Delavault P (2012) PrCYP707A1, an ABA catabolic gene, is a key component of *Phelipanche ramosa* seed germination in response to the strigolactone analogue GR24. J Exp Bot 63:5311–5322

Michell MR (1915) The embryo sac and embryo of *Striga lutea*. Bot Gaz 59:124–135

Musselman LJ, Mann WF (1976) A survey of surface characteristics of seeds of Scrophulariaceae and Orobanchaceae using scanning electron microscopy. Phytomorphology 26:370–378

Musselman LJ, Parker C (1981) Surface features of *Striga* seeds (Scrophulariaceae). Adansonia 20:431–438

Okonkwo SNC, Nwoke FIO (1978) Initiation, development and structure of the primary haustorium in *Striga gesnerioides* (Scrophulariaceae). Ann Bot 42:455–463

Okonkwo SNC, Raghavan V (1982) Studies on the germination of seeds of the root parasites *Alectra vogelii* and *Striga gesnerioides*. I. Anatomical changes in the embryos. Am J Bot 69:1636–1645

Plakhine D, Tadmor Y, Ziadne H, Joel DM (2012) A maternal tissue is involved in stimulant reception by seeds of the parasitic plant *Orobanche*. Ann Bot 109:979–986

Portnoy VH, Katzir N, Joel DM (1997) Species identification of soil-borne *Orobanche* seeds by DNA fingerprinting. Pest Biochem Physiol 58:49–54

Privat G (1960) Recherches sur les phanérogames parasites (étude d'*Orobanche hederae* DUBY). Ann Sci Nat Bot Ser 12:721–871

Raghavan V, Okonkwo SNC (1982) Studies on the germination of seeds of the root parasites *Alectra vogelii* and *Striga gesnerioides*. II. DNA synthesis and development of the quiescent center in the radicle. Am J Bot 69:1646–1656

Riopel JL, Baird WV (1987) Morphogenesis of the early development of primary haustoria in *Striga asiatica*. In: Musselman LJ (ed) Parasitic weeds in agriculture. CRC, Boca Raton, pp 107–125

Sauerborn J, Dörr I, Abbasher A, Thomas H, Kroschel J (1996) Electron microscopic analysis of the penetration process of *Fusarium nygamai*, a hyperparasite of *Striga hermonthica*. Biol Control 7:53–59

Saunders AR (1933) Studies in phanerogamic parasitism, with particular reference to *Striga lutea* Lour. S Afr Dept Agric Sci Bull 128

Talbot MJ, Offler CE, McCurdy DW (2002) Transfer cell wall architecture: a contribution towards understanding localized wall deposition. Protoplasma 219:197–209

Teryokhin ES (1997) Weed broomrapes: systematic, ontogenesis, biology, evolution. Aufstieg, Landshut

Teryokhin ES, Kravtsova TI (1987) On the importance of the seed structural features for the systematics of Orobanchaceae of the flora of the USSR. Bot Zh SSSR 72:15–24

Teryokhin ES, Shibakyna GB, Serafimovitch NB, Kravtzova TY (1993) Determination of broomrapes of the USSR Flora. Nauka, St. Petersburg

Thompson RD, Hueros G, Becker H-A, Maitz M (2001) Development and functions of seed transfer cells. Plant Sci 160:775–783

Ungurean L (1986) Recherches sur le tégument séminal dans le cadre des espèces de la section Trionychon Wallr. du genre *Orobanche* L. Rev Roum Biol Biol Vég 31:121–125

Velasco L, Goffman FD, Pujadas-Salva A (2000) Fatty acids and tocochromanols in seeds of *Orobanche*. Phytochemistry 54:295–300

Visser JH, Dörr I, Kollmann R (1977) On the parasitism of *Alectra vogelii* Benth. (Scrophulariaceae). Early development of the primary haustorium and initiation of the stem. Zeitschr Pflanzenphysiol 84:213–222

Whitney PJ, Carsten C (1981) Chemotropic response of broomrape radicles to host root exudates. Ann Bot 48:919–921

Williams CN (1961) Tropism and morphogenesis of *Striga* seedlings in the host rhizosphere. Ann Bot 25:406–415

Yoshida S, Shirasu K (2009) Multiple layers of incompatibility to the parasitic witchweed, *Striga hermonthica*. New Phytol 183:180–189

Chapter 10
Induction of Germination

Koichi Yoneyama, Carolien Ruyter-Spira, and Harro Bouwmeester

10.1 Introduction

In general, plant seeds germinate when they are exposed to appropriate temperature, humidity, and oxygen (Fig. 10.1). In addition to these environmental factors, light also plays a pivotal role in the regulation of germination in many species. In contrast, a host-derived signal—called **germination stimulant**—is needed for the germination of the seeds of obligate root parasitic plants of the Orobanchaceae, including species of *Orobanche*, *Phelipanche*, *Cistanche*, *Striga*, and *Alectra* (Fig. 10.1). This is of great importance for the obligate parasites since they will not be able to survive for more than just a few days after germination unless they reach their host. Hence, seeds of these root parasites will only germinate within the host rhizosphere so that after germination they have a better chance to rapidly attach to the host roots. It has been known for a long time that seeds of root parasitic plants only germinate when they are in close vicinity of their host. Vaucher (1823) reported that host root-derived stimulants are necessary for *Orobanche* seeds to germinate. During the late 1940s to the early 1950s, Brown et al. demonstrated that various plant species produce germination stimulants for seeds of *Orobanche minor*

K. Yoneyama (✉)
Weed Science Center, Utsunomiya University, 350 Mine-machi, Utsunomiya 321-8505, Japan
e-mail: yoneyama@cc.utsunomiya-u.ac.jp

C. Ruyter-Spira
Laboratory of Plant Physiology, Wageningen University, Droevendaalsesteeg 1, 6708 PB Wageningen, The Netherlands

Plant Research International, 6700 AA Wageningen, The Netherlands

H. Bouwmeester
Laboratory of Plant Physiology, Wageningen University, Droevendaalsesteeg 1, 6708 PB Wageningen, The Netherlands

Centre for BioSystems Genomics, P. O. Box 98, 6700 AB Wageningen, The Netherlands
e-mail: harro.bouwmeester@wur.nl

D.M. Joel et al. (eds.), *Parasitic Orobanchaceae*,
DOI 10.1007/978-3-642-38146-1_10, © Springer-Verlag Berlin Heidelberg 2013

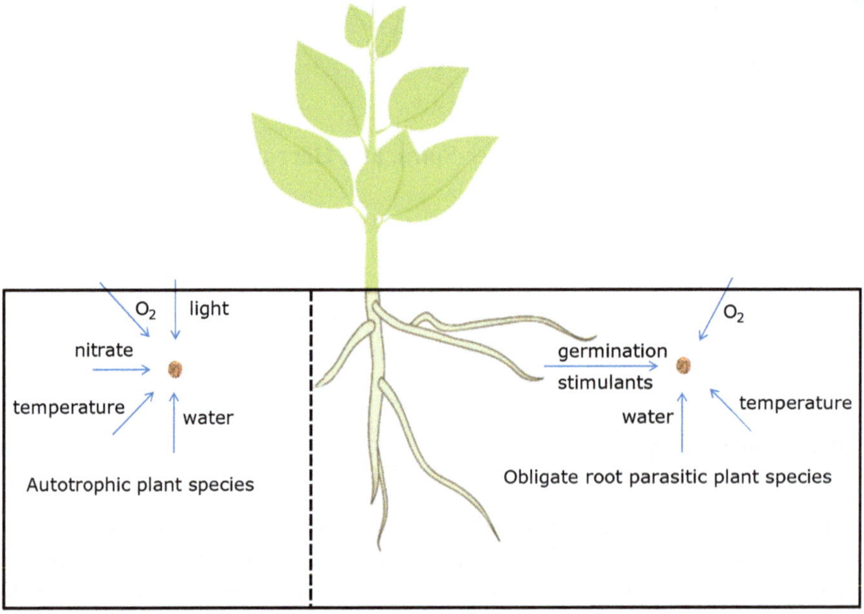

Fig. 10.1 Factors affecting germination. Factors affecting the germination of seeds of obligate root parasitic plants compared to the seeds of most autotrophic plant species

(Brown et al. 1951a, b, 1952a, b) and *Striga hermonthica* (Brown et al. 1949, 1952b). They also predicted, after partial purification of the active compounds, that these germination stimulants contain a lactone group since they were relatively stable in weakly acidic solutions but not in alkaline solutions. In 1966, this was proven to be correct as the structure of the first *Striga* germination stimulant, strigol, contained two lactone groups (Fig. 10.2) (Cook et al. 1966). Since then, many structural variants of strigol have been discovered. These compounds are now collectively called **strigolactones** (SLs) (Butler 1995).

Germination stimulants for root parasitic plants are isolated from plant root exudates, and the majority of the natural (so far identified) germination stimulants are the SLs (Xie et al. 2010). SLs are also host recognition signals for arbuscular mycorrhizal (AM) fungi with which >80 % of the land plants can have a symbiotic relationship (Akiyama et al. 2005). SLs induce hyphal branching in AM fungi, a process that precedes colonization of the host root and increases the chance to contact the host root (Akiyama et al. 2005; Besserer et al. 2006). Furthermore, it was recently unveiled that SLs, or their close derivatives, are plant hormones regulating shoot branching (Gomez-Roldan et al. 2008; Umehara et al. 2008; see Sect. 10.2.6), which probably means that SLs are ubiquitous in the plant kingdom. Indeed, we now know that also non-mycorrhizal plants produce and exude SLs. Therefore it is likely that the SLs are indeed the major germination stimulants for root parasitic plants in the rhizosphere of host plants. However, natural germination stimulants other than the SLs have also been identified (see Sect. 10.3). In this

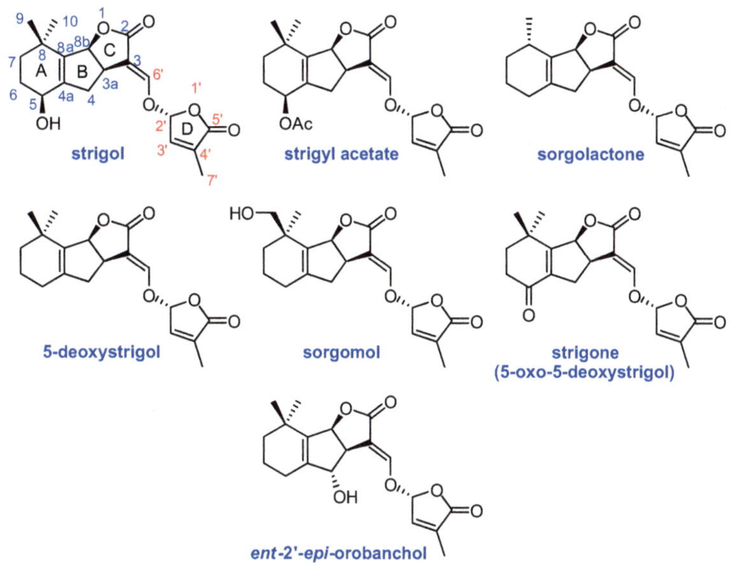

Fig. 10.2 Natural strigol-type strigolactones

chapter, the chemistry of germination stimulants including SL and non-SL type of compounds is discussed as well as other biological activities of SLs, their biosynthesis, and regulation. We also discuss the mechanisms of germination stimulation and the possibility to manipulate germination stimulation as a possible parasitic weed control strategy (Joel et al. 2007).

10.2 Strigolactones

10.2.1 Discovery of Strigolactones

Strigol and strigyl acetate, the first reported SLs, were isolated as *Striga asiatica* (syn. *S. lutea*) germination stimulants from root exudates of cotton (*Gossypium hirsutum*), a non-host of *Striga* (Fig. 10.2) (Cook et al. 1966, 1972). Although strigol was highly active in inducing *Striga* germination (at concentrations below 10^{-12} M), it was not immediately recognized as a host-derived germination stimulant until the identification of strigol in the root exudates of genuine *Striga* hosts, including sorghum (*Sorghum bicolor*), maize (*Zea mays*), and proso millet (common millet, *Panicum miliaceum*) (Siame et al. 1993). Furthermore, isolation of two additional SLs, sorgolactone and alectrol (=orobanchyl acetate, see below), from root exudates of sorghum (Hauck et al. 1992) and cowpea (*Vigna unguiculata*) (Müller et al. 1992), respectively, suggested that both host and non-host plants of root parasitic plants produce and exude SLs.

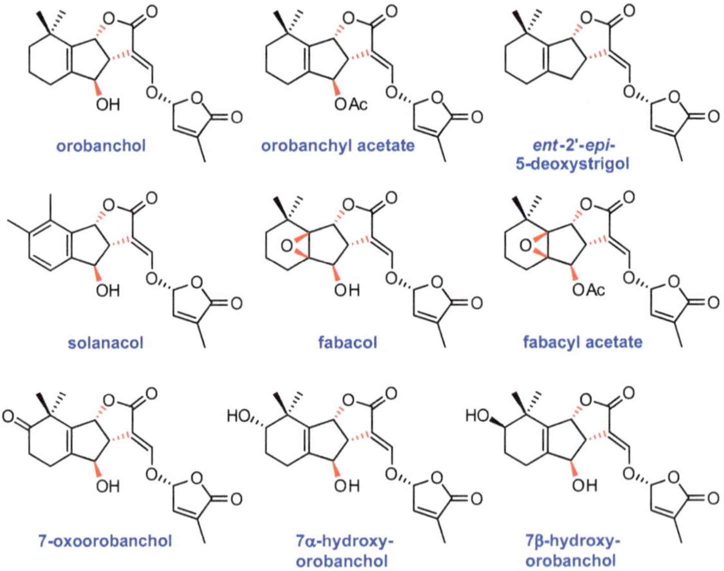

Fig. 10.3 Natural orobanchol-type strigolactones

Orobanchol was isolated from root exudates of red clover (*Trifolium pratense*), a host of *O. minor*, as the first *Orobanche* germination stimulant. This showed that SLs are natural germination cues for both *Striga* and *Orobanche* spp. (Yokota et al. 1998). In addition, red clover produces at least two other SLs, alectrol and a putative didehydro-orobanchol isomer. The first proposed structure for alectrol was rejected based on spectroscopic comparison with synthetic standards (Mori et al. 1998), and later alectrol was identified to be orobanchyl acetate (Fig. 10.3) (Matsuura et al. 2008; Xie et al. 2008b). The structures of orobanchol and orobanchyl acetate have recently been revised (Ueno et al. 2011). They were shown to be *ent*-SLs in which the orientation of the C-ring (Fig. 10.3) is opposite to that of other SLs such as strigol and sorgolactone (strigol-type SLs, see Sect. 10.2.2). 5-Deoxystrigol (Fig. 10.2) was first purified from root exudates of *Lotus japonicas* as a branching factor for AM fungi (Akiyama et al. 2005) but subsequently also shown to be a germination stimulant in root exudates of many monocot and dicot species (Awad et al. 2006; Xie et al. 2010; Yoneyama et al. 2008, 2009, 2010).

Root exudates of tobacco (*Nicotiana tabacum*) were found to contain several SLs including orobanchol, an orobanchol isomer, didehydro-orobanchol isomers, and tetradehydro-orobanchol (Fig. 10.3). The orobanchol isomer was identified as 2′-*epi*-orobanchol, the first 2′-*epi*-SL (Xie et al. 2007), but its actual structure is *ent*-2′-*epi*-orobanchol (Xie et al. 2012) (Fig. 10.2). The tetradehydro-orobanchol was purified and named solanacol, which is unique as it contains a phenyl ring (Fig. 10.3) (Xie et al. 2007). The stereochemistry of solanacol as an *ent*-SL was recently identified by synthesis (Chen et al. 2010).

One of the active compounds in sorghum root exudates was identified as sorgomol (Xie et al. 2008a) and subsequently also found in root exudates of maize, Chinese milk vetch (*Astragalus sinicus*), and white lupin (*Lupinus albus*) (Fig. 10.2) (Yoneyama et al. 2008). 7-Oxo-, 7α-, and 7β-hydroxyorobanchol and their acetates (see below) were isolated from root exudates of flax (*Linum usitatissimum*) (Xie et al. 2009b) and cucumber (*Cucumis sativus*) (Fig. 10.3) (X. Xie, unpublished data). In pea, fabacyl acetate is one of the major SLs having a unique epoxide group and the first reported *ent*-SL (Fig. 10.3) (Xie et al. 2009a). In addition to fabacyl acetate and fabacol (also identified in pea root exudates), there are several other *ent*-SLs including orobanchol, orobanchyl acetate (alectrol), solanacol (Chen et al. 2010), and *ent*-2′-*epi*-5-deoxystrigol (from rice; Xie et al. 2012). All of these SLs have an α-oriented C-ring and are classified as orobanchol-type SLs.

In addition, over ten novel SLs have been detected in root exudates of various plant species but their chemical structures remain to be characterized. For sure many others will be discovered in the future.

10.2.2 Structural Diversity of Strigolactones

All natural SLs characterized so far have the same structural features as shown in Figs. 10.2 and 10.3. The ABC part, composed of a tricyclic ring system, is connected to a 5-membered ring (the D-ring) via a vinyl ether bridge. The A-ring has one or two methyl groups. Additional substituents on the A and B rings and the stereochemistry of the C-ring and the enol-ether bridge between the C- and D-ring differentiate all the presently known SLs. It is likely that all of these SLs are derived from the simplest SLs 5-deoxystrigol and *ent*-2′-*epi*-5-deoxystrigol through hydroxylation, oxidation, decarboxylation, and esterification (Fig. 10.4). Allylic hydroxylation on C4 or C5 of 2′-*epi*-5-deoxystrigol or 5-deoxystrigol affords orobanchol or strigol, respectively. Hydroxylation at the homoallylic position, C9, results in formation of sorgomol, another monohydroxy-SL. Similar hydroxylation may also occur on the other homoallylic positions C6 and C7 affording 6- and 7-monohydroxy-SL, respectively. All these monohydroxy-SLs are or could potentially be acetylated and conjugated with sugars and amino acids, although so far such conjugates have not been reported. Sorgolactone could be formed via oxidation of the hydroxymethyl group of sorgomol to a carboxylic acid and subsequent decarboxylation (Fig. 10.4). It is likely that the phenyl ring in solanacol is formed through a series of hydroxylation and dehydration reactions coupled with migration of a methyl group from C8 to C7. Since solanacol is an orobanchol-type SL, the conversion pathway would be *ent*-2′-*epi*-5-deoxystrigol > orobanchol > 7-hydroxyorobanchol > didehydro-orobanchol(s) > solanacol. Therefore, 7α- and 7β-hydroxyorobanchol and their acetates, and 7-oxo derivatives are orobanchol-type SLs as shown in Fig. 10.3 (X. Xie, unpublished data). Among the three identified hydroxy-SLs, orobanchol seems to be the most common one in the

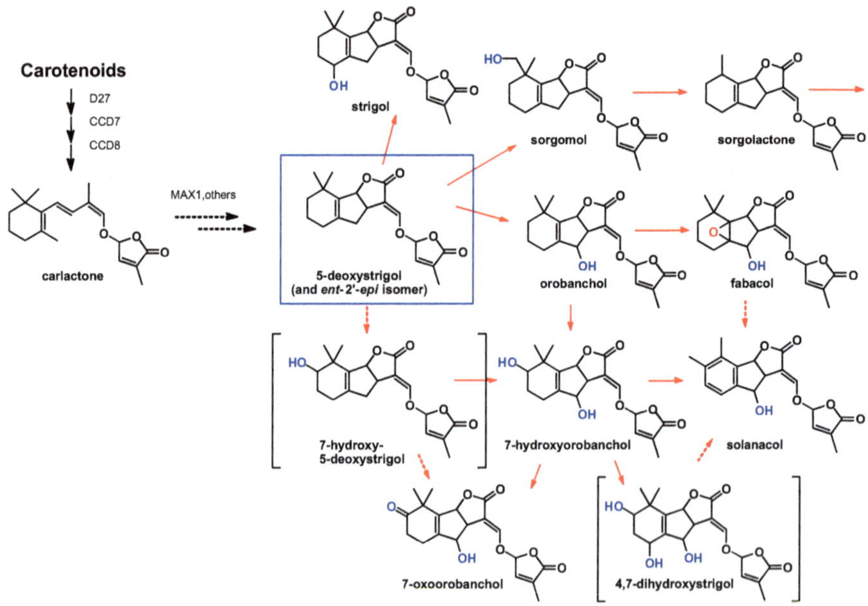

Fig. 10.4 Putative biosynthesis of natural strigolactones

plant kingdom. In addition, its acetate (alectrol) has been detected in various plant species that also produce orobanchol. Strigol can be acetylated (Cook et al. 1966) and oxidized to the 5-keto-SL, strigone, which was recently isolated from root exudates of *Houttuynia cordata* (Kisugi et al. 2013). Strigol was isolated from root exudates of cotton, sorghum, maize, and proso millet (Siame et al. 1993) and from cultured roots of a Chinese medicinal plant *Menispermum dauricum* (Yasuda et al. 2003). In contrast, sorgomyl acetate has so far not been detected in root exudates of any plant species.

The presence of strigol-type SLs and orobanchol-type SLs in plant root exudates suggests that at least some plant species produce both types of SLs. Recently, one of the major SLs in rice (*Oryza sativa*) was found to be *ent*-2′-*epi*-5-deoxystrigol (S. Yamaguchi unpublished data, Xie et al. 2012). It seems that the rate of the individual steps of the SL biosynthetic pathway is highly variable between plant species, resulting in the different mixtures of SLs that have been reported so far in root exudates. In addition, biosynthesis of orobanchol-type SLs may be regulated somewhat independently from that of strigol-type SLs because N and P starvation in Chinese milk vetch significantly promoted production of two strigol-type SLs, sorgomol and 5-deoxystrigol, but not that of orobanchyl acetate (Yoneyama et al. 2012; see Sect. 10.2.5). Finally, SLs occur in root exudates throughout the plant kingdom. They are also produced by tree roots (e.g., *Pinus* spp. and *Eucalyptus* spp.; K. Akiyama and X. Xie, unpublished data) and even by the bryophyte *Physcomitrella patens* (Proust et al. 2011) and liverworts (species of *Marchantia* and *Lunularia*; Delaux et al. 2012).

10.2.3 Biosynthesis of Strigolactones and the Genes Involved

Until 2005 the SLs were classified as sesquiterpene lactones (Bouwmeester et al. 2003). But in 2005, the biosynthetic origin of SLs was established (Matusova et al. 2005), by demonstrating that root exudates of carotenoid biosynthetic mutant seedlings of maize induced less *Striga* seed germination than wild-type plants. Treatment of wild-type plants with inhibitors of carotenoid biosynthesis similarly reduced germination stimulation activity of the root exudate. Fluridone, an inhibitor of carotenoid biosynthesis, blocking phytoene desaturase (PDS), strongly reduced SL production at a concentration of 10^{-8} M, far below that needed to induce chlorophyll bleaching. Although quantitative and/or qualitative changes in SL levels in the root exudates could not be demonstrated analytically, it was assumed that the reduction in germination stimulation activity was due to reduced production of SLs. Matusova et al. (2005) postulated that the SLs are derived—through cleavage—from the carotenoid pathway and should hence be classified as apocarotenoids (Fig. 10.4). In parallel, other groups were trying to identify an unknown, carotenoid-derived signaling molecule that had a profound effect on shoot branching. Genetic studies with mutants showed that two carotenoid cleavage dioxygenases and a P450 were responsible for the production of this signal (Booker et al. 2004, 2005; Sorefan et al. 2003; Stirnberg et al. 2002). These two research efforts merged when it was shown that the unknown signals are SLs (or derivatives) and vice versa—that the carotenoid cleavage dioxygenases, CCD7 and CCD8 (MAX3 and MAX4 in *Arabidopsis*, RMS5 and RMS1 in pea, D17/HTD1 and D10 in rice), are involved in the biosynthesis of the SLs (Fig. 10.4) (Gomez-Roldan et al. 2008; Umehara et al. 2008). MRM-LC-MS/MS analysis showed that *ccd8* and *ccd7* mutants in pea and rice produced no SLs or strongly reduced levels of SLs. Intriguingly, exogenous application of the synthetic SL analog GR24 complemented the branching phenotype of the mutants (Gomez-Roldan et al. 2008; Umehara et al. 2008). This work hence identified CCD7 and CCD8 as part of the SL biosynthetic pathway and the SLs as a new class of plant hormones that control (inhibit) tillering/branching in monocots/dicots, respectively. The highly branched phenotype of *Arabidopsis* plants mutated in *MAX1* (Booker et al. 2005; Stirnberg et al. 2002) could also be rescued by GR24 application (Gomez-Roldan et al. 2008; Umehara et al. 2008), but analytical proof for the involvement of MAX1 in SL biosynthesis was only provided 3 years later (Kohlen et al. 2011a). Interestingly, no *MAX1* orthologs have been identified in other species than *Arabidopsis,* although rice has five putative orthologs (Umehara et al. 2010), whereas orthologs for *MAX3 (CCD7)* and *MAX4 (CCD8)* have been identified in several other monocot and dicot species (Arite et al. 2007; Booker et al. 2004; Ledger et al. 2010; Lin et al. 2009; Proust et al. 2011; Snowden et al. 2005; Vogel et al. 2010; Zou et al. 2006). This suggests that *MAX1* in other species than *Arabidopsis* is redundant, which seems to be confirmed by the fact that rice contains five putative orthologs. A fourth step in the biosynthetic pathway, encoded by *DWARF27* (*D27*), was identified in 2009 (Fig. 10.4) (Lin et al. 2009). Although the precise function of this iron-containing protein at that time remained unknown, no SLs were detected in lines mutated in this gene (Lin et al. 2009). Recently, however, D27 was identified as a β-carotene isomerase that catalyzes the conversion of all-*trans*-β-carotene to 9-*cis*-β-carotene that is subsequently cleaved by CCD7 and CCD8 to form carlactone (Alder et al. 2012). These authors showed that the combined action of D27, CCD7, and CCD8 leads to the formation of carlactone. Carlactone already has the D-ring characteristic for strigolactones and can be envisaged to be converted by a number of oxidation reactions to 5-deoxystrigol and its isomer, possibly catalyzed by MAX1 (Fig. 10.4).

As described above, 5-deoxystrigol and *ent*-2′-*epi*-5-deoxystrigol are believed to be the first true SLs in the biosynthetic pathway, from which all other SLs can be derived by hydroxylation, decarboxylation, acetylation, and/or oxidation (Fig. 10.4) (Humphrey and Beale 2006; Matusova et al. 2005; Rani et al. 2008). However, the enzymes likely catalyzing these conversions so far remain unknown.

The precise tissue localization of SL biosynthesis also remains elusive. It is thought that a part of the total SL pool is synthesized in the roots (Ruyter-Spira et al. 2011) and that SLs are either exuded into the rhizosphere or transported to the shoot where they exert their inhibitory effect on shoot branching. However, biosynthesis of the SLs—of which we assume they function in the shoot—is not limited to the root system, as grafting studies indicated that interstock grafting with only a small part of the wild-type hypocotyl is sufficient to restore branching to near wild type in biosynthetic mutants (Foo et al. 2001). The expression of SL biosynthesis genes is also not limited to the root system (Booker et al. 2004, 2005; Sorefan et al. 2003), leaving unknown the exact origin of SLs that are present in the shoot. Still, transport through the plant is definitely required, and the xylem is likely to be involved, as orobanchol has been detected in the xylem sap of both *Arabidopsis* and tomato (Kohlen et al. 2011a).

10.2.4 Evolution of Strigolactones as Germination Stimulants

Putative orthologs of SL biosynthetic genes *CCD8* and *CCD7* are also found in bryophytes (Proust et al. 2011). The symbiotic interaction between angiosperms and AM fungi, in which SLs participate, is thought to date back over 400 million years and is believed to have played an important role in the migration of plants from water to land (Harrison 2005).

When parasitic Orobanchaceae evolved obligate parasitism, they apparently have adapted to the existing SL biosynthesis and signaling pathway according to the needs of their new life cycle, i.e., to use SLs as a (exogenous) germination trigger. This adaptation could have occurred by temporarily suspending or downregulating SL biosynthesis in the seeds or by evolving a higher SL requirement, as a consequence of other physiological changes. Also changes in SL perception may have evolved, that allow the parasitic plant seeds to respond to SL signals from outside (see also Sect. 10.2.6).

Two genes have been shown to be involved in SL perception/downstream signaling, *MAX2*, encoding an F-box protein (Stirnberg et al. 2002), and *DWARF14* (*D14*), encoding an α/β hydrolase (Arite et al. 2009). The Petunia equivalent of D14, DAD2, was recently crystallized and shown to hydrolyze GR24 (Hamiaux et al. 2012). It seems now that binding to and/or hydrolysis of a SL by D14 is required for SL downstream signaling and hence it will be of great interest to compare D14 homologs in root parasitic plant species with the crystal structure of Petunia DAD2 and to study whether there are differences in the catalytic cavity that correlate with differences in substrate and therefore possibly host specificity.

It is likely that SLs are involved in seed germination of other plant species as well. SLs can break dormancy of *Lactuca sativa* and *Avena fatua* (Westwood et al. 2010), and the SL downstream signaling mutant *max2* in *Arabidopsis* has been shown to be defective in light-dependent germination (Shen et al. 2007). Interestingly, *max1* seeds of *Arabidopsis* showed reduced germination compared with wild-type seeds under far-red followed by red light pulses. This phenotype could be rescued by GR24 application (Shen et al. 2007; Tsuchiya and McCourt 2009). Also under thermo-inhibitory conditions, SL biosynthetic mutants show reduced germination, which can be rescued by the application of the synthetic SL GR24 (Toh et al. 2012). The presence of MAX2 is required for this rescue. As described below (Sect. 10.3.1), forest-fire succession plant species germinate only after exposure to plant-derived smoke compounds, the karrikins, which partially resemble SLs (Chiwocha et al. 2009; Daws et al. 2008; and see Sect. 12.2). This fact supports the hypothesis that adaptation of the signaling pathway to break dormancy/induce germination to respond to different compounds (such as SLs or smoke-derived compounds) may have occurred several times in evolution. A key question is how in the seeds of parasitic plants SL perception that seems to be present in all plants was altered in such a way that they can respond to exogenous SLs (see Sect. 12.8). This should also shed light on a possible role of specificity of SL perception in host recognition (Fernández-Aparicio et al. 2011; Cardoso et al. 2011; Höniges et al. 2012).

10.2.5 Regulation of Strigolactone Biosynthesis

The production of SLs is affected by environmental conditions, particularly by nutrient deficiencies. This is in line with the observation that weedy *Striga* spp. are common particularly on poor soils, while the application of fertilizer has been reported to suppress their occurrence (Bouwmeester et al. 2007; Jamil et al. 2011a).

In red clover (*Trifolium pratense*), a low level of phosphate increased the exudation of orobanchyl acetate and orobanchol (Yoneyama et al. 2007b). A similar response was not obtained with low levels of nitrogen, potassium, calcium, or magnesium. The increased SL levels correlated well with the increased germination of *O. minor* seeds by these root exudates. Upon transfer of the plants to a nutrient solution with sufficient phosphate, SL exudation decreased within 24 h. A similar positive effect of low phosphate levels on SL production was also observed in tomato (López-Ráez et al. 2008a). However, the expression of the SL biosynthetic enzyme encoding *LeCCD7* and *LeCCD8* did not increase under phosphate starvation (López-Ráez et al. 2010). In contrast, the expression of *D10*, *D17*, and some *MAX1*-like *CYP711A* genes in rice was upregulated under phosphate starvation (Umehara et al. 2010; K. Yoneyama, unpublished data). Clearly, the many aspects of the mechanism by which SL production is upregulated under nutrient starvation are still unresolved.

In sorghum nitrogen deficiency also increased the exudation of 5-deoxystrigol and sorgomol (Yoneyama et al. 2007a). A combination of nitrogen and phosphate deficiencies, however, resulted in a minor additional increase only in 5-deoxystrigol exudation. SL content was not only enhanced in the exudates but also, to a comparable extent, in the root tissues as was also found in tomato (López-Ráez et al. 2008a). This suggests that shortage of nitrogen and/or phosphate directly induces SL biosynthesis rather than just their secretion. Also in rice, phosphate starvation dramatically increased the exudation of orobanchol, *(ent)*-2′-*epi*-5-deoxystrigol, and three unidentified methoxy-5-deoxystrigol isomers into the rhizosphere, more so than nitrogen starvation (Jamil et al. 2011a). Intriguingly, nitrogen and phosphate deficiency did not affect the concentration of 5-deoxystrigol in sorghum shoots (Yoneyama et al. 2007a). It is still unclear whether SLs present in the shoot are synthesized in situ or are imported from the roots. Nevertheless, in *Arabidopsis* and tomato, the SL content in the xylem sap increased upon phosphate starvation, suggesting that there is upregulation of SL transport to the shoot upon phosphate starvation (Kohlen et al. 2011a).

In general, it seems that in legumes, SL production is promoted only under low phosphate but not under nitrogen deficiency, whereas in other mycotrophic plants, nitrogen deficiency also promotes SL production (K. Yoneyama, unpublished data). This may be explained by the fact that while leguminous mycotrophic plants obtain nitrogen from nitrogen-fixing bacteria and phosphate from mycorrhizal fungi, nonleguminous mycotrophic plants depend on AM symbiosis for both nitrogen and phosphate supply. Non-mycotrophic plants such as *Arabidopsis*, spinach (*Spinacia oleracea*), and white lupin exude SLs at much lower levels than mycotrophic plants, and nitrogen and phosphate deficiency hardly affects their SL exudation (Kohlen et al. 2011a; K. Yoneyama, unpublished data). Nevertheless, in the non-mycotrophic *Arabidopsis,* there is a small but detectable increase in the concentration of SLs in the root exudates upon phosphate starvation (Kohlen et al. 2011a), although this increase is much smaller than in the above mentioned mycotrophic species.

The increase in SL production in response to nutrient shortage is clearly an adaptive strategy to promote AM colonization. Indeed, after AM colonization tomato SL exudation decreases again (López-Ráez et al. 2011). It is unclear whether this is a consequence of signaling or just of improved phosphate availability.

By using a split-root system, phosphate fertilization to one half of the split root was shown to strongly downregulate SL production also in the other half, indicating the existence of systemic signaling (Balzergue et al. 2011; K. Yoneyama, unpublished data).

Other environmental factors may also affect SL production. Weerasuriya et al. (1993) demonstrated a higher germination stimulant activity for *Striga* seeds in root exudates of sorghum and proso millet when grown under short days. However, a stimulating effect of light on SL production was recently reported for tomato roots (Koltai et al. 2011).

In several ABA mutants of tomato, SL production was reduced (López-Ráez et al. 2010), whereas application of GA strongly reduced SL production in rice (Ito et al. 2010). Hence, evidence is emerging that SLs interact with all or many of the other plant hormones and vice versa (Kohlen et al. 2011b).

10.2.6 Other Biological Functions of Strigolactones

Besides their ecological relevance as rhizosphere signaling molecules, SLs also have an endogenous signaling function as a plant hormone (Fig. 10.5). As already mentioned, SLs are involved in the suppression of axillary bud outgrowth, in establishing root system architecture (Kapulnik et al. 2011; Koltai et al. 2010a; Ruyter-Spira et al. 2011), and also in seed germination (Tsuchiya et al. 2010; Toh et al. 2012), light signaling (Shen et al. 2007; Tsuchiya et al. 2010), and reproductive development of plants (Kohlen et al. 2012). Evidence for additional roles is accumulating at a fast pace.

Application of GR24 stimulated primary root growth in 7-day-old *Arabidopsis* plants, which was accompanied by an increased size of the meristem and transition zones (Ruyter-Spira et al. 2011). The contribution of the transition zone expansion to the total increase in root length was relatively high. The transition zone size of SL biosynthesis mutant *max4* was significantly smaller compared to wild-type plants. These results suggest that SLs play an important role in defining the boundary and transition of the meristem and elongation zones. Interestingly, specific expansion of the transition zone is also observed in radicles of germinating *Arabidopsis* seeds (Sliwinska et al. 2009). It is not unlikely that this process is mediated by SLs and possibly it is at the basis of germination of most plant species. It could also imply that a SL-mediated expansion of the transition zone contributes to the mechanism underlying SL dependency of parasitic plant seed germination.

Environmentally regulated SL production most likely offers the plant evolutionary benefits. For instance, when plants are exposed to limiting phosphate conditions, architectural changes result in increased exploration of the rhizosphere for phosphate and reduced investment of resources into the shoot (Fig. 10.5). In the first place this is achieved by reducing the number of shoot branches (Cline 1997;

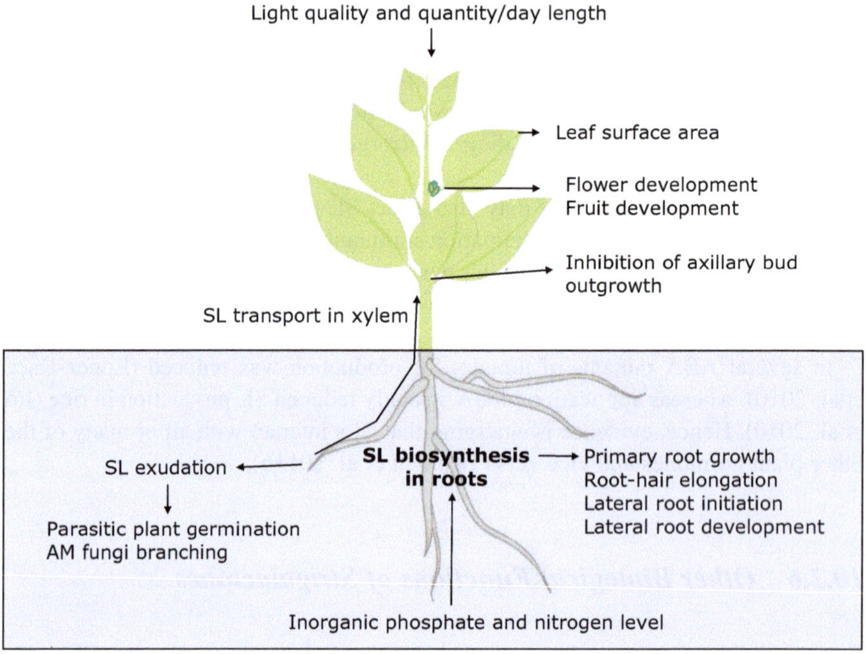

Fig. 10.5 Biological functions of strigolactones. SLs are predominantly produced in the plants' root system. Their biosynthesis is influenced by light conditions and by inorganic phosphate and nitrogen levels in the soil. SLs are exuded from the plant roots into the rhizosphere where they stimulate both hyphal branching of AM fungi and germination of parasitic plant seeds. Inside plant roots they have a hormonal function and regulate root architecture. SLs are also transported to the shoot where they inhibit axillary bud outgrowth and affect flower, fruit, and leaf development

Troughton 1977). It was recently demonstrated that the upregulation of SL biosynthesis under phosphate starvation is responsible for this response (Jamil et al. 2011a; Kohlen et al. 2011a; López-Ráez et al. 2008a; Umehara et al. 2008; Yoneyama et al. 2007b). Both rice and *Arabidopsis* mutants impaired in SL biosynthesis or signaling are unable to reduce shoot outgrowth under these conditions (Kohlen et al. 2011a; Umehara et al. 2010).

The root system architecture also changes under phosphate limiting conditions (Al-Ghazi et al. 2003; López-Bucio et al. 2002; Ma et al. 2003; Nacry et al. 2005; Sánchez-Calderón et al. 2005). In *Arabidopsis* it was shown that these changes are accompanied by an elevated level and perception of auxin in the roots (Al-Ghazi et al. 2003; López-Bucio et al. 2002; Pérez-Torres et al. 2008). This enhanced auxin functioning leads to a reduction in primary root growth while the outgrowth of lateral roots near the soil surface is stimulated (Al-Ghazi et al. 2003). It is believed that by this response the plant is able to exploit phosphate-rich areas that are usually found in the top layers of the soil (Al-Ghazi et al. 2003). As mentioned above, SLs are involved in shaping root system architecture which makes it likely that they are also involved in these adaptations to phosphate starvation (Fig. 10.5). Indeed, in

Arabidopsis the SL biosynthetic mutant *max4* showed a delay in lateral root development compared with wild-type plants (Ruyter-Spira et al. 2011). Combined, these changes in plant architecture under limited phosphate conditions lead to an increase in the root to shoot ratio, enabling the plant to better cope with its environment (Bonser et al. 1996), a response to which SLs make an important contribution.

10.3 Non-strigolactone Germination Stimulants

Various other plant compounds and microbe-derived compounds have also been reported to stimulate germination of parasitic plant seeds. Although most of them are several orders of magnitude less active than SLs, they may play a role in the stimulation of seed germination under natural conditions.

10.3.1 Plant-Derived Germination Stimulants

Dihydrosorgoleone (Fig. 10.6) was first identified as an in vitro germination stimulant for *Striga asiatica* (Chang et al. 1986). It is released from the roots of sorghum (*Sorghum bicolor*), the host of *S. asiatica*, and is rapidly oxidized in the rhizosphere to the phytotoxic allelochemical sorgoleone, which blocks photosynthetic and respiratory electron transport and blocks *p*-hydroxyphenylpyruvate dioxygenase (HPPD), an enzyme of the plastoquinone biosynthetic pathway (Dayan et al. 2009). The oily droplets exuded from sorghum root hairs are a 1:1 mixture of (dihydro)sorgoleone and a resorcinol, the latter being proposed to delay oxidation of dihydrosorgoleone (Lynn and Chang 1990). Since sorghum cultivars with different *Striga* resistance/tolerance were found to produce similar amounts of sorgoleone, it is not likely that dihydrosorgoleone plays an important role in germination stimulation of *Striga* seeds in the fields (Hess et al. 1992). Several SLs including sorgomol, strigol, sorgolactone, and 5-deoxystrigol have later been identified in sorghum root exudates (Xie et al. 2010), and sorghum cultivars susceptible to *Striga* seem to produce larger amounts of 5-deoxystrigol than resistant ones (Yoneyama et al. 2010, see Sect. 21.2.1).

Dehydrocostus lactone (Fig. 10.6) was recently identified in the root exudates of sunflower (*Helianthus annuus*) as the natural germination stimulant for *Orobanche cumana*, which is a sunflower-specific root parasite (Joel et al. 2011; see Sect. 18.2.3). Sunflower, like other plant species, produces SLs (K. Yoneyama and H. Bouwmeester, unpublished data), but *O. cumana* seeds are much less responsive to SLs than the seeds of other *Orobanche* spp. For example, 3×10^{-7} M of the synthetic SL, GR24, is required to elicit maximum germination of *O. cumana* seeds, whereas high *Phelipanche ramosa* germination is induced by a concentration of 10^{-9} M (López-Ráez et al. 2008a; Matusova et al. 2005). Indeed,

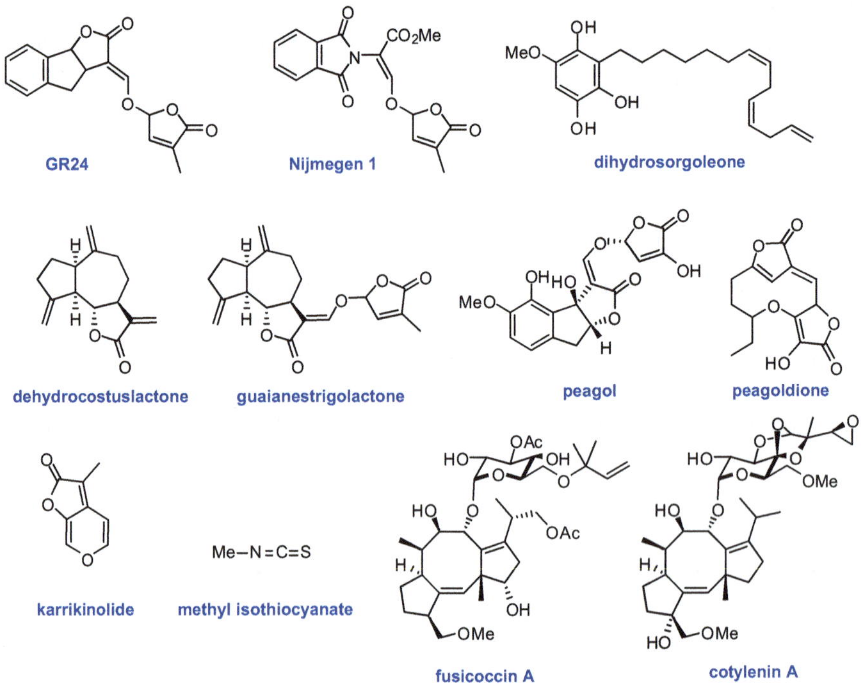

Fig. 10.6 Germination stimulants that are not strigolactones and the synthetic stimulants, the strigolactones GR24 and Nijmegen 1

phosphate starvation markedly decreased (rather than increased) the stimulatory activity of sunflower root exudates toward *O. cumana*, and fluridone did not inhibit the production of the germination stimulant in both shoots and roots of young sunflower plants, indicating that the *O. cumana* stimulant is not one of the SLs that are exuded from sunflower (Joel et al. 2011). Instead, guaianolide sesquiterpene lactones such as dehydrocostus lactone, which occur commonly in plant organs of Asteraceae species, were shown to elicit, in vitro, seed germination in *O. cumana* but not in other *Orobanche* and *Phelipanche* species (Pérez de Luque et al. 2000). In the synthetic modification of guaianolide sesquiterpene lactones, the introduction of another lactone group, yielding the so-called guaiane-SLs, renders them active as germination stimulants for other *Orobanche* and *Phelipanche* spp. as well (Fig. 10.6) (Macías et al. 2009).

Ethylene: In some *Striga* species including *S. hermonthica* and *S. asiatica*, seed germination can be elicited by the plant hormone ethylene and its precursors, 1-aminocyclopropanecarboxylic acid (ACC) and methionine (Logan and Stewart 1991; Zehhar et al. 2002). In these species, ethylene may be formed in response to SL exposure and hence form an intermediate signaling molecule, which can also induce germination (Babiker et al. 1993a, b; Sugimoto et al. 2003). Inhibitors of ethylene biosynthesis (2-aminoethoxyvinyloxyglycine, AVG) and ethylene perception (silver ions and 1-methylcyclopropene, 1-MCP) reduce seed germination elicited by SLs (Berner et al. 1999). Other plant (and microbe)-derived compounds that induce ethylene biosynthesis can also act as germination stimulants for these root parasites. In contrast, *O. minor* seed germination is not induced by ethylene and by its precursors, and neither AVG nor 1-MCP affects SLs induced germination in this species (Yoneyama et al. 1998b).

The plant hormone jasmonic acid and its derivatives also elicit *Striga* and *Orobanche* seed germination, but their activity is low (Yoneyama et al. 1998a).

Peagol and peagoldione, which bear some structural similarities to the SLs (Fig. 10.6), have recently been identified as germination stimulants in pea (*Pisum sativum*) root exudates (Evidente et al. 2009). Peagoldione at 2×10^{-3} M induced germination in *P. aegyptiaca* (syn. *O. aegyptiaca*) and peagol at 5×10^{-4} M in seeds of *O. foetida*. Evidente et al. (2011) also identified polyphenols as germination stimulants in pea root exudates (Evidente et al. 2010) and soyasapogenol B and *trans*-22-dehydrocampesterol in common vetch (*Vicia sativa*) exudates. Soyasapogenol B induced germination of *O. minor* seeds only, whereas *trans*-22-dehydrocampesterol stimulated *P. aegyptiaca*, *O. crenata*, *O. foetida*, and *O. minor* seed germination (Evidente et al. 2011). These compounds are only active at 10^{-3} M or even higher concentrations and therefore may not be involved in seed germination stimulation in the field. It would be possible to evaluate their contribution to parasite seed germination in the field by determining their concentrations in the soil.

Smoke from the combustion of plant material has been known to stimulate germination of a wide range of pioneer nonparasitic plant species (Baldwin et al. 1994; Brown and van Staden 1997; Dixon et al. 1995; Keeley and Fotheringham 1997). The biological relevance of this is that germination of these species is induced when the vegetation is cleared by fire. A butenolide (karrikinolide; Fig. 10.6) was isolated as one of the potent germination stimulants from the less complex cellulose-derived smoke, and its presence in plant-derived smoke was confirmed (Flematti et al. 2004). This compound also promotes seedling growth in some plant species. Although karrikinolide and related compounds, the karrikins, are not plant metabolites, they may be considered as allelochemicals in a broad sense as they are naturally derived by fire from plant material. There are a few reports that karrikinolide ("smoke water") induces seed germination of root parasitic plants (Bar Nun & Mayer 2005; Daws et al. 2008). However, pure karrikinolide (provided by Flematti, University of Western Australia) did not induce seed germination of *S. hermonthica* or *O. minor* (K. Yoneyama & Y. Sugimoto, unpublished data). In addition, it did not induce hyphal branching in AM fungi (Akiyama et al. 2010) nor did it inhibit tillering in rice plants (S. Yamaguchi, pers. communication). Therefore, although there are some structural similarities between karrikins and SLs, the mechanisms of germination stimulation seem to be different for the two compound classes. Nevertheless, the F-box protein MAX2 was recently shown to be involved in signal transduction of both karrikins and SLs (Nelson et al. 2011). However, Waters and co-workers showed recently that D14-like, a protein that is closely related to D14 that is required for

SL signaling (Arite et al. 2009), is required for karrikin signaling (Waters et al. 2012). See Chap. 12 for further discussion on the karrikin signaling mechanism.

Isothiocyanates have been shown to induce seed germination of *P. ramosa* (Virtue et al. 2006; Zhelev 1987) and *P. aegyptiaca* (K. Yoneyama, unpublished data). Since Brassicaceae including oilseed rape (*Brassica napus*) and *Arabidopsis thaliana* release isothiocyanates, these compounds could also contribute to host recognition by root parasites including *P. ramosa*. In contrast, *O. minor* seed germination is not affected by isothiocyanates, which may be the explanation why this species is less sensitive to germination stimulation by *Arabidopsis* (Goldwasser & Yoder 2001). Among the isothiocyanates examined, methyl isothiocyanate (Fig. 10.6) exhibited the highest and most consistent germination of *P. ramosa* seeds at around 10^{-5} M (Virtue et al. 2006). Recently, 2-phenylethyl isothiocyanate was identified as a major germination stimulant for *P. ramosa* in the rhizosphere of oilseed rape (Auger et al. 2012). Glucosinolates, precursors of isothiocyanates, and other degradation products of glucosinolates such as nitriles did not induce *P. ramosa* seed germination.

10.3.2 Germination Stimulants of Microbial Origin

Some microbial metabolites have also been reported to induce *Striga* and *Orobanche* seed germination. In particular, cotylenins and fusicoccins, produced by *Cladosporium* spp. and *Fusicoccum amygdali*, respectively, induced germination of *S. hermonthica* and *O. minor* at a concentration of $<10^{-5}$ M (Fig. 10.6) (Yoneyama et al. 1998b). The structure–activity relationship of these fusicoccanes for the induction of germination in *P. ramosa* seeds has been reported (Evidente et al. 2006). Other microbial metabolites, which were shown to elicit *S. hermonthica* seed germination, seemed to function by inducing ethylene biosynthesis (Yoneyama et al. 1998b; see also Sects. 22.4.1 and 26.3.1).

10.4 Can Germination Be a Target in the Control of Parasitic Weeds?

Parasitic weeds cause severe problems in agriculture (see Chaps. 17 and 18). Many solutions to this problem have been proposed and studied (see Chaps. 21–26). Germination, the first critical step in the interaction with a host, is an important target for control. Indeed, several control strategies that are based on this principle are already being used, and possible new ones have been proposed (Bouwmeester et al. 2003; Cardoso et al. 2011; López-Ráez et al. 2008c; Sun et al. 2007, 2008; see Sect. 22.3.3).

10.4.1 Control Through Enhanced Germination

Eliciting suicidal germination, i.e., germination in the absence of host plants, is a potential way to reduce the seed bank of parasitic weeds in agricultural fields (Babalola et al. 2002) and is discussed in detail in Chap. 22. Suicidal germination may be achieved by the use of chemical analogs of natural germination stimulants, such as synthetic SL analogs and synthetic guaianolide sesquiterpene lactones (Fukui et al. 2011; Kondo et al. 2007; Mwakaboko and Zwanenburg 2011; Macías et al. 2009). The synthetic SLs GR24 and Nijmegen 1 (Fig. 10.6) (Nefkens et al. 1997; Wigchert et al. 1999; Zwanenburg et al. 2009) are active in low concentrations, indicating that they could be used to induce suicidal germination by treating the soil before the crop is sown. A prerequisite for this to work is that the synthetic compounds should be more stable in soil than the natural SLs.

Trap and catch cropping is another control strategy involving increased germination, which is induced by non-host species that produce germination stimulants and that hence induce suicidal parasite germination (Chittapur et al. 2001; see Sect. 24.2). When susceptible crops are harvested before the seeds of the parasite are produced, they are called catch crops (see Sect. 22.4.5; Bouwmeester et al. 2003; Sun et al. 2007). The effectiveness of either strategy could be increased by the use of crop cultivars that produce higher amounts of germination stimulants. This could be achieved through traditional selection or through genetic engineering of the overexpression of one or more of the SL biosynthetic enzymes such as CCD7, CCD8, MAX1, or D27.

10.4.2 Control Through Reduced Germination

The opposite strategy would be to reduce the germination of parasitic plant seeds. As SLs are derived from carotenoids (López-Ráez et al. 2008a; Matusova et al. 2005), herbicides that inhibit carotenoid biosynthesis could reduce SL production and hence lead to reduced parasite germination. Indeed, application of low doses of carotenoid biosynthesis inhibitors to rice reduced the exudation of SLs (without causing chlorophyll bleaching) and decreased *Striga* germination and infection in a pot experiment (Jamil et al. 2010). The fact that some of the inhibitors that were used are applied as herbicides and that only low concentrations are needed to reduce SL formation makes this control strategy a promising one to further examine in the field (Jamil et al. 2010). However, fluridone and norflurazon, carotenoid biosynthesis-inhibiting herbicides, promoted conditioning of *O. minor* and *S. asiatica* seeds and enhanced their sensitivity to SLs (Chae et al. 2004; Kusumoto et al. 2006). Fluridone, even in the absence of SLs, induced *S. asiatica* seed germination in vitro (Kusumoto et al. 2006) and stimulated germination of *P. aegyptiaca* seeds in pot experiments (D. Plakhine, unpublished data). Nonetheless, novel types of inhibitors targeting the SL biosynthetic enzymes like CCD7,

CCD8, and MAX1 are being developed as SL-specific biosynthesis inhibitors (Ito et al. 2010; Kitahata et al. 2011; Sergeant et al. 2009).

Nutrient-deficient conditions and particularly low phosphate and nitrogen will increase the exudation of SLs (see Sect. 10.2.5). The application of fertilizers may address this problem in several different ways. It would improve soil fertility and increase plant fitness and crop yield, but it would also reduce host SL production and may thus reduce *Striga* infestation. In a pot experiment with rice, the application of phosphate (and to a lesser extent nitrogen) significantly reduced *Striga* infection (Jamil et al. 2011a). Also in tomato, the application of phosphate resulted in decreased SL production and suppressed the infection by *P. aegyptiaca* in pots (Jain and Foy 1992; López-Ráez et al. 2008a; Yoneyama et al. 2001; But see Sect. 22.3.3). In sorghum, nitrogen application reduced *S. hermonthica* germination (Ayongwa et al. 2006). The application of fertilizers could be a useful method to reduce SL production in crop plants and hence reduce parasitic weed infection. However, the response of SL production to nutrient availability is plant species dependent. Fertilizer rate and composition should therefore be designed to match with crop species, soil fertility, and soil properties and possibly also parasitic weed species (Jamil et al. 2011a; Yoneyama et al. 2009). See further discussion in Sect. 22.3.3.

As described above, colonization by AM fungi can reduce SL exudation of tomato (López-Ráez et al. 2011). Indeed, several studies have shown that inoculation of sorghum and maize with AM fungi can decrease infestation by parasitic plants (Lendzemo et al. 2007). Also in pea inoculation with AM fungi reduced the germination-stimulating activity of the root exudates for *Orobanche* and *Phelipanche* spp. (Fernández-Aparicio et al. 2010; see also Sect. 26.3.1). AM colonization can also induce resistance to pathogens and other biotic stresses of the host plant by the induction of defense-related genes (Pozo and Azcón-Aguilar 2007; Taylor and Harrier 2003). This may partially explain the improved *Striga* resistance of AM-colonized crops. However, the positive effect of AM colonization on plant fitness (by improving the availability of mineral nutrients), the reduction in SL production, and the induction of plant defense genes make AM fungi an interesting tool for parasitic weed control that deserves further research.

Genetic variation for low *Striga* germination stimulant (LGS) production in sorghum has been described and was used to breed for *Striga*-resistant varieties and introduce them into high-yielding sorghum cultivars into several African countries (Ejeta 2007; see Sect. 21.2.1). Also in tomato, genetic variation for the induction of *P. aegyptiaca* germination has been described (El-Halmouch et al. 2006), making breeding for LGS feasible and attractive. In addition, different cultivars of tomato produce/exude largely different amounts of SLs (López-Ráez et al. 2008b, 2010). The tomato mutant *high pigment-2* (*hp-2*^{dg}), which is an important mutant line introgressed into commercial tomato cultivars for enhanced levels of carotenoids including lycopene, was less susceptible to *P. aegyptiaca* infection than the corresponding wild-type background, and this reduced susceptibility correlated well with a lower production of SLs (López-Ráez et al. 2008b). Also in rice there is a strong genetic variation for the amount of SLs exuded by the roots. This is indeed also reflected in the *Striga* infection rate, which is generally lower in rice varieties that exude less SLs (Jamil et al. 2011b). Overall, these results indicate that selecting programs to breed for cultivars with low germination stimulant production that may hence be less susceptible to parasitic weeds is a valid and promising strategy (see Chap. 21.2.1 and 21.2.2). Genetic engineering can possibly be applied to change SL biosynthesis by activating or inhibiting one or more SL biosynthetic genes (see Chap. 24).

A fast-neutron-mutagenized tomato mutant that was resistant against *P. aegyptiaca* infection was shown to produce strongly reduced levels of SLs (Dor et al. 2010). *ccd7* and *ccd8* mutants in several plant species produced much less SLs and exhibited lower infection by *Phelipanche* spp. or *Striga* (Gomez-Roldan et al. 2008; Umehara et al. 2008). These genes are interesting targets for knockdown approaches in species where mutants are not available. Indeed, in tomato SL production could be strongly reduced through a gene silencing strategy. RNAi constructs against *LeCCD7* and *LeCCD8* were both effective in reducing SL production which resulted in a strongly reduced infection by *P. ramosa* (Kohlen et al. 2012; Vogel et al. 2010). Such genetically engineered crops may in the future be an important component of a parasitic weed control strategy (but see Sect. 24.3.2). More knowledge of the SL biosynthetic pathway and its regulation is needed to design the best strategies to achieve control with least side effects.

Manipulation of the biosynthesis of SLs in order to reduce parasitism is not without risk. SLs have other roles in the plant as described above. Alterations of SL biosynthesis may have implications for the regulation of plant root and shoot architecture. Nevertheless, some of the transgenic tomato lines with reduced SL production displayed only minor effects on shoot branching whereas *P. ramosa* infection was strongly reduced (Kohlen et al. 2012). Another possible strategy to overcome this problem would be to modulate the transport of SLs into the rhizosphere instead of regulating its biosynthesis. So far, it is not known whether the release of SLs into the rhizosphere is an active or a passive process. Understanding the mechanism that regulates the transport of SLs into the rhizosphere or to the shoot may help to develop new cultivars with reduced parasite germination induction but with normal root and shoot architecture.

The ability to establish a symbiotic interaction with AM fungi may be compromised if the release of SLs into the rhizosphere is suppressed. Parasitic weeds are generally most damaging in areas where soils are poor in nutrients and low input agriculture is practiced (Rubiales et al. 2009). In these cases, AM fungal symbiosis is likely to play an important role in enhancing crop productivity. To our knowledge so far, cultivars selected for low induction of parasite germination have not been tested for their ability to establish AM fungal symbiosis. However, mycorrhizal colonization is only compromised slightly in SL mutants or transgenic SL knockdown lines of several plant species (Gomez-Roldan et al. 2008; Kohlen et al. 2012; Koltai et al. 2010b; Umehara et al. 2010). The biological activity of SLs differs according to the biological process in which they are acting and the structural variations in the general ABCD-ring backbone (Akiyama et al. 2010; Kim et al. 2010; Zwanenburg et al. 2009). Therefore, host plants that stimulate low seed germination may not necessarily be producers of low amounts of SLs. It is possible that a different combination of SLs is produced instead, with low germination stimulatory activity but still inducing sufficient AM fungi hyphal branching and having normal shoot and root architecture.

10.5 Concluding Remarks

In the past decade we experienced a strong increase in the knowledge about germination stimulants and particularly about the most prominent class of germination stimulants, the SLs. Many new germination stimulants, including SLs, were identified, new biological functions uncovered, and the physiological and biochemical regulation of their production at least partly uncovered. A key discovery of this past decade is that SLs—first only known as parasitic plant germination stimulants—are rhizosphere signaling molecules for AM fungi and are a new class of plant hormones that regulate root and shoot architecture. The fact that SLs are involved in so many different processes in plants complicates control strategies based on changing their production. On the other hand, we now know so much more about the regulation of the biosynthesis of these signaling molecules that we are better able to interfere in their production. It is clear that for optimal control strategies of the root parasitic weeds based on modification of germination stimulant secretion, more knowledge of the molecular basis of SL production and secretion is required, but in the future germination will definitely remain an important target in strategies for parasitic weed control.

Acknowledgements KY acknowledges grants from KAKENHI (18208010, 23338006) and Program for Promotion of Basic and Applied Researches for Innovations in Bio-oriented Industry. HB acknowledges funding by the Netherlands Organization for Scientific Research (NWO; VICI grant, 865.06.002 and Equipment grant, 834.08.001). He was co-financed by the Centre for BioSystems Genomics (CBSG) which is part of the Netherlands Genomics Initiative/Netherlands Organization for Scientific Research.

References

Akiyama K, Matsuzaki K, Hayashi H (2005) Plant sesquiterpenes induce hyphal branching in arbuscular mycorrhizal fungi. Nature 435:824–827

Akiyama K, Ogasawara S, Ito S, Hayashi H (2010) Structural requirements of strigolactones for hyphal branching in AM fungi. Plant Cell Physiol 51:1104–1117

Alder A, Jamil M, Marzorati M, Bruno M, Vermathen M, Bigler P, Ghisla S, Bouwmeester HJ, Beyer P, Al-Babili S (2012) The path from β-carotene to carlactone, a strigolactone-like plant hormone. Science 335:1348–1351

Al-Ghazi Y, Muller B, Pinloche S, Tranbarger TJ, Nacry P, Rossignol M, Tardieu F, Doumas P (2003) Temporal responses of *Arabidopsis* root architecture to phosphate starvation: evidence for the involvement of auxin signalling. Plant Cell Environ 26:1053–1066

Arite T, Iwata H, Ohshima K, Maekawa M, Nakajima M, Kojima M, Sakakibara H, Kyozuka J (2007) *DWARF10*, an *RMS1/MAX4/DAD1* ortholog, controls lateral bud outgrowth in rice. Plant J 51:1019–1029

Arite T, Umehara M, Ishikawa S, Hanada A, Maekawa M, Yamaguchi S, Kyozuka J (2009) D14, a strigolactone-insensitive mutant of rice, shows an accelerated outgrowth of tillers. Plant Cell Physiol 50:1416–1424

Auger B, Pouvreau J-B, Pouponneau K, Yoneyama K, Montiel G et al (2012) Germination stimulants of *Phelipanche ramosa* in the rhizosphere of *Brassica napus* are derived from the glucosinolate pathway. Mol Plant Microbe Interact 25:993–1004

Awad AA, Sato D, Kusumoto D, Kamioka H, Takeuchi Y, Yoneyama K (2006) Characterization of strigolactones, germination stimulants for the root parasitic plants *Striga* and *Orobanche*, produced by maize, millet and sorghum. Plant Growth Regul 48:221–227

Ayongwa GC, Stomph TJ, Emechebe AM, Kuyper TW (2006) Root nitrogen concentration of sorghum above 2% produces least *Striga hermonthica* seed stimulation. Ann Appl Biol 149:255–262

Babalola OO, Osir EO, Sanni AI (2002) Characterization of potential ethylene-producing rhizosphere bacteria of *Striga*-infested maize and sorghum. Afr J Biotechnol 1:67–69

Babiker AG, Ejeta G, Butler LG, Woodson WR (1993a) Ethylene biosynthesis and strigol-induced germination of *Striga asiatica*. Physiol Plant 88:359–365

Babiker AGT, Butler KG, Ejeta G, Woodson WR (1993b) Enhancement of ethylene biosynthesis and germination by cytokinins and 1-aminocyclopropane-1-carboxylic acid in *Striga asiatica* seeds. Physiol Plant 89:21–26

Baldwin IT, Staszak-Kozinski L, Davidson R (1994) Up in smoke: smoke-derived germination cues for postfire annual, Nicotiana attenuata Torr. ex. Watson. J Chem Ecol 20:2345–2371

Balzergue C, Puech-Pags V, Bécard G, Rochange SF (2011) The regulation of arbuscular mycorrhizal symbiosis by phosphate in pea involves early and systemic signalling events. J Exp Bot 62:1049–1060

Bar Nun N, Mayer AM (2005) Smoke chemicals and coumarin promote the germination of the parasitic weed *Orobanche aegyptiaca*. Isr J Plant Sci 53:97–101

Berner DK, Schaad NW, Völksch B (1999) Use of ethylene-producing bacteria for stimulation of *Striga* spp. seed germination. Biol Control 15:274–282

Besserer A, Puech-Pagès V, Kiefer P, Gomez-Roldan V, Jauneau A, Roy S, Portais J-C, Roux C, Bécard G, Séjalon-Delmas N (2006) Strigolactones stimulate arbuscular mycorrhizal fungi by activating mitochondria. PLoS Biol 4:1239–1247

Bonser AM, Lynch J, Snapp S (1996) Effect of phosphorus deficiency on growth angle of basal roots in *Phaseolus vulgaris*. New Phytol 132:281–288

Booker J, Auldridge M, Wills S, McCarty D, Klee H, Leyser O (2004) MAX3/CCD7 is a carotenoid cleavage dioxygenase required for the synthesis of a novel plant signalling molecule. Curr Biol 14:1232–1238

Booker J, Sieberer T, Wright W, Williamson L, Willett B, Stirnberg P, Turnbull C, Srinivasan M, Goddard P, Leyser O (2005) *MAX1* Encodes a cytochrome P450 family member that acts downstream of *MAX3/4* to produce a carotenoid-derived branch-inhibiting hormone. Dev Cell 8:443–449

Bouwmeester HJ, Matusova R, Zhongkui S, Beale MH (2003) Secondary metabolite signalling in host-parasitic plant interactions. Curr Opin Plant Biol 6:358–364

Bouwmeester HJ, Roux C, Lopez-Raez JA, Bécard G (2007) Rhizosphere communication of plants, parasitic plants and AM fungi. Trends Plant Sci 12:224–230

Brown NAC, van Staden J (1997) Smoke as a germination cue: a review. Plant Growth Regul 22:115–124

Brown R, Johnson AW, Robinson E, Todd AR (1949) The stimulant involved in the germination of *Striga hermonthica*. Proc R Soc Lond B Biol Sci 136:395–404

Brown R, Greenwood AD, Johnson AW, Long AG (1951a) The stimulant involved in the germination of *Orobanche minor* Sm. 1. Assay technique and bulk preparation of the stimulant. Biochem J 48:559–564

Brown R, Greenwood AD, Johnson AW, Long AG, Tyler GL (1951b) The stimulant involved in the germination of *Orobanche minor* Sm. 2. Chromatographic purification of crude concentrates. Biochem J 48:564–568

Brown R, Greenwood AD, Johnson AW, Lansdown AR, Long AG, Suderland N (1952a) The *Orobanche* germination factor. III. Concentration of the factor by counter current distribution. Biochem J 52:571–574

Brown R, Johnson AW, Robinson E, Tyler GL (1952b) The *Striga* germination factor. II. Chromatographic purification of crude concentrates. Biochem J 50:596–600

Butler LG (1995) Chemical communication between the parasitic weed *Striga* and its crop host. A new dimension in allelochemistry. In: Inderjit KM, Dakshini M, Enhelling FA (eds) Allelopathy, organisms, processes and applications. American Chemical Society, Washington, DC, pp 158–166

Cardoso C, Ruyter-Spira C, Bouwmeester HJ (2011) Strigolactones and root infestation by plant-parasitic *Striga*, *Orobanche* and *Phelipanche* spp. Plant Sci 180:414–420

Chae HS, Yoneyama K, Takeuchi Y, Joel DM (2004) Fluridone and norflurazon, carotenoid-biosynthesis inhibitors, promote seed conditioning and germination of the holoparasite *Orobanche minor*. Physiol Plant 120:328–337

Chang M, Netzly DG, Butler LG, Lynn DG (1986) Chemical regulation of distance: characterization of the first natural host germination stimulant for *Striga asiatica*. J Am Chem Soc 108:7858–7860

Chen VX, Boyer FD, Rameau C, Retailleau P, Vors JP, Beau JM (2010) Stereochemistry, total synthesis, and biological evaluation of the new plant hormone solanacol. Chem Eur J 16:13941–13945

Chittapur BM, Hunshal CS, Shenoy H (2001) Allelopathy in parasitic weed management: role of catch and trap crops. Allelopathy J 8:147–160

Chiwocha SDS, Dixon KW, Flematti GR, Ghisalbert EL, Merritt DJ, Nelson DC, Riseborough J-AM, Smith SM, Stevens JC (2009) Karrikins: a new family of plant growth regulators in smoke. Plant Sci 177:252–256

Cline M (1997) Concepts and terminology of apical dominance. Am J Bot 84:1064

Cook CE, Whichard LP, Turner B, Wall ME, Egley GH (1966) Germination of witchweed (*Striga lutea* Lour.): isolation and properties of a potent stimulant. Science 154:1189–1190

Cook CE, Whichard LP, Wall ME, Egley GH, Coggon P, Luhan PA, McPhail AT (1972) Germination stimulants. II. The structure of strigol – a potent seed germination stimulant for witchweed (*Striga lutea* Lour.). J Am Chem Soc 94:6198–6199

Daws MI, Pritchard HW, Van Staden J (2008) Butenolide from plant-derived smoke functions as a strigolactone analogue: evidence from parasitic weed seed germination. S Afr J Bot 74:116–120

Dayan FE, Howell JL, Weidenhamer JD (2009) Dynamic root exudation of sorgoleone and its *in planta* mechanism of action. J Exp Bot 60:2107–2117

Delaux P-M, Xie X, Timme RE, Puech-Pages V, Dunand C et al (2012) Origin of strigolactones in the green lineage. New Phytol 195:857–871

Dixon KW, Roche S, Pate JS (1995) The promotive effect of smoke derived from burnt native vegetation on seed germination of Western Australian plants. Oecologia 101:185–192

Dor E, Alperin B, Wininger S, Ben-Dor B, Somvanshi VS, Koltai H, Kapulnik Y, Hershenhorn J (2010) Characterization of a novel tomato mutant resistant to the weedy parasites *Orobanche* and *Phelipanche* spp. Euphytica 171:371–380

Ejeta G (2007) Breeding for *Striga* resistance in sorghum: exploitation of an intricate host-parasite biology. Crop Sci 47:S216–S227

El-Halmouch Y, Benharrat H, Thalouarn P (2006) Effect of root exudates from different tomato genotypes on broomrape (*O. aegyptiaca*) seed germination and tubercle development. Crop Prot 25:501–507

Evidente A, Andolfi A, Fiore M, Boari A, Vurro M (2006) Stimulation of *Orobanche ramosa* seed germination by fusicoccin derivatives: a structure–activity relationship study. Phytochemistry 67:19–26

Evidente A, Fernández-Aparicio M, Cimmino A, Rubiales D, Andolfi A, Motta A (2009) Peagol and peagoldione, two new strigolactone-like metabolites isolated from pea root exudates. Tetrahedron Lett 50:6955–6958

Evidente A, Cimmino A, Fernández-Aparicio M, Andolfi A, Rubiales D, Motta A (2010) Polyphenols, including the new peapolyphenols A-C, from pea root exudates stimulate *Orobanche foetida* seed germination. J Agric Food Chem 58:2902–2907

Evidente A, Cimmino A, Fernández-Aparicio M, Rubiales D, Andolfi A, Melck D (2011) Soyasapogenol B and *trans*-22-dehydrocampesterol from common vetch (*Vicia sativa* L.) root exudates stimulate broomrape seed germination. Pest Manag Sci 67:1015–1022

Fernández-Aparicio M, García-Garrido JM, Ocampo JA, Rubiales D (2010) Colonisation of field pea roots by arbuscular mycorrhizal fungi reduces *Orobanche* and *Phelipanche* species seed germination. Weed Res 50:262–268

Fernández-Aparicio M, Yoneyama K, Rubiales D (2011) The role of strigolactones in host specificity of *Orobanche* and *Phelipanche* seed germination. Seed Sci Res 21:55–61

Flematti GR, Ghisalberti EL, Dixon KW, Trengrove RD (2004) A compound from smoke that promotes seed germination. Science 305:977

Foo E, Turnbull CGN, Beveridge CA (2001) Long-distance signalling and the control of branching in the *rms1* mutant of pea. Plant Physiol 126:203–209

Fukui K, Ito S, Ueno K, Yamaguchi S, Kyozuka J, Asami T (2011) Inhibition of rice branching and promotion of *Striga* germination by debranone derivatives mimicking strigolactone function. Bioorg Med Chem Lett 21:4905–4908

Goldwasser Y, Yoder JI (2001) Differential induction of *Orobanche* seed germination by *Arabidopsis thaliana*. Plant Sci 160:951–959

Gomez-Roldan V, Fermas S, Brewer PB, Puech-Pagès V, Dun EA, Pillot J-P, Letisse F, Matusova R, Danoun S, Portais J-C, Bouwmeester H, Bécard G, Beveridge CA, Rameau C, Rochange SF (2008) Strigolactone inhibition of shoot branching. Nature 455:189–194

Hamiaux C, Drummond RSM, Janssen BJ, Ledger SE, Cooney JM, Newcomb RD, Snowden KC (2012) DAD2 is an α/β hydrolase likely to be involved in the perception of the plant branching hormone, strigolactone. Curr Biol 22:2032–2036

Harrison MJ (2005) Signalling in the arbuscular mycorrhizal symbiosis. Annu Rev Microbiol 59:19–42

Hauck C, Müller S, Schildknecht H (1992) A germination stimulant for parasitic flowering plants from *Sorghum bicolor*, a genuine host plant. J Plant Physiol 139:474–478

Hess DE, Ejeta G, Butler LG (1992) Selecting sorghum genotypes expressing a quantitative biosynthetic trait that confers resistance to *Striga*. Phytochemistry 31:493–497

Höniges A, Ardelean A, Xie X, Yoneyama K, Yoneyama K, Wegmann K (2012) Towards understanding *Orobanche* host-specificity. Rom Agric Res 29:313–322

Humphrey AJ, Beale MH (2006) Strigol: biogenesis and physiological activity. Phytochemistry 67:636–640

Ito S, Kitahata N, Umehara M, Hanada A, Kato A, Ueno K, Mashiguchi K, Kyozuka J, Yoneyama K, Yamaguchi S, Asami T (2010) A new lead chemical for strigolactone biosynthesis inhibitors. Plant Cell Physiol 51:1143–1150

Jain R, Foy CL (1992) Nutrient effects on parasitism and germination of Egyptian broomrape (*Orobanche aegyptiaca*). Weed Technol 6:269–275

Jamil M, Charnikhova T, Verstappen F, Bouwmeester H (2010) Carotenoid inhibitors reduce strigolactone production and *Striga hermonthica* infection in rice. Arch Biochem Biophys 504:123–131

Jamil M, Charnikhova T, Cardoso C, Jamil T, Ueno K, Verstappen F, Asami T, Bouwmeester HJ (2011a) Quantification of the relationship between strigolactones and *Striga hermonthica* infection in rice under varying levels of nitrogen and phosphorus. Weed Res 51:373–385

Jamil M, Rodenburg J, Charnikhova T, Bouwmeester HJ (2011b) Pre-attachment *Striga hermonthica* resistance of NERICA cultivars based on low strigolactone production. New Phytol 192:964–975

Joel DM, Hershenhorn J, Eizenburg H, Aly R, Ejeta G, Rich PJ, Ransom JK, Sauerborn J, Rubiales D (2007) Biology and management of weedy root parasites. In: Janick J (ed) Horticultural reviews. Wiley, London, pp 267–349

Joel DM, Chaudhuri SK, Plakhine D, Ziadna H, Steffens JC (2011) Dehydrocostus lactone is exuded from sunflower roots and stimulates germination of the root parasite *Orobanche cumana*. Phytochemistry 72:624–634

Kapulnik Y, Delaux P-M, Resnick N, Mayzlish-Gati E, Wininger S, Bhattacharya C, Séjalon-Delmas N, Combier J-P, Bécard G, Belausov E, Beeckman T, Dor E, Hershenhorn J, Koltai H (2011) Strigolactones affect lateral root formation and root-hair elongation in *Arabidopsis*. Planta 233:209–216

Keeley JE, Fotheringham CJ (1997) Trace gas emissions and smoke-induced seed germination. Science 276:1248–1250

Kim HI, Xie X, Kim HS, Chun JC, Yoneyama K, Nomura T, Takeuchi Y, Yoneyama K (2010) Structure-activity relationship of naturally occurring strigolactones in *Orobanche minor* seed germination stimulation. J Pestic Sci 35:344–347

Kisugi T, Xie X, Kim HI, Yoneyama K, Sado A, Akiyama K, Hayashi H, Uchida K, Yokota T, Nomura T, Yoneyama K (2013) Strigone, the first isolation and identification as a natural strigolactone from *Houttuynia cordata*. Phytochemistry 87:60–64

Kitahata N, Ito S, Kato A, Ueno K, Nakano T, Yoneyama K, Asami T (2011) Abamine as a basis for new designs of regulators of strigolactone production. J Pestic Sci 36:53–57

Kohlen W, Charnikhova T, Liu Q, Bours R, Domagalska MA, Beguerie S, Verstappen F, Leyser O, Bouwmeester H, Ruyter-Spira C (2011a) Strigolactones are transported through the xylem and play a key role in shoot architectural response to phosphate deficiency in nonarbuscular mycorrhizal host *Arabidopsis*. Plant Physiol 155:974–987

Kohlen W, Ruyter-Spira C, Bouwmeester HJ (2011b) Strigolactones. A new musician in the orchestra of plant hormones. Botany 89:827–840

Kohlen W, Charnikhova T, Lammers M, Pollina T, Tóth P, Haider I, Pozo MJ, de Maagd RA, Ruyter-Spira C, Bouwmeester HJ, López-Ráez JA (2012) The tomato CAROTENOID CLEAVAGE DIOXYGENASE8 (SlCCD8) regulates rhizosphere signalling, plant architecture and affects reproductive development through strigolactone biosynthesis. New Phytol 196:535–547

Koltai H, Dor E, Hershenhorn J, Joel D, Weininger S, Lekalla S, Shealtiel H, Bahattacharya C, Eliahu E, Resnick N, Barg R, Kapulnik Y (2010a) Strigolactones' effect on root growth and root-hair elongation may be mediated by auxin-efflux carriers. J Plant Growth Regul 29:129–136

Koltai H, LekKala SP, Bhattacharya C, Mayzlish-Gati E, Resnick N, Wininger S, Dor E, Yoneyama K, Yoneyama K, Hershenhorn J, Joel DM, Kapulnik Y (2010b) A tomato strigolactone-impaired mutant displays aberrant shoot morphology and plant interactions. J Exp Bot 61:1739–1749

Koltai H, Cohen M, Chesin O, Mayzlish-Gati E, Bécard G, Puech-Pagès V, Dor BB, Resnick N, Wininger S, Kapulnik Y (2011) Light is a positive regulator of strigolactone levels in tomato roots. J Plant Physiol 168:1993–1996

Kondo Y, Tadokoro E, Matsuura M, Iwasaki K, Sugimoto Y, Miyake H, Takikawa H, Sasaki M (2007) Synthesis and seed germination stimulating activity of some imino analogs of strigolactones. Biosci Biotechnol Biochem 71:2781–2786

Kusumoto D, Chae SH, Mukaida K, Yoneyama K, Joel DM, Takeuchi Y (2006) Effects of fluridone and norflurazon on conditioning and germination of *Striga asiatica* seeds. Plant Growth Regul 48:73–78

Ledger SE, Janssen BJ, Karunairetnam S, Wang T, Snowden KC (2010) Modified CAROTENOID CLEAVAGE DIOXYGENASE8 expression correlates with altered branching in kiwifruit (*Actinidia chinensis*). New Phytol 188:803–813

Lendzemo VW, Kuyper TW, Matusova R, Bouwmeester HJ, Van Ast A (2007) Colonization by arbuscular mycorrhizal fungi of sorghum leads to reduced germination and subsequent attachment and emergence of *Striga hermonthica*. Plant Signal Behav 2:58–62

Lin H, Wang R, Qian Q, Yan M, Meng X, Fu Z, Yan C, Jiang B, Su Z, Li J, Wang Y (2009) DWARF27, an iron-containing protein required for the biosynthesis of strigolactones, regulates rice tiller bud outgrowth. Plant Cell 21:1512–1525

Logan DC, Stewart GR (1991) Role of ethylene in the germination of the hemiparasite *Striga hermonthica*. Plant Physiol 97:1435–1438

López-Bucio J, Hernandez-Abreu E, Sanchez-Calderon L, Nieto-Jacobo MF, Simpson J, Herrera-Estrella L (2002) Phosphate availability alters architecture and causes changes in hormone sensitivity in the *Arabidopsis* root system. Plant Physiol 129:244–256

López-Ráez JA, Charnikhova T, Gómez-Roldán V, Matusova R, Kohlen W, De Vos R, Verstappen F, Puech-Pages V, Bécard G, Mulder P, Bouwmeester H (2008a) Tomato strigolactones are derived from carotenoids and their biosynthesis is promoted by phosphate starvation. New Phytol 178:863–874

López-Ráez JA, Charnikhova T, Mulder P, Kohlen W, Bino R, Levin I, Bouwmeester H (2008b) Susceptibility of the tomato mutant high *pigment-2dg* (*hp-2*^{dg}) to *Orobanche* spp. infection. J Agric Food Chem 56:6326–6332

López-Ráez JA, Matusova R, Cardoso C, Jamil M, Charnikhova T, Kohlen W, Ruyter-Spira C, Verstappen F, Bouwmeester H (2008c) Strigolactones: ecological significance and use as a target for parasitic plant control. Pest Manag Sci 64:471–477

López-Ráez JA, Kohlen W, Charnikhova T, Mulder P, Undas AK, Sergeant MJ, Verstappen F, Bugg TDH, Thompson AJ, Ruyter-Spira C, Bouwmeester H (2010) Does abscisic acid affect strigolactone biosynthesis? New Phytol 187:343–354

López-Ráez JA, Charnikhova T, Fernandez I, Bouwmeester HJ, Pozo MJ (2011) Arbuscular mycorrhizal symbiosis decreases strigolactone production in tomato. J Plant Physiol 168:294–297

Lynn DG, Chang M (1990) Phenolic signals in cohabitation: implications for plant development. Annu Rev Plant Physiol Plant Mol Biol 41:497–526

Ma Z, Baskin TI, Brown KM, Lynch JP (2003) Regulation of root elongation under phosphorus stress involves changes in ethylene responsiveness. Plant Physiol 131:1381–1390

Macías FA, García-Díaz MD, Pérez-de-Luque A, Rubiales D, Galindo JCG (2009) New chemical clues for broomrape-sunflower host–parasite interactions: synthesis of guaianestrigolactones. J Agric Food Chem 57:5853–5864

Matsuura H, Ohashi K, Sasako H, Tagawa N, Takano Y, Ioka Y, Nabeta K, Yoshihara T (2008) Germination stimulant from root exudates of *Vigna unguiculata*. Plant Growth Regul 54:31–36

Matusova R, Rani K, Verstappen FWA, Franssen MCR, Beale MH, Bouwmeester HJ (2005) The strigolactone germination stimulants of the plant-parasitic *Striga* and *Orobanche* spp. are derived from the carotenoid pathway. Plant Physiol 139:920–934

Mori K, Matsui J, Bando M, Kido M, Takeuchi Y (1998) Synthetic disproof against the structure proposed for alectrol, the germination stimulant from *Vigna unguiculata*. Tetrahedron Lett 39:6023–6026

Müller S, Hauck C, Schildknecht H (1992) Germination stimulants produced by *Vigna unguiculata* Walp cv Saunders Upright. J Plant Growth Regul 11:77–84

Mwakaboko AS, Zwanenburg B (2011) Strigolactone analogs derived from ketones using a working model for germination stimulants as a blueprint. Plant Cell Physiol 52:699–715

Nacry P, Canivenc G, Muller B, Azmi A, Van Onckelen H, Rossignol M, Doumas P (2005) A role for auxin redistribution in the responses of the root system architecture to phosphate starvation in *Arabidopsis*. Plant Physiol 138:2061–2074

Nefkens GHL, Thuring JWJF, Beenakkers MFM, Zwanenburg B (1997) Synthesis of a phthaloylglycine-derived strigol analogue and its germination stimulatory activity towards seeds of the parasitic weeds *Striga hermonthica* and *Orobanche crenata*. J Agric Food Chem 45:2273–2277

Nelson DC, Scaffidi A, Dun EA, Waters MT, Flematti GR, Dixon KW, Beveridge CA, Ghisalberti EL, Smith SM (2011) F-box protein MAX2 has dual roles in karrikin and strigolactone signalling in *Arabidopsis thaliana*. Proc Natl Acad Sci USA 108:8897–8902

Pérez de Luque AP, Galindo JCG, Macías FA, Jorrín J (2000) Sunflower sesquiterpene lactone models induce *Orobanche cumana* seed germination. Phytochemistry 53:45–50

Pérez-Torres C-A, Lopez-Bucio J, Cruz-Ramirez A, Ibarra-Laclette E, Dharmasiri S, Estelle M, Herrera-Estrella L (2008) Phosphate availability alters lateral root development in *Arabidopsis* by modulating auxin sensitivity via a mechanism involving the TIR1 auxin receptor. Plant Cell 20:3258–3272

Pozo MJ, Azcón-Aguilar C (2007) Unraveling mycorrhiza-induced resistance. Curr Opin Plant Biol 10:393–398

Proust H, Hoffmann B, Xie X, Yoneyama K, Schaefer DG, Yoneyama K, Nogué F, Rameau C (2011) Strigolactones regulate protonema branching and act as a quorum sensing-like signal in the moss *Physcomitrella patens*. Development 138:1531–1539

Rani K, Zwanenburg B, Sugimoto Y, Yoneyama K, Bouwmeester HJ (2008) Biosynthetic considerations could assist the structure elucidation of host plant produced rhizosphere signalling compounds (strigolactones) for arbuscular mycorrhizal fungi and parasitic plants. Plant Physiol Biochem 46:617–626

Rubiales D, Verkleij J, Vurro M, Murdoch AJ, Joel DM (2009) Parasitic plant management in sustainable agriculture. Weed Res 49:1–5

Ruyter-Spira C, Kohlen W, Charnikhova T, van Zeijl A, van Bezouwen L et al (2011) Physiological effects of the synthetic strigolactone analog GR24 on root system architecture in *Arabidopsis*: another belowground role for strigolactones? Plant Physiol 155:721–734

Sánchez-Calderón L, Lopez-Bucio J, Chacon-Lopez A, Cruz-Ramirez A, Nieto-Jacobo F, Dubrovsky JG, Herrera-Estrella L (2005) Phosphate starvation induces a determinate developmental program in the roots of *Arabidopsis thaliana*. Plant Cell Physiol 46:174–184

Sergeant MJ, Li J-J, Fox C, Brookbank N, Rea D, Bugg TD, Thompson AJ (2009) Selective inhibition of carotenoid cleavage dioxygenases: phenotypic effects on shoot branching. J Biol Chem 284:5257–5264

Shen H, Luong P, Huq E (2007) The F-box protein MAX2 functions as a positive regulator of photomorphogenesis in *Arabidopsis*. Plant Physiol 145:1471–1483

Siame BP, Weerasuriya Y, Wood K, Ejeta G, Butler LG (1993) Isolation of strigol, a germination stimulant for *Striga asiatica*, from host plants. J Agric Food Chem 41:1486–1491

Sliwinska E, Bassel GW, Bewley JD (2009) Germination of *Arabidopsis thaliana* seeds is not completed as a result of elongation of the radicle but of the adjacent transition zone and lower hypocotyl. J Exp Bot 60:3587–3594

Snowden KC, Simkin AJ, Janssen BJ et al (2005) The *Decreased apical dominance1/Petunia hybrida* CAROTENOID CLEAVAGE DIOXYGENASE8 gene affects branch production and plays a role in leaf senescence, root growth, and flower development. Plant Cell 17:746–759

Sorefan K, Booker J, Haurogné K et al (2003) *MAX4* and *RMS1* are orthologous dioxygenase-like genes that regulate shoot branching in *Arabidopsis* and pea. Genes Dev 17:1469–1474

Stirnberg P, van de Sande K, Leyser HMO (2002) *MAX1* and *MAX2* control shoot lateral branching in *Arabidopsis*. Development 129:1131–1141

Sugimoto Y, Ali AM, Yabuta S, Kinoshita H, Inanaga S, Itai A (2003) Germination strategy of *Striga hermonthica* involves regulation of ethylene biosynthesis. Physiol Plant 119:1–9

Sun Z, Matusova R, Bouwmeester HJ (2007) Germination of *Striga* and chemical signalling involved: a target for control methods. In: Gressel J, Ejeta G (eds) Integrating new technologies for *Striga* control: towards ending the witch-hunt. World Scientific, Singapore, pp 47–60

Sun Z, Has J, Walter MH et al (2008) Cloning and characterisation of a maize carotenoid cleavage dioxygenase (*ZmCCD1*) and its involvement in the biosynthesis of apocarotenoids with various roles in mutualistic and parasitic interactions. Planta 228:789–801

Taylor J, Harrier LA (2003) Expression studies of plant genes differentially expressed in leaf and root tissues of tomato colonised by the arbuscular mycorrhizal fungus *Glomus mosseae*. Plant Mol Biol 51:619–629

Toh S, Kamiya Y, Kawakami N, Nambara E, McCourt P, Tsuchiya Y (2012) Thermoinhibition uncovers a role for strigolactones in *Arabidopsis* seed germination. Plant Cell Physiol 53:107–117

Troughton A (1977) The effect of phosphorus nutrition upon the growth and morphology of young plants of *Lolium perenne* L. Ann Bot 41:85–92

Tsuchiya Y, McCourt P (2009) Strigolactones: a new hormone with a past. Curr Opin Plant Biol 12:556–561

Tsuchiya Y, Vidaurre D, Toh S, Hanada A, Nambara E, Kamiya Y, Yamaguchi S, McCourt P (2010) A small-molecule screen identifies new functions for the plant hormone strigolactone. Nat Chem Biol 6:741–749

Ueno K, Nomura S, Muranaka S, Mizutani M, Takikawa H, Sugimoto Y (2011) *Ent-2′-epi*-orobanchol and its acetate, as germination stimulants for *Striga gesnerioides* seeds isolated from cowpea and red clover. J Agric Food Chem 59:10485–10490

Umehara M, Hanada A, Yoshida S, Akiyama K, Arite T, Takeda-Kamiya N, Magome H, Kamiya Y, Shirasu K, Yoneyama K, Kyozuka J, Yamaguchi S (2008) Inhibition of shoot branching by new terpenoid plant hormones. Nature 455:195–200

Umehara M, Hanada A, Magome H, Takeda-Kamiya N, Yamaguchi S (2010) Contribution of strigolactones to the inhibition of tiller bud outgrowth under phosphate deficiency in rice. Plant Cell Physiol 51:1118–1126

Vaucher JP (1823) Mémoire sur la germination des orobanches. Mém Mus Hist nat Paris 10:261–273

Virtue JG, DeDear C, Potter MJ, Rieger M (2006) Potential use of isothiocyanates in branched broomrape eradication. In: Preston C, Watts JHW, Crossman ND (eds) 15th Australian weeds conference, Adelaide. pp 629–632

Vogel JT, Walter MH, Giavalisco P et al (2010) SlCCD7 controls strigolactone biosynthesis, shoot branching and mycorrhiza-induced apocarotenoid formation in tomato. Plant J 61:300–311

Waters MT, Nelson DC, Scaffidi A, Flematti GR, Sun YK, Dixon KW, Smith SM (2012) Specialisation within the DWARF14 protein family confers distinct responses to karrikins and strigolactones in Arabidopsis. Development 139:1285–1295

Weerasuriya Y, Siame BA, Hess D, Ejets G, Butler LG (1993) Influence of conditions and genotype on the amount of *Striga germination* stimulants exuded by root of several host crops. J Agric Food Chem 41:1492–1496

Westwood JH, Yoder JI, Timko MP, dePamphilis CW (2010) The evolution of parasitism in plants. Trends Plant Sci 15:227–235

Wigchert SCM, Kuiper E, Boelhouwer GJ, Nefkens GHL, Verkleij JAC, Zwanenburg B (1999) Dose–response of seeds of the parasitic weeds *Striga* and *Orobanche* toward the synthetic germination stimulants GR 24 and Nijmegen 1. J Agric Food Chem 47:1705–1710

Xie X, Kusumoto D, Takeuchi Y, Yoneyama K, Yamada Y, Yoneyama K (2007) 2′-Epi-orobanchol and solanacol, two unique strigolactones, germination stimulants for root parasitic weeds, produced by tobacco. J Agric Food Chem 55:8067–8072

Xie X, Yoneyama K, Kusumoto D, Yamada Y, Takeuchi Y, Sugimoto Y, Yoneyama K (2008a) Sorgomol, germination stimulant for root parasitic plants, produced by *Sorghum bicolor*. Tetrahedron Lett 49:2066–2068

Xie X, Yoneyama K, Kusumoto D, Yamada Y, Yokota T, Takeuchi Y, Yoneyama K (2008b) Isolation and identification of alectrol as (+)-orobanchyl acetate, a novel germination stimulant for root parasitic plants. Phytochemistry 69:427–431

Xie X, Yoneyama K, Harada Y, Fusegi N, Yamada Y, Ito S, Yokota T, Takeuchi Y, Yoneyama K (2009a) Fabacyl acetate, a germination stimulant for root parasitic plants from *Pisum sativum*. Phytochemistry 70:211–215

Xie X, Yoneyama K, Kurita J-Y, Harada Y, Yamada Y, Takeuchi Y, Yoneyama K (2009b) 7-Oxoorobanchyl acetate and 7-oxoorobanchol as germination stimulants for root parasitic plants from flax (*Linum usitatissimum*). Biosci Biotechnol Biochem 73:1367–1370

Xie X, Yoneyama K, Yoneyama K (2010) The strigolactone story. Annu Rev Phytopathol 48:93–117

Xie X, Yoneyama K, Kisugi T, Uchida K, Ito S, Akiyama K, Hayashi H, Yokota T, Nomura T, Yoneyama K (2012) Confirming stereochemical structures of strigolactones produced by rice and tobacco. Mol Plant 6:153–163

Yasuda N, Sugimoto Y, Kato M, Inanaga S, Yoneyama K (2003) (+)-Strigol, a witchweed seed germination stimulant, from *Menispermum dauricum* root culture. Phytochemistry 62:1115–1119

Yokota T, Sakai H, Okuno K, Yoneyama K, Takeuchi Y (1998) Alectrol and orobanchol, germination stimulants for *Orobanche minor*, from its host red clover. Phytochemistry 49:1967–1973

Yoneyama K, Ogasawara M, Takeuchi Y, Konnai M, Sugimoto Y, Seto H, Yoshida S (1998a) Effect of jasmonates and related compounds on seed germination of *Orobanche minor* Smith and *Striga hermonthica* (Del.) Benth. Biosci Biotechnol Biochem 62:1448–1450

Yoneyama K, Takeuchi Y, Ogasawara M, Konnai M, Sugimoto Y, Sassa T (1998b) Cotylenins and fusicoccins stimulate seed germination of *Striga hermonthica* (Del.) Benth and *Orobanche minor* Smith. J Agric Food Chem 46:1583–1586

Yoneyama K, Takeuchi Y, Yokota T (2001) Production of clover broomrape seed germination stimulants by red clover root requires nitrate but is inhibited by phosphate and ammonium. Physiol Plant 112:25–30

Yoneyama K, Xie X, Kusumoto D, Sekimoto H, Sugimoto Y, Takeuchi Y, Yoneyama K (2007a) Nitrogen deficiency as well as phosphorus deficiency in sorghum promotes the production and exudation of 5-deoxystrigol, the host recognition signal for arbuscular mycorrhizal fungi and root parasites. Planta 227:125–132

Yoneyama K, Yoneyama K, Takeuchi Y, Sekimoto H (2007b) Phosphorus deficiency in red clover promotes exudation of orobanchol, the signal for mycorrhizal symbionts and germination stimulant for root parasites. Planta 225:1031–1038

Yoneyama K, Xie X, Sekimoto H, Takeuchi Y, Ogasawara S, Akiyama K, Hayashi H, Yoneyama K (2008) Strigolactones, host recognition signals for root parasitic plants and arbuscular mycorrhizal fungi, from Fabaceae plants. New Phytol 179:484–494

Yoneyama K, Xie X, Yoneyama K, Takeuchi Y (2009) Strigolactones; structures and biological activities. Pest Manag Sci 65:467–470

Yoneyama K, Awad AA, Xie X, Yoneyama K, Takeuchi Y (2010) Strigolactones as germination stimulants for root parasitic plants. Plant Cell Physiol 51:1095–1103

Yoneyama K, Xie X, Kim HI, Kisugi T, Nomura T, Sekimoto H, Yokota T, Yoneyama K (2012) How do nitrogen and phosphorus deficiencies affect strigolactone production and exudation? Planta 235:1197–1207

Zehhar N, Ingouff M, Bouya D, Fer A (2002) Possible involvement of gibberellins and ethylene in *Orobanche ramosa* germination. Weed Res 42:464–469

Zhelev N (1987) The biological role of exogenic factors in broomrape germination. Rastenievudni Nauki 26:36–43 (in Bulgarian)

Zou J, Zhang S, Zhang W, Li G, Chen Z, Zhai W, Zhao X, Pan X, Xie Q, Zhu L (2006) The rice HIGH-TILLERING DWARF1 encoding an ortholog of *Arabidopsis MAX3* is required for negative regulation of the outgrowth of axillary buds. Plant J 48:687–698

Zwanenburg B, Mwakaboko AS, Reizelman A, Anilkumar G, Sethumadhavan D (2009) Structure and function of natural and synthetic signalling molecules in parasitic weed germination. Pest Manag Sci 65:478–491

Chapter 11
Germination Ecophysiology

Alistair J. Murdoch and Ermias Kebreab

11.1 Introduction

For parasitic weeds of the Orobanchaceae to emerge successfully above the soil surface and have the potential to produce viable seeds for the next generation, the seeds must first be after-ripened, conditioned, stimulated and germinate. The seedlings must then locate, attach to and penetrate a host (Kebreab and Murdoch 2001). Next to the fertilisation of the ovule, germination is the primary event in the life cycle of flowering plants and successfully consummates the mother plant's investment in its offspring. It is especially significant in annual parasitic plants such as *Orobanche*, *Phelipanche*, *Alectra* and *Striga* species, most of which disperse and regenerate exclusively by seed. Environmental influences during maturation on the mother plant and subsequently in the soil seed bank interact with genotype and are adapted to increase the probabilities that a seed germinates in "the right place and at the right time" and regenerates successfully, completing a full life cycle (Murdoch and Ellis 2000). The key question in this chapter is: what proportion of viable seeds will regenerate successfully? Models to quantify the probability of a successful outcome are described below, focusing on genotype and environment interactions in the individual seed for which

> To germinate or not to germinate? That is the question!

Seeds of many species are sensitive to environmental germination triggers to ensure seeds germinate at a time and depth of burial from which they may emerge.

A.J. Murdoch (✉)
School of Agriculture, Policy and Development, University of Reading, Earley Gate, P.O. Box 237, Reading, Berkshire RG6 6AR, UK
e-mail: a.j.murdoch@reading.ac.uk

E. Kebreab
Department of Animal Science, University of California, Davis, 2111 Meyer Hall, One Shields Avenue, Davis, CA 95616, USA

D.M. Joel et al. (eds.), *Parasitic Orobanchaceae*,
DOI 10.1007/978-3-642-38146-1_11, © Springer-Verlag Berlin Heidelberg 2013

Such factors may include depth-sensing factors such as light quality and quantity and the amplitude of diurnal temperature fluctuation and seasonal factors such as soil nitrogen status, temperature and soil atmospheric composition (Roberts 1973; Roberts et al. 1987; Murdoch and Ellis 2000). As pointed out, however, by Joel et al. (2006), "the most specialized and intriguing germination mechanisms belong to some parasitic angiosperms that do not germinate until they detect the presence of a host plant". Adaptations which help to ensure some seeds successfully complete their life cycle include their fecundity, the stimulation of germination by host root exudates (Chap. 10) and the preservation of viability for long periods (Teryokhin 1997).

In terms of fecundity, mature *Orobanche* or *Phelipanche* plants may produce more than a quarter of a million (Parker and Riches 1993) very small seeds (0.2–0.7 mm long × 0.2–0.5 mm wide; Budantsev 1993, see Chap. 8 and Sect. 9.1). Since seed production is density dependent (López-Granados and García-Torres 1991; Kebreab and Murdoch 2001), it is useful to evaluate seed production per unit area where, for *O. crenata* on a range of legume hosts in Syria, Schnell et al. (1994) reported 0–619,000 seeds m^{-2} in a dry year (1990, 209 mm rainfall) and 1,100–10.6 million seeds m^{-2} on the same hosts in a wetter year (1991, 316 mm). Typically, 10–25 % of seeds produced may be non-viable (Teryokhin 1997).

Fecundity is slightly lower in *Striga*, with average seed numbers per *S. hermonthica* plant on sorghum of 6,700–26,500 (Rodenburg et al. 2006) and 15,600–18,800 on millet (van Mourik et al. 2008). Individual plants may of course produce much higher or smaller outputs. The maximum *S. hermonthica* seed production per sorghum host plant was 757,000 equivalent to 1.57 million seeds m^{-2} (van Mourik 2007), but seed production varies with host and decreases with decrease in crop duration, season length and intercropping with non-hosts (van Mourik et al. 2008). Some genera in the Orobanchaceae have much larger seeds (see Sect. 9.1). For example, across nine species of *Pedicularis* in Tibet, there is a tendency for seed production per plant to decrease with increase in seed size (Guo et al. 2010). The facultative parasite, *Rhinanthus minor*, also has relatively large seeds (4.9 × 3.8 mm) and produced on average 168 seeds per plant in the UK (Westbury 2004).

11.2 Seed Survival in Dry Storage

Persistence in soil is only possible if seed viability is preserved. Seed storage behaviour is classified as **orthodox** if seed longevity increases in a quantifiable and predictable manner with decrease in temperature and decrease in seed water potential from about −20 MPa to about −350 MPa, whereas **recalcitrant** seeds cannot be dried without losing viability (Ellis and Roberts 1980; Murdoch and Ellis 2000).

O. crenata and *S. hermonthica* seeds may be added to the 81 % of 13,913 species, including 53 from the Orobanchaceae, which exhibit "orthodox" seed storage behaviour (Kebreab and Murdoch 1999b; Murdoch et al. 2000; Royal Botanic Gardens Kew 2008). None of the 57 species of the Orobanchaceae, for which information on seed storage behaviour is documented, is recorded as recalcitrant although storability is deemed "uncertain" for four relatively large-seeded species: *Melampyrum nemorosum*, *M. pratense* L., *Pedicularis sylvatica* and *Rhinanthus aristatus* (Royal Botanic Gardens Kew 2008). Uncertainty in these four cases merely relates to the fact that their seeds have been recorded as surviving for less than 1 year in the soil (Thompson et al. 1997) rather than to definitive studies of seed storage behaviour.

Kebreab and Murdoch (1999b) stored seeds of *O. minor*, *O. crenata* and *P. aegyptiaca* in 13 environments above saturated salt solutions with equilibrium relative humidities (e.r.h.) between 11 and 85 % and at temperatures from 20 to 60 °C. Survival of dry seeds corresponded to negative cumulative normal distributions. Viability after any time t (days), at temperature T (°C), and e.r.h. r (%) was quantified as follows:

$$v = K_{\mathrm{i}} - t/10^{K_{\mathrm{E}} - C_{\mathrm{W}}r - C_{\mathrm{H}}T - C_{\mathrm{Q}}T^2} \quad (11.1)$$

where v is viability expressed in normal equivalent deviates (NED = probit − 5), K_{i} is a seed lot constant, the estimated initial viability of the seed lot (NED), and K_{E}, C_{W}, C_{H} and C_{Q} are species constants. K_{E} is an extrapolated estimate of the seed longevity at 0 % e.r.h. and 0 °C, C_{W} is the sensitivity of longevity to changes in e.r.h. and C_{H} and C_{Q} are temperature coefficients (Kebreab and Murdoch 1999b).

Although the research was only carried out on single seed lots of each species, the sensitivities to changes in e.r.h. and temperature were similar in the three species and are also similar to those of non-parasitic plant species in different families. As a working hypothesis, therefore, it is suggested that the parameter values for (11.1) should be able to be used with some confidence to predict the relative effects of changes in e.r.h. and temperature on seed longevity for all species and all seed lots of Orobanchaceae exhibiting "orthodox" seed storage behaviour. Caveats are that (1) the seed lots come from a single provenance in the same year (i.e. seeds from different sources and years should not be mixed) and (2) the seed lot has always been stored air-dry and has not been subdivided and then remixed so that the seed lot constant, K_{i}, remains valid. If these caveats are met, the time taken for viability to drop by one NED or probit (e.g. from 98 to 84 % or from 84 to 50 %) is doubled for each 8.4 % drop in e.r.h. The temperature term in (11.1) is quadratic, and so the sensitivity to temperature varies with temperature such that the relative benefit to longevity of a decrease in storage temperature decreases with decrease in temperature (the Q_{10} decreases).

What actually differs between species is the value of the "species" constant, K_{E}, which was lower in *O. minor* than in the other two species. This difference means seeds of *O. minor* are predicted to survive only half as long as *O. crenata* and *P. aegyptiaca* (Table 11.1), which may be one of the reasons why the latter two species are more successful as weeds. The higher the initial viability of the seed lot,

Table 11.1 Predicted viability periods

Initial viability	Years for 5 % loss of viability		
% (K_i, in NED)	*Phelipanche aegyptiaca*	*Orobanche crenata*	*Orobanche minor*
98 (2.00)	21	22	10
89 (1.25)	9	10	5
69 (0.50)	5	5	3

The effect of initial viability [K_i, (11.1)] on the predicted period for 5 % viability loss at 5 °C and 55 % equilibrium relative humidity (from Kebreab and Murdoch (1999b) with permission)

the longer the period for a small (5 %) drop in viability (Table 11.1), a consequence of seed survival curves corresponding to negative cumulative normal distributions. In addition, the ability of seeds to survive for several years means that loss of viability is unlikely to limit the longevity of these seeds when in dry soil. The proportion of viable seeds after a period of time in dry storage will, however, vary widely between seed lots of a given species, and the use of the viability equation in the laboratory requires an accurate estimate of initial seed viability, the seed lot constant, K_i.

11.3 Seed Survival in Moist Storage

At water potentials above about −20 MPa and in fully imbibed seeds, (11.1) no longer applies (Roberts and Ellis 1989). As seeds imbibe, metabolism increases, damage may be repaired and microbial attack may be actively resisted. The effect of increasing water potential is thus completely reversed so that seed longevity increases with increase in moisture availability, provided the seeds remain dormant and air is present to allow respiration (Ibrahim et al. 1983; Murdoch and Ellis 2000). Imbibed tissues and hence seed longevity are, however, very vulnerable to high temperatures, and predicted periods for 50 % loss of viability of imbibed seeds of *S. hermonthica*, *O. crenata* and *O. cumana* were, respectively, 98, 60 and 30 h at 45 °C but only 8, 7 and 6 min at 60 °C (calculated from regression coefficients in Dawoud 1995). *S. hermonthica* is, therefore, more tolerant to high temperature and hence should be the most resistant to soil solarisation (see Sect. 22.4.2). At the lower temperatures, likely to characterise most moist soils, fully imbibed seeds of *Orobanche* and *Phelipanche* spp. have been shown to survive for at least 3 months at 30 °C and to show negligible loss of viability over 7 months at temperatures of 10–20 °C provided air is present (Kebreab and Murdoch 1999a, b; Murdoch and Ellis 2000). In comparison to drier seeds whose survival follows the viability equation (11.1), these responses are relatively poorly quantified, but seed survival curves do approximate to negative cumulative normal distributions (Dawoud 1995; Kebreab and Murdoch 1999a, b).

The differences between very dry (< −350 MPa), dry (−350 to −20 MPa) and hydrated (> −20 MPa) seeds may relate to the binding of water (Roberts and Ellis

1989). Water is freely available in hydrated seeds allowing respiration to occur. Conversely, in very dry seeds, water is strongly bound and has little chemical potential, and so its removal has little effect on longevity. In between these extremes, water is weakly bound, and the effects of changes in water potential on longevity can be predicted by (11.1) (Roberts and Ellis 1989).

Whether at high or low water potentials, orthodox seeds of *Orobanche* and *Phelipanche*, if dormant and thus non-germinating, may persist for long periods in the soil seed bank. The main risks to their longevity are probably predation and anaerobiosis. Circumstantial evidence suggesting longevity of 14–20 years in soil for *Orobanche* seed is probably true. For example, López-Granados and García-Torres (1999) reported 98.7 % loss of viability of *O. crenata* seed after 12 years soil burial in Spain. It is also salutary to reflect on the prediction that even such a high percentage seed loss could still leave 3,250 seeds out of 250,000 produced by a single plant. *Orobanche* seed banks exhibit, therefore, "long-term persistence" in the soil seed bank as defined by Thompson et al. (1997) (see Chaps. 19 and 22).

It has generally been assumed that *Striga hermonthica* may similarly survive in the soil for long periods although recent results have been more equivocal. For example, 52 % depletion of a naturally occurring soil seed bank was observed over two wet seasons in Gambia with mono-cropped, unfertilised sorghum (Murdoch and Kunjo 2003), and slow depletion was also reported in Ghana (Sprich 1994). By contrast, a total loss of *S. hermonthica* seed viability occurred after only 3 months burial in Western Kenya (Pieterse et al. 1996). In drier, unimodal rainfall areas[1] of Benin, *S. hermonthica* seed viability declined from 90 to 15 % over 6 months (Gbèhounou et al. 1996a), with almost total loss of viability over two wet seasons (Gbèhounou 1998). Reasons for the discrepancy were resolved by showing that burial of *S. hermonthica* seeds in small mesh packets, as in the Benin and Western Kenyan reports, leads to rapid loss of viability associated with microbial decay (van Mourik et al. 2005).

Several species of hemiparasitic Orobanchaceae only form transient soil seed banks such that their seeds survive for less than 1 year in the soil, for example, *Rhinanthus minor* (Westbury 2004), *Bartsia alpina* L. and the four larger-seeded species mentioned in Sect. 11.2, while others, for example, *Euphrasia stricta* and *E. officinalis*, form short-term persistent seed banks surviving for 1–5 years in the soil (Thompson et al. 1997). In this respect, these hemiparasitic plants do not differ from non-parasitic plants.

[1] One rainy season each year.

11.4 Dormancy and Quiescence

No matter how spectacular the potential longevity, as quantified by the parameter K_E in (11.1), the potential will never be realised if a seed germinates shortly after entering the soil seed bank. Dormancy and/or quiescence are therefore essential for seeds to persist in the soil.

Dormancy is "the failure of a viable seed to germinate given moisture, air and a suitable constant temperature for radicle emergence and seedling growth" (Murdoch 2004). If one of these three requirements for germination of non-dormant seeds is lacking, the seed is arguably quiescent rather than dormant because metabolism will be lower, and growth will resume as soon as the limiting factor is available. As defined, individual seeds either germinate or do not, implying dormancy and quiescence are quantal responses. Unless all or no seeds germinate, germination assays only distinguish two groups of seeds—those which germinate (non-dormant) and those which do not (dormant). Measuring the extent or depth of physiological dormancy in individual seeds is currently impossible, but the variation in percentage germination as a function of the dose of dormancy-breaking treatments reflects variation in dormancy among individuals in the seed population. Like the seed-to-seed variation in periods for which seeds survive, this variation is often well described by the normal distribution function and so can be quantified by the mean and standard deviation of the distribution.

The qualitative distinction between primary and secondary dormancy is useful. **Primary dormancy** (innate dormancy) develops on the mother plant (Murdoch and Ellis 2000). It prevents or reduces both precocious germination and the germination of seeds immediately after shedding. Based on research on various non-parasitic species, the extent of primary dormancy on dispersal from the mother plant is likely to vary widely between seed lots with strong effects of both genotype and the environment. **Secondary dormancy** is induced after shedding (Murdoch and Ellis 2000) and helps to ensure the long-term persistence of the soil seed bank.

Frustration with negative definitions—"failure to germinate"—has led to dormancy being redefined as "a characteristic of the seed that determines the conditions required for germination" (Finch-Savage and Leubner-Metzger 2006) albeit begging the follow-on questions: which characteristic and how do you measure it? This definition also leads to the question: do host root exudates terminate dormancy in obligate parasitic Orobanchaceae or just stimulate their germination? The answer "depends on where one chooses to draw the line between the processes of dormancy and germination". Exposure to the chemical(s) in host root exudates changes the seed, removing a physiological or metabolic block, so that it can germinate at a suitable temperature in water and, sensu stricto, should therefore be regarded as "the last step in the dormancy-breaking process rather than the first step in the germination process" (quotations from Finch-Savage and Leubner-Metzger (2006)). The question is like looking at two sides of a coin. The chemical stimulant relieves the final physiological block to germination, and thus, the dual effect is that it relieves the final block to germination rather like light for some light-sensitive seeds and in so doing, stimulates germination, i.e. provides the first step of germination.

Primary dormancy of fresh, viable seeds of *Orobanche*, *Phelipanche* and *Striga* spp. can be envisaged as composed of three distinct physiological blocks to germination, which need to be relieved sequentially for germination to occur.

First of all, release from primary dormancy generally requires a period of "dry" storage also known as dry after-ripening. After-ripened seeds are not, however, non-dormant in the sense that they will not germinate spontaneously. The mechanism of after-ripening in seeds probably relates to "degradation of mRNAs and proteins for positive regulators of dormancy and for negative regulators of germination", and there is a possibility of some gene expression even though the seeds are, by definition, quiescent (Finch-Savage and Leubner-Metzger 2006).

Secondly, the seeds must undergo a short period of imbibed storage (conditioning), during which their ability to overcome the third block increases. Thirdly, the final block to germination is a requirement for stimulation by one or more chemicals exuded by host plant roots (see Chap. 10).

The following discussion mainly concerns publications where an approximately optimal and standard stimulatory treatment has been used.

11.4.1 Relief of Primary Dormancy

After-Ripening Air-Dry, Quiescent Seeds. Quantitative studies of dormancy loss in *Orobanche* seeds in controlled dry storage conditions, other than at "room" temperature, are lacking, and this has not altered since Pieterse and Verkleij (1994) reported that "hardly any reliable data on the occurrence of primary dormancy in *Orobanche* seeds are available" and that "conclusions drawn with respect to primary dormancy are largely based on speculation". For example, Edwards (1972) reported that several months' storage was required while Saghir (1986) made the unreferenced remark that seeds may require "a period of up to 2 years for after-ripening to occur". Research in non-parasitic plants implies that loss of primary dormancy may occur both on and off the mother plant. Moreover, the loss of primary dormancy is a progressive process in the seed population, and the rate of the process varies in a predictable way as a function of the environment (Murdoch and Ellis 2000).

With respect to environmental factors, seed moisture content is critical, but the rate of loss of primary dormancy during after-ripening varies with both temperature and seed moisture content (see Bazin et al. 2011 and references cited therein). Moisture contents (fresh weight basis) of 5–18 % are required, the rate of dormancy alleviation during dry storage generally decreasing with decrease in moisture content (see Probert 2000). In modelling the life cycle of *O. crenata*, Grenz and Sauerborn (2007) specified a soil moisture threshold of 22.5 % below which both primary and secondary dormancy can be relieved although the soil type and water potential are not known. These processes can clearly occur for seeds in secondary dormancy during warm, dry Mediterranean summers (van Hezewijk et al. 1994b), and equally *O. crenata* could not succeed as a weed where warm, dry conditions never occur (Grenz and Sauerborn 2007).

The influence of temperature on this relief of dormancy was not included by Grenz and Sauerborn (2007) in their model. Circumstantial evidence from the literature suggests, however, that many seeds, including *S. hermonthica* (Murdoch et al. 2000), behave like those members of the Poaceae in which the logarithm of the mean dormancy period is a negative linear function of temperature, the Q_{10} for the relation being typically in the range 2.5–3.8. For example, in *S. hermonthica*, the rate of after-ripening decreases slightly with increase in temperature with a Q_{10} of approx. 3 (Murdoch et al. 2000; Murdoch and Kebreab 2005). If the quantitative effect of temperature on the rate of loss of primary dormancy during dry after-ripening is an approximately conserved trait over species from different families (Murdoch and Ellis 2000), then for species for which experimental evidence is lacking, it would be reasonable to assume a Q_{10} of 3 to predict how periods for loss of primary dormancy would vary with a change in temperature. It follows that there should be an approximate 81-fold increase in the rate of after-ripening at 60 °C as compared to 20 °C. However, it is also important to note that the Q_{10} for loss of viability increases approximately exponentially with increase in temperature (the temperature term in (11.1) being quadratic) compared to a linear temperature term for after-ripening. The net effect is that above certain temperatures, a significant fraction of the seed population may lose viability before it loses dormancy! No information is available on this transition for *Orobanche* or *Phelipanche*, but in the case of *S. hermonthica*, loss of viability becomes significant during after-ripening at about 60 °C and is detectable although not significant at 50 °C (compare Murdoch and Ellis 2000). It is therefore recommended that routine after-ripening for laboratory experiments should *not* be carried out above 40 °C.

The variation in after-ripening periods described in the literature is, however, not only accounted for by differences in storage temperature and moisture. Other factors could include differences within and between species, variation in seed maturation conditions, the different ways in which seeds are processed and differences in germination protocols. The paucity of literature on these topics especially for *Orobanche* and *Phelipanche* spp. may seem surprising given the importance of dormancy in determining the extent of parasitism. These species mainly occur, however, as serious weeds in agro-ecologies which are subject to seasonal drought during which after-ripening will occur naturally. The second phase of dormancy relief, more commonly known as "conditioning" or "preconditioning", is of greater significance in relation to weed control as it occurs in moist soil and immediately precedes germination.

Conditioning Imbibed Seeds. Imbibition is a physical process which activates metabolic pathways in viable seeds. Once imbibed, seeds enter a lag phase when new physiological mechanisms prepare for elongation of the embryonic axis. Completion of seed germination can be temporarily blocked by dormancy, or indeed, dormancy may be relieved during this lag phase. In the Orobanchaceae,

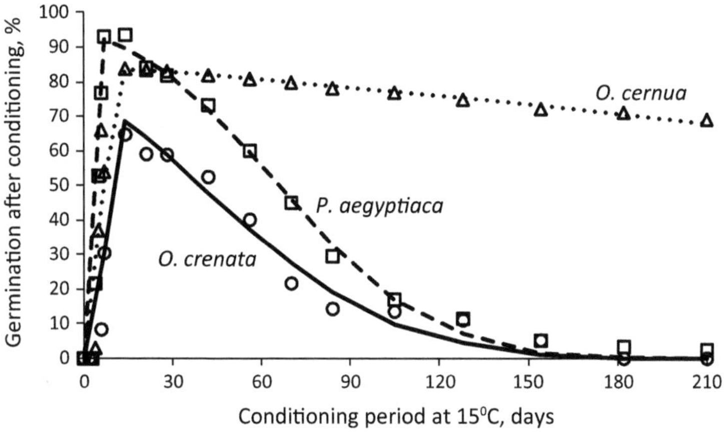

Fig. 11.1 Seed conditioning. Influence of the period of conditioning (at 15 °C) on the seed germination of three parasite species. Germination was stimulated with GR24 and assessed 10 days later (based on data published by Kebreab and Murdoch 1999a)

the process of conditioning has long been recognised as a beneficial seed pretreatment before laboratory experiments (Pieterse 1979). During conditioning, seeds become increasingly sensitive to germination stimulants (Joel et al. 1991; Matusova et al. 2004) over periods of up to about 14 days (Fig. 11.1), the period required decreasing with increase in temperature (Table 11.2). The extrapolated lowest temperatures at which the rate of loss of dormancy is zero, the **base temperature**, varied from −5.4 °C in *O. crenata* to 11.2 °C for *S. hermonthica* (Table 11.2; Vallance 1950). The effect is analogous to stratification in non-parasitic seeds in which seeds show a progressive increase in sensitivity to a subsequent treatment, which overcomes the final block to germination (Murdoch and Ellis 2000).

Respiration of *P. aegyptiaca* increased to a maximum after 3 days of conditioning with a second peak after 11 days and thereafter declined to low levels (Bar-Nun and Mayer 1993; Mayer and Bar-Nun 1994). A similar decline to low levels has also been reported for imbibed, non-germinating seeds of non-parasitic plants (Barton 1945; Ibrahim et al. 1983). Also like some non-parasitic plants (Roberts 1973; Roberts and Smith 1977), respiration prior to germination is at least in part via the cyanide-insensitive respiratory pathway, but this happens in the parasite seeds during early conditioning (Bar-Nun et al. 2003). Interestingly, the final percentage germination of *P. ramosa* did not differ significantly when seeds were conditioned for 14 days at 20 °C with oxygen partial pressures ranging from 1 to 21 % oxygen in the atmosphere (Gibot-Leclerc et al. 2004).

Gene expression occurs actively during conditioning (e.g. Joel et al. 2006) and is perhaps linked to gibberellin biosynthesis (Joel et al. 1991; Song et al. 2005) and other metabolic activities. Exogenous applications of GA_3 increase the subsequent

Table 11.2 Loss of dormancy during conditioning

	Phelipanche aegyptiaca	*Orobanche cernua*	*Orobanche crenata*	*Striga hermonthica*
Base temperature (°C)	0.55	3.42	−5.38	11.2
Approx. median thermal time (°C d)[a]	58.8	69.8	142.9	69.4
Approx. median conditioning periods (days) at:				
10 °C	6.2	10.6	9.3	–
15 °C	4.1	6.0	7.0	18.3
20 °C	3.0	4.2	5.6	7.9
25 °C	2.4	3.2	4.7	5.0
30 °C	2.0	2.6	4.0	3.7
35 °C	–	–	–	2.9

Base temperatures, median thermal times and predicted conditioning periods required at 10–35 °C for loss of dormancy during conditioning

Requirements to sensitise approx. 50 % of parasite seed populations to germination stimulants. Estimates based on parameter values for p_0 and p in (11.2) (Kebreab and Murdoch 1999a) and on comparable figures from Dzomeku and Murdoch (2007b)

[a]Thermal times for loss of dormancy to increase by three normal deviates or probits from −3 NED (0.1 %) to 0 NED (50 %) Thermal times for approx. complete loss of primary dormancy, i.e. from −3 to +3 NED (99.9 %), are twice those to 50 %

germination, whereas including an inhibitor of GA biosynthesis (e.g. 0.01 mg/L uniconazole) has the converse effect on seeds of *P. ramosa*, *P. aegyptiaca* and *O. minor* (Song et al. 2005). In addition, the concentration of ABA decreased from approx. 250 ng/g seed to 100 ng/g during 6 days of conditioning of *O. minor* in darkness at both 23 and 30 °C (Chae et al. 2004; see Sects. 9.2.1 and 9.4 regarding the timing of expression of an ABA catabolic gene in various seed tissues). Carotenoid biosynthesis inhibitors prevent ABA biosynthesis, and two out of three inhibitors tested by Kusumoto et al. (2006), fluridone and norflurazon, reduced the conditioning period required for germination of *Striga asiatica* seeds. Given the importance of both ABA and gibberellin in control of dormancy and germination of many species (e.g. Finch-Savage and Leubner-Metzger 2006; Shepherd et al. 2007), it is not surprising that changes in plant hormones during conditioning are linked to the increase in seed sensitivity to germination stimulants. ABA is catabolised during conditioning (Fig. 9.3c; see Sect. 9.2.1) and Lechat et al. (2012) showed significant up-regulation of the ABA catabolism gene, *PrCYP707A1*, after stimulating conditioned *P. ramosa* seeds with GR24 (Fig. 9.8b; see Sect. 9.4). Slavov et al. (2004) showed IAA was released by *P. ramosa* seeds after treating with germination stimulant.

Despite some conflicting and inconsistent reports, subsequent germination of *O. crenata* increased significantly with increase in pH levels during conditioning for 14 days at 20 °C from 36 % at pH 4 to about 50 % at pH 6–7 and 62 % at pH 8.5 (van Hezewijk et al. 1994a).

Recent research has suggested that while conditioning is a prerequisite for germination of *Striga* spp. and *O. crenata*, it is not in either *P. aegyptiaca* or *O. cumana* provided germination test periods are extended to about 14 days, because seeds of the latter two species achieve high germination percentages whether or not they have been conditioned prior to stimulation with GR24 (Plakhine et al. 2009; Plakhine and Joel 2010). Given the need to prolong the germination period, it could still be argued that conditioning occurred prior to germination; it was just that exposure to the germination stimulant occurred earlier. It should also be noted that conditioned seeds responded better to lower concentrations of stimulant. For example, 53 % of conditioned *P. aegyptiaca* seeds germinated with 10^{-8} M GR24 compared to only 25 % of the unconditioned ones (SED 2.56, Fig. 3 in Plakhine et al. 2009). So, while the earlier literature implied that conditioning was a prerequisite before exposure to stimulant, it is now more appropriate to say it is beneficial, but the sequence of dormancy-relieving treatments is less critical in some species than previously thought.

11.4.2 Secondary Dormancy

Prolonging conditioning beyond the optimum leads to an induction of secondary dormancy (Fig. 11.1). The term “wet dormancy” was introduced by Vallance (1950) for this process in *S. hermonthica*. Induction of secondary dormancy during prolonged moist aerobic treatments also occurs in non-parasitic species of other plant families, e.g. *Rumex* spp. and *Picea sitchensis* (Totterdell and Roberts 1979; Jones et al. 1997). Interestingly, the rate of induction of secondary dormancy in imbibed *R. crispus* seeds decreased with decrease in temperature (from 20 to 1.5 °C, Murdoch and Ellis 2000), whereas the converse occurs in *Orobanche* and *Striga* seeds in which rates of induction of secondary dormancy were respectively fastest during prolonged conditioning at 10 and 17.5 °C and decreased to a minimum above about 20 and 25 °C (Kebreab and Murdoch 1999a; Song et al. 2005; Dzomeku and Murdoch 2007a), which indicates dissimilarity in secondary dormancy mechanisms.

Relief of secondary dormancy is usually achieved in laboratory experiments by drying, reconditioning and restimulating the seeds (Kebreab and Murdoch 1999a). This observation implies some similarity in the mechanism of secondary dormancy to the requirement for after-ripening, which is acquired during maturation on the mother plant (see Sect. 11.4.1).

11.4.3 Modelling Conditioning

Although variation within homogeneous seed populations is normally distributed, non-normal responses are observed during conditioning because up to three processes are occurring during this pre-germination phase, as demonstrated for seeds of *O. cernua*, *O. crenata* and *P. aegyptiaca* (Kebreab and Murdoch 1999a) and *S. hermonthica* (Dzomeku and Murdoch 2007b). Provided that the germination tests were carried out identically, germination (G) after conditioning in water for t days ($0 \leq t \leq 210$ days) at temperature T (from $10 \leq T \leq 30$ °C) is then the product of the proportions of the seed population that have (a) lost primary dormancy (Φ_{d}^{-1}), (b) not entered secondary dormancy (Φ_{s}^{-1}) and (c) at 30 °C only retained viability (Φ_{v}^{-1}), according to the following equation:

$$G = \left[\Phi_{\mathrm{d}}^{-1}(K_{\mathrm{d}} + (p_0 + pT)t)\right]\left[\Phi_{\mathrm{s}}^{-1}(K_{\mathrm{i}} + (s_m + sr^T)t)\right] \times \left[\Phi_{\mathrm{v}}^{-1}(K_{\mathrm{i}} + \beta_{30}t)\right]/\Phi^{-1}(K_{\mathrm{i}}) \tag{11.2}$$

where Φ^{-1} indicates back-transformation from NED to proportions; K_{d} and K_{i} are respectively the initial levels of non-dormancy and viability in NED when $t = 0$ days; p_0 and p quantify a linear increase in the rate of loss of primary dormancy with increase in temperature; s_m, s and r are an asymptotic decrease in the rate of induction of secondary dormancy with increase in temperature; and β_{30} is the rate of loss of viability at 30 °C (Kebreab and Murdoch 1999a). Examples for seeds conditioned at 15 °C are illustrated in Fig. 11.1.

The counter-intuitive inference from (11.2) is that the sensitisation to germination stimulants associated with loss of dormancy during conditioning is both independent and concurrent with the desensitisation to germination stimulants associated with induction of secondary dormancy. The hypothesis of independence and concurrence is based on a similar idea proposed by Totterdell and Roberts (1979) and was used by Grenz and Sauerborn (2007) to model seed bank dynamics of *O. crenata*, and it is consistent with the mechanism of secondary dormancy suggested in Sect. 11.4.2. Vleeshouwers and Bouwmeester (2001) contended, however, that only one process could take place at a time.

Using a much larger set of data in which *S. hermonthica* seeds were conditioned not only at different temperatures but also at different water potentials and in different concentrations of urea, Dzomeku and Murdoch (2007b) preferred a sequential model in which seeds first lose primary dormancy and can only then have secondary dormancy induced in them. As a step to resolving this question, multiplicative and sequential models could be compared statistically on the same datasets.

Water stress and urea during conditioning both reduce subsequent germination of *S. hermonthica* (Dzomeku and Murdoch 2007a, b), while water stress reduced the subsequent germination rate of *P. ramosa* (Gibot-Leclerc et al. 2004) and also final germination percentages of *P. aegyptiaca*, *P. ramosa* and *O. minor* seeds especially if a −2 MPa treatment was applied at relatively low (13 °C) or high (28 °C) conditioning temperatures (Song et al. 2005). Interestingly, the rate of induction of secondary dormancy appeared not to be affected by decrease in water potential (to −2 MPa) during conditioning (Song et al. 2005). For the same three species, Song et al. (2006) demonstrated that germination after prolonged conditioning can be enhanced by the use of the growth promoters, gibberellin (GA_3,

10 mg L^{-1}), fluridone (10 mg L^{-1}) and brassinolide (1 mg L^{-1}), combinations of pairs of these chemicals being particularly effective.

The variation between species (and often also between populations of the same species) in the effects of conditioning is particularly interesting. For example, *P. aegyptiaca* lost dormancy most rapidly and *O. crenata* most slowly (Table 11.2, Fig. 11.1), perhaps reflecting the absolute conditioning requirement of the latter (Plakhine et al. 2009). Lowest temperatures for loss of dormancy during conditioning vary between species, *O. cernua* showing relatively little induction of secondary dormancy compared to *P. aegyptiaca* and *O. crenata* (Fig. 11.1). Similarly, Song et al. (2005) found that *O. minor* was much less susceptible to secondary dormancy compared to *P. aegyptiaca* and *P. ramosa*. Interestingly, Gibot-Leclerc et al. (2004) found induction of secondary dormancy in *P. ramosa* to a greater extent at 30 °C than at lower temperatures, whereas the asymptotic decrease in the rate of induction of secondary dormancy with increase in temperature in (11.2) was based on a minimal induction of secondary dormancy at 30 °C (Kebreab and Murdoch 1999a). A further contrast is found in Song et al.'s (2005) results for *P. ramosa*, which showed little difference in the rate of induction of secondary dormancy over the temperature range 13–28 °C and at water stress levels of 0 to −2 MPa.

It is therefore clear that responses to conditioning are not at all conserved in different species, base temperatures, thermal times and rates of induction of secondary dormancy all varying markedly between species and often also between seed lots of the same species (Table 11.2, Fig. 11.1). Conditioning is therefore likely to be a highly adaptive trait in parasitic seeds. Further research to explore this adaptation in different seed lots of the same species and in different species could assist in predicting the potential spread of these species with climate change.

11.5 From Relief of Dormancy to the Initiation of Germination

If the seed ultimately fails to germinate, adaptations to maintain viability, avoid predation and prevent germination in unfavourable conditions will have been with no advantage as far as the individual seed is concerned. Using the banking analogy, some seeds need to be transferred from a deeply dormant "savings" account to a less or non-dormant "current" account. These latter seeds need to be at a low enough level of dormancy so as to be capable of responding to the chemicals exuded from host plant roots.

11.5.1 Annual Cycles in Dormancy

Annual cycles in which physiologically based dormancy is relieved and induced during the course of a year occurred in buried seeds of *O. crenata* in Syria, Egypt and Spain (Van Hezewijk et al. 1994b; López-Granados and García-Torres 1999) as they do in many non-parasitic annuals with persistent soil seed banks in both temperate and tropical soil environments (Baskin and Baskin 1985, 1998; Benech-Arnold and Sanchez 1995; Murdoch and Ellis 2000). *O. crenata* seeds, which after-ripened in the soil in summer, would then be exposed to some moisture, lose their primary dormancy and become able to germinate in response to germination stimulant in the autumn or winter (Van Hezewijk et al. 1994b). Secondary dormancy was then induced at low winter temperatures (compare Sect. 12.2) followed by its being progressively relieved during warm dry summer conditions, a similar pattern recurring over at least 6 years (López-Granados and García-Torres 1999). Evidence for a comparable annual dormancy cycle in other species of *Orobanche* is not known, but it is speculated that differences in dormancy cycles are likely due, for example, to the obligate requirement for conditioning in *O. crenata*, but not in *P. aegyptiaca* or *O. cumana* (Plakhine et al. 2009; Plakhine and Joel 2010) and differences in rates of induction of secondary dormancy (Fig. 11.1).

This annual dormancy cycle is not found in quiescent *Orobanche* seeds stored dry in controlled constant environments, but in *S. hermonthica*, Gbèhounou (1998) suggested that there was an endogenous dormancy cycle in dry-stored seeds. While such cycles have sometimes been suggested in other species, the absence of control or monitoring of temperature and humidity during laboratory storage means the cycle needs confirmation. For seeds buried in the soil, changes in dormancy were, however, evident ensuring the seeds were ready to germinate at the end of the dry season (Gbèhounou et al. 1996b).

11.6 Germination

For species which respond to germination stimulants, the percentage germination achievable varies with the amounts and type of germination stimulants exuded by the host (see Chap. 10). In the field, the concentration sensed by the seed is a function of soil moisture and proximity of the seed to active host roots, and these are also the most important factors determining whether or not germination occurs. Many factors influence the proportion of seeds which germinate at any given stimulant concentration. Promotion of suicidal germination (see Sect. 22.4.5) by application of stimulants to soil has achieved some success in glasshouse experiments (Kgosi et al. 2012), but not so far in commercial agricultural systems. Research supplementing synthetic stimulants with growth promoters may improve the reliability of the treatment (Song et al. 2006; see Sect. 10.4.1).

Whether the goal is the stimulation of suicidal germination or prediction of infestations (Sect. 23.4), research has quantified germination responses in relation to both **germination rate** (germination per unit time) and final **percentage germination**.

Soil fertility and in particular the presence of ammonium ions or urea tend to be inhibitory (Pieterse 1996). In studies of five weedy parasitic *Orobanche* species, Westwood and Foy (1999) showed that while ammonium ions inhibit germination slightly, their main effect was to reduce radicle elongation after germination, approximately 80 % reduction in radicle length occurring after 10 days in 25 mM ammonium sulphate compared to the water control. Temperature and moisture are the main environmental factors that determine germination of non-dormant seeds in the field. Water stress often limits germination (Benech-Arnold and Sanchez 1995), while temperature not only influences germination through its effects on seed viability and dormancy but also affects the germination process itself.

11.6.1 Effects of Fluctuating and Constant Temperatures

Relief of dormancy and promotion of germination by fluctuating temperatures in non-parasitic plants have been interpreted ecologically as depth and seasonal sensing mechanisms of small seeds, which increase the likelihood of successful regeneration (Roberts et al. 1987). Seeds of the small-seeded, parasitic members of the Orobanchaceae need to detect their proximity to the roots of host plants rather than the soil surface, and so a positive response to fluctuating temperatures might actually be selected against, as it would indicate proximity to the soil surface where density of host roots might be lower.

Racovitza (1959) suggested that germination of seeds of *P. ramosa* may be enhanced by a 6–8 °C fluctuation of temperature. Strictly, however, to demonstrate unequivocally that alternating temperatures affect germination, the constant temperature controls should include the mean, minimum and maximum temperatures of the alternating temperature regime. Comparing germination at a 15/25 °C (12 h/12 h) alternation with the mean temperature of the regime (20 °C) as a constant temperature control, Van Hezewijk (1994) found a deleterious effect of the alternating temperature regime on germination of *O. crenata*.

The use of a temperature gradient plate by Kebreab and Murdoch (1999c) allowed testing a total of 78 different combinations of daily minimum and maximum temperature regimes, each with either 8 or 16 h/day at the maximum/minimum temperature. Germination after up to 30 days was compared with 13 constant temperature controls for non-dormant seeds of *P. aegyptiaca*, *O. cernua*, *O. crenata* and *O. minor* and for *P. aegyptiaca* and *O. crenata* seeds with secondary dormancy. The expected absence of a positive response to temperature fluctuation was confirmed, and indeed, as observed by Van Hezewijk (1994), alternating temperatures tended to be deleterious. Alternating temperatures never increased germination compared to the corresponding mean temperatures, and at wider amplitudes, alternating temperatures decreased germination percentage. Four characteristics of the temperature regime (whether constant or alternating) appeared to affect germination percentage: (a) the mean temperature, (b) the maximum temperature, (c) the temperature range or amplitude and (d) the period spent at the maximum temperature each day, the latter two modifying the effect of maximum temperature (Kebreab and Murdoch 1999c).

Variation in germination responses could be accounted for by a multiplicative probability model which implied that there were two cardinal temperatures determining whether or not an individual seed would germinate in a given regime: (a) a mean temperature, which had to be exceeded, and (b) a maximum temperature, which must not be exceeded. These two limits were modelled on the bases that they were independent in individual seeds and were normally distributed in the seed population. In general, a very sharp decline in germination was noted in temperature regimes in which one or both temperatures exceeded 25 °C, indicating that the adverse effect of alternating temperatures was mainly due to the maximum temperature (Kebreab and Murdoch 1999c). Eizenberg et al. (2003) reported a similarly dramatic effect in seeds of *O. cumana* whose germination decreased from 88 to 38 % with only a 3 °C increase in mean temperature at alternating temperatures of 26/18 °C and 29/21 °C, respectively (14 h/10 h).

The acceptance of a model with two cardinal temperatures implies that individual seeds vary in the temperature range over which they will germinate. Integration of these ranges gives an overall range of temperature, final germination percentage varying within that range (Table 11.3). The overall range itself varies between species and may also vary between populations of a given species.

It is also important to recognise that the optimal temperature for maximum final germination percentage is much lower than the optimal temperature for rate of germination. Water stress narrows the temperature range for germination (Kebreab and Murdoch 2000). For example, germination of *P. aegyptiaca* exceeded 90 % from 11 to 29 °C in water but only 70–75 % over the range of 17–21 °C at −1.25 MPa (Kebreab and Murdoch 2000).

11.6.2 Thermal Time and the Rate of Germination in the Laboratory and Field

The success of germination in the field is often linked to the time it takes for seeds to germinate or the reciprocal of time, the rate of germination. Progress of germination in a seed population has commonly been described by **thermal time** or **hydrothermal time** models, the latter taking account of both temperature and water potential (see Kebreab and Murdoch 1999d; Bradford 2002; Allen et al. 2007). Thermal time models have been developed to simulate seed bank dynamics (Grenz et al. 2008) and subsequent development (Manschadi 1999; Ephrath and Eizenberg 2010).

Table 11.3 Optimum temperatures and thermal times for germination of *Phelipanche* and *Orobanche* species

	Temperature range for germination (°C)	Temperature for maximum final germination (°C)		Temperature for maximum rate of germination (°C)	Median thermal time for germination (°C d)	Standard deviation of thermal time for germination (°C d)
		No dormancy	Secondary dormancy			
P. aegyptiaca	5–32	18–21	17.5–19.5	29	52.4	15.2
O. crenata	8–29	18	17.5	29	52.0	20.4
O. cernua	5–32	15	–	29	47.0	15.6
O. minor	8–29	17–20	–	26–29	37.6	10.4
O. cumana	5–32	–	–	26	46.4	10.5

Based on temperature gradient experiments (Kebreab and Murdoch 1999c), germination tests continued for 30–50 days according to species and temperature (Kebreab and Murdoch 1998, 1999d). Median thermal times in °C days (°C d) are estimates based on a common base temperature of 4.9 °C

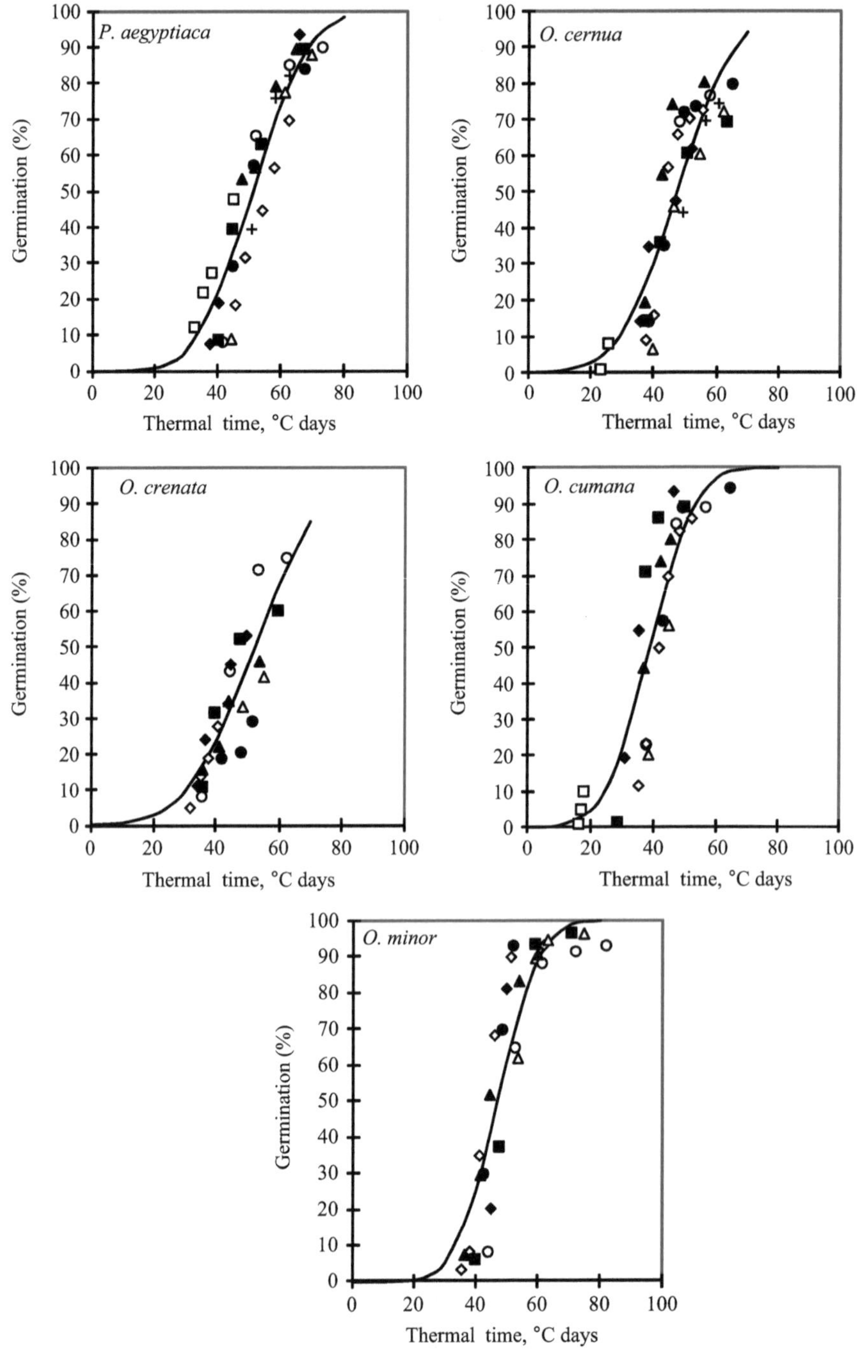

Fig. 11.2 Thermal times for progress of germination of different parasites. Progress of germination at various temperatures: 5 °C (*open squares*), 8 °C (*open diamonds*), 11 °C (*open triangles*),

Thermal time for progress of germination is approximately normally distributed within a seed lot (Covell et al. 1986; Ellis et al. 1986; Kebreab and Murdoch 1999d, Fig. 11.2) and is always calculated above the estimated base temperature at which, by extrapolation, the rate of germination is predicted to be zero. In *S. hermonthica*, the base and optimum temperatures for rate of germination were 22.7 and 40 °C in seeds conditioned at 20 °C compared with 18.9 and 32–35 °C for seeds conditioned at 30 °C (Aflakpui et al. 1998).

The variation in base and optimal temperatures with conditioning temperature is particularly interesting because Ellis and Butcher (1988) found that base temperature varied little between autumn and spring sown onion seeds and suggested it was probably a species characteristic, i.e. every individual seed of every seed lot of a given species has the same base temperature. In studies of germination rates of *P. aegyptiaca*, *O. crenata*, *O. cernua*, *O. minor* and *O. cumana*, separate base temperatures for each species were tabulated by Kebreab and Murdoch (1998), and further analysis showed that a base temperature of 4.9 °C was not only constant within each seed population, but it did not differ significantly between all five species (Kebreab and Murdoch 1998). Supporting evidence comes from Van Hezewijk et al. (1991), who reported a base temperature of approx. 5 °C for *O. crenata*, and Grenz et al. (2008) who used 5 °C for *O. cumana*. Eizenberg et al. (2012) achieved excellent predictions of attachments of *O. cumana* to sunflower roots using the base temperature of the sunflower host (4 °C) while noting that models of attachment are only applicable above base temperatures for growth of both parasite (5 °C) and host (see Sect. 23.4). The germination (*G* in NED or probits) after t_g (days) of non-dormant seeds of these five species when germinated at a mean temperature, *T*, with 10^{-5} M GR24 at a water potential of 0 MPa (water) can thus be predicted as follows:

$$\text{Probit}\ (G) = [t_g(T - T_b) - \theta_{T(50)}]/\sigma_{\theta_T} \qquad (11.3)$$

where T_b is the base temperature, $\theta_{T(50)}$ is the median thermal time to germination in water (Table 11.3) and σ_{θ_T} is the standard deviation of thermal times (°C d, Table 11.3, Fig. 11.2).

Base temperature and thermal time both increase, however, as a linear function of the decrease in water potential (Kebreab and Murdoch 1999d). The effect of conditioning temperature on base temperature of *S. hermonthica* (Aflakpui et al. 1998) emphasises the caution needed. It is not known how these effects of water potential differ between seed lots or species. The highest water potential at which seeds of *P. aegyptiaca* were predicted to be unable to germinate (the base water potential at which the rate of germination is predicted to be zero) varied, however,

Fig. 11.2 (continued) 14 °C (*open circles*), 17 °C (*filled squares*), 20 °C (*filled diamonds*), 23 °C (*filled triangles*), 26 °C (*filled circles*) and 29 °C (*plus symbols*), using a common base temperature of 4.9 °C for all species. Normal distributions were fitted according to (11.3) (reanalysis of data in Kebreab and Murdoch 1998)

between individual seeds (mean −1.96 MPa and standard deviation 0.33 MPa at 20 °C). The median base water potential was approx. −2 MPa at 14–23 °C and increased at both higher and lower temperatures, meaning *P. aegyptiaca* seeds were more susceptible to water stress when also subjected to either low or high temperature stress.

11.7 Conclusion

Adaptations to holo- and hemiparasitism are seen in many chapters of this book, not least in seed characteristics. All aspects of the seed cycle can be seen as maximising the likelihood of successful attachment to a host plant, but variations between species have also been evident. These variations in combination with those regarding the response to germination stimulation facilitate adaptations to different agroecologies and different hosts, since germination is the primary event driving infestations of annual weeds. The concepts, results and models considered in this chapter have important ramifications for integrated weed management (see Chaps. 22 and 23). As an example, changes in the dormancy of buried seeds underpin delayed sowing practices which exploit induction of secondary dormancy, and so there is clearly little point in using such approaches for species such as *O. cernua* and *O. minor* in which secondary dormancy is induced slowly.

Despite the small size of the seeds, they also seem well adapted to form a persistent soil seed bank, survival of at least some seeds until conditions are right for germination being assured by their fecundity, small size (avoiding predation) and their ability to maintain viability while preventing germination by dormancy.

References

Aflakpui GKS, Gregory PJ, Froud-Williams RJ (1998) Effect of temperature on seed germination rate of *Striga hermonthica* (Del.) Benth. Crop Prot 17:129–133

Allen PS, Benech-Arnold RL, Batlla D, Bradford KJ (2007) Modeling of seed dormancy. In: Bradford KJ, Nonogaki H (eds) Annual plant reviews: seed development, dormancy and germination, vol 27. Blackwell, Oxford, pp 72–112

Bar-Nun N, Mayer AM (1993) Preconditioning and germination of *Orobanche* seeds: respiration and protein synthesis. Phytochemistry 34:39–45

Bar-Nun N, Plakhine D, Joel DM, Mayer AM (2003) Changes in the activity of the alternative oxidase in *Orobanche* seeds during conditioning and their possible physiological function. Phytochemistry 64:235–241

Barton LV (1945) Respiration and germination studies of seeds in moist storage. Ann N Y Acad Sci 46:185–200

Baskin JM, Baskin CC (1985) The annual dormancy cycle in buried weed seeds: a continuum. Bioscience 35:492–498

Baskin CC, Baskin JM (1998) Seeds: ecology, biogeography and evolution of dormancy and germination. Academic, San Diego, CA, p 666

Bazin J, Batlla D, Dussert S, El-Maarouf-Bouteau H, Bailly C (2011) Role of relative humidity, temperature, and water status in dormancy alleviation of sunflower seeds during dry after-ripening. J Exp Bot 62:627–640

Benech-Arnold RL, Sanchez RA (1995) Modelling weed seed germination. In: Kigel J, Galili G (eds) Seed development and germination. Marcel-Dekker, New York, pp 545–566

Bradford KJ (2002) Applications of hydrothermal time to quantifying and modeling seed germination and dormancy. Weed Sci 50:248–260

Budantsev LY (1993) Determinator of broomrapes of the USSR flora. Nauka, St Petersburg, Russia

Chae SH, Yoneyama K, Takeuchi Y, Joel DM (2004) Fluridone and norflurazon, carotenoid-biosynthesis inhibitors, promote seed conditioning and germination of the holoparasite *Orobanche minor*. Physiol Plant 120:328–337

Covell S, Ellis RH, Roberts EH, Summerfield RJ (1986) The influence of temperature on seed germination rate in grain legumes I. A comparison of chickpea, lentil, soyabean and cowpea at constant temperatures. J Exp Bot 37:705–715

Dawoud DA (1995) Autecological studies on phytoparasites of the genera *Alectra*, *Orobanche* and *Striga* with special reference to their early developmental stages. PLITS 13:1–112

Dzomeku IK, Murdoch AJ (2007a) Effects of prolonged conditioning on dormancy and germination of *Striga hermonthica*. J Agron 6:29–36

Dzomeku IK, Murdoch AJ (2007b) Modelling effects of prolonged conditioning on dormancy and germination of *Striga hermonthica*. J Agron 6:235–249

Edwards WGH (1972) *Orobanche* and other plant parasite factors. In: Harborne JB (ed) Phytochemical ecology. Academic, New York, pp 235–248

Eizenberg H, Plakhine D, Hershenhorn J, Kleifeld Y, Rubin B (2003) Resistance to broomrape (*Orobanche* spp.) in sunflower (*Helianthus annuus* L.) is temperature dependent. J Exp Bot 54:1305–1311

Eizenberg H, Hershenhorn J, Achdari G, Ephrath JE (2012) A thermal time model for predicting parasitism of *Orobanche cumana* in irrigated sunflower - field validation. Field Crops Res 137:49–55

Ellis RH, Butcher PD (1988) The effects of priming and 'natural' differences in quality amongst onion seed lots on the response of the rate of germination to temperature and identification of the characteristics under genotypic control. J Exp Bot 39:935–950

Ellis RH, Roberts EH (1980) Improved equations for the prediction of seed longevity. Ann Bot 45:13–30

Ellis RH, Covell S, Roberts EH, Summerfield RJ (1986) The influence of temperature on seed germination rate in grain legumes II. Intraspecific variation in chickpea (*Cicer arietinum* L.). J Exp Bot 37:1503–1515

Ephrath JE, Eizenberg H (2010) Quantification of the dynamics of *Orobanche cumana* and *Phelipanche aegyptiaca* parasitism in confectionery sunflower. Weed Res 50:140–152

Finch-Savage WE, Leubner-Metzger G (2006) Seed dormancy and the control of germination. New Phytol 171:501–523

Gbèhounou G (1998) Seed ecology of *Striga hermonthica* in the Republic of Benin: host specificity and control potentials. PhD thesis, Vrije Universiteit, Amsterdam, 126 pp

Gbèhounou G, Pieterse AH, Verkleij JAC (1996a) The decrease in seed germination of *Striga hermonthica* in Benin in the course of the rainy season is due to a dying-off process. Experientia 52:264–267

Gbèhounou G, Pieterse AH, Verkleij JAC (1996b) A study on primary dormancy of *Striga hermonthica* seeds in the northern provinces of the Republic of Benin. In: Moreno MT, Cubero JI, Berner D, Joel D, Musselman LJ, Parker C (eds) Advances in parasitic plant research. Junta de Andalucia, Consejería de Agricultura y Pesca, Sevilla, pp 273–282

Gibot-Leclerc S, Corbineau F, Sallé G, Côme D (2004) Responsiveness of *Orobanche ramosa* L. seeds to GR 24 as related to temperature, oxygen availability and water potential during preconditioning and subsequent germination plant growth regulation. Plant Growth Regul 43:63–71

Grenz J, Sauerborn J (2007) Mapping the potential geographical range of the parasitic weed *Orobanche crenata*. Agric Ecosyst Environ 122:275–281

Grenz JH, Istoc VA, Manschadi AM, Sauerborn J (2008) Interactions of sunflower (*Helianthus annuus*) and sunflower broomrape (*Orobanche cumana*) as affected by sowing date, resource supply and infestation level. Field Crops Res 107:170–179

Guo H, Mazer SJ, Du G (2010) Geographic variation in seed mass within and among nine species of *Pedicularis* (Orobanchaceae): effects of elevation, plant size and seed number per fruit. J Ecol 98:1232–1242

Ibrahim AE, Roberts EH, Murdoch AJ (1983) Viability of lettuce seeds. II. Survival and oxygen uptake in osmotically controlled storage. J Exp Bot 34:631–640

Joel DM, Back A, Kleifeld Y, Gepstein S (1991) Seed conditioning and its role in *Orobanche* seed germination: inhibition by paclobutrazol. In: Wegmann K, Musselman LJ (eds) Progress in *Orobanche* research. Proceedings of the international workshop on *Orobanche* research. Eberhard-Karls-Universität, Tübingen, pp 147–156

Joel DM, Bar H, Mayer AM, Welbaum G, Westwood J (2006) Characterization of a dioxygenase gene with a potential role in steps leading to germination of the root parasite *Orobanche aegyptiaca*. In: Adkins S, Ashmore S, Navie S (eds) Seeds: biology, development and ecology. CAB International, Wallingford, UK, pp 296–306

Jones SK, Ellis RH, Gosling PG (1997) Loss and induction of conditional dormancy in seeds of Sitka spruce maintained moist at different temperatures. Seed Sci Res 7:351–358

Kebreab E, Murdoch AJ (1998) Thermal time models for rate of germination of five *Orobanche* species. In: Wegmann K, Musselman LJ, Joel, DM (eds) Current problems of *Orobanche* research. Proceedings of the 4th international workshop on *Orobanche* research, Albena, Bulgaria. Institute for Wheat and Sunflower, General Toshevo, Bulgaria, pp 59–64

Kebreab E, Murdoch AJ (1999a) A quantitative model for loss of primary dormancy and induction of secondary in imbibed seeds of *Orobanche* spp. J Exp Bot 50:211–219

Kebreab E, Murdoch AJ (1999b) Effect of moisture and temperature on the longevity of *Orobanche* seeds. Weed Res 39:199–211

Kebreab E, Murdoch AJ (1999c) Modelling the germination *Orobanche* seeds at a wide range of alternating and constant temperatures. Ann Bot 84:549–557

Kebreab E, Murdoch AJ (1999d) Modelling the effects of water stress and temperature on germination rate of *Orobanche aegyptiaca* seeds. J Exp Bot 50:655–664

Kebreab E, Murdoch AJ (2000) The effect of water stress on temperature range for germination of *Orobanche aegyptiaca* seeds. Seed Sci Res 10:127–133

Kebreab E, Murdoch AJ (2001) Simulation of integrated control strategies for *Orobanche* spp. based on a life cycle model. Exp Agric 37:37–51

Kgosi RL, Zwanenburg B, Mwakaboko AS, Murdoch AJ (2012) Strigolactone analogues induce suicidal germination of *Striga* spp in soil. Weed Res 52:197–203

Kusumoto D, Chae SH, Mukaida K, Yoneyama K, Yoneyama K, Joel DM, Takeuchi Y (2006) Effects of fluridone and norflurazon on conditioning and germination of *Striga asiatica* seeds. Plant Growth Regul 48:73–78

Lechat MM, Pouvreau JB, Péron T, Gauthier M, Montiel G, Véronési C, Todoroki Y, Le Bizec B, Monteau F, Macherel D, Simier P, Thoiron S, Delavault P (2012) *PrCYP707A1*, an ABA catabolic gene, is a key component of *Phelipanche ramosa* seed germination in response to the strigolactone analogue GR24. J Exp Bot 63:5311–5322

López-Granados F, García-Torres L (1991) Estimation of *Orobanche crenata* seed production. In: Wegmann K, Musselman LJ (eds) Progress in *Orobanche* research. Proceedings of the international workshop on *Orobanche* research. Eberhard-Karls-Universität, Tübingen, pp 92–98

López-Granados F, García-Torres L (1999) Longevity of Crenate Broomrape (*Orobanche crenata*) seed under soil and laboratory conditions. Weed Sci 47:161–166

Manschadi AM (1999) Modelling the growth and development of Faba Bean (*Vicia faba* L.) infested with the parasitic weed *Orobanche crenata* Forsk. APIA, Germany, p 128

Matusova R, van Mourik T, Bouwmeester HJ (2004) Changes in the sensitivity of parasitic weed seeds to germination stimulants. Seed Sci Res 14:335–344

Mayer AM, Bar-Nun N (1994) Metabolism during conditioning and germination of *Orobanche aegyptiaca*. In: Pieterse AH, Verkleij JAC, ter Borg S (eds) Biology and management of *Orobanche*. Proceedings of the 3rd international workshop on *Orobanche* and related *Striga* research. Royal Tropical Institute, Amsterdam, pp 146–156

Murdoch AJ, Kunjo EM (2003) Depletion of natural soil seedbanks of *Striga hermonthica* in West Africa under different integrated management regimes. In: Bekker RM, Forcella F, Grundy AC, Jones NE, Marshall EJP, Murdoch AJ (eds) Seedbanks: determination, dynamics and management. Aspects of Applied Biology, vol 69. Association of Applied Biologists, Warwickshire, pp 261–268

Murdoch AJ (2004) Seed dormancy. In: Goodman RM (ed) Encyclopedia of plant and crop science. Marcell Dekker, New York, pp 1130–1133

Murdoch AJ, Ellis RH (2000) Dormancy, viability and longevity. In: Fenner M (ed) Seeds: the ecology of regeneration and plant communities, 2nd edn. CAB International, Wallingford, pp 183–214

Murdoch AJ, Kebreab E (2005) Temperature dependence of *Orobanche* germination and implications for the northward spread of *Orobanche* within Europe. In: COST Action 849, Parasitic plant management in sustainable agriculture, WG1+4 Workshop on means for limiting *Orobanche* propagation and dispersal in agricultural fields, Newe-Yaar Research Center, Israel, pp 15–16

Murdoch AJ, Sonko L, Kebreab E (2000) Population responses to temperature for loss and induction of seed dormancy and consequences for predictive empirical modelling. In: Viémont J-D, Crabbé J (eds) Dormancy in plants: from whole plant behaviour to cellular control. Proceedings of the 2nd international symposium on plant dormancy, Angers, France. CAB International, Wallingford, pp 57–68

Parker C, Riches CR (1993) Parasitic weeds of the world – biology and control. CABI, Wallingford

Pieterse AH (1979) The broomrapes (Orobanchaceae) – a review. Abstr Trop Agric 5:9–35

Pieterse AH (1996) The effect of nitrogen on *Orobanche* and *Striga* – state of the art. In: Moreno MT, Cubero JI, Berner D, Joel D, Musselman LJ, Parker C (eds) Advances in parasitic plant research. Junta de Andalucia, Consejería de Agricultura y Pesca, Sevilla, pp 273–282

Pieterse AH, Verkleij JAC (1994) Germination ecology of *Striga* and *Orobanche* – an overview. In: Pieterse AH, Verkleij JAC, ter Borg S (eds) Biology and management of *Orobanche*. Proceedings of the 3rd international workshop on *Orobanche* and related *Striga* Research. Royal Tropical Institute, Amsterdam, pp 36–48

Pieterse AH, Verkleij JAC, den Hollander NG, Odhiambo GD, Ransom JK (1996) Germination and viability of *Striga hermonthica* seeds in western Kenya in the course of the long rainy season. In: Moreno MT, Cubero JI, Berner D, Joel D, Musselman LJ, Parker C (eds) Advances in parasitic plant research. Junta de Andalucia, Consejería de Agricultura y Pesca, Sevilla, pp 457–464

Plakhine D, Joel DM (2010) Ecophysiological consideration of *Orobanche cumana* germination. HELIA 33(52):13–18

Plakhine D, Ziadna H, Joel DM (2009) Is seed conditioning essential for *Orobanche* germination? Pest Manag Sci 65:492–496

Probert RJ (2000) The role of temperature in the regulation of seed dormancy and germination. In: Fenner M (ed) Seeds: the ecology of regeneration and plant communities, 2nd edn. CAB International, Wallingford, pp 261–292

Racovitza A (1959) Nouvelles observations sur la germination des graines d'*Orobanche ramosa* L. Journal d'Agriculture Tropical Botanique Appliqué 6:111–114

Roberts EH (1973) Oxidative processes and the control of seed germination. In: Heydecker W (ed) Seed ecology. Butterworths, London, pp 189–218

Roberts EH, Ellis RH (1989) Water and seed survival. Ann Bot 63:39–52

Roberts EH, Smith RD (1977) Dormancy and the pentose phosphate pathway. In: Khan AA (ed) The physiology and biochemistry of seed dormancy and germination. Elsevier North-Holland Biomedical Press, Amsterdam, pp 385–411

Roberts EH, Murdoch AJ, Ellis RH (1987) The interaction of environmental factors on seed dormancy. In: Proceedings of 1987 British crop protection conference – weeds, pp 687–694

Rodenburg J, Bastiaans L, Kropff MJ, Van Ast A (2006) Effects of host plant genotype and seedbank density on *Striga* reproduction. Weed Res 46:251–263

Royal Botanic Gardens Kew (2008) Seed information database. Version 7.1. Available from: http://data.kew.org/sid/. Accessed February 2012

Saghir AR (1986) Dormancy and germination of *Orobanche* seeds in relation to control methods. In: ter Borg S (ed) Proceedings of workshop on biology and control of *Orobanche*, Wageningen, pp 25–34

Sauerborn J, Kroschel J (1996) Underrated methods of weed control and their use in the agriculture of developing countries. In: Proceedings of the 2nd international weed science congress, Copenhagen, Denmark, pp 611–621

Schnell H, Sauerborn J, Linke KH (1994) Crop rotation in Syria and its impact on *Orobanche crenata* seed bank. In: Pieterse AH, Verkleij JAC, ter Borg S (eds) Biology and management of *Orobanche*. Proceedings of the 3rd international workshop on *Orobanche* and related *Striga* research. Royal Tropical Institute, Amsterdam, pp 644–651

Shepherd SK, Murdoch AJ, Dunwell JM, Lutman PJ (2007) A model of seed dormancy in wild oats (*Avena fatua*) for investigating genotype × environment interactions. In: Adkins S, Ashmore S, Navie S (eds) Seeds: biology, development and ecology. CAB International, Wallingford, UK, pp 342–353

Slavov S, van Onckelen H, Batchvarova R, Atanassov A, Prinsen E (2004) IAA production during germination of *Orobanche* spp. seeds. J Plant Physiol 161:847–853

Song WJ, Zhou WJ, Jin ZL, Cao DD, Joel DM, Takeuchi Y, Yoneyama K (2005) Germination response of *Orobanche* seeds subjected to conditioning temperature, water potential and growth regulator treatments. Weed Res 45:467–476

Song WJ, Zhou WJ, Jin ZL, Zhang D, Yoneyama K, Takeuchi Y, Joel DM (2006) Growth regulators restore germination of *Orobanche* seeds that are conditioned under water stress and suboptimal temperature. Aust J Agric Res 57:1195–1201

Sprich H (1994) Bedeutung der Fruchtfolge zur Ertragssicherung in getreidebetonten Produktionssystemen der Guinea-Savanne unter besonderer Berücksichtigung des parasitischen Unkrautes *Striga hermonthica* (Del.) Benth. PLITS 12(3), cited by Sauerborn and Kroschel (1996)

Teryokhin E (1997) Weed broomrapes – systematics, ontogenesis, biology and evolution. Aufstieg, Germany

Thompson K, Bakker JP, Bekker RM (1997) Soil seed banks of North West Europe: methodology, density, longevity. Cambridge University Press, Cambridge, p 276

Totterdell S, Roberts EH (1979) Effects of low temperatures on the loss of innate dormancy and the development of induced dormancy in seeds of *Rumex obtusifolius* L. and *Rumex crispus* L. Plant Cell Environ 2:131–137

Vallance KB (1950) Studies on the germination of the seeds of *Striga hermonthica*. I. The influence of moisture-treatment, stimulant-dilution and after-ripening on germination. Ann Bot 14:347–363

Van Hezewijk MJ, Verkleij JAC, Pieterse AH (1991) Temperature dependence of germination in *Orobanche crenata*. In: Wegmann K, Musselman LJ (eds) Progress in *Orobanche* research. Proceedings of the international workshop on *Orobanche* research. Eberhard-Karls-Universität, Tübingen, pp 125–133

Van Hezewijk MJ (1994) Germination ecology of *Orobanche crenata* – implications for cultural control measures. PhD Dissertation, Vrije Universiteit, Amsterdam

Van Hezewijk MJ, Verkleij JAC, Pieterse AH (1994a) The effect of pH on germination of *Orobanche crenata*. In: Pieterse AH, Verkleij JAC, ter Borg S (eds) Biology and management of *Orobanche*. Proceedings of the 3rd international workshop on *Orobanche* and related *Striga* research. Royal Tropical Institute, Amsterdam, pp 173–179

Van Hezewijk MJ, Linke K-H, Lopez-Granados F, Al-Menoufi OA, Garcia-Torres L, Saxena MC, Verkleij JAC, Pieterse AH (1994b) Seasonal changes in germination response of buried seeds of *Orobanche crenata* Forsk. Weed Res 34:369–376

Van Mourik TA (2007) *Striga hermonthica* seed bank dynamics: process quantification and modelling. PhD Thesis, Wageningen University, 123 pp

Van Mourik TA, Stomph TJ, Murdoch AJ (2005) Why high seed densities within buried mesh bags may overestimate depletion rates of soil seed banks. J Appl Ecol 42:299–305

Van Mourik TA, Bianchi FJJA, Van der Werf W, Stomph TJ (2008) Long-term management of *Striga hermonthica*: strategy evaluation with a spatio-temporal population model. Weed Res 48:329–339

Vleeshouwers LM, Bouwmeester HJ (2001) A simulation model for seasonal changes in dormancy and germination of weed seeds. Seed Sci Res 11:77–92

Westbury DB (2004) *Rhinanthus minor* L. J Ecol 92:906–927

Westwood JH, Foy CL (1999) Influence of nitrogen on germination and early development of broomrape (*Orobanche* spp.). Weed Sci 47:2–7

Chapter 12
Are Karrikin Signaling Mechanisms Relevant to Strigolactone Perception?

David C. Nelson

12.1 Introduction

The mechanisms underlying host-induced germination of Orobanchaceae have remained elusive for lack of a genetically tractable system. Recent investigations using *Arabidopsis thaliana* to explore the mode of action of karrikins, a class of germination stimulants found in smoke that are structurally related to strigolactones (Fig. 12.1), have revealed that karrikin and strigolactone signaling involve common genetic components. Here we discuss the potential significance of these findings for understanding parasitic weed germination.

12.2 Karrikins, Germination Stimulants Found in Smoke

Fire is destructive, but it also provides a prime opportunity for plant species that have evolved post-fire germination mechanisms to rapidly recolonize a burnt area. While heat from fire causes release of physical dormancy for seeds of some species (e.g., fire-mediated serotiny), the chemical signals found in smoke can also trigger germination (De Lange and Boucher 1990). Smoke application in the absence of heat, either as an aerosol or as a smoke-water solution, is a highly effective germination stimulant (Roche et al. 1997). Seeds of over 1,200 species from 80 genera, in a diverse array of ecosystems, are known to positively respond to smoke (Chiwocha et al. 2009; Dixon et al. 2009).

The bioactive chemicals in smoke remained unknown until 2004, when a butenolide now known as KAR_1, 3-methyl-2*H*-furo[2,3-*c*]pyran-2-one, was

D.C. Nelson (✉)
Department of Genetics, University of Georgia, 120 E. Green St, Athens, GA 30602, USA
e-mail: dcnelson@uga.edu

D.M. Joel et al. (eds.), *Parasitic Orobanchaceae*,
DOI 10.1007/978-3-642-38146-1_12, © Springer-Verlag Berlin Heidelberg 2013

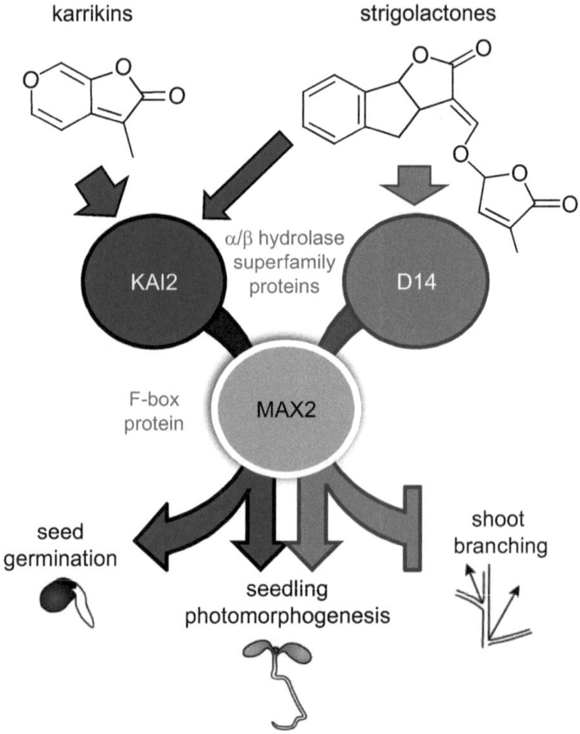

Fig. 12.1 Model of karrikin and strigolactone signaling in *Arabidopsis thaliana*. *Arrows* indicate signal transduction based upon current genetic evidence but do not necessarily imply direct interactions

discovered (Flematti et al. 2004; van Staden et al. 2004). Several germination-promoting compounds with similar structures were later found in smoke, forming a chemical family collectively known as karrikins (Flematti et al. 2009; Nelson et al. 2012).

12.3 Regulation of Plant Development by Karrikins and Strigolactones

Karrikins share partial structural similarity with strigolactones, a class of phytohormones that activate parasitic weed germination (see Sect. 10.2), but the effectiveness of these germination stimulants can vary highly between species (Nelson et al. 2012). The invasive weed *Brassica tournefortii*, for example, can respond to <1 nM KAR_1 (Stevens et al. 2007) but is up to four orders of magnitude less sensitive to the synthetic strigolactone GR24 (Nelson et al. 2009). Conversely, common broomrape (*Orobanche minor*) seems to be entirely insensitive to KAR_1, while 1 nM GR24 is sufficient to induce substantial germination (Nelson et al. 2009). These distinct chemical preferences make sense from an ecological viewpoint: karrikins and strigolactones are opposite indicators of the presence of nearby plants that would be potential competitors or a food supply, respectively, for autotrophic and auxotrophic seedlings.

In contrast to these extreme examples, *Arabidopsis thaliana* germination is enhanced by both karrikins and GR24, making it a suitable model system for studying both signaling pathways (Nelson et al. 2009). *Arabidopsis* seed is about 10- to 100-fold more sensitive to karrikins than GR24 treatment (Nelson et al. 2009; Waters et al. 2012). It is currently unclear if endogenous strigolactones have a significant role in regulation of *Arabidopsis* seed germination. Strigolactone-deficient *max1*, *max3*, and *max4* mutants appear to have normal seed germination and dormancy (Nelson et al. 2011; Shen et al. 2012). However, another group has reported that strigolactone-deficient mutants exhibit reduced germination under supraoptimal temperatures or limited light exposure (Toh et al. 2012; Tsuchiya et al. 2010).

Intriguingly, karrikin application and strigolactone application also influence early seedling development (Nelson et al. 2010, 2011; Waters et al. 2012; Tsuchiya et al. 2010). GR24 is more effective than either KAR_1 or KAR_2 in enhancing light-dependent inhibition of hypocotyl elongation in *Arabidopsis thaliana* (Nelson et al. 2010).

Although karrikins and strigolactones produce similar effects on some aspects of *Arabidopsis* development and gene expression, these signals are distinguished by plants. Karrikins cannot repress the increased axillary shoot-branching phenotype of strigolactone-deficient *max3* and *max4* mutants in *Arabidopsis thaliana* and pea (Nelson et al. 2011). Also, while karrikins enhance light-induced expansion of cotyledons, GR24 inhibits this growth (Nelson et al. 2011; Waters et al. 2012). The full extent of the overlap between karrikins and strigolactones in regulating plant development remains to be determined.

12.4 Karrikin and Strigolactone Responses Are *MAX2*-Dependent

A genetic screen to investigate how karrikins control seed germination was performed in *Arabidopsis thaliana* (Nelson et al. 2011). Two allelic mutants with karrikin-insensitive germination also had increased seed dormancy, decreased seedling photomorphogenesis, delayed leaf senescence, and more axillary shoot-branching phenotypes. The latter three phenotypes had been attributed to mutations of the *MAX2/ORE9/PPS* gene through three independent genetic screens, and sequencing revealed that the karrikin-insensitive mutants were frameshift alleles of *MAX2* (Woo et al. 2001; Shen et al. 2007; Stirnberg et al. 2002; Nelson et al. 2011).

Remarkably, shoot-branching responses to strigolactone are dependent on *MAX2* orthologs in *Arabidopsis*, pea, and rice (Umehara et al. 2008; Gomez-Roldan et al. 2008). *Arabidopsis max2* mutants are completely insensitive to both karrikin and strigolactone treatments in seed germination and seedling photomorphogenesis assays. Furthermore, several early transcriptional response markers (*STH7*, *KUF1*, *KUOX1*, *DLK2*) are insensitive to karrikin and strigolactone treatment in *max2* (Nelson et al. 2011; Waters et al. 2012). Therefore, *MAX2* has a central role in mediating both karrikin and strigolactone signaling in *Arabidopsis thaliana*.

This discovery was a significant step toward understanding smoke-induced seed germination, but it remained unclear how MAX2 alone could manage distinct responses to karrikins and strigolactones. A recent publication has shed light on this mystery, with the characterization of *KAI2* and *D14* (Waters et al. 2012).

12.5 *KAI2* and *D14* Are Required for Specific Responses to Karrikins and Strigolactones

The *DWARF14* (*D14/D88/HTD2*) gene was originally identified in rice mutants with strigolactone-insensitive, high-tillering phenotypes. *D14* encodes an α/β hydrolase superfamily protein (Arite et al. 2009; Gao et al. 2009; Liu et al. 2009). Reverse genetic characterization of the *D14* ortholog in *Arabidopsis thaliana*, *AtD14*, was undertaken to determine if its role in strigolactone response was conserved in dicots (Waters et al. 2012). An *Atd14* loss of function allele produced a GR24-insensitive, increased axillary-branching phenotype—similar to *max2*. Unlike *max2*, however, *Atd14* had wild-type seed dormancy and seedling development. *Atd14* retained normal germination responses to GR24 application, but *Atd14* seedlings were less sensitive to GR24. All responses to karrikins were unaffected by the loss of *AtD14*.

Forward and reverse genetic approaches in *Arabidopsis* led to the identification of two mutant alleles of *KAI2*, a paralog of *D14*. The *kai2* mutants had enhanced seed dormancy and reduced seedling photomorphogenesis that was phenotypically similar to *max2*, but in contrast to *d14*. *kai2* seed had no germination response to karrikins or GR24. *kai2* mutants had normal axillary branching, unlike *max2* and *d14*. Hypocotyl elongation of *kai2* seedlings was karrikin insensitive but still partially responsive to strigolactones.

KAI2 and *D14* therefore collectively regulate the same developmental responses that are controlled by *MAX2* in *Arabidopsis*, including seed germination, seedling photomorphogenesis, and axillary shoot branching (Fig. 12.1). *KAI2* is essential during seed germination and mediates both karrikin and strigolactone signals. *D14* is required for control of shoot branching and is strigolactone-specific. *kai2 d14* double mutants have the same phenotypes as *max2* seedlings, demonstrating that *KAI2* and *D14* have partially redundant roles during seedling development (Waters et al. 2012). *D14* and *KAI2* expression patterns in *Arabidopsis* suggest that the role of these genes in development may partly be due to transcriptional regulation: in seed *KAI2* is 100-fold more highly expressed than *D14*, while in seedlings expression of *D14* is ~2- to 3-fold greater than *KAI2* (Waters et al. 2012).

12.6 Common Elements of Karrikin, Strigolactone, and Gibberellin Signaling

MAX2 encodes an F-box protein that is highly conserved among land plants (Waters et al. 2011). F-box proteins classically act as adapters that confer specificity to E3 ubiquitin-protein ligase complexes. The proteins targeted by the F-box protein are polyubiquitinated and subsequently degraded by the 26S proteasome (Somers and Fujiwara 2009). Therefore, the inability to degrade one or more MAX2 targets may cause the phenotypes of *max2*, including insensitivity to karrikins and strigolactones.

F-box proteins are implicated in the direct perception or early signal transduction of several plant hormones, including auxins, jasmonates, ethylene, and gibberellins (reviewed in Somers and Fujiwara 2009; Sun 2011; Kendrick and Chang 2008). It is remarkable that the genes currently known to be required for karrikin and strigolactone responses are similar in type to gibberellin signaling pathway components. The GID1 gibberellin receptors are classified as α/β hydrolase superfamily proteins, as are KAI2 and D14. Upon binding gibberellin the GID1 protein undergoes a conformational change that promotes its association with DELLA proteins, which have partially redundant roles as repressors of gibberellin-regulated development. The recognition of the gibberellin-GID1-DELLA complex by the F-box protein SLY1 leads to polyubiquitination and degradation of DELLA, resulting in gibberellin growth responses (Sun 2011).

The common features of α/β hydrolases and F-box proteins as key components of gibberellin and karrikin/strigolactone signaling suggest that further mechanistic parallels may be present. This leads to the question, are KAI2 and D14 receptors for karrikins and strigolactones? What are the targets of MAX2, and are they degraded following karrikin or strigolactone perception? Answering these questions will provide the next key steps in deciphering karrikin- and strigolactone-induced germination.

12.7 D14/DAD2 Is a Candidate Receptor for Strigolactones

The characterization of DAD2, an ortholog of D14 in petunia, provides evidence that supports a role for D14 as a strigolactone receptor (Hamiaux et al. 2012). As DAD2 protein has the canonical catalytic triad of α/β hydrolases (S96, H246, and D217), it was tested for enzymatic activity. DAD2 slowly hydrolyzed GR24 to two products in vitro, the formyl tricyclic lactone of GR24 (i.e., the ABC ring) and an unknown compound with a molecular mass of 270. The products of hydrolysis were ineffective inhibitors of bud growth; if these products reach the same tissues as GR24, this result would indicate that strigolactone and not its metabolites are the bioactive signals. Differential scanning fluorimetry revealed that DAD2 undergoes a shift in thermal stability in the presence of GR24. DAD2 hydrolytic activity is required for strigolactone responses in vivo, as the DAD2-S96A variant does not complement the *dad2*-branching phenotype. Active site mutations of DAD2 abolished enzymatic activity on GR24 and GR24-activated protein destabilization (Hamiaux et al. 2012).

Cumulatively these data suggest that strigolactone signaling is transduced by a conformational change triggered in DAD2 during GR24 hydrolysis rather than by the products of hydrolysis themselves. Preliminary evidence from yeast two-hybrid experiments suggested that GR24 promotes the association of DAD2 and one of the two MAX2 paralogs in petunia (Hamiaux et al. 2012). As GR24 also enhanced β-galactosidase reporter activity in control yeast carrying DAD2 bait *without* MAX2 prey, the biological significance of the putative DAD2–MAX2 interaction must be evaluated through further studies.

The 2.15 Å resolution X-ray crystal structure of DAD2 is a critical resource for identifying residues that confer ligand specificity (Hamiaux et al. 2012). Strigolactone was not detected within the hydrophobic ligand-binding pocket, although modeling predicts sufficient space for it.

Current evidence favors the role of D14 as a strigolactone receptor. It follows from the phenotypes of *kai2* mutants and the close homology of KAI2 and D14 that KAI2 is a possible receptor that recognizes both karrikins and strigolactones. Competitive binding assays with karrikins and strigolactones will be required to provide compelling evidence for this hypothesis.

12.8 What Can *Arabidopsis thaliana* Tell Us About Parasitic Weed Germination?

We now return to the central question of this chapter: what is the basis of host-triggered germination in the Orobanchaceae? To begin, a few observations should be considered:

1. **A lactone moiety is common to karrikins, strigolactones, and sesquiterpene lactones, yet preference for different lactones as germination stimulants can be observed even between closely related parasite species.** For example, natural and synthetic sesquiterpene lactones (e.g., costunolide and dihydroparthenolide) stimulate *Striga asiatica* germination at nM concentrations (Fischer et al. 1989, 1990), and *Orobanche cumana* seed specifically responds to dehydrocostus lactone exuded from the sunflower host (Joel et al. 2011). Differential responses to three strigolactones have been described for 15 parasitic weeds (Fernandez-Aparicio et al. 2011b).

2. **Close relatives of the Orobanchaceae have smoke- or KAR-responsive germination.** Smoke-responsive germination occurs in the Lamiaceae, e.g., *Lavandula* spp., *Thymus vulgaris*, *Nepeta rtanjensis*, and *Salvia* spp. (Moreira et al. 2010; Todorovic et al. 2007; Keeley and Fotheringham 1998). *Mimulus nasutus* (Phrymaceae) seeds positively respond to karrikins (Nelson et al., unpublished data).
3. **Genes required for germination responses to karrikin and strigolactone in *Arabidopsis* are well conserved in the Orobanchaceae.** MAX2 orthologs in *Striga hermonthica*, *Phelipanche aegyptiaca*, and *Triphysaria versicolor* are 58 % identical to *A. thaliana* MAX2 and have 73 % identity to the 200 amino acids of the highly conserved C-terminus (Waters et al. 2011). KAI2 and D14 orthologs are also present in the transcriptome assemblies produced by the Parasitic Plant Genome Project for these three parasitic weeds (Westwood et al. 2010). In each species a single copy of *D14* and four to five paralogs of *KAI2* have been identified (Nelson et al., unpublished data).

Karrikin-responsive germination is observed in diverse clades of angiosperms and may indicate a widespread capacity in seeds for perception of lactone-type molecules. If so, specialization of this mechanism in parasitic plants seems a likely means by which host-specific germination could have evolved. We hypothesize that a MAX2-dependent signaling pathway controls germination in the Orobanchaceae in a similar manner to *Arabidopsis*, except that the capacity for dual signal recognition has narrowed to exclude karrikins. This signaling adaptation could have arisen in the Orobanchaceae through mutation of the ligand-binding site of an ancestral receptor that recognized both karrikins and strigolactones. This idea is particularly attractive given the observed increase in *KAI2* gene copy number in parasites; multiple subfunctionalized paralogs of this candidate receptor could explain the range of host-specific germination stimulants recognized by closely related parasitic weeds. If there are independent receptors for karrikins and strigolactones, however, then inactivation of the karrikin receptor or a shift in expression of the two types of receptors could also produce strigolactone-specific germination. Either of these scenarios would be expected to closely accompany the evolution of obligate parasitism.

Testing these hypotheses will require multiple approaches. Reverse genetic techniques such as RNAi can be used to test directly if *MAX2*, *KAI2*, or *D14* orthologs in Orobanchaceae are required for host-triggered germination. The development of transformation methods for parasitic weeds now makes these experiments feasible (Tomilov et al. 2007; Fernandez-Aparicio et al. 2011a; Ishida et al. 2011). A second approach could involve reconstructing strigolactone-specific germination in *Arabidopsis thaliana* through cross-species complementation. This would involve, for instance, the introduction of parasitic weed orthologs of *KAI2* into *Arabidopsis kai2* mutants, which have karrikin- and strigolactone-insensitive seed germination. If parasitic weed *KAI2* conferred strigolactone-specific germination responses, a molecular evolutionary analysis of KAI2 orthologs in

Orobanchaceae and other angiosperms could identify amino acids that are critical to ligand specialization.

12.9 Conclusion

It is now evident that strigolactone and karrikin signaling pathways are intrinsically linked. Therefore, research on karrikins can provide insights that extend beyond fire ecology. If the Orobanchaceae and *Arabidopsis thaliana* utilize similar mechanisms to respond to strigolactones and other lactone-based germination stimulants, then the well-developed genetic resources of *Arabidopsis* can be exploited to rapidly unlock the secrets of a critical agricultural problem.

References

Arite T, Umehara M, Ishikawa S, Hanada A, Maekawa M, Yamaguchi S, Kyozuka J (2009) *d14*, a strigolactone-insensitive mutant of rice, shows an accelerated outgrowth of tillers. Plant Cell Physiol 50:1416–1424

Chiwocha SDS, Dixon KW, Flematti GR, Ghisalberti EL, Merritt DJ, Nelson DC, Riseborough J-AM, Smith SM, Stevens JC (2009) Karrikins: a new family of plant growth regulators in smoke. Plant Sci 177:252–256

De Lange JH, Boucher C (1990) Autecological studies on *Audouinia capitata* (Bruniaceae). I. Plant-derived smoke as a seed germination cue. S Afr J Bot 56:700–703

Dixon KW, Merritt DJ, Flematti GR, Ghisalberti EL (2009) Karrikinolide – a phytoreactive compound derived from smoke with applications in horticulture, ecological restoration, and agriculture. Acta Hortic 813:155–170

Fernandez-Aparicio M, Rubiales D, Bandaranayake PC, Yoder JI, Westwood JH (2011a) Transformation and regeneration of the holoparasitic plant *Phelipanche aegyptiaca*. Plant Methods 7:36

Fernandez-Aparicio M, Yoneyama K, Rubiales D (2011b) The role of strigolactones in host specificity of *Orobanche* and *Phelipanche* seed germination. Seed Sci Res 21:55–61

Fischer NH, Weidenhamer JD, Bradow JM (1989) Dihydroparthenolide and other sesquiterpene lactones stimulate witchweed germination. Phytochemistry 28:2315–2317

Fischer NH, Weidenhamer JD, Riopel JL, Quijano L, Menelaou MA (1990) Stimulation of witchweed germination by sesquiterpene lactones: a structure-activity study. Phytochemistry 29:2479–2483

Flematti GR, Ghisalberti EL, Dixon KW, Trengove RD (2004) A compound from smoke that promotes seed germination. Science 305:977

Flematti GR, Ghisalberti EL, Dixon KW, Trengove RD (2009) Identification of alkyl substituted 2*H*-furo[2,3-*c*]pyran-2-ones as germination stimulants present in smoke. J Agric Food Chem 57:9475–9480

Gao Z, Qian Q, Liu X, Yan M, Feng Q, Dong G, Liu J, Han B (2009) Dwarf 88, a novel putative esterase gene affecting architecture of rice plant. Plant Mol Biol 71:265–276

Gomez-Roldan V, Fermas S, Brewer PB, Puech-Pages V, Dun EA, Pillot JP, Letisse F, Matusova R, Danoun S, Portais JC, Bouwmeester H, Becard G, Beveridge CA, Rameau C, Rochange SF (2008) Strigolactone inhibition of shoot branching. Nature 455:189–194

Hamiaux C, Drummond RS, Janssen BJ, Ledger SE, Cooney JM, Newcomb RD, Snowden KC (2012) DAD2 is an alpha/beta hydrolase likely to be involved in the perception of the plant branching hormone, strigolactone. Curr Biol 22:2032–2036

Ishida JK, Yoshida S, Ito M, Namba S, Shirasu K (2011) *Agrobacterium rhizogenes*-mediated transformation of the parasitic plant *Phtheirospermum japonicum*. PLoS One 6(10):e25802

Joel DM, Chaudhuri SK, Plakhine D, Ziadna H, Steffens JC (2011) Dehydrocostus lactone is exuded from sunflower roots and stimulates germination of the root parasite *Orobanche cumana*. Phytochemistry 72:624–634

Keeley JE, Fotheringham CJ (1998) Smoke-induced seed germination in California chaparral. Ecology 79:2320–2336

Kendrick MD, Chang C (2008) Ethylene signalling: new levels of complexity and regulation. Curr Opin Plant Biol 11:479–485

Liu W, Wu C, Fu Y, Hu G, Si H, Zhu L, Luan W, He Z, Sun Z (2009) Identification and characterization of *HTD2*: a novel gene negatively regulating tiller bud outgrowth in rice. Planta 230:649–658

Moreira B, Tormo J, Estrelles E, Pausas JG (2010) Disentangling the role of heat and smoke as germination cues in Mediterranean Basin flora. Ann Bot 105:627–635

Nelson DC, Riseborough JA, Flematti GR, Stevens J, Ghisalberti EL, Dixon KW, Smith SM (2009) Karrikins discovered in smoke trigger *Arabidopsis* seed germination by a mechanism requiring gibberellic acid synthesis and light. Plant Physiol 149:863–873

Nelson DC, Flematti GR, Riseborough JA, Ghisalberti EL, Dixon KW, Smith SM (2010) Karrikins enhance light responses during germination and seedling development in *Arabidopsis thaliana*. Proc Natl Acad Sci USA 107:7095–7100

Nelson DC, Scaffidi A, Dun EA, Waters MT, Flematti GR, Dixon KW, Beveridge CA, Ghisalberti EL, Smith SM (2011) F-box protein MAX2 has dual roles in karrikin and strigolactone signalling in *Arabidopsis thaliana*. Proc Natl Acad Sci USA 108:8897–8902

Nelson DC, Flematti GR, Ghisalberti EL, Dixon KW, Smith SM (2012) Regulation of seed germination and seedling growth by chemical signals from burning vegetation. Annu Rev Plant Biol 63:107–130

Roche S, Koch JM, Dixon KW (1997) Smoke enhanced seed germination for mine rehabilitation in the southwest of Western Australia. Restor Ecol 5:191–203

Shen H, Luong P, Huq E (2007) The F-box protein MAX2 functions as a positive regulator of photomorphogenesis in *Arabidopsis*. Plant Physiol 145:1471–1483

Shen H, Zhu L, Bu Q, Huq E (2012) MAX2 affects multiple hormones to promote photomorphogenesis. Mol Plant 5:750–762

Somers DE, Fujiwara S (2009) Thinking outside the F-box: novel ligands for novel receptors. Trends Plant Sci 14:206–213

Stevens JC, Merritt DJ, Flematti GR, Ghisalberti EL, Dixon KW (2007) Seed germination of agricultural weeds is promoted by the butenolide 3-methyl-2*H*-furo[2,3-*c*]pyran-2-one under laboratory and field conditions. Plant Soil 298:113–124

Stirnberg P, van De Sande K, Leyser HM (2002) *MAX1* and *MAX2* control shoot lateral branching in *Arabidopsis*. Development 129:1131–1141

Sun TP (2011) The molecular mechanism and evolution of the GA-GID1-DELLA signalling module in plants. Curr Biol 21:338–345

Todorovic SLA, Zivkovic S, Giba Z, Grubisic D, Misic D (2007) Basic seed germination characteristics of the endemic species *Nepeta rtanjensis* (Lamiaceae). Plant Species Biol 22:205–210

Toh S, Kamiya Y, Kawakami N, Nambara E, McCourt P, Tsuchiya Y (2012) Thermoinhibition uncovers a role for strigolactones in *Arabidopsis* seed germination. Plant Cell Physiol 53:107–117

Tomilov A, Tomilova N, Yoder JI (2007) *Agrobacterium tumefaciens* and *Agrobacterium rhizogenes* transformed roots of the parasitic plant *Triphysaria versicolor* retain parasitic competence. Planta 225:1059–1071

Tsuchiya Y, Vidaurre D, Toh S, Hanada A, Nambara E, Kamiya Y, Yamaguchi S, McCourt P (2010) A small-molecule screen identifies new functions for the plant hormone strigolactone. Nat Chem Biol 6:741–749

Umehara M, Hanada A, Yoshida S, Akiyama K, Arite T, Takeda-Kamiya N, Magome H, Kamiya Y, Shirasu K, Yoneyama K, Kyozuka J, Yamaguchi S (2008) Inhibition of shoot branching by new terpenoid plant hormones. Nature 455:195–200

van Staden J, Jager AK, Light ME, Burger BV (2004) Isolation of the major germination cue from plant-derived smoke. S Afr J Bot 70:654–659

Waters MT, Smith SM, Nelson DC (2011) Smoke signals and seed dormancy: where next for MAX2? Plant Signal Behav 6:1418–1422

Waters MT, Nelson DC, Scaffidi A, Flematti GR, Sun YK, Dixon KW, Smith SM (2012) Specialisation within the DWARF14 protein family confers distinct responses to karrikins and strigolactones in *Arabidopsis*. Development 139:1285–1295

Westwood JH, Yoder JI, Timko MP, dePamphilis CW (2010) The evolution of parasitism in plants. Trends Plant Sci 15:227–235

Woo HR, Chung KM, Park JH, Oh SA, Ahn T, Hong SH, Jang SK, Nam HG (2001) ORE9, an F-box protein that regulates leaf senescence in *Arabidopsis*. Plant Cell 13:1779–1790

Chapter 13
Changing Host Specificities: By Mutational Changes or Epigenetic Reprogramming?

Toby J.A. Bruce and Jonathan Gressel

13.1 Introduction

In the agricultural sense, a species is not a weed until it competes with a crop or, in the case of a parasitic weed, until it repeatedly attacks a crop and affects the quality or quantity of yield. Thus, just as crops are the rare species domesticated from wild species, weeds too are among the few wild species that were (inadvertently) domesticated from the wild. Weeds bear a syndrome of characters that make them quite different from their wild relatives and are often unable to exist outside of agricultural ecosystems (Warwick and Stewart 2005). As with all syndromes, a plant need not possess all the characteristics of weediness, just a sufficient number that renders the plant competitive with the crop.

One of the special weediness traits of parasitic plants is host specificity. Most of the weedy parasitic Orobanchaceae thrive less well on wild hosts, and conversely many parasitic *Orobanchaceae* species attacking nonagricultural species rarely attack domesticated crops, or when they do, they cause little damage. The most virulent parasitic *Orobanchaceae* weeds such as *Striga hermonthica* and *Orobanche cumana* are hardly known to attack nonagricultural species, a sign that they are highly domesticated weeds. Nothing is black and white in weediness, and there is typically a continuum, and indeed the far less virulent *O. minor* attacks both weeds and wild species. Some previously non-weedy Orobanchaceae such as *O. foetida* seem to be in an evolutionary flux, recently evolving strains[1] that are

[1] Various authors refer to the intraspecific variants as strains, races, clades, pathovars, subspecies, etc., and these can be functionally interchangeable terms in the context of this chapter.

T.J.A. Bruce
Department of Biological Chemistry, Rothamsted Research, Harpenden, Herts AL5 2JQ, UK
e-mail: toby.bruce@rothamsted.ac.uk

J. Gressel (✉)
Plant Sciences, Weizmann Institute of Science, Rehovot 76100, Israel
e-mail: jonathan.gressel@weizmann.ac.il

D.M. Joel et al. (eds.), *Parasitic Orobanchaceae*,
DOI 10.1007/978-3-642-38146-1_13, © Springer-Verlag Berlin Heidelberg 2013

agricultural weeds in Tunisia (Román et al. 2007; Vaz Patto et al. 2008). In Spain, it still attacks only weedy legumes while in the same fields *O. crenata* concurrently attacks legume crops, but not the legume weeds.

Crop-parasitic weeds continue to spread. In France there are various wild species-specific *Orobanche* spp., but a decade ago French weed specialists thought their country is immune to the crop-specific parasites found in Spain and Italy. Now, *Phelipanche ramosa* is marching north, attacking three French "crop" species (one is *Cannabis sativa*) with host specificity among strains (Brault et al. 2007). *P. ramosa* is also spreading in Germany (Kohlschmid et al. 2011). Evolution continues to the east as well; the indigenous parasite on some wild species, *O. pubescens* is now damaging parsley fields in Israel, while a European parasite of wild species, *O. amethystea* is now attacking vetch (Joel and Eisenberg 2002). Such evolution of expanded host range often occurs in fields with a huge parasite seedbank, e.g., the first attack of commercial anemones occurred in a field rotated from faba beans that had been heavily infested with *O. crenata* (Dor et al. 2008). Similarly, *P. aegyptiaca* has suddenly parasitized various Malvaceae weeds as well as sunflower in Israel (Yaacoby et al. 2011).

What evolutionary forces allowed these dynamic changes, and how? While there is ample evidence that most host specificities are due to classical genetic variation, there are cases of rapid changes in intraspecific host specificity where this is less clear. Such rapid adaptations may be genetic (which may not be fast enough to explain huge sudden population changes), or are epigenetically imprinted (Bruce et al. 2007), or possibly due to alternative splicing. Each case may be due to a different cause.

The aim of this chapter is to summarize some of the evidence for intraspecific host specificity, the evolutionary changes in this specificity, and mainly to discuss the type of evidence needed to differentiate between epigenetic and classical changes in host specificity, and to speculate on how such knowledge might affect management practices.

13.2 Static Evidence for Intraspecific Variation in Host Specificity

There is considerable evidence for strain variation leading to host specificity within species (Table 13.1A). Molecular tools typically suggest that this is genetic due to banding differences in AFLP or similar separation techniques. The use of these tools in *Orobanche* species and the need for codominant markers such as microsatellites for analyzing genetic drift are extensively reviewed by Satovic et al. (2009; see Sect. 19.3). There are discrepancies; self-pollinating *Striga gesnerioides* has very little genetic diversity by AFLP; only 6 % of 1,200 AFLP bands are polymorphic (Dube and Belzile 2010). Still, there are clear host strain specificities (Botanga and Timko 2006), and that group could correlate them with

Table 13.1 Evidence for intraspecific variation in host specificity and evidence for host change

Species	Type of evidence	Genetic evidence	Reference
(A) Static intraspecific variation of host specificity			
S. gesnerioides	Host preference: cowpea vs weed; geographic	AFLP clustering	Botanga and Timko (2006)
S. hermonthica	Considerable in-field variation	Isozymes	Olivier et al. (1998)
O. minor	Clades of this generalist species with host (carrot, clover) specificity	Bioassay, ISSR, SCAR	Thorogood et al. (2008, 2009a, b)
P. ramosa	Host-specific strains on tobacco, oilseed rape, and *Cannabis*	Bioassay correlated with RAPD	Brault et al. (2007)
P. ramosa	Strains with different virulence towards tobacco	Bioassay correlated with ISSR	Buschmann et al. (2005)
(B) Changes in intraspecific host range			
O. cumana	New races evolve that overcome genetic resistance in sunflower	Crop breeding only; no DNA/genetic evidence	Molinero-Ruiz et al. (2009)
O. foetida	Recent jump from weed to both faba bean and chickpea (with specificity between crops)	RAPD clustering between crops (no data on weedy form)	Román et al. (2007); Vaz Patto et al. (2008)
S. hermonthica	Lag in ability to infest when crop hosts are changed	Observational	Olivier et al. (1998)
O. minor	Multiple northward shifts to new hosts with clade specificity for each	SCAR	Thorogood et al. (2009b)
O. pubescens	Wild hosts to parsley	Observational	Joel and Eisenberg (2002)
O. amethystea	Wild hosts to vetch	Observational	Joel and Eisenberg (2002)
O. crenata	Faba beans to commercial anemones	Observational	Dor et al. (2008)
P. aegyptiaca	Tomatoes to sunflowers and weedy Malvaceae	Observational	Yaacoby et al. (2011)

O. Orobanche, *S.* Striga, *P.* Phelipanche, *AFLP* amplified fragment length polymorphism, *ISSR* intersimple sequence repeat, *SCAR* sequence characterized amplified region

slight AFLP banding changes. But using the same primers and different populations, Dube and Belzile (2010) could not find strain differences.

Epigenetic changes such as DNA methylation would not be revealed by conventional AFLP analyses, which only indicate differences in nucleotide sequences, and thus this technology cannot differentiate between mutations and epigenetic changes.

The physiological cause in such variation is often due to adaptation to the specific germination stimulant from the host (Thorogood et al. 2009c; Fernandez-Aparicio et al. 2011). If artificially germinated parasite seeds are planted next to a nonhost, they will often attach and infest (e.g., Westwood 2000). Can species of parasite evolve to recognize different host germination stimulants? Presently only *O. cumana* recognizes dehydrocostus lactone, a specific stimulant exuded by sunflower roots (Joel et al. 2011; see Sect. 10.3.1). What are the drivers of such evolution? Specificity is often multilayered, in some cases the parasite will germinate, but not form haustoria or attach, or attach and not develop, or develop somewhat and stop (Shen et al. 2006; Thorogood and Hiscock 2010; see Sect. 3.15). While this can be the basis of specificity, it is also the basis of resilient breeding for crop resistance, i.e., having genes that confer resistance at each level (see Chaps. 7 and 21).

13.3 Evidence for Rapid Dynamic Intraspecific Changes in Host Specificity

The above studies were all made at single static time points, and it is thus difficult to ascertain just how, under what conditions and controls, did changes in host specificities evolve, both from wild host-generalist parasites of wild species to weedy strains of parasites, each of specific crops. Changes in any parasitic mechanism that allows changing host preference, such as changes in stimulant recognition (see Chaps. 4 and 10), may also lead to a parallel selection against it in host populations (see Chap. 19). The selection may be slower in agricultural crops because of the way they are bred and maintained.

The evidence for rapid change is typically observational (Table 13.1B). For example, "When a sorghum field infested with *S. hermonthica* is replaced by pearl millet, we usually observe that the new crop is almost free of the parasitic weed during the first season of cropping. But the new crop can become highly infested after a few years of cropping" (Olivier et al. 1998). Such fields are cultivated by hand, so there would be little movement of *Striga* seed by cultivation or harvesting machinery, yet the whole field changes over. Unfortunately, such observational data have yet to have had parallel molecular genetic scrutiny, except as in the static sense, as described in the previous section. Could these changes in host specificity be due to a uniform distribution of a rare mutant throughout the field, or is some other, possibly epigenetic adaptation at play? Others report informally that if one reverts back to the first crop, or to maize, there is another lag before either is infested. *S. hermonthica* is an obligate outcrosser, with considerable genetic variability, yet in East Africa this species is rarely seen on weedy grasses bordering the fields, so the ability to adapt is far less than that of its interbreeding presumed progenitor, *Striga aspera*.

13.4 Critically Differentiating Between Classical Genetic Evolution and Epigenetic Adaptation

Epigenetic adaptation involves a change in gene expression and hence phenotype caused by mechanisms other than changes in the DNA sequence (Russo et al. 1996). Epigenetic changes involve higher-level alterations to the DNA structure, for example, by methylation of cytosine residues in DNA or by histone modification. Crucially, DNA methylation in plants is not necessarily reset at meiosis as generally occurs in animals, and so these marks can be passed on to the following generations. This means that they can play an important role in plant fitness and have adaptive value. Histone modifications are not usually transferred to the next generation, even in plants. Epigenetic mechanisms allow an organism to respond to the environment through changes in gene expression (Jaenisch and Bird 2003; Bossdorf et al. 2008) and allow more rapid and flexible changes than classical genetic evolution, and thus may be the cause of changes in host specificity. However, such changes are reversible unlike mutations that cause fixed changes to the DNA sequence itself. The rapid and reversible changes in host specificity could be explained by epigenetic changes, most likely alterations in DNA methylation at loci involved in host recognition and colonization. This hypothesis could be tested by analyzing DNA methylation patterns in weed strains with different host affinities. If the genes involved in host recognition and colonization are known, these could be focused on. Alternatively whole-genome DNA methylation patterns could be compared between strains. A crucial step would be to eliminate conventional sequence-based variation between strains as a cause of the differences between strains. This would require some knowledge of the genes associated with host affinity as well as nucleotide sequencing to show that there is no difference in the order of base pairs between strains with different host affinities. If the genes associated with host affinity are not well known, it may be possible to apply statistical procedures to measure the relative contribution of nucleotide sequence and epigenetic differences to the observed host affinity phenotype.

Alternative splicing of mRNA is also becoming an area of interest for those studying rapid adaptations to stress (Mastrangelo et al. 2012), and germinating next to a nonhost is a life or death stress for a parasite germling. Thus, alternative splicing might also facilitate the rapid adaptation to new hosts in some cases.

Analysis of DNA methylation patterns relies increasingly on sequencing-based profiling methods. These methods have recently been compared by Harris et al. (2010). The most frequently used approaches involve bisulfate conversion (Fig. 13.1a) or enrichment (Fig. 13.1b) of methylated DNA. Methylated cytosine sequencing (MethylC-seq) and reduced representation bisulfite sequencing (RRBS) use bisulfate conversion. Methylated DNA immunoprecipitation sequencing (MeDIP-seq) and methylated DNA-binding domain sequencing (MBD-seq) use enrichment of methylated DNA. Results from these four different approaches are comparable (Harris et al. 2010), but differences in coverage, resolution, quantitative accuracy, efficiency, and cost were noted. However, all four sequencing-based profiling methods are expensive. The pros and cons of each are summarized in Table 13.2.

Methylation-sensitive enzyme digestion (Fig. 13.1c) is another approach that has been used to ascertain DNA methylation pattern changes in plants. The metAFLP technique used to investigate DNA methylation changes in *Deschampsia antarctica* allowed investigation of differences in DNA methylation levels within populations of plants (Chwedorzewska and Bednarek 2011). Methylation-sensitive amplified

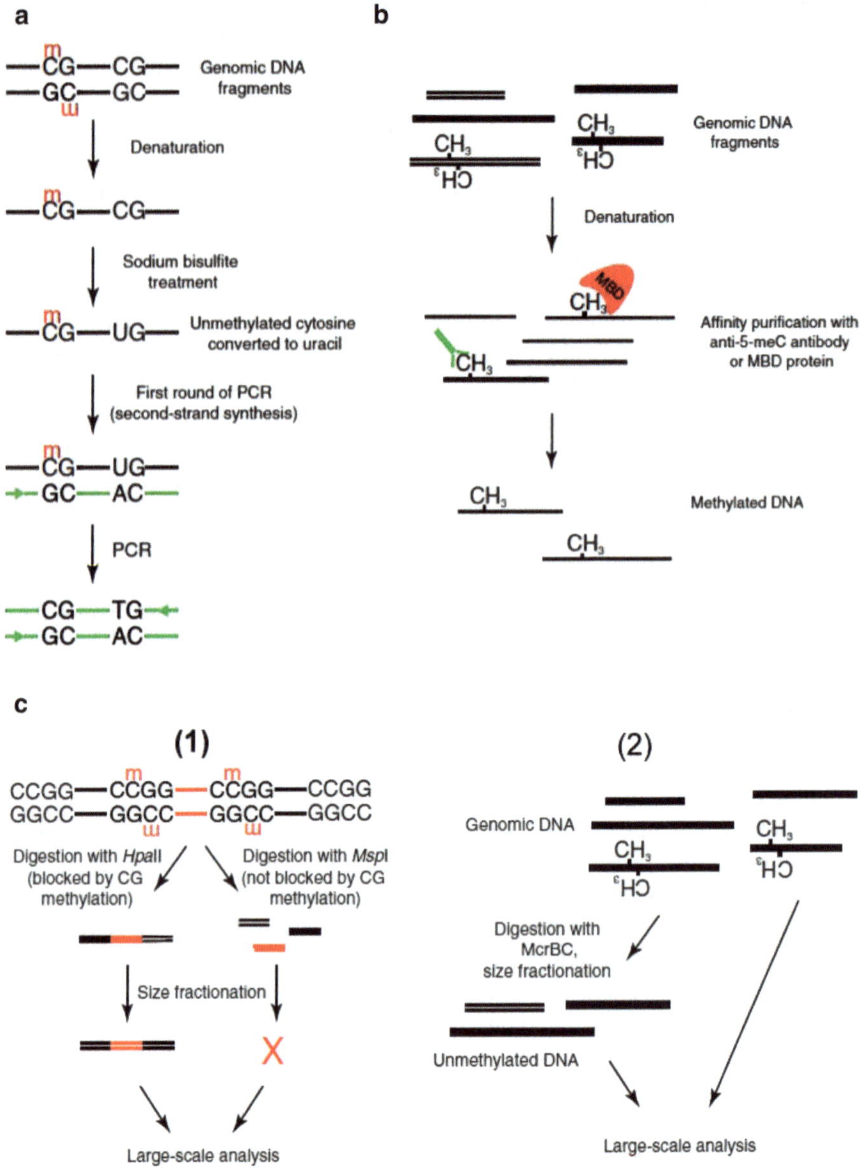

Fig. 13.1 The three main approaches to measuring cytosine methylation of DNA to ascertain epigenetic modifications. (**a**) Bisulfite conversion: DNA is denatured and then treated with sodium bisulfite to convert unmethylated cytosine to uracil, which is converted to thymine by PCR. This technique can reveal the methylation status of every cytosine residue, and it is amenable to parallel sequencing methods. Comparison of sequence information between the reference genome and bisulfite-treated DNA can provide single-nucleotide resolution information about cytosine methylation patterns. Following bisulfite conversion, the DNA strands are no longer complementary, and primers are designed to assay the methylation (m) status of a specific strand. (**b**) Affinity

fragment length polymorphism (MSAP) was used to demonstrate epigenetic reprogramming and the reversible phenotypic response of alligator weed (*Alternanthera philoxeroides*) to drought stress (Gao et al. 2010). MSAP was also used to investigate the association between browsing damage and epigenetic characteristics of individual *Viola cazorlensis* plants (Herrera and Bazaga 2011). These methods could easily be adapted for use with the weedy Orobanchaceae. Bisulfite sequencing requires prior detailed information on genome sequences, as it is also the case for other affinity-based methods. For non-model plants without genome information, any variant of the MSAP method is, for the time being, the most sensible choice.

If indeed the phenomenon of gene-for-gene resistance defining host specificity, as found with *S. gesnerioides* and cowpea strains (Li and Timko 2009), is general, then the differences in host specificity can be governed by the changed expression of a single gene. One must remember that such changing expression can be due to any of the factors described: a mutation in a structural gene or controlling element, alternative splicing, or epigenetic methylation.

Because a certain portion of the remembered epigenetic changes or alternative mRNA splicing revert in each generation, such a reversion may confound the interpretation of crossbreeding experiments to ascertain if the change is due to Mendelian genetics. Additionally, alternative splicing of defense genes is implicated in response to many abiotic and biotic stresses (Chung et al. 2010), but has yet to be researched in relation to changing host specificities of parasites. The stimulated germination of a parasite near a nonhost is surely a life or death stress on the parasite. Thus, to be certain in a diagnosis of the cause of change in host specificity, it is necessary to ascertain changes in DNA methylation patterns for epigenetics, DNA sequence changes for Mendelian genetics, and mRNA sequence changes for alternative splicing.

Fig. 13.1 (continued) enrichment of methylated DNA: affinity-based methods use methylated DNA-binding antibodies or proteins to enrich the experimental DNA sample for subsequent analysis. Genomic DNA is denatured and then affinity purified with either an antibody (*green*) or a methyl-binding domain (MBD, *red*) protein that can be attached to a column. (**c**) Methylation-sensitive enzyme digestion: restriction enzymes are used to generate DNA fragments for methylation analysis. Some restriction enzymes are methylation sensitive (i.e., digestion is impaired or blocked by methylated DNA). When used in conjunction with an isoschizomer that has the same recognition site but is methylation insensitive, information about methylation status can be obtained. Two examples are the following: (1) a methylated (m) region of genomic DNA digested with HpaII (*left*) or MspI (*right*); smaller fragments are discarded (red X), enriching for methylated DNA in the HpaII-treated sample, relative to the MspI treated; (2) genomic DNA is treated with McrBC, which cuts methylated (CH3) DNA; smaller fragments are discarded, enriching for unmethylated DNA (figure modified from figures in Zilberman and Henikoff (2007), with permission of the publisher; the text from: http://www.neb.com/nebecomm/tech_reference/epigenetics/epigenetics_technology.asp, with permission)

Table 13.2 Affinity-based methods use methylated DNA-binding proteins or antibodies to enrich the experimental DNA sample for subsequent analysis

Method	Description	Advantages	Disadvantages
Sodium bisulfite conversion	Treatment of denatured DNA (i.e., single-stranded DNA) with sodium bisulfite leads to deamination of unmethylated cytosine residues to uracil, leaving 5-mC intact. The uracils are amplified as thymines, and 5-mC residues are amplified as cytosines in PCR. Comparison of sequence information between the reference genome and bisulfite-treated DNA can provide single-nucleotide resolution information about cytosine methylation patterns	Resolution at the nucleotide level Works on 5-mC-containing DNA Automated analysis	Requires micrograms of DNA input Harsh chemical treatment of DNA can lead to its damage Potentially incomplete conversion of DNA Cannot distinguish between 5-mC and 5-hmC Multistep protocol
Sequence-specific enzyme digestion	Restriction enzymes are used to generate DNA fragments for methylation analysis. Some restriction enzymes are methylation sensitive (i.e., digestion is impaired or blocked by methylated DNA). When used in conjunction with an isoschizomer that has the same recognition site but is methylation insensitive, information about methylation status can be obtained. Additionally, the use of methylation-dependent restriction enzymes (i.e., requires methylated DNA for cleavage to occur) can be used to fragment DNA for sequencing analysis	High enzyme turnover Well studied Easy to use Availability of recombinant enzymes	Determination of methylation status is limited by the enzyme recognition site Overnight protocols Lower throughput

(continued)

Table 13.2 (continued)

Method	Description	Advantages	Disadvantages
Methylated DNA immunoprecipitation	Fragmented genomic DNA (restriction enzyme digestion or sonication) is denatured and immunoprecipitated with antibodies specific for 5-mC. The enriched DNA fragments can be analyzed by PCR for locus-specific studies or by microarrays (MeDIP-chip) and massively parallel sequencing (MeDIP-seq) for whole-genome studies	Relatively fast Compatible with array-based analysis Applicable for high-throughput sequencing	Dependent on antibody specificity May require more than one 5-mC for antibody binding Requires DNA denaturation Resolution depends on the size of the immunoprecipitated DNA and for microarray experiments, depends on probe design Data from repeat sequences may be overrepresented
Methylated DNA-binding proteins	Instead of relying on antibodies for DNA enrichment, affinity-based assays use proteins that specifically bind methylated or unmethylated CpG sites in fragmented genomic DNA (restriction enzyme digestion or sonication). The enriched DNA fragments can be analyzed by PCR for locus-specific studies or by microarrays and massively parallel sequencing for whole-genome studies	Well studied Does not require denaturation Compatible with array-based analysis Applicable for high-throughput sequencing	May require high DNA input May require a long protocol Requires salt elutions Does not give single base methylation resolution data

From: http://www.neb.com/nebecomm/tech_reference/epigenetics/epigenetics_technology.asp (with permission)

13.5 Does It Matter to Parasite Management Whether Classical Genetic Evolution or Epigenetic Adaptation?

If changes in host specificity are due to epigenetic adaptations, the weed is able to adapt to changing host abundance more rapidly than with classical mutation-based shifts in host affinity. This has consequences for weed management. For example, crop rotation is less likely to reduce weed seed bank levels if the other crops in the rotation are the ones which the weed can adapt to epigenetically.

Acknowledgments Helpful discussions with Daniel Joel contributed extensively to the text on host specificity, and Avi Levy made useful suggestions on the text.

References

Bossdorf O, Richards CL, Pigliucci M (2008) Epigenetics for ecologists. Ecol Lett 11:106–115

Botanga CJ, Timko MP (2006) Phenetic relationships among different races of *Striga gesnerioides* (Willd.) Vatke from West Africa. Genome 49:1351–1365

Brault M, Betsou F, Jeune B, Tuquet C, Sallé G (2007) Variability of *Orobanche ramosa* populations in France as revealed by cross infestations and molecular markers. Environ Exp Bot 61:272–280

Bruce TJA, Matthes MC, Napier JA, Pickett JA (2007) Stressful "memories" of plants: evidence and possible mechanisms. Plant Sci 173:603–608

Buschmann H, Gonsior G, Sauerborn J (2005) Pathogenicity of branched broomrape (*Orobanche ramosa*) populations on tobacco cultivars. Plant Pathol 54:650–656

Chung HS, Cooke TF, Depew CL, Patel LC, Ogawa N, Kobayashi Y, Howe GA (2010) Alternative splicing expands the repertoire of dominant JAZ repressors of jasmonate signaling. Plant J 63:613–622

Chwedorzewska KJ, Bednarek PT (2011) Genetic and epigenetic studies on populations of *Deschampsia antarctica* Desv. from contrasting environments on King George Island. Pol Polar Res 32:15–26

Dor E, Eizenberg H, Joel DM, Levitin E, Hershenhorn J (2008) First report of *Orobanche crenata* parasitism on ornamental anemone (*Anemone coronaria*) in Israel. Plant Dis 92:655

Dube MP, Belzile FJ (2010) Low genetic variability of *Striga gesnerioides* populations parasitic on cowpea might be explained by a recent origin. Weed Res 50:493–502

Fernandez-Aparicio M, Yoneyama K, Rubiales D (2011) The role of strigolactones in host specificity of Orobanche and Phelipanche seed germination. Seed Sci Res 21:55–61

Gao LX, Geng YP, Li B, Chen JK, Yang J (2010) Genome-wide DNA methylation alterations of *Alternanthera philoxeroides* in natural and manipulated habitats: implications for epigenetic regulation of rapid responses to environmental fluctuation and phenotypic variation. Plant Cell Environ 33:1820–1827

Harris RA, Wang T, Coarfa C, Nagarajan RP et al (2010) Comparison of sequencing-based methods to profile DNA methylation and identification of monoallelic epigenetic modifications. Nat Biotechnol 28:1097–1105

Herrera CM, Bazaga P (2011) Untangling individual variation in natural populations: ecological, genetic and epigenetic correlates of long-term inequality in herbivory. Mol Ecol 20:1675–1688

Jaenisch R, Bird A (2003) Epigenetic regulation of gene expression: how the genome integrates intrinsic and environmental signals. Nat Genet 33:245–254

Joel DM, Eisenberg H (2002) Three *Orobanche* species newly found on crops in Israel. Phytoparasitica 30:187–190

Joel DM, Chaudhuri SK, Plakhine D, Ziadna H, Steffens J (2011) Dehydrocostus lactone is exuded from sunflower roots and stimulates germination of the root parasite *Orobanche cumana*. Phytochemistry 72:624–634

Kohlschmid E, Muller-Stover D, Sauerborn J (2011) Spreading of the parasitic weed *Phelipanche ramosa* in German agriculture (in German). Ges Pflanz 63:69–74

Li JX, Timko MP (2009) Gene-for-gene resistance in *Striga*-cowpea associations. Science 325:1094

Mastrangelo AM, Marone D, Laidò G, De Leonardis AM, De Vita P (2012) Alternative splicing: enhancing the plant ability to cope with stress via transcriptome plasticity. Plant Sci 185:40–49

Molinero-Ruiz ML, Garcia-Ruiz R, Melero-Vara JM, Dominguez J (2009) *Orobanche cumana* race F: performance of resistant sunflower hybrids and aggressiveness of populations of the parasitic weed. Weed Res 49:469–478

Olivier A, Glaszmann JC, Lanaud C, Leroux GD (1998) Population structure, genetic diversity and host specificity of the parasitic weed *Striga hermonthica* (Scrophulariaceae) in Sahel. Plant Syst Evol 209:33–45

Román B, Satovic Z, Alfaro C, Moreno MT, Kharrat M, Perez-de-Luque A, Rubiales D (2007) Host differentiation in *Orobanche foetida* Poir. Flora 202:201–208

Russo VEA, Martienssen RA, Riggs AD (1996) Epigenetic mechanisms of gene regulation. Cold Spring Harbor Laboratory Press, Cold Spring Harbor, NY

Satovic D, Joel DM, Rubiales D, Cubero JI, Román B (2009) Population genetics in weedy species of *Orobanche*. Australas Plant Pathol 38:228–234

Shen H, Ye W, Hong L, Huang H, Wang Z, Deng X, Yang Q, Xu Z (2006) Progress in parasitic plant biology: host selection and nutrient transfer. Plant Biol 8:175–185

Thorogood CJ, Hiscock SJ (2010) Compatibility interactions at the cellular level provide the basis for host specificity in the parasitic plant *Orobanche*. New Phytol 186:572–575

Thorogood CJ, Rumsey FJ, Harris SA, Hiscock SJ (2008) Host-driven divergence in the parasitic plant *Orobanche minor* Sm. (Orobanchaceae). Mol Ecol 17:4289–4303

Thorogood CJ, Rumsey FJ, Hiscock SJ (2009a) Host-specific races in the holoparasitic angiosperm *Orobanche minor:* implications for speciation in parasitic plants. Ann Bot 103:1005–1014

Thorogood CJ, Rumsey FJ, Harris SA, Hiscock SJ (2009b) Gene flow between alien and native races of the holoparasitic angiosperm *Orobanche minor* (Orobanchaceae). Plant Syst Evol 282:31–42

Thorogood CJ, Rumsey FJ, Hiscock SJ (2009c) Seed viability determination in parasitic broomrapes (*Orobanche* and *Phelipanche*) using fluorescein diacetate staining. Weed Res 49:461–468

Vaz Patto MC, Diaz-Ruiz R, Satovic Z, Roman B, Pujadas-Salva AJ, Rubiales D (2008) Genetic diversity of Moroccan populations of *Orobanche foetida*: evolving from parasitising wild hosts to crop plants. Weed Res 48:179–186

Warwick SI, Stewart CN Jr (2005) Crops come from wild plants: how domestication, transgenes, and linkage together shape ferality. In: Gressel J (ed) Crop ferality and volunteerism. CRC, Boca Raton, FL, pp 9–30

Westwood JH (2000) Characterization of the *Orobanche-Arabidopsis* system for studying parasite-host interactions. Weed Sci 48:742–748

Yaacoby T, Goldwasser Y, Rubin B (2011) Malvaceae weeds as hosts for *Phelipanche aegyptiaca* in Israel. In: Proceedings of the 11th world congress on parasitic plants, Martina Franca, Italy, p 124

Zilberman D, Henikoff S (2007) Genome-wide analysis of DNA methylation patterns. Development 134:3959–3965

Chapter 14
Phylogenetic Relationships and Evolutionary Trends in Orobanchaceae

Gerald M. Schneeweiss

14.1 Introduction

Parasitism within angiosperms has independently evolved about 12 times (Fig. 14.1; see Chap. 1). The majority of these parasitic lineages comprise small genera or families with one or few genera and usually not more than a dozen species each (Nickrent 2012; Fig. 14.1). The highest taxon diversity can be found in Santalales, which currently comprises 3 nonparasitic and 17 parasitic families (Barkman et al. 2007; Nickrent et al. 2010), and the Orobanchaceae in Lamiales. Whereas molecular data often dramatically changed our understanding of the phylogenetic position of enigmatic parasitic families, such as Rafflesiaceae or Apodanthaceae, they corroborated the phylogenetic position of Orobanchaceae in the vicinity of Scrophulariaceae (Olmstead et al. 2001; Bremer et al. 2002).

14.2 Phylogenetic Relationships

14.2.1 Circumscription of Orobanchaceae

In its traditional circumscription,[1] the Orobanchaceae exclusively included holoparasitic genera with unilocular ovaries (but bilocular in *Lathraea*) and parietal placentation. In contrast, members of the Scrophulariaceae, even if they were parasitic (mostly hemiparasitic), usually have bi- to trilocular syncarpous ovaries

[1] Circumscription refers to the definition of the limits of a taxonomic group.

G.M. Schneeweiss (✉)
Department of Systematic and Evolutionary Botany, University of Vienna, Rennweg 14, 1030 Vienna, Austria
e-mail: gerald.schneeweiss@univie.ac.at

D.M. Joel et al. (eds.), *Parasitic Orobanchaceae*,

DOI 10.1007/978-3-642-38146-1_14,

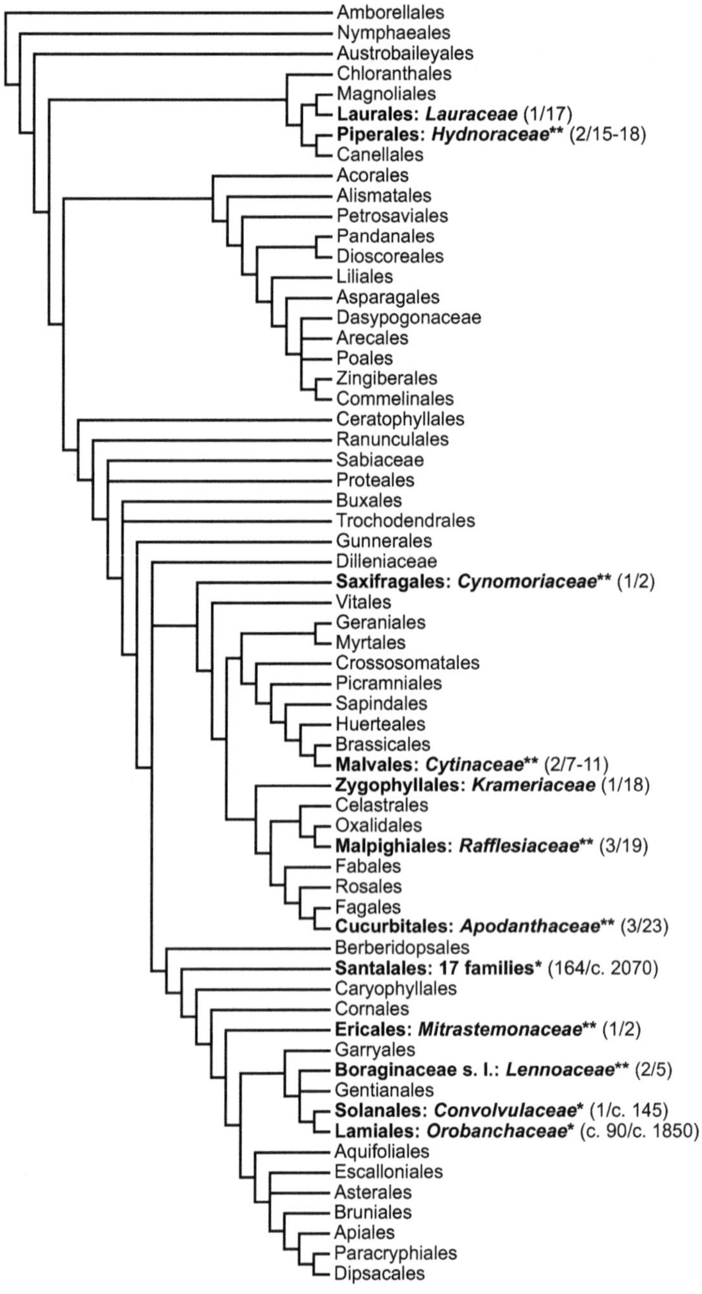

Fig. 14.1 Assignment of at least partly parasitic plant families to orders (indicated in bold) and their phylogenetic position within the angiosperm tree of life (Angiosperm Phylogeny Group 2009; Filipowicz and Renner 2010; Moore et al. 2010). Number of parasitic genera/species in parentheses (for details in Santalales, see Nickrent 2012). Families containing both hemi- and holoparasites are indicated by an *asterisk*; exclusively holoparasitic ones by *two asterisks*. The phylogenetic position of Cynomoriaceae in Saxifragales (Nickrent et al. 2005) is still contentious (Barkman et al. 2007; Zhang et al. 2009)

(Bentham 1876; Beck-Mannagetta 1891; Wettstein 1891). The lack of clear-cut differences between Orobanchaceae and Scrophulariaceae (Boeshore 1920) led some authors to merge both families (Hallier 1903; Takhtajan 1997; Teryokhin et al. 2003), but most authors kept the two families separate (Bentham 1876; Beck-Mannagetta 1891, 1930; Wettstein 1891; Novopokrovskij and Cvelev 1958; Webb 1972; Zhang and Cvelev 1998).

Recent molecular phylogenetic results show that the Scrophulariaceae is a non-monophyletic family (Olmstead et al. 2001; Bremer et al. 2002; Oxelman et al. 2005) and that all root-parasitic genera, irrespective of their former assignment, now belong to the same exclusively parasitic clade, thus corresponding to the family circumscription used by Bellini (1907; as Rhinanthaceae) and by Fischer (2004). This clade is sister to the nonparasitic genus *Lindenbergia*, and together they constitute the greatly expanded family—the Orobanchaceae (Nickrent et al. 1998; Young et al. 1999; Olmstead et al. 2001; Wolfe et al. 2005; Bennett and Mathews 2006; Tank et al. 2006; Schäferhoff et al. 2010; McNeal et al. 2013). Only recently, a clade of *Rehmannia* and *Triaenophora*, two small eastern Asian genera, has been identified as sister group to Orobanchaceae (Albach et al. 2009; Xia et al. 2009; Schäferhoff et al. 2010), which taxonomically can be accommodated by extending the circumscription of Orobanchaceae to include *Rehmannia* and *Triaenophora*.

14.2.2 Major Groups Within the Orobanchaceae

Several studies have addressed phylogenetic relationships within Orobanchaceae (dePamphilis et al. 1997; Wolfe and dePamphilis 1998; Young et al. 1999; Bennett and Mathews 2006; Park et al. 2008; McNeal et al. 2013). They use different taxon sampling and different molecular markers, which sometimes are burdened with parasite-specific (aberrant evolution of plastid genomes of holoparasites [see Chap. 15]; dePamphilis and Palmer 1990; Wolfe and dePamphilis 1998; Young and dePamphilis 2005) or marker-specific problems (in the widely used internal transcribed spacer (ITS) region of 35S ribosomal RNA; Álvarez and Wendel 2003). Therefore, our current understanding of phylogenetic relationships within Orobanchaceae is still incomplete and subject to further amendment. Nevertheless, a certain consensus has been reached concerning several major clades (Fig. 14.2), even if their morphological characterization and the identification of synapomorphies are still largely lacking. The descriptions of nearly all genera can be found in Fischer (2004). These clades may become subfamilies or tribes in future classifications, but for the time being we will use informal names derived from prominent genera within each group.

14.2.2.1 *Lindenbergia* Clade

The *Lindenbergia* clade (clade I of Bennett and Mathews 2006) consists of the Afro-Asian genus *Lindenbergia* (Hjertson 1995; Fig. 14.3a, b). Its 15 species are the

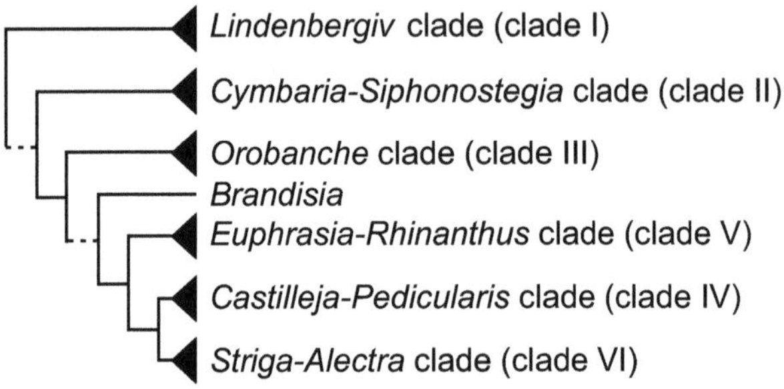

Fig. 14.2 Phylogenetic relationships of major clades of Orobanchaceae. Clade numbers used by Bennett and Mathews (2006) are given in parentheses. Uncertain branches are indicated by *dashed lines*

only nonparasitic genus in Orobanchaceae. The close relationship to parasitic genera is supported by the shared mode of corolla aestivation (Fischer 2004) and permanently open stomata (Wolfe et al. 2005). Most molecular phylogenetic analyses indicate that the *Lindenbergia* clade is sister to the remainder of the family (dePamphilis et al. 1997; Nickrent et al. 1998; Young et al. 1999; Bennett and Mathews 2006; Park et al. 2008; McNeal et al. 2013). This agrees with the presence of tricolporate pollen with a reticulate exine (Minkin and Eshbaugh 1989; Hjertson 1995) and the nonparasitic lifestyle of *Lindenbergia* not found elsewhere in Orobanchaceae, but frequent in other families of Lamiales.

14.2.2.2 *Cymbaria–Siphonostegia* Clade

The *Cymbaria–Siphonostegia* clade (clade II of Bennett and Mathews 2006) includes the small hemiparasitic genera *Bungea*, *Cymbaria* (Fig. 14.3c), *Monochasma*, *Schwalbea* and *Siphonostegia* (Figs. 14.3d and 14.4a). A close relationship among those genera was previously suggested (Bentham 1876; Wettstein 1891) and acknowledged by recognition of tribe Cymbarieae. This tribe is characterized by the presence of bracteoles, a tubular and only weakly dorsiventral calyx, a strongly two-lipped corolla, and anthers with two equal mostly rounded thecae (Fischer 2004). A molecular synapomorphy of this clade is a unique intron in the phytochrome A gene (Bennett and Mathews 2006). Wettstein (1891) contemplated merging *Schwalbea* with *Siphonostegia*, which is now supported by molecular data (Fig. 14.4a).

14.2.2.3 *Orobanche* Clade

The exclusively holoparasitic *Orobanche* clade (clade III of Bennett and Mathews 2006) consists of *Orobanche* (Fig. 14.3e), *Phelipanche* (Fig. 14.3f) and related genera (Fig. 14.4b), all members of the traditional Orobanchaceae. This extratropical clade represents the first transition from hemi- to holoparasitism within the family. A summary of molecular phylogenetic hypotheses on relationships among genera within this clade can be found in Park et al. (2008).

Orobanche in its traditional circumscription (Beck-Mannagetta 1930) includes four sections: *Gymnocaulis*, *Myzorrhiza*, *Trionychon*, and *Orobanche*. These have morphological differences in the absence/presence of bracteoles, calyx shape, differences in modes of fruit dehiscence, seed surface and placentation patterns (Beck-Mannagetta 1890, 1930; Teryokhin et al. 2003; Plaza et al. 2004; Domina and Colombo 2005) and possess different chromosome base numbers: $x = 19$ in *Orobanche* versus $x = 12$ in the others (Schneeweiss et al. 2004b). This has led some authors to treat the sections as separate genera (Holub 1977, 1990; Teryokhin et al. 2003). Molecular phylogenetic data corroborate their distinctness (Wolfe et al. 2005; Bennett and Mathews 2006). Furthermore, although *Orobanche* sensu lato is supported as a clade (Bennett and Mathews 2006; Park et al. 2008; McNeal et al. 2013), it additionally includes the genus *Phelypaea* characterized by large single flowers with red to orange corollas (Beck-Mannagetta 1930), rendering *Orobanche* non-monophyletic (Schneeweiss et al. 2004a). As all lineages of *Orobanche* sensu lato can be readily distinguished morphologically and show different evolutionary trajectories with respect to chromosome number evolution and polyploidy (Schneeweiss et al. 2004b), genome size evolution (Weiss-Schneeweiss et al. 2006), and genome evolution (Park et al. 2007b; Piednoël et al. 2012), it appears sensible to adopt earlier narrow generic circumscriptions for *Orobanche* (species numbers in Fig. 14.4b follow Uhlich et al. 1995) and to treat the following as separate genera: *Aphyllon* (*O.* sect. *Gymnocaulis*), *Myzorrhiza* (*O.* sect. *Myzorrhiza*), *Phelipanche* (*O.* sect. *Trionychon*), *Boulardia* (*O. latisquama* of *O.* sect. *Orobanche*), *Orobanche* (*O.* sect. *Orobanche*), and *Phelypaea*. This nomenclature has been recommended for adoption in the applied literature as well (Joel 2009). The monotypic *Necranthus* from northeastern Turkey (Gilli 1968) has been suggested to be conspecific with *O. gamosepala* (Teryokhin 2001), but this still awaits corroboration from molecular data.

Molecular data strongly suggest a close relationship between *Epifagus* and *Conopholis*. *Epifagus* was the first holoparasitic plant whose entire plastid genome was sequenced (Wolfe et al. 1992; see Sect. 15.3). Prior to having molecular data, a close relationship between these genera had not been suggested. Fischer (2004) had even assigned them to different tribes. The closest relative of *Epifagus* and *Conopholis* is either *Kopsiopsis*, at times considered a section of *Orobanche* or as part of *Boschniakia* (Beck-Mannagetta 1890 *versus* Bentham 1876), or *Boschniakia* sensu stricto (Bennett and Mathews 2006; McNeal et al. 2013; versus Wolfe et al. 2005; Park et al. 2008). East Asian *Boschniakia* has been designated as *Xylanche* (Beck-Mannagetta 1930), but appears to be closely related to *Boschniakia* sensu stricto (McNeal et al. 2013). Since the putative differentiating characters, carpel number and presence of bracteoles, are unstable (Zhang and Cvelev 1998), the final taxonomic assessment of *Xylanche* requires further data. The recently described *Eremitilla* (Yatskievych and Contreras Jiménez 2009), which has a number of characters previously unknown in Orobanchaceae such as five-ribbed ovaries, also belongs here as supported by as yet unpublished molecular data. *Mannagettaea* and *Cistanche* are members of the *Orobanche* clade with phylogenetic affinity to the subclade including *Orobanche* (Fig. 14.4b).

Fig. 14.3 (continued)

Fig. 14.3 Representatives of major clades of Orobanchaceae. (**a**) *Lindenbergia indica*. (**b**) *L. philippensis*. (**c**) *Cymbaria mongolica*. (**d**) *Siphonostegia chinensis*. (**e**) *Orobanche alba*. (**f**) *Phelipanche purpurea* subsp. *purpurea*. (**g**) *Castilleja pallescens* var. *inverta*. (**h**) *C. rhexifolia*. (**i**) *Pedicularis rostrato-capitata*. (**j**) *Euphrasia minima*. (**k**) *Rhinanthus glacialis*. (**l**) *Striga gesnerioides*. (**m**) *Alectra sessiliflora* (photos: **a**, **l**, **m**—Jeffrey Morawetz; **b**—Susann Wicke; **c**—Josef Buchner; **d**—David E. Boufford; **e**, **f**, **i**, **j**, **k**—Michaela Sonnleitner; **g**, **h**—Dave Tank)

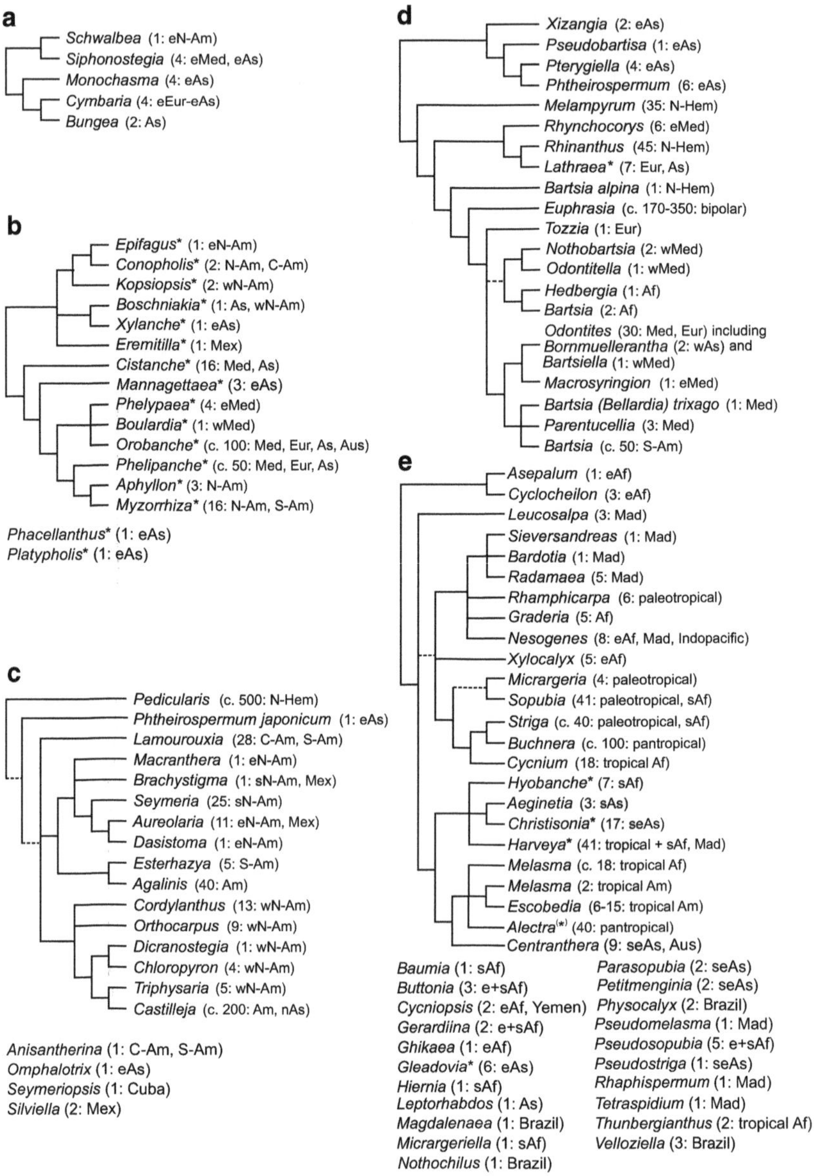

Fig. 14.4 Phylogenetic relationships of genera within (**a**) the *Cymbaria–Siphonostegia* clade, (**b**) the *Orobanche* clade, (**c**) the *Castilleja-Pedicularis* clade, (**d**) the *Euphrasia-Rhinanthus* clade, and (**e**) the *Striga–Alectra* clade of Orobanchaceae. Species numbers and geographical distribution (Nickrent 2012) are given in parentheses. Exclusively holoparasitic genera are indicated with an *asterisk*, those with both hemi- and holoparasites with an *asterisk* in parentheses. Uncertain branches—*dashed lines*. Genera, whose phylogenetic affinity has not been tested yet by molecular data, are listed below the phylogenetic tree of the clade they likely belong to. Major geographical regions: *Af* Africa, *Am* America, *As* Asia, *Aus* Australia, *C-Am* Central America, *Eur* Europe, *Mad* Madagascar, *Med* Mediterranean, *Mex* Mexico, *N-Am* North America, *N-Hem* northern hemisphere, *S-Am* South America

14.2.2.4 *Brandisia*

Despite its parasitic habit, the East Asian *Brandisia* (13 species) has never been associated with parasitic Scrophulariaceae (Fischer 2004). This genus consists of shrubs and vines with a densely tomentose indumentum, bracteate thyrsic inflorescences, a five-lobed calyx, and funnel-shaped corollas (Fischer 2004). Molecular data firmly place it in Orobanchaceae (Oxelman et al. 2005; Bennett and Mathews 2006; Albach et al. 2009; McNeal et al. 2013), where it may constitute a separate lineage sister to the remaining clades (Fig. 14.2).

14.2.2.5 *Castilleja–Pedicularis* Clade

The exclusively hemiparasitic *Castilleja–Pedicularis* clade (clade IV of Bennett and Mathews 2006; Fig. 14.4c) contains two of the largest genera within the family, *Pedicularis* (ca. 500 species; Fig. 14.3i) and *Castilleja* (ca. 200 species; Fig. 14.3g, h). The most comprehensive molecular phylogenetic study of the monophyletic *Pedicularis* (Ree 2005) indicates that vegetative characters, such as phyllotaxis, tend to better reflect phylogenetic relationships than floral characters, which have been used predominantly for infra-generic taxonomy. *Phtheirospermum* is not monophyletic because *Ph. japonicum*, the single representative in earlier molecular phylogenetic studies (Bennett and Mathews 2006), is more closely related to *Pedicularis* than to the remaining *Phtheirospermum* species (Dong et al. 2013) that belong to the *Euphrasia–Rhinanthus* clade described below. Morphological synapomorphies supporting the inclusion of *Pedicularis* and of *Phtheirospermum japonicum* in the *Castilleja–Pedicularis* clade remain to be identified.

The remaining genera of the *Castilleja–Pedicularis* clade fall into two subclades (Fig. 14.4c). The first subclade is mainly American (Fig. 14.4c) and corresponds to subtribe Castillejinae (tribe "Castillejeae" in Fischer 2004), characterized by laterally compressed and anteriorly or laterally cleft calyces, strongly two-lipped corollas, and anthers with unequal thecae that are unequally attached (Fischer 2004; Tank et al. 2009). Recent molecular phylogenetic analyses of this subtribe (Tank and Olmstead 2008) resulted in several changes in genus delimitation (Tank et al. 2009). Specifically, the monotypic genera *Clevelandia* and *Ophiocephalus* phylogenetically nest within *Castilleja* (including the sometimes distinguished *Gentrya*) and were merged with it. Furthermore, the three subgenera of *Cordylanthus* were raised to the generic level (Tank et al. 2009): *Chloropyron*, *Cordylanthus*, and *Dicranostegia*. *Orthocarpus* and *Triphysaria* are unchanged in their circumscription. Flower morphology, anther morphology, position of seed hilum, and chromosome base number are important diagnostic characters (Tank et al. 2009).

The second subclade likely corresponds to tribe Gerardieae as circumscribed by Fischer (2004). It is characterized by plain corolla lobes, anthers with two separate thecae, bilocular ovaries and loculicid capsules. Whereas the precise position of

Lamourouxia is uncertain (Ernst 1972; Bennett and Mathews 2006), the other genera form a well-supported tightly knit group. *Agalinis*, including the central North American *Tomanthera* (Neel and Cummings 2004; Pettengill and Neel 2008), is closely related to *Esterhazya* (Bennett and Mathews 2006; McNeal et al. 2013). Since no molecular phylogenetic information is available for South American *Agalinis* species, which are sometimes treated as members of a separate genus *Virgularia* (Pennell 1928), the precise circumscription of *Agalinis* remains to be established. *Brachystigma*, *Aureolaria* and its close relative *Dasistoma*, *Seymeria* as well as *Macranthera* form a clade distinct from *Agalinis* and *Esterhazya*. A denser sampling will be necessary to ascertain monophyly of *Aureolaria* and *Seymeria*.

14.2.2.6 *Euphrasia–Rhinanthus* Clade

The genera in the *Euphrasia–Rhinanthus* clade (clade V of Bennett and Mathews 2006, the *Bartsia–Melampyrum* clade of Tank et al. 2006; Fig. 14.4d) largely correspond to tribe Rhinantheae in the circumscription of Fischer (2004). Its members lack bracteoles, have usually strongly two-lipped corollas with the upper lip often being galeate (helmet-shaped), have anthers with two separate and usually equal thecae and have bilocular ovaries and mostly loculicid capsules. As many of these features are shared with *Pedicularis* of the *Castilleja–Pedicularis* clade (see Sect. 14.2.2.5), the identification of morphological differential characters for this clade is still needed. The earliest branch comprises the Chinese endemic genera *Pterygiella* and its segregate *Xizangia*, the Chinese endemic *Pseudobartsia* and the East Asian *Phtheirospermum* (Dong et al. 2011, 2013) except *P. japonicum* belonging to the *Castilleja-Pedicularis* clade (see Sect. 14.2.2.5). The largest genus within the *Euphrasia-Rhinanthus* clade is *Euphrasia* (Fig. 14.3j) with ca. 350 species, which reached its conspicuous distribution by multiple long-distance dispersals from Eurasia (Gussarova et al. 2008). Molecular data confirm monophyly of *Melampyrum* and of *Rhinanthus*. *Rhinanthus* (Fig. 14.3k) together with *Rhynchocorys* are closely related to *Lathraea* (Těšitel et al. 2010; Scheunert et al. 2012; Fig. 14.4d), which is the sole holoparasitic member of the *Euphrasia–Rhinanthus* clade and is an independent transition from hemi- to holoparasitism. *Odontites* in its current circumscription (Bolliger 1996) is paraphyletic by including *Bartsiella* and *Bornmuellerantha* (Fig. 14.4d), and these genera have been recently merged again; the precise position of *Macrosyringion* with respect to a thus extended *Odontites* is still uncertain (Scheunert et al. 2012). *Bartsia* consists of at least four lineages, *B. alpina*, African *Bartsia* species (*B.* sect. *Longiflorae*; Molau 1990), *B. trixago* (=*Bellardia t.*), and South American *Bartsia* species (Fig. 14.4d). The African *Bartsia* species are closely related to and have been recently merged with the African *Hedbergia* (Scheunert et al. 2012). South American *Bartsia* species and *B. trixago* are more closely related to *Parentucellia*, *Odontites* or *Tozzia* than to *B. alpina* (Wolfe et al. 2005; Bennett and Mathews 2006; Těšitel et al. 2010; Scheunert et al. 2012); taxonomically, this has been recently accounted for by merging (the investigated) South American *Bartsia* species, *Parentucellia* and *B. trixago* into *Bellardia* (Scheunert et al. 2012).

14.2.2.7 *Striga–Alectra* Clade

The *Striga–Alectra* clade (clade VI of Bennett and Mathews 2006; *Alectra–Sopubia* clade of Tank et al. 2006), centred in subtropical and tropical areas, is the taxonomically most challenging clade and combines several currently recognized tribes (Fischer 2004) as well as the small families (Marais 1981) Cyclocheilaceae and Nesogenaceae (Bremer et al. 2002; Oxelman et al. 2005; Morawetz et al. 2010). Although circumscription of the clade appears to be established (Wolfe et al. 2005; Bennett and Mathews 2006; McNeal et al. 2013), its precise phylogenetic structure and composition is still insufficiently known as evident from the list of unstudied genera (Fig. 14.4e). Inclusion of those as yet unstudied genera may well change circumscription and delimitation of groups discussed here (Fig. 14.4e). This will be of particular interest in the context of the multiple evolution of holoparasitism within this clade and the presence of the major weedy species in subtropical regions (*Striga*, *Alectra*; see Sect. 14.3.2).

The earliest diverging subclade consists of *Cyclocheilon* and *Asepalum* (Morawetz et al. 2010), which both are woody and lack a conspicuous calyx (Demissew 2004). Earlier phylogenetic studies (Bremer et al. 2002; Oxelman et al. 2005) showed that *Cyclocheilon* actually belongs to Orobanchaceae without allowing more detailed positioning within the family, and clearly recognition of a separate family Cyclocheilaceae (Marais 1981; Demissew 2004) is obsolete.

Striga (Fig. 14.3l) is closely related to *Buchnera* and to *Cycnium* (Young et al. 1999; Wolfe et al. 2005; Bennett and Mathews 2006; Morawetz et al. 2010; McNeal et al. 2013). Current evidence suggests that these genera are monophyletic and together are sister to a subclade containing *Sopubia* and *Micrargeria* (Fischer et al. 2012; McNeal et al. 2013). Denser sampling is necessary to test this hypothesis and the relationships of the above genera to *Xylocalyx* and the subclade including *Graderia*, *Nesogenes*, *Rhamphicarpa*, and the closely related Madagascan genera *Bardotia*, *Radamaea*, and *Sieversandreas* (Fig. 14.4e). These genera have been assigned to three different tribes (Fischer 2004) or, in case of *Nesogenes*, to a separate family Nesogenaceae, characterized by one-loculed ovaries (Marais 1981; Harley 2004), none of which finds support from molecular data.

The following subclade represents the largest diversification of holoparasitic species outside the *Orobanche* clade. It consists of three lineages with unclear relationships: (1) hemiparasitic *Aeginetia*, which includes *Ae. indica* and the closely related holoparasitic *Christisonia*; (2) holoparasitic *Hyobanche*, whose monophyly is largely uncontested (Randle and Wolfe 2005; Wolfe et al. 2005; Morawetz and Wolfe 2009; Morawetz et al. 2010; but see Bennett and Mathews 2006); and (3) nearly exclusively holoparasitic *Harveya* in an extended circumscription to include *H. alba* (Morawetz and Randle 2010), which has been suggested to be designated as separate genus *Paraharveya* (Fischer 2004), and *H. alectroides* (Fischer et al. 2012), originally described as separate genus *Parastriga* (Mildbraed 1930).

A further subclade includes *Alectra* (Fig. 14.3m), *Melasma*, and *Escobedia* and may thus correspond to tribe Escobedieae (Fischer 2004). Whereas circumscription of *Alectra* is corroborated by molecular data (Morawetz and Wolfe 2009; Morawetz et al. 2010), *Melasma* consists of two phylogenetically and geographically distinct lineages, the American one being more closely related to *Escobedia* than to African *Melasma* species (Morawetz and Wolfe 2009; Morawetz et al. 2010), but further data will be necessary before a revised generic classification can be presented. Morphological evidence suggests that *Alectra orobanchoides* is a holoparasite (Morawetz et al. 2010), indicating an independent transition from hemi- to holoparasitism. Whereas *Centranthera* possibly is sister group to the subclade including *Alectra*, the precise position of *Leucosalpa* within the *Striga–Alectra* clade (Fischer et al. 2012) still needs to be established.

14.3 Phylogenetic Relationships of Weedy Taxa

Whereas the majority of Orobanchaceae species are of no economic concern, a few genera of the *Orobanche* clade and the *Striga–Alectra* clade (Fig. 14.4b, e) include weedy species that attack various crops, where they can cause substantial yield losses. A detailed account on host range, extent of the damage caused, and their relevance in different regions can be found in Chap. 18.

14.3.1 Orobanche *and* Phelipanche

Several recent studies have addressed phylogenetic relationships of and within *Orobanche* and *Phelipanche* (*Orobanche* clade) using DNA sequence (Manen et al. 2004; Schneeweiss et al. 2004a; Carlón et al. 2005, 2008; Park et al. 2008) and fingerprinting (Joel et al. 1998; Román et al. 2003) as well as proteomic data (Castillejo et al. 2009). The agreement of groups identified by molecular data with those distinguished in traditional taxonomy (Beck-Mannagetta 1930; Teryokhin et al. 2003) is limited (Manen et al. 2004; Schneeweiss et al. 2004a).

Overall, molecular data confirm the gross phylogenetic position of the most important weedy species *O. cernua* sensu lato (incl. *O. cumana*), *O. crenata*, *Phelipanche aegyptiaca*, *P. mutelii* and *P. ramosa* suggested by traditional morphology-based taxonomy (Beck-Mannagetta 1930; Teryokhin et al. 2003). Whereas *O. cernua* sensu lato is well supported as a distinct lineage (Schneeweiss et al. 2004a; Carlón et al. 2005, 2008), its taxon composition is still unclear. This concerns forms of *O. cernua* growing on solanaceous crops, some of which (those with strongly developed annual tubers) have been distinguished as *O. cernua* subsp. *rajahmundrica* (Teryokhin et al. 1996), and *O.* (*cernua* var.) *cumana* parasitizing wild and cultivated Asteraceae, particularly sunflower (Parker 2009; see Sect. 18.2.3). *O. crenata* is very closely related to the morphologically similar southwest Asian *O. owerini* (Schneeweiss et al. 2004a). Both nest within the taxonomically daunting *O. minor* aggregate (Schneeweiss et al. 2004a), which includes *O. minor*, itself a widespread, but only locally important weed mainly on forage legumes (Parker 2009; see Sect. 18.2.6).

Accounts on weedy *Phelipanche* are strongly hampered by our poor understanding of species diversity, morphological taxon delimitation and the influence of host plants on parasite morphology. The latter is also true for *Orobanche*. Consequently, taxonomic assignments of weedy populations need to be viewed with caution and might change considerably, once a sound taxonomic framework has been established. *Phelipanche aegyptiaca*, *P. mutelii*, and *P. ramosa* are part of the species rich but mostly unresolved main diversification within the genus (Schneeweiss et al. 2004a; Carlón et al. 2005, 2008) corresponding to *Phelipanche* sect. *Phelipanche* (Teryokhin et al. 2003). There is evidence that within this large assemblage, *P. aegyptiaca* (excluding *P. hirtiflora*) is closely related to *P. mutelii* (Carlón et al. 2008). Several species with branched inflorescences and medium-sized flowers have been at times subsumed under the name *P. mutelii*, and only recently, the name was assigned to a taxon widespread in the southern Mediterranean (Carlón et al. 2008). *Phelipanche mutelii* is taxonomically usually associated with *P. ramosa*, but *P. mutelii* appears to be more closely related to *P. lavandulacea*, a morphologically distinct species of no economic concern, than to *P. ramosa* (Schneeweiss et al. 2004a; Carlón et al. 2008). Further data on *P. aegyptiaca* are urgently needed to assess whether it is actually distinct from *P. mutelii* (in the circumscription of Carlón et al. 2008) or not and whether populations currently subsumed under *P. aegyptiaca* actually belong to the same taxon. In some regions, *P. ramosa* populations are morphologically clearly distinct from co-occurring wild races (e.g. Pujadas Salvà 2002). This appears not be the case in the eastern Mediterranean region, where species diversity is high, but insufficiently known. Usually, *P. ramosa* is associated with the widespread Mediterranean *Ph. nana*. Although molecular data do not disagree with a close relationship, the lack of sufficient molecular distinction also from morphologically and ecologically clearly different species (e.g. *P. reuteriana*; Carlón et al. 2008) renders meaningful taxonomic conclusions impossible.

DNA sequence data are insufficient to delimit closely related species, negatively affecting assessment of the distinctness (or lack thereof) of weedy species from their usually not precisely known wild congeners. A possible solution might be the application of much more variable fingerprinting techniques. Indeed, such markers have already been widely applied in *Orobanche* and *Phelipanche*, for instance, for identification purposes (Joel et al. 1996, 1998; Paran et al. 1997; Portnoy et al. 1997; Zeid et al. 1997; Benharrat et al. 2002; Román et al. 2007a; see Chap. 19 and Sect. 20.3) or for assessing intraspecific diversity in the context of potential host-specific parasite races (Gagne et al. 2000; Román et al. 2001, 2002, 2007b, c; Brault et al. 2007; Thorogood et al. 2008, 2009a; Satovic et al. 2009; Vaz Patto et al. 2009). Although these are important contributions towards a better understanding of species differentiation, they fall short in terms of geographic and taxonomic coverage. An approach combining molecular techniques, including novel methods like next-generation sequencing-based RAD sequencing (Baird et al. 2008; Emerson et al. 2010), with cross-inoculation experiments, as applied in *O. minor* and relatives of northwestern Europe (Thorogood et al. 2008, 2009a, b), employing a geographically and taxonomically representative set of accessions will be necessary to cut the Gordian knot of *Orobanche* and *Phelipanche* taxonomy.

14.3.2 Striga *and* Alectra

The most important weedy species of the *Striga–Alectra* clade can be found in *Striga* (see Chap. 17 and Sect. 18.3). Parker and Riches (1993) suggest that *S. hermonthica*, which parasitizes many cereal crops and sugarcane, "must surely be the most important parasitic weed species on a world scale". Despite this economic importance and although *S. asiatica* has become a model parasitic plant, little is known about phylogenetic relationships within the genus. In their monograph on African species, Mohamed et al. (2001) neither suggest any infrageneric classification nor do they provide a hypothesis on genus-wide relationships despite the cladistic analysis of morphological data published previously (Mohamed et al. 1996). Only a handful of species has been included in molecular phylogenetic studies, usually with scarce and often only partially overlapping taxon coverage. According to those studies, *S. gesnerioides* (including *S. orobanchoides*), a major parasite on cowpea, is related to the non-pest *S. bilabiata* subsp. *bilabiata* (Wolfe et al. 2005; Bennett and Mathews 2006; Morawetz and Wolfe 2009; Morawetz et al. 2010), and *S. hermonthica* is related to *S. passargei* (Wolfe and dePamphilis 1998), a locally important weedy species on maize. Relationships to each other and to other investigated species, including the weedy species *S. asiatica*, remain uncertain. No molecular phylogenetic information is available for the other species, although sometimes of local importance (Parker and Riches 1993: 9–18): *S. angustifolia*, *S. aspera*, *S. densiflora*, *S. forbesii* and *S. latericea*. Fingerprinting techniques have been applied to selected *Striga* species aiming at assessing genetic diversity or at race identification and differentiation (Botanga et al. 2002; Gethi et al. 2005; Botanga and Timko 2006; Dube and Belzile 2010), the latter potentially of taxonomic relevance.

Several *Alectra* species are of economic importance. Of those, *A. vogelii*, a pest on cowpea in different regions of Africa, is closely related to *A. orobanchoides* (Morawetz and Wolfe 2009), which despite its wide host range is only of limited relevance as a pest (Parker and Riches 1993: 86). A group of African species, including *A. sessiliflora*, which can attack crop species, but so far has not had any wider economic importance (Parker and Riches 1993: 82), is sister group to a small clade of exclusively New World species (Morawetz and Wolfe 2009), which includes *A. aspera*, a local pest on sugarcane (listed as *A. fluminensis* by Parker and Riches 1993: 87). No molecular phylogenetic data are available for *A. picta*, a minor weedy species in several regions of Africa, but morphological and geographical similarity with *A. vogelii* strongly suggests that both are closely related or even conspecific.

14.4 Evolutionary Trends: Some Examples

14.4.1 Parasitism

Parasitism has evolved several times independently within angiosperms. However, only Orobanchaceae includes all forms of parasitism, i.e. facultative and obligate hemiparasitism and holoparasitism (Kuijt 1969). Parasitism has evolved within Orobanchaceae only once, suggesting that mechanisms and processes leading to haustorium formation and host penetration (Westwood et al. 2010) follow a common pathway established once in the evolution of the family. Since in holoparasitic plants the genetic basis and consequently the machinery for photosynthesis usually become partially or completely lost (dePamphilis 1995), the evolution from hemi- to holoparasitism is considered irreversible (but see the possible exception of hemiparasitic *Harveya obtusifolia*, which deeply nests within its holoparasitic congeners; Morawetz et al. 2010). Consequently, the observed occurrence of holoparasitism in different clades indicates independent transitions from hemi- to holoparasitism (Tank et al. 2006). See further discussion on evolution of the Orobanchaceae in Chap. 15.

14.4.2 Host Range Evolution

Host range, i.e. the number and phylogenetic diversity of successfully attackable host plants, differs vastly among different parasites. Host range evolution in Orobanchaceae has not been formally tested yet. This is due to the lack of a comprehensive phylogenetic framework but also to the lack of precise and quantifiable data on host range. Circumstantial evidence suggests that in Orobanchaceae (but not necessarily in other parasitic lineages; Kuijt 1969), holoparasitic groups tend to have narrower host ranges than hemiparasitic members (Heide-Jørgensen 2008). This ecological specialization may be an important driver of speciation as suggested for animal parasites (Huyse et al. 2005).

Generalists in animal parasites have non-predictable resources, whereas specialists tend to exploit predictable resources (Ward 1992). A similar pattern was found in *Orobanche* and *Phelipanche*, which show a significant association of host range width (narrow versus wide) and host longevity (perennial versus annual; Schneeweiss 2007). As in both genera most weedy species share wide host ranges and utilize short-lived hosts, thus being at least facultatively annuals themselves, it is tempting to speculate that features allowing fast life cycles and utilization of wide host ranges may be important preadaptations for a parasite to become weedy (see Sect. 19.3.3). It remains to be tested whether such patterns hold also for other Orobanchaceae.

It can be expected that host range is influenced by the parasite's host recognition, and species with narrow host range are expected to respond to more specific host cues than a generalist parasite. Chemical germination stimulants probably play an important role (Yoneyama et al. 2010), but their precise function in determining host specificity is not yet known (see Chap. 13). Host range may also be affected by the availability of host species, and narrow and wide host ranges might be advantageous in little or more diverse communities, respectively. Testing this hypothesis will require not only more ecological data but also a need to take into account community change over ecological and even more so evolutionary time scales. Possible ways that a parasite may quickly evolve to a new host include epigenetic modifications and alternative splicing of mRNA and are discussed in Chap. 13.

14.4.3 Polyploidy and Horizontal Gene Transfer

A striking feature of *Orobanche* and related genera is that they possess high chromosome base numbers ranging from $x = 12$ in *Aphyllon*, *Myzorrhiza*, and *Phelipanche* to $x = 41$ in *Kopsiopsis* (Schneeweiss et al. 2004b; H. Weiss-Schneeweiss unpublished data). Other holoparasitic groups follow the same pattern ($x = 15$ in *Aeginetia* and $x =$ ca. 21 in *Lathraea* and *Phacellanthus*; Fedorov 1969; Schneeweiss and Weiss 2003). However, an association of polyploidy and holoparasitism still needs to be formally tested (see Sect. 15.2.2).

Parasitic plants are suitable vectors for the transfer of genetic material between host and parasite (Davis and Wurdack 2004) and vice versa (Mower et al. 2004; Davis et al. 2005). Orobanchaceae are no exception. Horizontal gene transfer of a plastid region has occurred from *Phelipanche* into *Orobanche* (Manen et al. 2004; Park et al. 2007a), even if the horizontally acquired genes likely do not reside in the plastid genome. *Striga hermonthica* has a horizontally acquired nuclear gene from its grass host (Yoshida et al. 2010). In *Rafflesia*, some of the genes horizontally acquired from its host have replaced the parasite's own gene activity (Xi et al. 2012), a process likely also occurring in Orobanchaceae. Horizontal gene transfer not only manifests itself on an evolutionary time scale but probably is also relevant on an ecological time scale, considering that interplant transfer of genetic material occurs regularly (Westwood et al. 2009; Bock 2010) and has been suggested for *Hyobanche* (Randle and Wolfe 2005). See further discussion of horizontal gene transfer in Sect. 15.5.

14.5 Outlook

Orobanchaceae provides a unique opportunity to study the evolution of parasitism concerning, for instance, host recognition and its effects on host range, macromolecular trafficking between plants or genome evolution after loss of photosynthesis. Molecular data were instrumental in establishing a sound phylogenetic framework identifying the well-supported main groups. Since these only partly agree with morphologically defined groups, a morphological re-evaluation of the entire family allowing synapomorphies to be identified is urgently needed. Furthermore, the position and boundaries of numerous genera are still insufficiently understood, and there are plenty unsolved issues concerning relationship among and discrimination of species (including the economically important weedy species; see Sect. 14.3 and Chap. 18). Targeted sampling and application of the steadily growing molecular toolbox will help solving such issues in the near future.

Acknowledgments Two anonymous reviewers are acknowledged for their cogent comments and suggestions. My research on *Orobanche* and related genera would not have been possible without the help of numerous colleagues. My work is financially supported by the Austrian Science Foundation (FWF grants P14352, P19404) and the Faculty of Life Sciences of the University of Vienna.

References

Albach DC, Yan K, Jensen SR, Li H-Q (2009) Phylogenetic placement of *Triaenophora* (formerly Scrophulariaceae) with some implications for the phylogeny of Lamiales. Taxon 58:749–756

Álvarez I, Wendel JF (2003) Ribosomal ITS sequences and plant phylogenetic inference. Mol Phylogenet Evol 29:417–434

Angiosperm Phylogeny Group (2009) An update of the Angiosperm Phylogeny Group classification for the orders and families of flowering plants: APG III. Bot J Linn Soc 161:105–121

Baird NA, Etter PD, Atwood TS, Currey MC, Shiver AL, Lewis ZA, Selker EU, Cresko WA, Johnson EA (2008) Rapid SNP discovery and genetic mapping using sequenced RAD markers. PLoS One 3(10):e3376. doi:10.1371/journal.pone.0003376

Barkman TJ, McNeal JR, Lim SH, Coat G, Croom HB, Young ND, dePamphilis CW (2007) Mitochondrial DNA suggests at least 11 origins of parasitism in angiosperms and reveals genomic chimerism in parasitic plants. BMC Evol Biol 7:248. doi:10.1186/1471-2148-7-248

Beck-Mannagetta G (1890) Monographie der Gattung *Orobanche*. Theodor Fischer, Cassel

Beck-Mannagetta G (1891) Orobanchaceae. In: Engler A, Prantl K (eds) Die natürlichen Pflanzenfamilien nebst ihren Gattungen und wichtigeren Arten, insbesondere den Nutzpflanzen. Wilhelm Engelmann, Leipzig, pp 123–132

Beck-Mannagetta G (1930) Orobanchaceae. In: Engler A (ed) Das Pflanzenreich. Regni vegetabilis conspectus, 4, vol 261. Wilhelm Engelmann, Leipzig, pp 1–348

Bellini R (1907) Criteri per una nuova classificazione della Personatae (Scrophulariaceae et Rhinantaceae [sic]). Ann Bot 6:131–145

Benharrat H, Veronesi C, Theodet C, Thalouarn P (2002) *Orobanche* species and population discrimination using intersimple sequence repeat (ISSR). Weed Res 42:470–475

Bennett J, Mathews S (2006) Phylogeny of the parasitic plant family Orobanchaceae inferred from phytochrome A. Am J Bot 93:1039–1051

Bentham G (1876) Ordo CXV. Scrophularineae. Ordo CXVI. Orobanchaceae. In: Bentham G, Hooker JD (eds) Genera plantarum, vol 2. Reeve, London, pp 913–980, 980–986

Bock R (2010) The give-and-take of DNA: horizontal gene transfer in plants. Trends Plant Sci 15:11–22

Boeshore I (1920) The morphological continuity of Scrophulariaceae and Orobanchaceae. Contrib Bot Lab Univ Pennsylv 5:139–177

Bolliger M (1996) Monographie der Gattung *Odontites* (Scrophulariaceae) sowie der verwandten Gattungen *Macrosyringion*, *Odontitella*, *Bornmuellerantha* und *Bartsiella*. Willdenowia 26:37–168

Botanga CJ, Timko MP (2006) Phenetic relationships among different races of *Striga gesnerioides* (Willd.) Vatke from West Africa. Genome 49:1351–1365

Botanga CJ, Kling JG, Berner DK, Timko MP (2002) Genetic variability of *Striga asiatica* (L.) Kuntz based on AFLP analysis and host-parasite interaction. Euphytica 128:375–388

Brault M, Betsou F, Jeune B, Tuquet C, Sallé G (2007) Variability of *Orobanche ramosa* populations in France as revealed by cross infestations and molecular markers. Environ Exp Bot 61:272–280

Bremer B, Bremer K, Heidari N, Erixon P, Olmstead RG, Anderberg AA, Källersjö M, Barkhordarian E (2002) Phylogenetics of asteroids based on 3 coding and 3 non-coding chloroplast DNA markers and the utility of non-coding DNA at higher taxonomic levels. Mol Phylogenet Evol 24:274–301

Carlón L, Gómez Casares G, Laínz M, Moreno Moral G, Sánchez Pedraja Ó, Schneeweiss GM (2005) Más, a propósito de algunas *Orobanche* L. y *Phelipanche* Pomel (Orobanchaceae) del oeste del Paleártico. Doc Jardin Bot Atl Gijón 3:1–71

Carlón L, Gómez Casares G, Laínz M, Moreno Moral G, Sánchez Pedraja Ó, Schneeweiss GM (2008) Más, a propósito de algunas *Phelipanche* Pomel, *Boulardia* F. W. Schultz y *Orobanche* L. (Orobanchaceae) del oeste del Paleártico. Doc Jardin Bot Atl Gijón 6:1–128

Castillejo MA, Fernández-Aparicio M, Satovic Z, Rubiales D (2009) Comparative proteomic analysis of *Orobanche* and *Phelipanche* species inferred from seed proteins. Weed Res 49 (Supp 1):81–87

Davis CC, Wurdack KJ (2004) Host-to-parasite gene transfer in flowering plants: phylogenetic evidence from Malpighiales. Science 305:676–678

Davis CC, Fritsch PW, Li J, Donoghue MJ (2002) Phylogeny and biogeography of *Cercis* (Fabaceae): evidence from nuclear ribosomal ITS and chloroplast *ndhF* sequence data. Syst Bot 27:289–302

Demissew S (2004) Cyclocheilaceae. In: Kadereit JW (ed) Flowering plants, dicotyledons, Lamiales (except Acanthaceae including Avicenniaceae), vol 7, The families and genera of vascular plants. Springer, Berlin, pp 60–62

dePamphilis CW (1995) Genes and genomes. In: Press MC, Graves JD (eds) Parasitic plants. Chapman & Hall, London, pp 177–205

dePamphilis CW, Palmer JD (1990) Loss of photosynthetic and chlororespiratory genes from the plastid genome of a parasitic flowering plant. Nature 348:337–339

dePamphilis CW, Young ND, Wolfe AD (1997) Evolution of plastid gene *rps2* in a lineage of hemiparasitic and holoparasitic plants: many losses of photosynthesis and complex patterns of rate variation. Proc Natl Acad Sci USA 94:7367–7372

Domina G, Colombo P (2005) Seed micromorphology and its variation in Sicilian *Orobanche* (Magnoliopsida). Flora Mediterr 15:99–107

Dong LN, Wortley AH, Wang H, Li DZ, Lu L (2011) Efficiency of DNA barcodes for species delimitation: a case in *Pterygiella* Oliv. (Orobanchaceae). J Syst Evol 49:189–202

Dong LN, Wang H, Wortley AH, Lu L, Li DZ (2013) Phylogenetic relationships in the *Pterygiella* complex (Orobanchaceae) inferred from molecular and morphological evidence. Bot J Linn Soc 171:491–507

Dube MP, Belzile FJ (2010) Low genetic variability of *Striga gesnerioides* populations parasitic on cowpea might be explained by a recent origin. Weed Res 50:493–502

Emerson KJ, Merz CR, Catchen JM, Hohenlohe PA, Cresko WA, Bradshaw WE, Holzapfel CM (2010) Resolving postglacial phylogeography using high-throughput sequencing. Proc Natl Acad Sci USA 107:16196–16200
Ernst WR (1972) Floral morphology and systematics of *Lamourouxia* (Scrophulariaceae: Rhinanthoideae). Smithsonian Contrib Bot 6:1–63
Fedorov AA (1969) Khromosomiye chisla tsvetkovykh rastenij. Nauka, Leningrad
Filipowicz N, Renner SS (2010) The worldwide holoparasitic *Apodanthaceae* confidently placed in the Cucurbitales by nuclear and mitochondrial gene trees. BMC Evol Biol 10:219. doi:10.1186/1471-2148-10-219
Fischer E (2004) Scrophulariaceae. In: Kadereit JW (ed) Flowering plants, dicotyledons, Lamiales (except Acanthaceae including Avicenniaceae), vol 7, The families and genera of vascular plants. Springer, Berlin, pp 333–432
Fischer E, Schäferhoff B, Müller KF (2012) The new monotypic genus *Bardotia* (Orobanchaceae) from Madagascar and remarks on the phylogenetic relationships of the African and Madagascan genera *Parastriga*, *Radamaea*, *Rhamphicarpa* and *Sieversandreas*. Phytotaxa 46:19–33
Gagne G, Roeckel-Drevet P, Grezes-Besset B, Shindrova P, Ivanov P, Grand-Ravel C, Vear F, Charmet G, Nicolas P (2000) Amplified fragment length polymorphism (AFLP) as suitable markers to study *Orobanche cumana* genetic diversity. J Phytopathol 148:457–459
Gethi JG, Smith ME, Mitchell SE, Kresovich S (2005) Genetic diversity of *Striga hermonthica* and *Striga asiatica* populations in Kenya. Weed Res 45:64–73
Gilli A (1968) *Necranthus*: a new genus of Orobanchaceae from Turkey. Notes R Bot Gard Edinb 28:297–298
Gussarova G, Popp M, Vitek E, Brochmann C (2008) Molecular phylogeny and biogeography of the bipolar *Euphrasia* (Orobanchaceae): recent radiations in an old genus. Mol Phylogenet Evol 48:444–460
Hallier H (1903) Ueber die Abgrenzung und Verwandtschaft der einzelnen Sippen bei den Scrophularineen. Bull l'Herbier Boissier II 3:181–207
Harley RM (2004) Nesogenaceae. In: Kadereit JW (ed) Flowering plants, dicotyledons, Lamiales (except Acanthaceae including Avicenniaceae), vol 7, The families and genera of vascular plants. Springer, Berlin, pp 293–295
Heide-Jørgensen HS (2008) Parasitic flowering plants. Brill, Leiden
Hjertson ML (1995) Taxonomy, phylogeny and biogeography of *Lindenbergia* (Scrophulariaceae). Bot J Linn Soc 119:265–321
Holub J (1977) New names in Phanerogamae 6. Folia Geobot Phytotax 12:417–432
Holub J (1990) Some taxonomic and nomenclatural changes within *Orobanche* s. l. (Orobanchaceae). Preslia 62:193–198
Huyse T, Poulin R, Théron A (2005) Speciation in parasites: a population genetics approach. Trends Parasitol 21:469–475
Joel DM (2009) The new nomenclature of *Orobanche* and *Phelipanche*. Weed Res 49(Supp 1):6–7
Joel DM, Portnoy V, Katzir N (1996) Identification of single tiny seeds of *Orobanche* using RAPD analysis. Plant Mol Biol Rep 14:243–248
Joel DM, Portnoy VH, Katzir N (1998) Use of DNA fingerprinting for soil-borne seed identification. Aspects Appl Biol 51:23–27
Kuijt J (1969) The biology of parasitic flowering plants. University of California Press, Berkeley
Manen JF, Habashi C, Jeanmonod D, Park JM, Schneeweiss GM (2004) Phylogeny and intraspecific variability of holoparasitic *Orobanche* (Orobanchaceae) inferred from plastid *rbcL* sequences. Mol Phylogenet Evol 33:482–500
Marais W (1981) Two new gamopetalous families, Cyclocheilaceae and Nesogenaceae, for extra-Australian 'Dicrastylidaceae'. Kew Bull 35:797–812
McNeal JR, Bennett JR, Wolfe AD, Mathews S (2013) Phylogeny and origins of holoparasitism in Orobanchaceae. Am J Bot 100:971–983
Mildbraed J (1930) New species and a new genus from East Africa. J Arnold Arboretum 11:50–55

Minkin JP, Eshbaugh WH (1989) Pollen morphology of the Orobanchaceae and rhinanthoid Scrophulariaceae. Grana 28:1–18

Mohamed KI, Mussleman LJ, Aigbokhan EI, Berner DK (1996) Evolution and taxonomy of agronomically important *Striga* species. In: Moreno MT, Cubero JI, Berner D, Joel D, Musselman LJ, Parker C (eds) Advances in parasitic plant research. Consejería de Agricultura y Pesca, Junta de Andalucía, Cordoba, pp 53–73

Mohamed KI, Musselman LJ, Riches CR (2001) The genus *Striga* (Scrophulariaceae) in Africa. Ann Mo Bot Gard 88:60–103

Molau U (1990) The genus *Bartsia* (Scrophulariaceae-Rhinanthoideae). Opera Bot 102:1–99

Moore MJ, Soltis PS, Bell CD, Burleigh JG, Soltis DE (2010) Phylogenetic analysis of 83 plastid genes further resolves the early diversification of eudicots. Proc Natl Acad Sci USA 107:4623–4628

Morawetz JJ, Randle CP (2010) The status of *Harveya alba* (Orobanchaceae). Kew Bull 65:1–2

Morawetz JJ, Wolfe AD (2009) Assessing the monophyly of *Alectra* and its relationship to *Melasma* (Orobanchaceae). Syst Bot 34:561–569

Morawetz JJ, Randle CP, Wolfe AD (2010) Phylogenetic relationships within the tropical clade of Orobanchaceae. Taxon 59:416–426

Mower JP, Stefanović S, Young GJ, Palmer JD (2004) Gene transfer from parasitic to host plants. Nature 432:165–166

Neel MC, Cummings MP (2004) Section-level relationships of North American *Agalinis* (Orobanchaceae) based on DNA sequence analysis of three chloroplast gene regions. BMC Evol Biol 4:15. doi:10.1186/1471-2148-4-15

Nickrent DL (2012) The parasitic plant connection. http://www.parasiticplants.siu.edu. Cited 5 Oct 2012

Nickrent DL, Duff RJ, Colwell AE, Wolfe AD, Young ND, Steiner KE, dePamphilis CW (1998) Molecular phylogenetic and evolutionary studies of parasitic plants. In: Soltis DE, Soltis PS, Doyle JJ (eds) Molecular systematics of plants II. Kluwer, Boston, pp 211–241

Nickrent DL, Der JP, Anderson FE (2005) Discovery of the photosynthetic relatives of the "Maltese mushroom" *Cynomorium*. BMC Evol Biol 5:38. doi:10.1186/1471-2148-5-38

Nickrent DL, Malécot V, Vidal-Russell R, Der JP (2010) A revised classification of Santalales. Taxon 59:538–558

Novopokrovskij IV, Cvelev NN (1958) Semejstvo Orobanchaceae. In: Shishkin BK, Bobov EG (eds) Flora SSSR, vol 23. Akad Nauk SSSR, Moscow, pp 19–115

Olmstead RG, dePamphilis CW, Wolfe AD, Young ND, Elisons WJ, Reeves PA (2001) Disintegration of the Scrophulariaceae. Am J Bot 88:348–361

Oxelman B, Kornhall P, Olmstead RG, Bremer B (2005) Further disintegration of Scrophulariaceae. Taxon 54:411–425

Paran I, Gidoni D, Jacobsohn R (1997) Variation between and within broomrape (*Orobanche*) species revealed by RAPD markers. Heredity 78:68–74

Park JM, Manen JF, Schneeweiss GM (2007a) Horizontal gene transfer of a plastid gene in the non-photosynthetic flowering plants *Orobanche* and *Phelipanche* (Orobanchaceae). Mol Phylogenet Evol 43:974–985

Park JM, Schneeweiss GM, Weiss-Schneeweiss H (2007b) Diversity and evolution of Ty1-*copia* and Ty3-*gypsy* retroelements in the non-photosynthetic flowering plants *Orobanche* and *Phelipanche* (Orobanchaceae). Gene 387:75–86

Park JM, Manen JF, Colwell AE, Schneeweiss GM (2008) A plastid gene phylogeny of the non-photosynthetic parasitic *Orobanche* (Orobanchaceae) and related genera. J Plant Res 121:365–376

Parker C (2009) Observations on the current status of *Orobanche* and *Striga* problems worldwide. Pest Manag Sci 65:453–459

Parker C, Riches CR (1993) Parasitic weeds of the world. Biology and control. CAB, Wallingford

Pennell FW (1928) *Agalinis* and allies in North America – I. Proc Acad Nat Sci Phila 80:339–449

Pettengill JB, Neel MC (2008) Phylogenetic patterns and conservation among North American members of the genus *Agalinis* (Orobanchaceae). BMC Evol Biol 8:264. doi:10.1186/1471-2148-8-264

Piednoël M, Aberer AJ, Schneeweiss GM, Macas J, Novak P, Gundlach H, Temsch EM, Renner SR (2012) Next-generation sequencing reveals the impact of repetitive DNA across phylogenetically closely related genomes of Orobanchaceae. Mol Biol Evol 29:3601–3611

Plaza L, Fernández I, Juan R, Pastor J, Pujadas A (2004) Micromorphological studies on seeds of *Orobanche* species from the Iberian Peninsula and the Balearic Islands, and their systematic significance. Ann Bot 94:167–178

Portnoy VH, Katzir N, Joel DM (1997) Species identification of soil-borne *Orobanche* seeds by DNA fingerprinting. Pestic Biochem Phys 58:49–54

Pujadas Salvà AJO (2002) *Orobanche* L. In: López Sáez JA, Catalán P, Sáez LI (eds) Plantas parásitas de la Península Ibérica e Islas Baleares. Mundi Prensa, Madrid, pp 345–451

Randle CP, Wolfe AD (2005) The evolution and expression of RBCL in holoparasitic sister-genera *Harveya* and *Hyobanche* (Orobanchaceae). Am J Bot 92:1575–1585

Ree RH (2005) Phylogeny and the evolution of floral diversity in *Pedicularis* (Orobanchaceae). Int J Plant Sci 166:595–613

Román B, Rubiales D, Torres AM, Cubero JI, Satovic Z (2001) Genetic diversity in *Orobanche crenata* populations from southern Spain. Theor Appl Genet 103:1108–1114

Román B, Satovic Z, Rubiales D, Torres AM, Cubero JI, Katzir N, Joel DM (2002) Variation among and within populations of the parasitic weed *Orobanche crenata* from Spain and Israel revealed by Inter Simple Sequence Repeat markers. Phytopatholgy 92:1262–1266

Román B, Alfaro C, Torres AM, Moreno MT, Satovic Z, Pujadas A, Rubiales D (2003) Genetic relationships among *Orobanche* species as revealed by RAPD analysis. Ann Bot 91:637–642

Román B, González Verdejo CI, Satovic Z, Madrid MD, Cubero JI, Nadal S (2007a) Detecting *Orobanche* species by using cpDNA diagnostic markers. Phytoparasitica 35:129–135

Román B, Hernández R, Pujadas-Salvá AJ, Cubero JI, Rubiales D, Satovic Z (2007b) Genetic diversity in two variants of *Orobanche gracilis* Sm. [var. *gracilis* and var. *deludens* (Beck) A. Pujadas] (Orobanchaceae) from different regions of Spain. Electron J Biotechnol 10:6. doi:10.2225/vol10-issue2-fulltext-6

Román B, Satovic Z, Alfaro C, Moreno MT, Kharrat M, Pérez-de-Luque A, Rubiales D (2007c) Host differentiation in *Orobanche foetida* Poir. Flora 202:201–208

Satovic Z, Joel DM, Rubiales D, Cubero JI, Román B (2009) Population genetics in weedy species of *Orobanche*. Australas Plant Pathol 38:228–234

Schäferhoff B, Fleischmann A, Fischer E, Albach DC, Borsch T, Heubl G, Müller KF (2010) Towards resolving Lamiales relationships: insights from rapidly evolving chloroplast sequences. BMC Evol Biol 10:352. doi:10.1186/1471-2148-10-352

Scheunert A, Fleischmann A, Olano-Marín C, Bräuchler C, Heubl G (2012) Phylogeny of tribe Rhinantheae (Orobanchaceae) with a focus on biogeography, cytology and re-examination of generic concepts. Taxon 61:1269–1285

Schneeweiss GM (2007) Correlated evolution of life history and host range in the non-photosynthetic parasitic flowering plants *Orobanche* and *Phelipanche* (Orobanchaceae). J Evol Biol 20:471–478

Schneeweiss GM, Weiss H (2003) Polyploidy in *Aeginetia indica* L. (Orobanchaceae). Cytologia 68:15–17

Schneeweiss GM, Colwell AE, Park JM, Jang CG, Stuessy TF (2004a) Phylogeny of holoparasitic *Orobanche* (Orobanchaceae) inferred from nuclear ITS sequences. Mol Phylogenet Evol 30:465–478

Schneeweiss GM, Palomeque T, Colwell AE, Weiss-Schneeweiss H (2004b) Chromosome numbers and karyotype evolution in holoparasitic *Orobanche* (Orobanchaceae) and related genera. Am J Bot 91:439–448

Takhtajan A (1997) Diversity and classification of flowering plants. Columbia University Press, New York

Tank DC, Olmstead RG (2008) From annuals to perennials: phylogeny of subtribe Castillejinae (Orobanchaceae). Am J Bot 95:608–625

Tank DC, Beardsley PM, Kelchner SA, Olmstead RG (2006) Review of the systematics of Scrophulariaceae s.l. and their current disposition. Austr Syst Bot 19:289–307

Tank DC, Egger JM, Olmstead RG (2009) Phylogenetic classification of subtribe Castillejinae (Orobanchaceae). Syst Bot 34:182–197

Teryokhin ES (2001) O rode *Necranthus* Gilli (Scrophulariaceae). Nov sist vysshich rastenij 33:205–207

Teryokhin ES, Chubarov SI, Raju CA (1996) Two modes of the vegetative propagation in *Orobanche cernua* (Orobanchaceae). In: Moreno MT, Cubero JI, Berner D, Joel D, Musselman LJ, Parker C (eds) Advances in parasitic plant research. Consejería de Agricultura y Pesca, Junta de Andalucía, Sevilla, pp 243–248

Teryokhin ES, Shibakina GV, Serafimovich NV, Kravtsova TI (2003) Opredelitel̆ zarazichov'ich flor'i SSR (s atlasom plodov i semjan). Nauka, St Petersburg

Těšitel J, Řiha P, Svobodová Š, Malinová T, Štech M (2010) Phylogeny, life history evolution and biogeography of the rhinanthoid Orobanchaceae. Folia Geobot 45:347–367

Thorogood CJ, Rumsey FJ, Harris SA, Hiscock SJ (2008) Host-driven divergence in the parasitic plant *Orobanche minor* Sm. (Orobanchaceae). Mol Ecol 17:4289–4303

Thorogood CJ, Rumsey FJ, Harris SA, Hiscock SJ (2009a) Gene flow between alien and native races of the holoparasitic angiosperm *Orobanche minor* (Orobanchaceae). Plant Syst Evol 282:31–42

Thorogood CJ, Rumsey FJ, Hiscock SJ (2009b) Host-specific races in the holoparasitic angiosperm *Orobanche minor*: implications for speciation in parasitic plants. Ann Bot 103:1005–1014

Uhlich H, Pusch J, Barthel KJ (1995) Die Sommerwurzarten Europas. Westarp Wissenschaften, Magdeburg

Vaz Patto MC, Fernández-Aparicio M, Satovic Z, Rubiales D (2009) Extent and pattern of genetic differentiation within and between European populations of *Phelipanche ramosa* revealed by amplified fragment length polymorphism analysis. Weed Res 49(Supp 1):48–55

Ward SA (1992) Assessing functional explanations of host-specificity. Am Nat 139:883–891

Webb DA (1972) CLIV. Scrophulariaceae. In: Tutin TG, Heywood VH, Burges NA, Moore DM, Valentine DH, Walters SM, Webb DA (eds) Flora Europaea, vol 3. Cambridge University Press, Cambridge, pp 202–281

Weiss-Schneeweiss H, Greilhuber J, Schneeweiss GM (2006) Genome size evolution in holoparasitic *Orobanche* (Orobanchaceae) and related genera. Am J Bot 93:148–156

Westwood JH, Roney JK, Khatibi PA, Stromberg VK (2009) RNA translocation between parasitic plants and their hosts. Pest Manag Sci 65:533–539

Westwood JH, Yoder JI, Timko MP, dePamphilis CW (2010) The evolution of parasitism in plants. Trends Plant Sci 15:227–235

Wettstein R (1891) Scrophulariaceae. In: Engler A, Prantl K (eds) Die natürlichen Pflanzenfamilien nebst ihren Gattungen und wichtigeren Arten, insbesondere den Nutzpflanzen 4/3b. Wilhelm Engelmann, Leipzig, pp 39–107

Wolfe AD, dePamphilis CW (1998) The effect of relaxed functional constraints on the photosynthetic gene *rbcL* in photosynthetic and nonphotosynthetic parasitic plants. Mol Biol Evol 15:1243–1258

Wolfe KH, Morden CW, Palmer JD (1992) Function and evolution of a minimal plastid genome from a nonphotosynthetic parasitic plant. Proc Natl Acad Sci USA 89:10648–10652

Wolfe AD, Randle CP, Liu L, Steiner KE (2005) Phylogeny and biogeography of Orobanchaceae. Folia Geobot 40:115–134

Xi Z, Bradley RK, Wurdack KJ, Wong KM, Sugumaran M, Bomblies K, Rest JS, Davis CC (2012) Horizontal transfer of expressed genes in a parasitic flowering plant. BMC Genomics 13:227. doi:10.1186/1471-2164-13-227

Xia Z, Wang YZ, Smith JF (2009) Familial placement and relations of *Rehmannia* and *Triaenophora* (Scrophulariaceae s.l.) inferred from five gene regions. Am J Bot 96:519–530
Yatskievych G, Contreras Jiménez JL (2009) A new genus of holoparasitic Orobanchaceae from Mexico. Novon 19:266–276
Yoneyama K, Awad AA, Xie X, Yoneyama K, Takeuchi Y (2010) Strigolactones as germination stimulants for root parasitic plants. Plant Cell Pysiol 51:1095–1103
Yoshida S, Maruyama S, Nozaki H, Shirasu K (2010) Horizontal gene transfer by the parasitic plant *Striga hermonthica*. Science 328:1128
Young ND, dePamphilis CW (2005) Rate variation in parasitic plants: correlated and uncorrelated patterns among plastid genes of different functions. BMC Evol Biol 5:16. doi:10.1186/1471-2148-5-16
Young ND, Steiner KE, dePamphilis CW (1999) The evolution of parasitism in Scrophulariaceae/ Orobanchaceae: plastid gene sequences refute an evolutionary transition series. Ann Mo Bot Gard 86:876–893
Zeid M, Madkour M, Koraiem Y, Nawar A, Soliman M, Zaitoun F (1997) Molecular studies on *Orobanche*. J Phytopathol 145:351–355
Zhang ZY, Cvelev [Tzvelev] NN (1998) Orobanchaceae. In: Wu ZY, Raven PH (eds) Flora of China, vol 18. Missouri Botanical Garden Press, St. Louis & Science Press, Beijing, pp 229–243
Zhang ZH, Li CQ, Li J (2009) Phylogenetic placement of *Cynomorium* in Rosales inferred from sequences of the inverted repeat region of the chloroplast genome. J Syst Evol 47:297–304

Chapter 15
Genomic Evolution in Orobanchaceae

Susann Wicke

15.1 Introduction

The broomrape family, Orobanchaceae, is widely recognized as *the* model group to study genomic evolution in parasitic plants, especially because it is the only parasitic family to include the entire range of evolutionary transitional stages, from a fully autotrophic via semi-heterotrophic to completely holo-heterotrophic lifestyle (Westwood et al. 2010). Orobanchaceae, encompassing an estimated number of 2,000 species,[1] are confidently placed in the large and diverse group of Lamiales, which contains a great number of species with highly specialized life forms including desiccation tolerance, carnivory, and parasitism (Schäferhoff et al. 2010). The parasitic lifestyle has brought about numerous morphological and developmental changes. Substantial progress has been made during the past few years in uncovering basic genetic reconfigurations and signalling pathways necessary in establishing a haustorial connection to another plant. Nevertheless, little is known about the evolution of nuclear and mitochondrial genes and genomes in Orobanchaceae, even 20 years after the first plastid genome of a non-photosynthetic member of the family has been sequenced. This is especially astonishing given the great advance in molecular biological methods and sequencing technologies over the past 5–10 years. Reasons therefore are manifold—as usual. Owing to the great diversity of the family, evolutionary studies based upon molecular data are restricted to only a few members of either species-rich and commonly distributed genera or to members of considerable ecological importance such as *Orobanche*,

[1] This chapter uses the most recent taxonomic changes outlined in Chap. 14.

S. Wicke (✉)
Department for Systematic and Evolutionary Botany, University of Vienna, Rennweg 14, 1030 Vienna, Austria

Current address: Institute for Evolution and Biodiversity, University of Muenster, Huefferstr. 1, 48149 Muenster, Germany
e-mail: susann.wicke@uni-muenster.de

D.M. Joel et al. (eds.), *Parasitic Orobanchaceae*,

DOI 10.1007/978-3-642-38146-1_15,

Phelipanche, and *Striga*. Additionally, *in vitro* cultivation of obligate parasites is often difficult or requires permits in some countries, hampering genetic and reverse genetic approaches. Finally, the trend for rather large genomes renders Orobanchaceae challenging objects for genomic surveys.

This chapter summarizes the current knowledge of the genomic evolution in Orobanchaceae. The following sections provide overviews about nuclear, mitochondrial, and plastid genomics and about horizontal DNA transfer. Subsequently, a short concluding paragraph outlines some prospects on where genomics in the broomrape family may be headed in the next few years (see also Sect. 4.5 for the evolution of parasite-specific functions).

15.2 The Nuclear Genome

15.2.1 Nuclear Genes

Nothing is known about the evolution of nuclear coding regions in the Orobanchaceae, and thus very little is known about the molecular basis of parasite-specific life stages. Recently, a hydroxylase (PRCYP707A1) functioning in the abscisic acid (ABA) catabolic pathway in *Phelipanche ramosa* has been identified as playing a major role during germination of seeds after exposure to germination stimulants (Lechat et al. 2012). The same study also identified two heat shock proteins and a few more transcripts associated with ABA cascades. Even though those transcripts have not been characterized in detail, preliminary lines of evidence link those transcripts to proteins active during seed germination in *Arabidopsis thaliana*. Experiments with the holoparasitic *Orobanche minor* have shown that the involvement of phytochromes (PHY) during germination, shoot elongation, and anthocyanin content differs from that observed in photosynthetic plants, suggesting reconfigured regulatory cascades involving at least PHY proteins A and B (Takagi et al. 2009). Relative to *Arabidopsis*, 26 amino acids are substituted in PHYA of *O. minor* (Trakulnaleamsai et al. 2005). Of these, some substitutions perhaps hold the potential to alter protein function, thereby contributing to unusual light responses in the holoparasite compared to autotrophic plants (Trakulnaleamsai et al. 2005; Takagi et al. 2009). Given that expression patterns as well as cellular localizations of PHYA in *O. minor* are comparable to autotrophic plants and that the chromatophore-binding site in PHYA is highly conserved, the reported amino acid changes may also represent results of coevolution with rapidly evolving PHYA-interacting photosynthesis genes.

In *Striga hermonthica*, Yoshida et al. (2010) found 589 assembled fragments of expressed genes (unigenes) that are not similar to known plant genes, implying that at least some of these may be specific to parasitism. Sequencing of cDNAs from the facultative hemiparasite *Triphysaria versicolor* revealed an up-regulation of more than a hundred unigenes during early haustorium initiation. These fragments were assigned similarity to proteins functioning in quinone detoxification, transcription and regulatory processes, membrane transport, and the citric acid cycle (Matvienko et al. 2001). Two of these transcripts, a quinone oxidoreductase (QR1) and a protein associated to plant signalling pathways (*TvPirin*), have been further characterized as essential for haustorium initiation after contact to host roots or exposure to haustorium-inducing chemicals (Bandaranayake et al. 2010; see also Sects. 4.4 and 4.5). However, not much is known about the molecular evolution of those genes, and future studies will have to show whether genes relevant for the development of lateral haustoria of *Triphysaria* are also essential for the induction of terminal haustoria (see Sect. 4.4). Deep sequencing of ultrathin slices of host-parasite interface tissue of *T. versicolor* furthermore revealed the differential expression of a β-expansin gene (TvEXPB1) when the parasite was grown in the presence of different hosts (Honaas et al. 2013). However, it still needs to be elucidated whether cell wall modifying proteins and their differential expression are common among other Orobanchaceae.

Using three Orobanchaceae species differing in their extent of heterotrophy, the ongoing large-scale transcriptome-sequencing approach of the *Parasitic Plant Genome Project* (PPGP, Westwood et al. 2010, 2012) aims, among other aspects, at discovering and studying genes that are exclusive to specific ontogenetic stages. In a first study, Wickett et al. (2011) found that expression of nuclear-encoded photosynthesis subunits in aboveground tissue is considerably reduced in the obligate hemiparasite *S. hermonthica* compared to the facultative hemiparasite *T. versicolor*. No expression of nuclear-encoded photosynthesis genes was detected in *P. aegyptiaca*, where these genes might have become pseudogenes or have already been deleted from the genome. In contrast, genes for chlorophyll synthesis were still expressed in *Phelipanche* (Wickett et al. 2011), corroborating results of the detection of trace amounts of chlorophyll a in some holoparasites of Orobanchaceae (*Epifagus*, *Myzorrhiza cooperi* [syn. *Orobanche cooperi*], *Aphyllon uniflora* [syn. *O. uniflora*]) and other families (Cummings and Welschmeyer 1998).

A survey of the evolution of the small ribosomal RNA subunit (SSU) found that parasitic plants possess significantly elevated nucleotide substitution rates (Nickrent and Duff 1996). However, comparative studies across parasitic and myco-heterotrophic plants (see Sect. 1.8) did not show a significant acceleration in the SSU evolution in several Orobanchaceae holoparasites, and the pattern of rate acceleration across lineages remains widely elusive (Lemaire et al. 2011).

15.2.2 Chromosome Numbers

Chromosome numbers have been the focus of several studies on hemiparasitic Orobanchaceae, although the vast majority of these reports lacked an explicit evolutionary context. Chromosome numbers and ploidy are highly variable in the family and apparently do not correlate with genome size. The current knowledge of chromosome numbers and genome sizes in Orobanchaceae is graphically

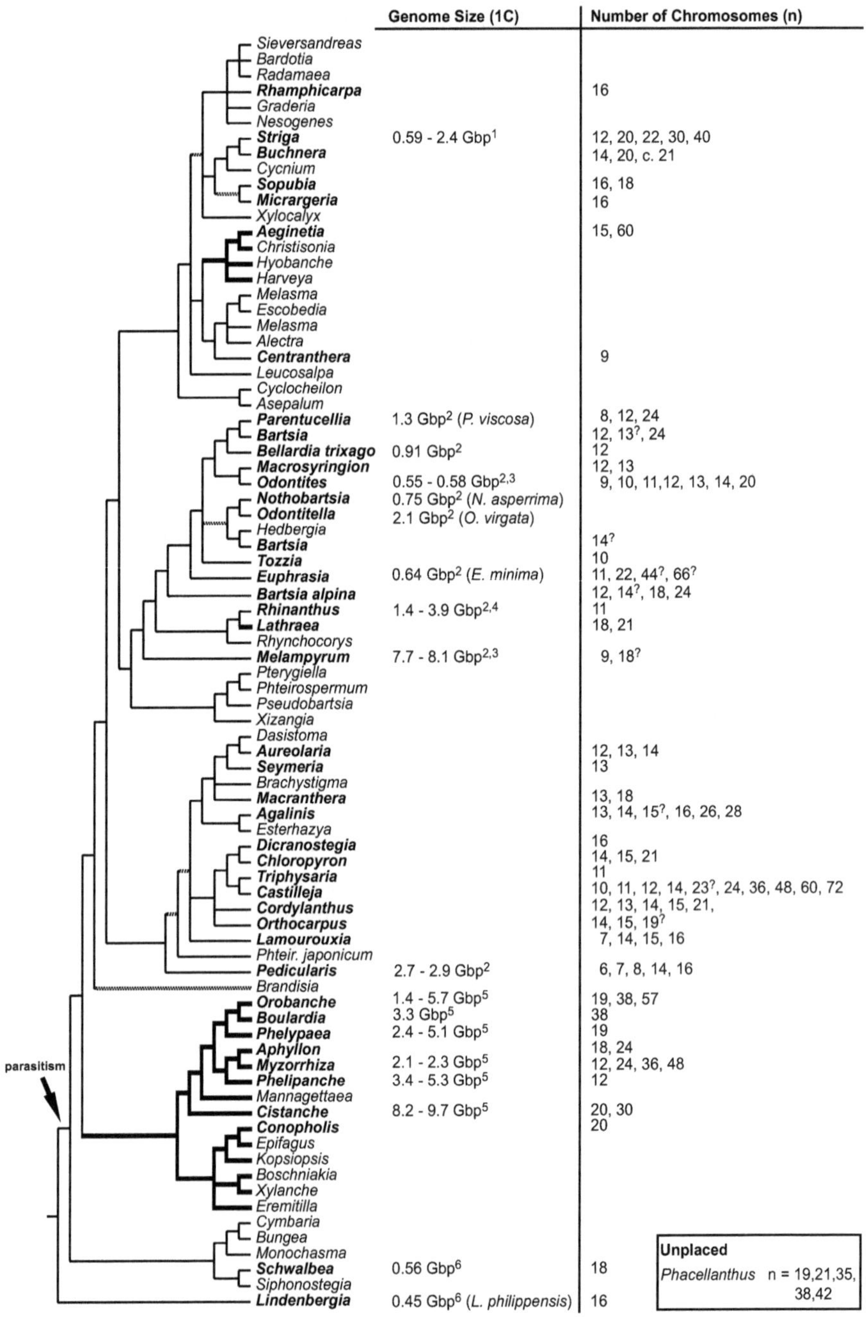

Fig. 15.1 Evolution of chromosome number and genome size in Orobanchaceae. *Arrow*, the origin of parasitism. *Thin branches*, autotrophic and photosynthetic heterotrophs; *thick branches*, non-photosynthetic heterotrophs; *dashed branches*, uncertain placement. Tree topology after

summarized in Fig. 15.1. *Lindenbergia* and *Schwalbea*, members of the first branching lineages, have $n = 16$ (Hjertson 1995) and $n = 18$ (Kondo et al. 1981), respectively. Being mostly diploid with the exception of one tetraploid species, the closely related sister group of Orobanchaceae, *Rehmannia*, harbours $n = 14$ chromosomes (Albach et al. 2007). In the light of current data, there seems to be a slight trend towards higher chromosome numbers in the exclusively holoparasitic *Orobanche* clade (see Sect. 14.2.2.3), although polyploidy appears to be common in some of the hemiparasites as well (e.g. *Castilleja*, *Striga*, *Euphrasia*; Tank et al. 2009; Kondo et al. 1981; Barker et al. 1988; Iwo et al. 1993). Except for *Phelipanche* with $n = 12$, most genera of the holoparasite clade have $n = 19$ or more chromosomes (Fig. 15.1; Schneeweiss et al. 2004). Plots of synonymous substitutions of selected expressed genes revealed unambiguously that *Phelipanche* must have undergone at least one whole-genome duplication after the split from hemiparasitic ancestors (Wickett et al. 2011), corroborating previous hypotheses of ancient duplication events in the *Orobanche* clade (Schneeweiss et al. 2004). Differences in chromosome number indicate that lineages within this clade might have undergone at least one or more rounds of polyploidization. Inconsistent chromosome morphologies imply that these events might have even occurred independently (Schneeweiss et al. 2004). It will be interesting to see whether independent changes and ancient polyploidization also occurred in other holoparasitic groups, especially in the light of the severe ecological changes accompanying the transition to the non-photosynthetic lifestyle. A change of ploidy level in *O. transcaucasica* apparently coincides with a shifted host range, suggesting that, at least in this particular case, genome duplication favours ecological differentiation from its progenitors (Schneeweiss et al. 2004).

It is unclear whether several independent rounds of polyploidization and dysploidization (i.e. reduction of ploidy/chromosome numbers) in the hemiparasites of the Old and New World led to the great diversity of chromosome number, which ranges from $n = 8$ to $n = 20$ (Fig. 15.1). Changes of ploidy level are a substantial basis for speciation among angiosperms, allowing offsprings to settle into new niches (Wood et al. 2009). Niching due to host range shifts as a result of polyploidization or dysploidization may thus be a more significant aspect for speciation in parasitic lineages. At this point, such putative correlations between the degrees of parasitism and chromosomal evolution including ploidy are, however, hypothetical at best. Caryological studies in most of the tropical and/or Asian hemi- and holoparasitic lineages (i.e. nearly 50 % of all Orobanchaceae genera) are still lacking.

Fig. 15.1 (continued) Schneeweiss (see Chap. 14). *Question mark* indicates uncertain chromosome counts. Chromosome data: IPCN (Goldblatt and Johnson 1979), Fedorov (1969), and Moore (1982). Genome size data: [1]Yoshida et al. (2010); Estep et al. (2012); [2]Castro et al. 2012; [3]Hanson et al. (2002); [4]Zonneveld et al. (2005); Nagl and Fusenig (1979); [5]Weiss-Schneeweiss et al. (2006); [6]Piednoël et al. (2012).

15.2.3 Genome Size

In line with the rampant, even though uncorrelated occurrence of polyploidy, genome sizes vary greatly within Orobanchaceae. *Lindenbergia* (1C = 0.45 Gbp), *Schwalbea* (1C = 0.56 Gbp; Piednoël et al. 2012), and *Odontites* (1C = 0.55–0.56 Gbp; Hanson et al. 2002) possess small genomes, the sizes of which are comparable to those of poplar and rice. Other photosynthetic Orobanchaceae such as *Rhinanthus* and *Melampyrum* have considerably larger genomes, with that of *Melampyrum* reaching almost three times the size of the human genome (Fig. 15.1; Hanson et al. 2002). The smallest genome of the holoparasitic *Orobanche* clade is found in *O. cumana* (Weiss-Schneeweiss et al. 2006). The *Orobanche* genus contains some polyploids (e.g. *O. transcaucasica*, *O. gracilis*) that may exceed 1C = 5.5 Gbp. Among diploids, *O. crenata* with 1C = 2.8 Gbp ranks among the biggest according to currently available measurements (Fig. 15.1; Weiss-Schneeweiss et al. 2006). In contrast to *Orobanche*, *Phelipanche* has larger genomes on average (Weiss-Schneeweiss et al. 2005), but the largest genomes in Orobanchaceae have been described so far for species of *Cistanche* with 1C = 8.7 Gbp in *C. phelypaea* (Weiss-Schneeweiss et al. 2006). Even larger ones may occur in other *Cistanche* species (N. Ataei, D. Quandt, and H. Weiss-Schneeweiss, unpublished data). Nevertheless, compared to the (cryptically) photosynthetic heterotrophs of other parasitic angiosperm families such as *Cuscuta* (1C = 0.57–32.1 Gbp, McNeal et al. 2007a) or members of Santalales (1C = 0.3–80.2 Gbp, Martin 1983; Hanson et al. 2001; Zonneveld 2010), the range of genome sizes is rather moderate in Orobanchaceae, a fact that contributes to its status as the 'model family' among parasitic plants.

As in most plant genomes, the abundance of repetitive DNA contributes substantially to genome size differences in Orobanchaceae. The five economically most important species of *Striga* show considerable genomic variation with respect to the 14 largest genus-specific repeat families residing in the genomes with more than a few hundred copies (Estep et al. 2012). Those repetitive DNAs account for 10–19 % of the nuclear genomes of *Striga* species, but they are not strictly correlated with genome size. They belong to classes commonly found in angiosperm genomes with transposable elements being the most abundant. The analyses of repeat classes point towards a ploidy series in the genus *Striga* (Estep et al. 2012). Interestingly, the variability among different populations of single species of *Striga*, e.g. *S. asiatica* or *S. gesnerioides*, is moderate or even low, respectively (Botanga et al. 2002; Botanga and Timko 2005, 2006; see Sect. 19.2).

The genomes of seven holoparasitic broomrapes and two photosynthetic Orobanchaceae were characterized employing a whole-genome shotgun pyrosequencing approach (Piednoël et al. 2012). The proportion of repeat DNA sequences is low in the small-sized genomes of the nonparasite *Lindenbergia* and the hemiparasite *Schwalbea* with repetitive elements accounting for no more than 30 % of the genomes. As implied by chromosomal and genome size data, divergent dynamics of genome evolution exist in the sister groups *Orobanche* and *Phelipanche*.

This hypothesis is corroborated by differing quantities of genus-specific clusters of transposable elements (Piednoël et al. 2012). The proportion of long and short interspersed nuclear elements (LINE and SINE, respectively) seems to be generally lower in *Phelipanche* than in *Orobanche*. LINEs contribute to the increase in genome size (as do many retrotransposons) in that they autonomously copy themselves. *Phelipanche* spp. may have evolved a more sophisticated machinery for silencing transposable elements, which results in a more stable genomic and chromosomal evolution. Control and regulatory mechanisms for transposable elements are lineage-specific and contribute widely to genome stability (e.g. He et al. 2012). It will be interesting to see whether genome size evolution is related to host range and/or to the degree of parasitism. For instance, in some plants, nutrient limitation leaves behind genomic signatures (Acquisti et al. 2009a, b), but obligate parasites may not be affected by those limitations in the same way because of the host-provided nutrient supply.

Polyploid Orobanchaceae tend to a reduction of the monoploid genome size (1Cx value) after events of polyploidization, which is in congruence with several nonparasitic polyploid angiosperm lineages (Leitch and Bennett 2004). In most cases, 1Cx values from polyploids are smaller than those of diploid relatives (Weiss-Schneeweiss et al. 2005). Although the genetic mechanisms are still poorly understood, genome-size reduction may be selected for because of, e.g. biophysical (e.g. chromosome pairing in meiosis and mitosis) and biochemical reasons ('biochemical economy') (Leitch and Bennett 2004; Leitch and Leitch 2012). Perhaps there is a trade-off between genomic plasticity that comes with genome size and nutritional constraints. An obligate parasitic way of life might favour moderately to large-sized genomes irrespective of the ability to carry out photosynthesis, enhancing chances of sub- or neofunctionalization of duplicated genes that contribute to host specificity and host adaptation, leading eventually to speciation within parasite lineages.

Several other Orobanchaceae groups may have had comparable scenarios of frequent increase and decrease of chromosome number and genome size like those observed in the *Orobanche* clade. Independent events of polyploidization have also been hypothesized for some other lineages (e.g. *Euphrasia, Lathraea*) based upon duplications of the phytochrome A gene (Bennett and Mathews 2006).

15.3 The Plastid Genome

The plastid chromosome (plastome) is the best understood cellular genome in angiosperms. The plastome normally has a highly conserved structure with a large and a small single-copy region (LSC and SSC, respectively) that are separated from each other by two large and virtually identical inverted repeats. Plastomes encode a large set of subunits for the photosynthesis apparatus including genes for photosystems I and II, the cytochrome complex, an ATP synthase, and an NAD(P)H complex as well as few genes involved in photosynthetic energy gain (*rbcL*, *ccsA*, *cemA*) or lipid synthesis (*accD*). Several proteins for the genetic apparatus are solely plastid encoded including several ribosomal protein genes, a plastid-encoded polymerase complex, as well as few others involved in either transcript maturation (*matK*) or protein turnover (*infA*, *clpP*, photosystem assembly factors *ycf3*, *ycf4*). The essential function of the two largest plastid genes (*ycf1*, *ycf2*) is as yet unknown, but both reading frames are conserved among photosynthetic and non-photosynthetic land plants (Wicke et al. 2011). Based on protein-domain comparisons, both proteins probably function in housekeeping processes rather than having a metabolic function (Wolfe 1994; Boudreau et al. 1997; Drescher et al. 2000). The plastid genome normally also harbours two sets of four ribosomal RNA genes as well as 30 tRNA genes, the latter of which enable the delivery of all codons due to (extended) wobbling and superwobbling (Lagerkvist 1978; Rogalski et al. 2008; Alkatib et al. 2012).

Due to its compact nature and its prime role in photosynthesis, the evolution of the plastid genome of non-photosynthetic plants has received attention early on. Already in 1990, dePamphilis and Palmer reported the loss of all genes for the plastid NAD(P)H dehydrogenase complex from the plastome of the holoparasite *Epifagus virginiana*. Soon the complete plastid genome sequence of *E. virginiana* was described (Wolfe et al. 1992b). Massive gene loss led to an extraordinary structure of the plastome which is reduced to less than half the size of that of photosynthetic relatives. Nevertheless, the relative order of genes in LSC, SSC, and the inverted repeats remains largely colinear to photosynthetic plants (Fig. 15.2). Besides *ndh* genes, most genes involved in light and dark reaction of photosynthesis are completely absent from the plastome; only a few photosynthesis-related genes reside in the plastome as pseudogenes (e.g. Ψ*rbcL*, Ψ*atpA*). Furthermore, several genes encoding proteins of the genetic apparatus are (functionally) lost including tRNA genes, the plastid-encoded polymerase complex, and some ribosomal protein genes (Morden et al. 1991; Wolfe et al. 1992b). Comparable dramatic reductions of plastid DNAs occur in a variety of parasitic plants, including *Cuscuta* species (Funk et al. 2007; McNeal et al. 2007b), mistletoes (Nickrent and García 2009), green algae (Knauf and Hachtel 2002; de Koning and Keeling 2006) as well as myco-heterotrophic plant lineages, including non-photosynthetic orchids (Logacheva et al. 2011; Delannoy et al. 2011) and achlorophyllous Ericaceae (Braukmann and Stefanović 2012). As in some of the other parasites, residing plastid genes of the translation apparatus in *Epifagus* evolved significantly faster than those of nonparasitic relatives (Wolfe et al. 1992a). Nevertheless, the retained plastid genes of *E. virginiana* are transcribed, mature, and are translated into functional RNAs and proteins (Morden et al. 1991; Wolfe et al. 1992a; Ems et al. 1995; Lohan and Wolfe 1998; also see Wimpee et al. 1991, 1992).

In terms of gene losses, other holoparasitic Orobanchaceae lineages possess considerably different plastomes than *E. virginiana*, indicating that reductive evolution of plastid DNA is a highly lineage-specific process within Orobanchaceae (and presumably within other parasitic plant lineages as well). Extensive restriction-mapping experiments suggested that *Conopholis americana* has an even smaller plastid genome (ca. 42kb) than the closely related *E. virginiana*,

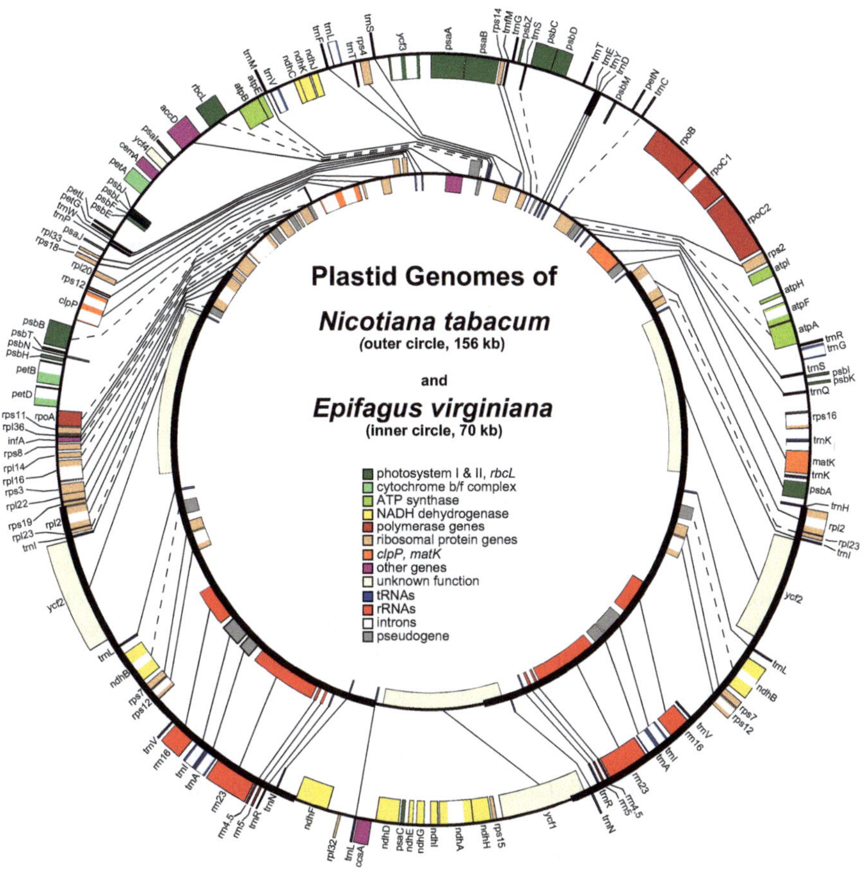

Fig. 15.2 Comparison of plastid genome structure of *Epifagus virginiana* (*inner circle*) and the nonparasitic plant *Nicotiana tabacum* (*outer circle*). The large inverted repeats are indicated as thickened chromosomal segments relative to the large and the small single-copy regions. Genes are coloured according to their function with the name of genes depicted on the *outer circle*. Pseudogenes are coloured in *grey*. *Thin lines* from the *Nicotiana* to the *Epifagus* genome indicate structural reorganizations due to massive gene loss. genes No line between *Nicotiana* and *Epifagus* indicates that the region has been lost in the latter; a *dashed* connection indicates pseudogenization in *Epifagus*. Genome maps were drawn using OGDRAW (Lohse et al. 2007) based upon Shinozaki et al. (1986) and Wolfe et al. (1992b)

mainly due to the loss of one large inverted repeat (Downie and Palmer 1992; Colwell 1994). Other large deletions are comparable to those of *Epifagus*, implying that functional reduction is similar in both species (Colwell 1994). Conversely, restriction mapping and PCR screens suggest that the plastid genome of the holoparasite *Lathraea clandestina* is ca. 100–110 kb in size with gene synteny mostly colinear to *Epifagus* and most photosynthetic plants (Delavault et al. 1996). Reductive evolution of the plastid genome in *Lathraea* has obviously not proceeded as far as in *E. virginiana*. Still, some losses have also affected the plastid genome of

Lathraea: most prominently all SSC-located *ndh* genes have been deleted (Delavault et al. 1996). This is in line with data from other holoparasitic and myco-heterotrophic lineages with minimally reduced plastomes (Funk et al. 2007; McNeal et al. 2007b; Wickett et al. 2008; Delannoy et al. 2011; Logacheva et al. 2011).

Further support for the hypothesis of a highly lineage-specific reductive plastome evolution comes from studies using a broad taxon sampling, but focusing only on a few plastome regions. Because of its prime function during photosynthetic carbon fixation, most data is available for the plastid gene *rbcL*, which encodes the large subunit of RuBisCO. Some non-photosynthetic lineages (e.g. *Lathraea*, *Harveya*, *Myzorrhiza* [syn. *Orobanche sect. M.*]) preserve an intact reading frame for *rbcL* (Delavault et al. 1995, 1996; Wolfe and dePamphilis 1997, 1998; Randle and Wolfe 2005). In *Lathraea*, however, *rbcL* is transcribed by a nuclear-encoded polymerase rather than by the normally used plastome-encoded polymerase, which also transcribes most of the other photosynthesis genes (Lusson et al. 1998). Accordingly, it is not surprising that plastome-encoded polymerase-specific promoter regions are lost in *Epifagus* (Morden et al. 1991) and also in some other parasites (Krause et al. 2003; Berg et al. 2004). Several non-photosynthetic species harbour only a pseudogene copy (e.g. *Aphyllon* [syn. *Orobanche sect. Gymnocaulis*], *Hyobanche*, most *Orobanche* s. str.; Wolfe and dePamphilis 1997; Delavault and Thalouarn 2002; Manen et al. 2004; Young and dePamphilis 2005), and several lines of evidence indicate that the *rbcL*-gene region is deleted from the plastomes of *Phelipanche* (Manen et al. 2004; Park et al. 2007a; Wicke et al. in prep.). Low levels of *rbcL* expression have been detected in holoparasitic *Harveya* and *Lathraea* (Lusson et al. 1998; Randle and Wolfe 2005). *Myzorrhiza corymbosa* maintains functional upstream and downstream untranslated regulatory elements, which is indicative of maintained transcription activity (Wolfe and dePamphilis 1997); expression data is, however, lacking. The function of the translated polypeptide transcribed from *rbcL* has not been investigated in holoparasites. A function of RuBisCO that is unrelated to photosynthesis has been speculated (e.g. Wolfe and dePamphilis 1997; Leebens-Mack and dePamphilis 2002), corroborated by findings that link RuBisCO to amino acid synthesis and a glycolysis-bypassing pathway (Tolbert 1997; Schwender et al. 2004). The transcript of an *rbcL* pseudogene has been detected in *Hyobanche* (Randle and Wolfe 2005), implying that regulation of the largely nuclear-encoded transcription (and transcript processing) machineries lacks behind plastid DNA evolution, at least in this particular case.

Evolutionary analyses of *rps2* and *matK* show low rates of nucleotide substitution of these genes in *rbcL*-preserving lineages (dePamphilis et al. 1997; Young and dePamphilis 2005), suggesting that *rps2* and *matK* are under purifying selection and, thus, still functional. Furthermore, *Orobanche minor* retains most plastid tRNAs, although some only as pseudogenes (Lohan and Wolfe 1998), and it retains several DNA fragments that are deleted from the plastome of *Epifagus* and *Conopholis*. Taken together, this and the previously mentioned studies point towards less-reduced plastid genomes in *Harveya*, *Hyobanche*, *Myzorrhiza*, and

Orobanche. The reasons for these lineage-specific reductions are not yet understood, but the *time* since transition to holoparasitism seems to play a most relevant role. In general, older holoparasitic Orobanchaceae lineages, such as *Epifagus*, have greater reductions than younger ones, such as *Lathraea* and the *Harveya*/*Hyobanche* lineages (Leebens-Mack and dePamphilis 2002; also consider phylogenetic relationships depicted in Fig. 15.1).

Besides time, factors influencing the rate of gene loss and physical reduction of the plastome are still widely elusive. However, there seems to be a strong correlation between the deletion of a plastomic gene region and its physical proximity to indispensable genes of housekeeping or metabolic function (Wicke et al. submitted). The localization of a gene that has become dispensable after the loss of photosynthesis is apparently also protected by its location within an operon that encodes genes of various functional complexes (Wicke et al. submitted). Given the high gene density of plastid chromosomes, both effects are not mutually exclusive for the protection of physical gene deletion. A complex interaction of species-specific repair and recombination rates may further be elucidated as important factors in "regulating" how plastid genome reduction proceeds in holoparasites (Wicke et al. in prep.).

Structural reorganization of the plastid DNAs (e.g. inversions) in parasites occurs to a considerably smaller extent (if at all), compared with the amount of segmental DNA losses. The only reports of large-scale structural changes come from Colwell (1994) and Downie and Palmer (1992), who revealed the independent loss of one of the inverted repeats in and its *Conopholis* and *S. asiatica*, respectively. Outside Orobanchaceae, the highly reduced plastome of the underground orchid *Rhizanthella* has severe alterations around the inverted repeats (Delannoy et al. 2011), whereas the less dramatically reduced genome of the closely related bird's-nest orchid *Neottia nidus-avis* shows no genomic rearrangements (Logacheva et al. 2011). Small inversions were reported in plastomes of some species of *Cuscuta*, but no large-scale plastomic reconfigurations were found (Krause 2011). Thus, the generally high degree of structural conservation reported for the majority of angiosperm plastomes (Wicke et al. 2011) appears to be upheld in Orobanchaceae holoparasites for a long duration after the loss of photosynthesis. In contrast, ongoing research suggests that the relaxation of functional constraints and subsequent gene loss rapidly commence after the transition to a (obligate) heterotrophic way of life (Wicke et al. submitted).

15.4 The Mitochondrial Genome

Unlike plastomes, mitochondrial genomes (chondriomes) are highly susceptible to the incorporation of both horizontally transferred DNA and DNA from other cellular genomes (Won and Renner 2003; Bergthorsson et al. 2004; Davis et al. 2005; Knoop et al. 1996, 2011). Additionally, the high variability of size and gene content of plant chondriomes makes them difficult targets for phylogenetic and comparative evolutionary studies (Knoop et al. 2011).

As of this writing, no complete sequence of a chondriome of a parasitic flowering plant is available. Only few studies focused on the molecular evolution of mitochondrial genes, although several genes (e.g. *matR*, *atp1*, and *coxI*) have been employed to infer the placement of parasitic plant lineages in the tree of flowering plants. Despite the fact that DNA evolves normally more slowly than nuclear and plastid DNAs (Wolfe et al. 1987), some holoparasitic lineages (e.g. Apodanthaceae, Rafflesiaceae) exhibit elevated nucleotide substitution rates in the mitochondrial small ribosomal RNA (mtSSU) as well as in the mitochondrial genes *coxI*, *atp1*, *matR*, and in exons B and C of *nad1* (Duff and Nickrent 1997; Nickrent et al. 2004; Barkman et al. 2004, 2007; Filipowicz and Renner 2010). However, no rate acceleration has been found in holoparasitic Orobanchaceae genera such as *Epifagus*, *Orobanche*, and *Boschniakia* (supposedly *Kopsiopsis*; Mower et al. 2004; Barkman et al. 2007). Hemiparasitic Orobanchaceae (e.g. *Lamourouxia*, *Agalinis*, *Pedicularis*, *Hedbergia*, *Parentucellia*, *Bartsia*, *Buchnera*) also appear to evolve at similar evolutionary rates as *Lindenbergia* (Mower et al. 2004).

Little is known about the evolution of macro- and microstructural changes, such as (small) insertions, deletions, and inversions in coding and non-coding mitochondrial DNAs of parasitic plants. Duff and Nickrent (1997) reported a slight increase of indel events in the mtSSU of non-asterid parasite lineages. In contrast, mtSSU in *Epifagus* shows only an insignificant length variation (2 nt) compared to photosynthetic relatives. Remarkably, parasitic plants frequently possess an intron in the *coxI* gene. The *coxI* intron is found in ten of the at least 12 independently evolved angiosperm lineages with a parasitic lifestyle (Barkman et al. 2007). The source of the intron remains largely unclear. While an initial acquisition of the *coxI* intron via horizontal homing from a fungus seems likely for some angiosperms (Vaughn et al. 1995; Adams et al. 1998; Cho and Palmer 1999; Seif et al. 2005, cp. Cusimano et al. 2008), most gains in parasites seem to have occurred by horizontal plant-to-plant transfers (Sanchez-Puerta et al. 2008; see below). The close interaction between a parasitic plant and its host further supports the hypothesis of a plant donor of parasite *coxI* introns.

15.5 Horizontal DNA Transfer

The identification of true horizontal DNA transfers from one plant to another and its origin can be very problematic (critically reviewed in Renner and Bellot 2012). Parasite–host systems appear to be especially prone to horizontal gene/DNA transfer (HGT) (see Sect. 6.5.2). In plants, the uptake and incorporation of DNA from another species occur more frequently in mitochondrial DNA, although comparative data of functional and non-functional HGT is still widely lacking for the nuclear genome. Prominent cases of HGT involving mitochondrial genes concern the *atp1* region of parasitic plants of the Rafflesiaceae and Apodanthaceae (Davis and Wurdack 2004; Nickrent et al. 2004). A chimeric copy consisting of parasite-specific and horizontally acquired genic parts of *atp1* was found in the mitochondria in *Pilostyles thurberi* (Apodanthaceae; Barkman et al. 2007). Copies of host *atp1* appear to have also been independently transferred to species of the *Bartsia* clade of Orobanchaceae, and to *Cuscuta* (Mower et al. 2004, 2010). Other mitochondrial genes involved in HGT have not yet been identified in Orobanchaceae.

The transfer of macromolecules such as RNAs predominantly from the host into the parasite was reported for *Triphysaria versicolor* (Tomilov et al. 2008) and *Phelipanche aegyptiaca* (Aly et al. 2009; see Sect. 6.5.1). *Phelipanche* seems to also take up a host protein (Aly et al. 2011). A similar phenomenon was found in *Cuscuta* (reviewed by LeBlanc et al. 2012); comparable data from other parasitic plant families are currently lacking. Although the cellular components involved in RNA trafficking in host–parasite systems supposedly differ according to haustorial anatomy (see Sect. 3.9.3), current data suggest that parasites may have access to a great variety of host RNAs, including those encoding proteins that are functionally located in host plastids (e.g. *rbcS*, LeBlanc et al. 2012). Mobile RNAs may eventually also be incorporated into the nuclear genome of the parasite. Such cases have already been reported for an expressed nuclear gene of unknown function in *Striga hermonthica* (Yoshida et al. 2011) and also for *Rafflesia* (Xi et al. 2012).

Another case of HGT involves fragments of the plastid regions *rbcL*, *rps2*, and *trnL-F*, which appear to have been transmitted from *Orobanche* into some species of *Phelipanche* (Park et al. 2007a). Current data on plastid genome evolution in both genera suggest, however, that the horizontally acquired fragments from at least two out of the five originally studied *Phelipanche* species do not reside in the plastid genome (Wicke et al. submitted). Those fragments may be located in either the nuclear or the mitochondrial genome, as hypothesized earlier (Park et al. 2007a). Regardless of the location of the horizontally acquired copies, putative HGT remains highly interesting as it might involve transmission via a host plant as the vector, even though a vertical transmission may, however, also be considered (Park et al. 2007b).

15.6 Conclusions

Orobanchaceae possess highly dynamic genomes, which is in part due to the rampant occurrence of polyploidy as implied by genome-size data and chromosomal evolution. Transcript-profiling and transcriptome-sequencing projects have already identified several genes that are candidates for being newly recruited in parasite-specific developmental pathways, and large-scale sequencing in combination with basic genetic work has revealed complex patterns of macromolecule trafficking and signalling in selected host–parasite systems. Studies of plastid genes and genomes of members of Orobanchaceae have additionally brought to light the first insights into the complexity and differential dynamics in the process of plastid genome reduction after the loss of photosynthesis.

Ongoing and future research and research networks will allow the stepwise elucidation of the physiological evolution of the Orobanchaceae from the autotrophic to holo-heterotrophic lifestyle.

Despite the great progress that has already been achieved, our understanding of genomic evolution in Orobanchaceae is still hampered by a substantial lack of data on, for instance, chromosome numbers, genome sizes, gene content and organization, gene expression, and epigenetic variation (see Chap. 13). Further basic and comparative research are needed, including large-scale transcriptome and genome sequencing, to determine basic parasite-specific genetics and to elucidate the complex (co-)evolution of the parasites and their hosts.

Orobanchaceae, the largest and most diverse family of parasitic angiosperms has already proven to be highly suitable for studying the functional basis of a parasitic lifestyle in higher plants. Several projects are currently underway that will shed further light on genomic evolution as well as on the extent of horizontal gene transfer. Besides crucial physiological and ultrastructural works, genome surveys utilizing the rapidly developing sequencing technologies and large-scale proteomic approaches should be a key element in understanding the evolution of parasitism and all adaptations that come with the parasitic way of life, e.g. haustorium formation, host recognition, and nutrient acquisition. The sequencing of reference genomes of Orobanchaceae species is inevitable (though challenged by its genome size) and will eventually be an important step towards finding effective ways for control of the weedy species (see Chaps. 21 and 24).

The availability of reverse genetic approaches is another key point that will bring parasitic plant research to another level. *Agrobacterium*-mediated transformation systems have already been established successfully for the hemiparasites *Triphysaria versicolor* (Tomilov et al. 2007), *Phtheirospermum japonicum* (Ishida et al. (2011), and for the holoparasite *Phelipanche aegyptiaca* (Fernandez-Aparicio et al. 2011). Thus, the broomrape family provides a unique framework to experimentally test the function of putative parasite-specific genetic elements and to study physiological evolution across close relatives with differing degrees of heterotrophy.

Acknowledgments I thank Gerald Schneeweiss for the helpful comments on this manuscript and I am grateful to Kai Müller for critical discussion. I also thank Najibeh Ataei, Dietmar Quandt, and Hanna Weiss-Schneeweiss for sharing their unpublished data. Support from the Austrian Science Fund (FWF grant P19404 to G. Schneeweiss), the University of Vienna, and the German Academic Exchange Service (DAAD) for own research is highly appreciated.

References

Acquisti C, Elser JJ, Kumar S (2009a) Ecological nitrogen limitation shapes the DNA composition of plant genomes. Mol Biol Evol 26:953–956

Acquisti C, Kumar S, Elser JJ (2009b) Signatures of nitrogen limitation in the elemental composition of the proteins involved in the metabolic apparatus. Proc R Soc B 276:2605–2610

Adams KL, Clements MJ, Vaughn JC (1998) The *Peperomia* mitochondrial *coxI* group I intron: timing of horizontal transfer and subsequent evolution of the intron. J Mol Evol 46:689–696

Albach DC, Li H-Q, Zhao N, Jensen SR (2007) Molecular systematics and phytochemistry of *Rehmannia* (Scrophulariaceae). Biochem Syst Ecol 35:293–300

Alkatib S, Fleischmann TT, Scharff LB, Bock R (2012) Evolutionary constraints on the plastid tRNA set decoding methionine and isoleucine. Nucleic Acids Res 40:6713–6724

Aly R, Cholakh H, Joel DM, Leibman D, Steinitz B, Zelcer A, Naglis A, Yarden O, Gal-On A (2009) Gene silencing of mannose 6-phosphate reductase in the parasitic weed *Orobanche aegyptiaca* through the production of homologous dsRNA sequences in the host plant. Plant Biotechnol J 7:487–498

Aly R, Hamamouch N, Abu-Nassar J, Wolf S, Joel DM, Eizenberg H, Kaisler E, Cramer C, Gal-On A, Westwood JH (2011) Movement of protein and macromolecules between host plants and the parasitic weed *Phelipanche aegyptiaca* Pers. Plant Cell Rep 30:2233–2241

Bandaranayake PCG, Filappova T, Tomilov AA, Tomilova NB, Jamison-McClung D, Ngo Q, Inoue K, Yoder JI (2010) A single-electron reducing quinone oxidoreductase Is necessary to induce haustorium development in the root parasitic plant *Triphysaria*. Plant Cell 22:1404–1419

Barker WR, Kiehn M, Vitek E (1988) Chromosome numbers in Australian *Euphrasia* (Scrophulariaceae). Plant Syst Evol 158:161–164

Barkman TJ, Lim S-H, Salleh KM, Nais J (2004) Mitochondrial DNA sequences reveal the photosynthetic relatives of *Rafflesia*, the world's largest flower. Proc Natl Acad Sci USA 101:787–792

Barkman TJ, McNeal JR, Lim SH, Coat G, Croom HB, Young ND, dePamphilis CW (2007) Mitochondrial DNA suggests at least 11 origins of parasitism in angiosperms and reveals genomic chimerism in parasitic plants. BMC Evol Biol 7:248

Bennett JR, Mathews S (2006) Phylogeny of the parasitic plant family Orobanchaceae inferred from phytochrome A. Am J Bot 93:1039–1051

Berg S, Krause K, Krupinska K (2004) The *rbcL* genes of two *Cuscuta* species, *C. gronovii* and *C. subinclusa*, are transcribed by the nuclear-encoded plastid RNA polymerase (NEP). Planta 219:541–546

Bergthorsson U, Richardson AO, Young GJ, Goertzen LR, Palmer JD (2004) Massive horizontal transfer of mitochondrial genes from diverse land plant donors to the basal angiosperm *Amborella*. Proc Natl Acad Sci USA 101:17747–17752

Botanga CJ, Timko MP (2005) Genetic structure and analysis of host and non-host interactions of *Striga gesnerioides*. Phytopathology 95:1166–1173

Botanga CJ, Timko MP (2006) Phenetic relationships among different races of *Striga gesnerioides* (Willd.) Vatke from West Africa. Genome 49:1351–1365

Botanga CJ, Kling JG, Berner DK, Timko MP (2002) Genetic variability of *Striga asiatica* (L.) Kuntz based on AFLP analysis and host-parasite interaction. Euphytica 128:375–388

Boudreau E, Turmel M, Goldschmidt-Clermont M, Rochaix J-D, Sivan S, Michaels A, Leu S (1997) A large open reading frame (orf1995) in the chloroplast DNA of *Chlamydomonas reinhardtii* encodes an essential protein. Mol Gen Genet 253:649–653

Braukmann T, Stefanović S (2012) Plastid genome evolution in mycoheterotrophic Ericaceae. Plant Mol Biol 79:5–20

Castro M, Castro S, Loureiro J (2012) Genome size variation and incidence of polyploidy in Scrophulariaceae sensu lato from the Iberian Peninsula. AoB Plants 2012: pls037. doi:10.1093/aobpla/pls037

Cho KY, Palmer JD (1999) Multiple acquisitions via horizontal transfer of a group I intron in the mitochondrial *cox1* gene during evolution of the Araceae family. Mol Biol Evol 16:1155–1165
Colwell AE (1994) Genome evolution in a non-photosynthetic plant, *Conopholis americana*. Ph. D. Thesis, Washington University. Division of Biology and Biomedical Sciences, St. Louis, WA, USA
Cummings MP, Welschmeyer NA (1998) Pigment composition of putatively achlorophyllous angiosperms. Plant Syst Evol 210:105–111
Cusimano N, Zhang L-B, Renner SS (2008) Reevaluation of the cox1 group I intron in Araceae and angiosperms indicates a history dominated by loss rather than horizontal transfer. Mol Biol Evol 25:265–276
Davis C, Wurdack KJ (2004) Host-to-parasite gene transfer in flowering plants: phylogenetic evidence from Malpighiales. Science 305:676–678
Davis CC, Anderson WR, Wurdack KJ (2005) Gene transfer from a parasitic flowering plant to a fern. Proc R Soc B 272:2237–2242
de Koning AP, Keeling PJ (2006) The complete plastid genome sequence of the parasitic green alga *Helicosporidium sp.* is highly reduced and structured. BMC Biol 4:12
Delannoy E, Fujii S, des Francs CC, Brundrett M, Small I (2011) Rampant gene loss in the underground orchid *Rhizanthella gardneri* highlights evolutionary constraints on plastid genomes. Mol Biol Evol 28:2077–2086
Delavault PM, Thalouarn P (2002) The obligate root parasite *Orobanche cumana* exhibits several *rbcL* sequences. Gene 297:85–92
Delavault PM, Sakanyan V, Thalouarn P (1995) Divergent evolution of two plastid genes, *rbcL* and *atpB*, in a non-photosynthetic parasitic plant. Plant Mol Biol 29:1071–1079
Delavault PM, Russo NM, Lusson NA, Thalouarn P (1996) Organization of the reduced plastid genome of *Lathraea clandestina*, an achlorophyllous parasitic plant. Physiol Plant 96:674–682
dePamphilis CW, Palmer JD (1990) Loss of photosynthetic and chlororespiratory genes from the plastid genome of a parasitic flowering plant. Nature 348:337–339
dePamphilis CW, Young ND, Wolfe AD (1997) Evolution of plastid gene rps2 in a lineage of hemiparasitic and holoparasitic plants: many losses of photosynthesis and complex patterns of rate variation. Proc Natl Acad Sci USA 94:7367–7372
Downie SR, Palmer JD (1992) Restriction site mapping of the chloroplast DNA inverted repeat – a molecular phylogeny of the Asteridae. Ann Mo Bot Gard 79:266–283
Drescher A, Ruf S, Calsa T, Carrer H, Bock R (2000) The two largest chloroplast genome-encoded open reading frames of higher plants are essential genes. Plant J 22:97–104
Duff RJ, Nickrent DL (1997) Characterization of mitochondrial small-subunit ribosomal RNAs from holoparasitic plants. J Mol Evol 45:631–639
Ems SC, Morden CW, Dixon CK, Wolfe KH, dePamphilis CW, Palmer JD (1995) Transcription, splicing and editing of plastid RNAs in the nonphotosynthetic plant *Epifagus virginiana*. Plant Mol Biol 29:721–733
Estep MC, Gowda BS, Huang K, Timko MP, Bennetzen JL (2012) Genomic characterization for parasitic weeds of the genus *Striga* by sample sequence analysis. Plant Genome 5:30–41
Fedorov AA (ed) (1969) Khromosomnye chisla tsetkovykh rasteny (Chromosome numbers of flowering plants). Izdatel'stvo Nauka, Leningrad
Fernandez-Aparicio M, Rubiales D, Bandaranayake PCG, Yoder J, Westwood J (2011) Transformation and regeneration of the holoparasitic plant *Phelipanche aegyptiaca*. Plant Methods 7:36
Filipowicz N, Renner SS (2010) The worldwide holoparasitic Apodanthaceae confidently placed in the Cucurbitales by nuclear and mitochondrial gene trees. BMC Evol Biol 10:219
Funk H, Berg S, Krupinska K, Maier U, Krause K (2007) Complete DNA sequences of the plastid genomes of two parasitic flowering plant species, *Cuscuta reflexa* and *Cuscuta gronovii*. BMC Plant Biol 7:45
Goldblatt P, Johnson DE (1979) Index to plant chromosome numbers. Missouri Botanical Garden, St. Louis. http://www.tropicos.org/Project/IPCN. Accessed August 2012

Hanson L, McMahon KA, Johnson MAT, Bennett MD (2001) First nuclear DNA C-values for another 25 angiosperm families. Ann Bot 88:851–858
Hanson L, Leitch IJ, Bennett MD (2002) Unpublished data from the Jodrell Laboratory, Royal Botanic Gardens, Kew. Accessed via the Kew C-Value Database at data.kew.org/cvalues/ in August 2012
He F, Zhang X, Hu J-Y, Turck F, Dong X, Goebel U, Borevitz JO, de Meaux J (2012) Widespread interspecific divergence in cis-regulation of transposable elements in the *Arabidopsis* genus. Mol Biol Evol 29:1081–1091
Hjertson ML (1995) Taxonomy, phylogeny and biogeography of *Lindenbergia* (Scrophulariaceae). Bot J Linn Soc 119:265–321
Honaas L, Wafula E, Yang Z, Der J, Wickett N, Altman N, Taylor C, Yoder J, Timko M, Westwood J, dePamphilis (2013) Functional genomics of a generalist parasitic plant: Laser microdissection of host-parasite interface reveals host-specific patterns of parasite gene expression. BMC Plant Biol 13:9
Ishida JK, Yoshida S, Ito M, Namba S, Shirasu K (2011) *Agrobacterium rhizogenes*-mediated transformation of the parasitic plant *Phtheirospermum japonicum*. PLoS One 6(10):e25802. doi:10.1371/journal.pone.0025802
Iwo GA, Husaini SWH, Olaniyan GO (1993) Cytological observations and distribution of *Striga* species in central part of Nigeria. Feddes Repertorium 104:497–501
Knauf U, Hachtel W (2002) The genes encoding subunits of ATP synthase are conserved in the reduced plastid genome of the heterotrophic alga *Prototheca wickerhamii*. Mol Genet Genomics 267:492–497
Knoop V, Unseld M, Marienfeld J, Brandt P, Sunkel S, Ullrich H, Brennicke A (1996) Copia-, gypsy- and LINE-like retrotransposon fragments in the mitochondrial genome of *Arabidopsis thaliana*. Genetics 142:579–585
Knoop V, Volkmar U, Hecht J, Grewe F (2011) Mitochondrial genome evolution in the plant lineage. In: Kempten F (ed) Advances in plant biology – mitochondrial genomes. Springer Science & Business, New Yorlk, pp 3–29
Kondo K, Segawa M, Musselman LJ, Mann WF (1981) Comparative ecological study of the chromosome races in certain root parasitic plants of the southeastern U.S.A. Bol Soc Broteriana 53:793–807
Krause K (2011) Piecing together the puzzle of parasitic plant plastome evolution. Planta 234:647–656
Krause K, Berg S, Krupinska K (2003) Plastid transcription in the holoparasitic plant genus *Cuscuta*: parallel loss of the *rrn16* PEP-promoter and of the *rpoA* and *rpoB* genes coding for the plastid-encoded RNA polymerase. Planta 216:815–823
Lagerkvist U (1978) "Two out of three": an alternative method for codon reading. Proc Natl Acad Sci USA 75:1759–1762
LeBlanc M, Kim G, Westwood JH (2012) RNA trafficking in parasitic plant systems. Front Plant Sci 3:203
Lechat M-M, Pouvreau J-B, Péron T et al (2012) *PrCYP707A1*, an ABA catabolic gene, is a key component of *Phelipanche ramosa* seed germination in response to the strigolactone analogue GR24. J Exp Bot 63:5311–5322
Leebens-Mack JH, dePamphilis CW (2002) Power analysis of tests for loss of selective constraint in cave crayfish and nonphotosynthetic plant lineages. Mol Biol Evol 19:1292–1302
Leitch IJ, Bennett MD (2004) Genome downsizing in polyploid plants. Biol J Linn Soc Lond 82:651–663
Leitch AR, Leitch IJ (2012) Ecological and genetic factors linked to contrasting genome dynamics in seed plants. New Phytol 194:629–646
Lemaire B, Huysmans S, Smets E, Merckx V (2011) Rate accelerations in nuclear 18S rDNA of mycoheterotrophic and parasitic angiosperms. J Plant Res 124:561–576
Logacheva MD, Schelkunov MI, Penin AA (2011) Sequencing and analysis of plastid genome in mycoheterotrophic orchid *Neottia nidus-avis*. Genome Biol Evol 3:1296–1303

Lohan AJ, Wolfe KH (1998) A subset of conserved tRNA genes in plastid DNA of nongreen plants. Genetics 150:425–433

Lohse M, Drechsel O, Bock R (2007) Organellar Genome DRAW (OGDRAW): a tool for the easy generation of high-quality custom graphical maps of plastid and mitochondrial genomes. Curr Genet 52:267–274

Lusson NA, Delavault PM, Thalouarn P (1998) The *rbcL* gene from the non-photosynthetic parasite *Lathraea clandestina* is not transcribed by a plastid-encoded RNA polymerase. Curr Genet 34:212–215

Manen J-F, Habashi C, Jeanmonod D, Park J-M, Schneeweiss GM (2004) Phylogeny and intraspecific variability of holoparasitic *Orobanche* (Orobanchaceae) inferred from plastid *rbcL* sequences. Mol Phylogenet Evol 33:482–500

Martin NJ (1983) Nuclear DNA variation in the Australasian Loranthaceae. In: Calder M, Berhnhardt P (eds) Biology of mistletoes. Academic Press, New York, pp 277–293

Matvienko M, Torres MJ, Yoder JI (2001) Transcriptional responses in the hemiparasitic plant *Triphysaria versicolor* to host plant signals. Plant Physiol 127:272–282

McNeal JR, Arumugunathan K, Kuehl J, Boore J, dePamphilis C (2007a) Systematics and plastid genome evolution of the cryptically photosynthetic parasitic plant genus *Cuscuta* (Convolvulaceae). BMC Biol 5:55

McNeal JR, Kuehl J, Boore J, de Pamphilis C (2007b) Complete plastid genome sequences suggest strong selection for retention of photosynthetic genes in the parasitic plant genus *Cuscuta*. BMC Plant Biol 7:57

Moore DM (1982) Flora Europaea Check-List and chromosome index. Cambridge University Press, Cambridge UK

Morden CW, Wolfe KH, dePamphilis CW, Palmer JD (1991) Plastid translation and transcription genes in a nonphotosynthetic plant – Intact, missing and pseudogenes. EMBO J 10:3281–3288

Mower JP, Stefanovic S, Young GJ, Palmer JD (2004) Gene transfer from parasitic to host plants. Nature 432:165–166

Mower JP, Stefanovic S, Hao W, Gummow J, Jain K, Ahmed D, Palmer J (2010) Horizontal acquisition of multiple mitochondrial genes from a parasitic plant followed by gene conversion with host mitochondrial genes. BMC Biol 8:150

Nagl W, Fusenig HP (1979) Types of chromatin organization in plant nuclei. Plant Syst Evol Suppl 2:221–223

Nickrent DL, Duff RJ (1996) Molecular studies or parasitic plants using ribosomal RNA. In: Moreno MT, Cubero JI, Berner D, Joel D, Musselman LJ, Parker C (eds) Advances in parasitic plant research. Junta de Andalucia, Dirección General de Investigación Agraria, Cordoba, Spain, pp 28–52

Nickrent DL, García M (2009) On the brink of holoparasitism: plastome evolution in dwarf mistletoes (*Arceuthobium*, Viscaceae). J Mol Evol 68:603–615

Nickrent DL, Blarer A, Qiu Y-L, Vidal-Russell R, Anderson FE (2004) Phylogenetic inference in Rafflesiales: the influence of rate heterogeneity and horizontal gene transfer. BMC Evol Biol 4:40

Park J-M, Manen J-F, Schneeweiss GM (2007a) Horizontal gene transfer of a plastid gene in the non-photosynthetic flowering plants *Orobanche* and *Phelipanche* (Orobanchaceae). Mol Phyl Evol 43:974–985

Park J-M, Schneeweiss GM, Weiss-Schneeweiss H (2007b) Diversity and evolution of *Ty1-copia* and *Ty3-gypsy* retroelements in the non-photosynthetic flowering plants *Orobanche* and *Phelipanche* (Orobanchaceae). Gene 387:75–86

Piednoël M, Aberer AJ, Schneeweiss GM, Macas J, Novak P, Gundlach H, Temsch EM, Renner SS (2012) Next-generation sequencing reveals the impact of repetitive DNA across phylogenetically closely related genomes of Orobanchaceae. Mol Biol Evol 29:3601–3611

Randle CP, Wolfe AD (2005) The evolution and expression of RBCL in holoparasitic sister-genera *Harveya* and *Hyobanche* (Orobanchaceae). Am J Bot 92:1575–1585

Renner SS, Bellot S (2012) Horizontal gene transfer in eukaryotes: fungi-to-plant and plant-to-plant transfers of organellar DNA. In: Bock R, Knoop V (eds) Genomics of chloroplasts and mitochondria. Springer, Heidelberg, pp 223–235

Rogalski M, Karcher D, Bock R (2008) Superwobbling facilitates translation with reduced tRNA sets. Nat Struct Mol Biol 15:192–198

Sanchez-Puerta MV, Cho Y, Mower JP, Alverson AJ, Palmer JD (2008) Frequent, phylogenetically local horizontal transfer of the *cox1* group I intron in flowering plant mitochondria. Mol Biol Evol 25:1762–1777

Schäferhoff B, Fleischmann A, Fischer E, Albach D, Borsch T, Heubl G, Müller KF (2010) Towards resolving Lamiales relationships: insights from rapidly evolving chloroplast sequences. BMC Evol Biol 10:352

Schneeweiss GM, Palomeque T, Colwell AE, Weiss-Schneeweiss H (2004) Chromosome numbers and karyotype evolution in holoparasitic *Orobanche* (Orobanchaceae) and related genera. Am J Bot 91:439–448

Schwender J, Goffman F, Ohlrogge JB, Shachar-Hill Y (2004) Rubisco without the Calvin cycle improves the carbon efficiency of developing green seeds. Nature 432:779–782

Seif E, Leigh J, Liu Y, Roewer I, Forget L, Lang BF (2005) Comparative mitochondrial genomics in zygomycetes: bacteria-like RNase P RNAs, mobile elements and a close source of the group I intron invasion in angiosperms. Nucleic Acids Res 33:734–744

Shinozaki K, Ohme M, Tanaka M et al (1986) The complete nucleotide sequence of the tobacco chloroplast genome: its gene organization and expression. EMBO J 5:2043–2049

Takagi K, Okazawa A, Wada Y, Mongkolchaiyaphruek A, Fukusaki E, Yoneyama K, Takeuchi Y, Kobayashi A (2009) Unique phytochrome responses of the holoparasitic plant *Orobanche minor*. New Phytol 182:965–974

Tank DC, Egger JM, Olmstead RG (2009) Phylogenetic classification of subtribe Castillejinae (Orobanchaceae). Syst Bot 34:182–197

Tolbert NE (1997) The C2 oxidative photosynthetic carbon cycle. Annu Rev Plant Physiol Plant Mol Biol 48:1–25

Tomilov AA, Tomilova NB, Yoder JI (2007) *Agrobacterium tumefaciens* and *Agrobacterium rhizogenes* transformed roots of the parasitic plant *Triphysaria versicolor* retain parasitic competence. Planta 225:1059–1071

Tomilov AA, Tomilova NB, Wroblewski T, Michelmore R, Yoder JI (2008) Trans-specific gene silencing between host and parasitic plants. Plant J 56:389–397

Trakulnaleamsai C, Okazawa A, An C-I, Kajiyama S, Fukusaki E, Yoneyama K, Takeuchi Y, Kobayashi A (2005) Isolation and characterization of a cDNA encoding phytochrome A in the non-photosynthetic parasitic plant, *Orobanche minor* Sm. Biosci Biotechnol Biochem 69:71–78

Vaughn JC, Mason MT, Sper-Whitis GL, Kuhlman P, Palmer JD (1995) Fungal origin by horizontal transfer of a plant mitochondrial group I intron in the chimeric *coxI* gene of *Peperomia*. J Mol Evol 41:563–572

Weiss-Schneeweiss H, Greilhuber J, Schneeweiss GM (2006) Genome size evolution in holoparasitic *Orobanche* (Orobanchaceae) and related genera. Am J Bot 93:148–156

Westwood JH, Yoder JI, Timko MP, dePamphilis CW (2010) The evolution of parasitism in plants. Trends Plant Sci 15:227–235

Westwood JH, dePamphilis CW, Das M, Fernández-Aparicio M, Honaas LA, Timko MP, Wafula EK, Wickett NJ, Yoder JI (2012) The Parasitic Plant Genome Project: new tools for understanding the biology of *Orobanche* and *Striga*. Weed Sci 60:295–306

Wickett NJ, Zhang Y, Hansen SK, Roper JM, Kuehl JV, Plock SA, Wolf PG, dePamphilis CW, Boore JL, Goffinet B (2008) Functional gene losses occur with minimal size reduction in the plastid genome of the parasitic liverwort *Aneura mirabilis*. Mol Biol Evol 25:393–401

Wickett NJ, Honaas LA, Wafula EK, Das M, Huang K, Wu B, Landherr L, Timko MP, Yoder J, Westwood JH, dePamphilis CW (2011) Transcriptomes of the parasitic plant family Orobanchaceae reveal surprising conservation of chlorophyll synthesis. Curr Biol 21:2098–2104

Wimpee CF, Wrobel R, Garvin D (1991) A divergent plastid genome in Conopholis americana, an achlorophyllous parasitic plant. Plant Mol Biol 17(166):161

Wimpee CF, Morgan R, Wrobel RL (1992) Loss of transfer RNA genes from the plastid 16S–23S ribosomal RNA gene spacer in a parasitic plant. Curr Genet 21:417–422

Wolfe KH (1994) Similarity between putative ATP-binding sites in land plant plastid ORF2280 proteins and the *FtsH*/CDC48 family of ATPases. Curr Genet 25:379–383

Wolfe AD, dePamphilis CW (1997) Alternate paths of evolution for the photosynthetic gene *rbcL* in four nonphotosynthetic species of *Orobanche*. Plant Mol Biol 33:965–977

Wolfe AD, dePamphilis CW (1998) The effect of relaxed functional constraints on the photosynthetic gene *rbcL* in photosynthetic and nonphotosynthetic parasitic plants. Mol Biol Evol 15:1243–1258

Wolfe KH, Li WH, Sharp PM (1987) Rates of nucleotide substitution vary greatly among plant mitochondrial, chloroplast, and nuclear DNAs. Proc Natl Acad Sci USA 84:9054–9058

Wolfe KH, Morden CW, Ems SC, Palmer JD (1992a) Rapid evolution of the plastid translational apparatus in a nonphotosynthetic plant: loss or accelerated sequence evolution of tRNA and ribosomal protein genes. J Mol Evol 35:304–317

Wolfe KH, Morden CW, Palmer JD (1992b) Function and evolution of a minimal plastid genome from a nonphotosynthetic parasitic plant. Proc Natl Acad Sci USA 89:10648–10652

Won H, Renner SS (2003) Horizontal gene transfer from flowering plants to *Gnetum*. Proc Natl Acad Sci USA 100:10824–10829

Wood TE, Takebayashic N, Abrahamsen MS, Mayrose I, Greenspoond PB, Rieseberg LH (2009) The frequency of polyploid speciation in vascular plants. Proc Natl Acad Sci USA 106:13875–13879

Xi Z, Bradley R, Wurdack K, Wong KM, Sugumaran M, Bomblies K, Rest J, Davis C (2012) Horizontal transfer of expressed genes in a parasitic flowering plant. BMC Genomics 13:227

Yoshida S, Ishida JK, Kamal N, Ali A, Namba S, Shirasu K (2010) A full-length enriched cDNA library and expressed sequence tag analysis of the parasitic weed, *Striga hermonthica*. BMC Plant Biol 10:55

Young ND, dePamphilis CW (2005) Rate variation in parasitic plants: correlated and uncorrelated patterns among plastid genes of different function. BMC Evol Biol 5:16

Zonneveld BJM (2010) New record holders for maximum genome size in Eudicots and Monocots. J Bot 2010:4 pages

Zonneveld BJM, Leitch IJ, Bennett MD (2005) First nuclear DNA amounts in more than 300 angiosperms. Ann Bot 96:229–244

Chapter 16
Ecology of Hemiparasitic Orobanchaceae with Special Reference to Their Interaction with Plant Communities

Duncan D. Cameron and Gareth K. Phoenix

16.1 Introduction

Members of the family Orobanchaceae have colonised the majority of terrestrial biomes on a global scale, from the high arctic where *Pedicularis* species are found on the Svalbard archipelago to the tropics with *Striga* species in Africa. Whilst the parasitic syndrome in plants has been recognised for centuries, it is only relatively recently that their importance in agricultural (Parker and Riches 1993) and natural (Gibson and Watkinson 1992) ecosystems has been highlighted. The majority of the physiological research has focused mainly on parasitic weeds over the past 30 years. Recent work has focused on understanding host–parasite relations between these economically important parasites and their range of crop hosts (see Chaps. 6 and 7; Gurney et al. 2003; Goldwasser et al. 2000), whilst the ecology of non-agriculturally significant parasitic plants has not been considered in such detail. This is despite parasitic plants representing significant components of natural and semi-natural ecosystems around the globe. As a consequence their potential role in the functioning of these systems is not fully resolved. A number of studies have shown the potential of parasitic plants to shape the structure of the communities they inhabit (Cameron et al. 2005, 2009; Westbury and Dunnett 2000, 2007; Marvier 1998; Davies et al. 1997; Pennings and Callaway 1996; Gibson and Watkinson 1992). However, there is a great deal of variability in the response of the host community to the parasite. There is also a debate as to the mechanisms through which the parasites effect changes in communities and ecosystems (Irving and Cameron 2009; Cameron et al. 2005). This chapter reviews the current knowledge on the ecology of the family Orobanchaceae with special reference to community-level effects.

D.D. Cameron (✉) • G.K. Phoenix
Department of Animal and Plant Sciences, The University of Sheffield, Alfred Denny Building, Western Bank, Sheffield S10 2TN, UK
e-mail: d.cameron@shef.ac.uk; g.phoenix@shef.ac.uk

D.M. Joel et al. (eds.), *Parasitic Orobanchaceae*,
DOI 10.1007/978-3-642-38146-1_16,

16.2 Interactions Between Parasitic Plants and Their Hosts at the Individual Scale

Community-level effects are based on the interactions of the parasite and host at the level of the individual plant. Understanding these interactions is therefore essential to understanding community-level effects of the Orobanchaceae.

16.2.1 Host Range

Many of the hemiparasitic Orobanchaceae have the capacity to utilise a wide range of hosts. The facultative hemiparasite *Rhinanthus minor*, for example, has a host range of 50 species (Gibson and Watkinson 1989; but see comments at the end of this section) and the obligate hemiparasite *Striga asiatica* has a host range of approximately 42 non-agricultural host species (Cochrane and Press 1997). To a far lesser extent, extreme host specificity has also been recorded. One such example is the holoparasitic species *Orobanche minor* where sub-speciation events have led to the parasite expanding its host range through co-evolution with potential host species. This has resulted in individual *O. minor* subspecies specialising on a specific host species or small group of related species (Thorogood et al. 2009). Perhaps the most notable exception among the non-weedy Orobanchaceae is *Epifagus grandifolia*, which only parasitises *Fagus grandifolia* (American beech) (Musselman and Press 1995).

Not all of the host species are equally beneficial for the growth and development of the parasite (Hwangbo 2000). There is a broad trend that grasses and legumes are the best hosts for the facultative hemiparasite *Rhinanthus minor*, in terms of parasite growth and fecundity, whilst the forbs (non-leguminous, herbaceous species) make the poorest hosts. Association with some forbs can even be lethal to the parasite (Cameron et al. 2005; Seel and Press 1993; Hwangbo 2000). The reasons underlying differences in host quality remained unclear until recently. *R. minor* performed better when attached to hosts with high foliar nitrogen content (Seel et al. 1993). Therefore, the poorest hosts may contain fewer resources for the parasite, although the ability of the host to limit the parasite from being a sink for resources could also account for variation in host quality. Likewise, potential hosts are not equally beneficial for *Striga*; for example, the tropical grass *Tripsacum dactyloides* has durable, non-host resistance to *Striga hermonthica* (Gurney et al. 2003).

Whilst the hypothesis that the differential responses of potential hosts to parasites can be based on host- and non-host resistance and tolerance has been extensively studied in *Striga* and *Orobanche* species (see Chap. 7), the physiology of host–parasitic plant interactions is less well studied in non-agricultural parasites (see Chap. 6; Cameron and Seel 2007; Rümer et al. 2007; Cameron et al. 2006). Resistance or susceptibility to *R. minor* is conserved within functional groups.

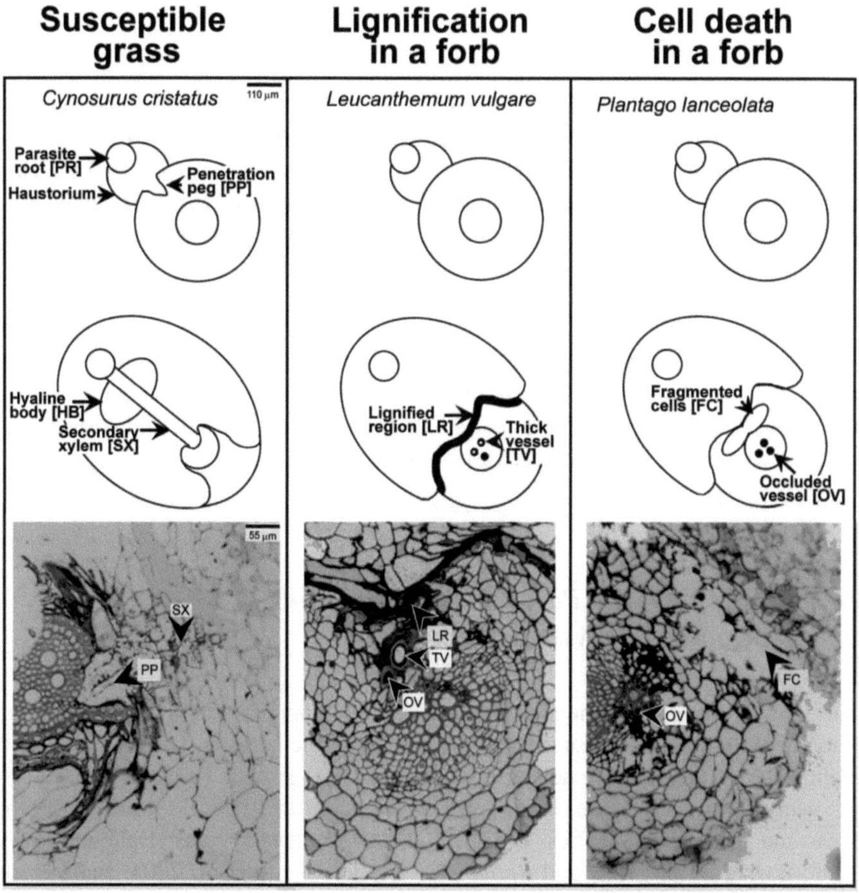

Fig. 16.1 Haustorium penetration in susceptible *vs* resistant hosts. Ontogeny of haustoria formed by *Rhinanthus minor* on the susceptible grass *Cynosurus cristatus*, compared to the resistance reaction in the forbs *Leucanthemum vulgare* and *Plantago lanceolata* (from Cameron and Seel 2007, with permission)

Grasses (Poaceae) are susceptible to the parasite, and although the cortical tissues of the grasses become lignified in response to parasites (Rümer et al. 2007), this does not prevent access to the host vasculature (Figs. 3.10a and 16.1; Rümer et al. 2007; see Chaps. 4 and 7) and does not impede parasite acquisition of host resources (Cameron and Seel 2007). In contrast, forbs have dramatic resistance responses to the parasite, encapsulating the parasite endophytic tissues in lignin (Fig. 16.1, Rümer et al. 2007) and/or dying cells upon contact with the parasite (Fig. 16.1; Cameron et al. 2006) preventing resource transfer from host to parasite (Cameron and Seel 2007). Legumes (Fabaceae) are also infested by *R. minor* but do not seem to undergo reduction in growth and fecundity as a result of infestation and yet are not resistant to this parasite (Jiang et al. 2008; Rümer et al. 2007;

Cameron et al. 2006; but see Sect. 7.3). The reasons behind such tolerance are unresolved (Jiang et al. 2008).

Together, these observations provide an explanation for the differential responses of grasses, forbs and legumes in terms of growth, fecundity and metabolism, to infestation by *R. minor* (Cameron et al. 2009), providing a mechanism through which *R. minor* can influence plant communities (see Sect. 16.3.1). This also raises critical questions regarding the identification of the host range of a parasitic plant. For example, it was necessary to trace the roots of *R. minor* to host roots in the field to determine the parasite host range (Gibson and Watkinson 1989). They inferred that haustoria on roots of any given species reveal it to be a host for the parasite even though *R. minor* may form wart haustoria or meta-haustoria not only on any plant root (Keith et al. 2004; see Sect. 3.15) but also on twigs, stones and even the plant pots. It is necessary that these haustoria be identified as functional before a given species can be defined as a host, as not all haustoria are successful at penetrating the host vasculature.

16.2.2 Effect of the Parasite on the Host in Compatible Associations

The growth and development of the host can be suppressed as a consequence of resource withdrawal after successful attachment of the parasite. Different parasitic plant genera typically have deleterious effects on host growth. Few examples out of the many studied cases are *Striga* species (Kuiper et al. 1998; Taylor et al. 1996; Press and Stewart 1987; Press et al. 1987a), *Orobanche* species (Dale and Press 1998; Barker et al. 1996) and *Rhinanthus* species (Keith et al. 2004; Jiang et al. 2003; Davies and Graves 2000; Hwangbo 2000; Seel and Press 1996). Parasites can also significantly suppress the reproductive output of infested hosts, e.g. *S. hermonthica* (Olivier 1996) and *R. minor* (Seel et al. 1993).

There is a puzzle, however, in that parasite-induced host damage is often greater than expected based on resource removal alone (i.e. the biomass achieved by the parasite does not compensate for the loss of host biomass, e.g. Gurney et al. 2002). Reduction in host productivity is potentially due to resource abstraction by the parasite (Press et al. 1987b) but can also result from parasite-induced suppression of host photosynthesis (Press et al. 1987a; Graves et al. 1989). The induced suppression of host biomass was non-source-sink, and the 'missing biomass' was attributed to reduced host photosynthetic competency (Cameron et al. 2008).

In addition to direct negative effects on growth and fecundity of some hosts, parasitic plants can also indirectly reduce the ability of their hosts to compete with their neighbours as a result of resource loss and/or suppression of metabolic processes (Cameron et al. 2005; Matthies 1995, 1996; Gibson and Watkinson 1991). If neighbouring plants are either uninfested or are able to prevent parasitism, then the infested host plants may be out competed in field conditions. The parasite is likely to induce shifts in host community structure when the net effects of suppression of host competition restructure competitive hierarchies (see Sect. 16.3).

16.2.3 Interaction with Resource Availability

Increased mineral nutrient availability often lessens the impact of parasitic plants such as *Striga*, (Cechin and Press 1993b, 1994; Patterson 1987), *Orobanche* (Egley 1971) and *Rhinanthus* species (Matthies and Egli 1999) on their hosts. Elevated availability of nitrogen in the soil hinders *Striga* emergence (Farina et al. 1985; Last 1961) and biomass (Cechin and Press 1993a) and suppresses success of *Orobanche ramosa* (syn. *Phelipanche ramosa*) and *Orobanche minor* (Westwood and Foy 1999; see Sects. 10.2.5 and 22.3.3).[1]

Elevated levels of phosphorus have the same suppressant effects on the *Rhinanthus*-host association as elevated nitrogen has on the *Striga*-host association. The parasite performs poorly and the damage to the host is lessened (Davies and Graves 2000). Negative effects of high nutrients on parasitic plants may lie in the fact that parasitic plants are mainly restricted to nutrient poor environments (Pennings and Callaway 2002). This seems to be related to competition with the host for light as hemiparasites perform well in low productivity environments where competition for light is also likely to be low (Pennings and Callaway 2002).

Elevated nitrogen seems to suppress *Striga* and *Orobanche* through its toxicity as these parasites have a low sensitivity to ammonium and nitrate (Egley 1971; Westwood and Foy 1999). It does not, however, appear that elevated phosphorus is damaging to *R. minor* in the same way; instead, the parasites receive less from their hosts as phosphorus stimulates host growth and hence increases its sink strength relative to the parasite (Davies and Graves 2000).

16.3 Orobanchaceae in Plant Communities: Multiple Impacts, Multiple Consequences

The impacts of parasites on host species in natural plant communities lead to a cascade of effects operating through changes in the competitive ability of the hosts; shifts in the competitive balance between hosts, non-hosts and parasites; and ultimately shifts in community structure and diversity, vegetation zonation and cycling of plant populations (Phoenix and Press 2005). Furthermore, community impacts also operate through a range of trophic levels, including other parasites, herbivores, mycorrhizal fungi and soil bacteria, and through impacts on the abiotic environment, including soil nutrient cycling, water availability and physical

[1] However, the required levels of N may be far too high for common agricultural use.

structure of the vegetation. Given this impressive array of *modi operandi*, it is perhaps unsurprising that the Orobanchaceae provide excellent examples of keystone species in certain plant ecosystems and may act as allogenic (driven by the abiotic components) and autogenic (driven by the biotic components) 'ecosystem engineers' (Press and Phoenix 2005).

16.3.1 Differential Resistance Underpins Community Consequences

One of the single most important factors in determining the community level is the identity of the co-occurring species that are parasitised. As with root parasites of most other plant families, members of the Orobanchaceae tend to have a broad host range (see Sect. 16.2.1). Despite this typical broad host range, the differential resistance of potential hosts ensures that only a small subset of plants is parasitised. It is this bias towards a small number of poorly defended host species that allows some parasitic plants to impact on the structure of plant communities. There is, therefore, now growing realisation that shifts in plant community composition may result from the differential resistance inherent in hosts, rather than active 'host selection' (Gibson and Watkinson 1989) by parasites acting as a 'discriminate consumers' (Press and Phoenix 2005).

16.3.2 Parasites Reduce Whole Community Productivity

Since reductions in productivity of hosts are so great and not countered by the gain in parasite biomass, parasitism often reduces the biomass of the whole community. In fact, losses in total productivity of European grasslands resulting from *Rhinanthus* parasitism range from 8 % to a considerable 73 % (Davies et al. 1997). This reaches an extreme situation with certain crops in agriculture (see Chap. 17). Indeed, it is this disproportionate and considerable effect in contrast to the abundance of the parasite that makes hemiparasitic Orobanchaceae excellent examples of keystone species in plant communities (Paine 1969). The reasons for such a large range of biomass reduction are not fully understood. It seems likely that parasite-induced suppression of host photosynthesis is a component of reduced productivity (Cameron et al. 2008). The reduction in biomass may be much stronger under nutrient limited conditions (Matthies and Egli 1999). Reductions in above-ground community biomass may also be smaller in plant communities with greater diversity of plant functional types. The mechanism is not known but it is proposed that the greater functional diversity allows a larger number of better-defended hosts whose productivity can compensate for the loss of biomass from susceptible hosts (Joshi et al. 2000).

16.3.3 Impacts on Floristic Diversity

There are many examples of parasitic Orobanchaceae altering community composition by reducing host productivity and competitive ability, and so ultimately releasing non-host species from competition and increasing non-host abundance. Perhaps the most widely reported example is the parasitism of poorly defended grasses by *Rhinanthus* species in European grasslands and the resulting promotion of forb abundance (Westbury and Dunnett 2000, 2007; Davies et al. 1997). *Rhinanthus* is thus an effective tool in the restoration of agriculturally improved, species poor, grasslands to a floristically rich pasture or hay meadow (Pywell et al. 2004; Westbury 2004) where it is desirable to reduce grass abundance due to their expansion as a result of agricultural improvement. Since grasses are least well defended, they have the greatest declines in cover upon introduction of the parasite. This approach promotes an increase in abundance of the better-defended forbs. In this example, hosts are derived from dominant species (grasses) and the end result is therefore an increase in biodiversity (Press 1998). This situation may be facilitated by the fact that community dominants are simply more likely to be parasitised due to their root abundance and greater chance of encounter by the parasite (Davies et al. 1997). A promotion of biodiversity, however, is not always the outcome: where hosts are derived from subordinate species, further reductions in their abundance occurs and community dominant species may conversely benefit as seen in sand dune systems with *R. minor* (Gibson and Watkinson 1989).

Intriguingly, annual hemiparasites such as *Rhinanthus alectorolophus* may increase diversity by their very absence. In this case, by creating open ground resulting from their dieback, safe sites can be created for establishment of invasive species and hence increase community diversity (Joshi et al. 2000). Such sites may benefit from nutrient addition via decaying parasite biomass (see Sect. 16.3.4).

Beyond these immediate diversity effects, there is also further potential for positive feedback to the parasite since a more floristically rich community may enhance the productivity of the parasite. Again, this has been seen in the work of Joshi et al. (2000) studying *Rhinanthus alectorolophus*. Such a response may occur if greater floristic diversity provides greater opportunity for a 'mixed diet' that may promote parasite productivity (Marvier 1998) or a greater diversity may simply increase the chance of the parasite finding a poorly defended host species (Press and Phoenix 2005). Though yet to be proven, a positive community feedback could be initiated with the parasite promoting community diversity. This in turn would promote parasite productivity, increasing community diversity.

16.3.4 Cycling in Community Composition and 'Waves of Change'

Negative feedbacks between parasite and host community can result in cycling of community composition. Parasite reduction of the abundance of its main host species can result in a subsequent decline in its own abundance. This then allows recovery of the main host species, and once again a return of the parasite due to the increase in its favoured hosts. This can establish population cycling similar to that seen in classic predator–prey systems (Krebs et al. 1995). This feedback also explains why some parasites appear to move through vegetation in waves, as seen with *Rhinanthus* species in grasslands. Here parasitised patches will have a decline in the main host grass species, and therefore subsequent decline in *Rhinanthus* due to the low abundance of hosts. The annual *Rhinanthus* then establishes in neighbouring patches where the main grass hosts are still abundant. The overall effect is a highly spatio-temporally dynamic system (Press and Phoenix 2005; Gibson and Watkinson 1992). Indeed, Kelly (1989) suggested that short-lived hemiparasites may exist in a community as 'shifting clouds', depleting hosts in a given area then invading new territory only to re-invade previously occupied areas once the host community has recovered. Cameron et al. (2009) used mathematical modelling approaches to investigate this hypothesis for *R. minor*. The model was parameterised with pairwise interaction coefficients derived from pot studies where a range of host species (forbs and grasses) were grown with and without the parasite at two nutrient levels. The model suggested that the community dynamics induced by *R. minor* conformed to *a rock-paper-scissors* game (intransitive dynamics) where the community cycles between grass- and forb-rich phases over time (Fig. 16.2a). Spatially, the community exists as contiguous patches across a continuum of grass-rich to forb-rich with the parasite causing a reduction in grasses and a relative increase in forbs, eventually leading to localised extirpation of the parasite due to a lack of hosts (Fig. 16.2b–d). Over time, and in the absence of the parasite, grasses begin to out-compete forbs in uninfested patches only for the parasite to re-invade once there are sufficient grass hosts. This supports the notion that hemiparasite populations may form 'shifting clouds' in natural populations and explains shifts in host community structure of seemingly unpredictable magnitude following introduction of hemiparasites. The extent to which intransitive community dynamics would result from infestation by a parasite species with a long-term persistent seedbank is, however, questionable.

16.3.5 Competition and the Case for Facilitation

The parasite is also in direct competition with its surrounding host and non-host species. This competition can be strongest for light in the case of hemiparasitic Orobanchaceae. The performance of *Rhinanthus serotinus* (syn. *R. angustifolius*)

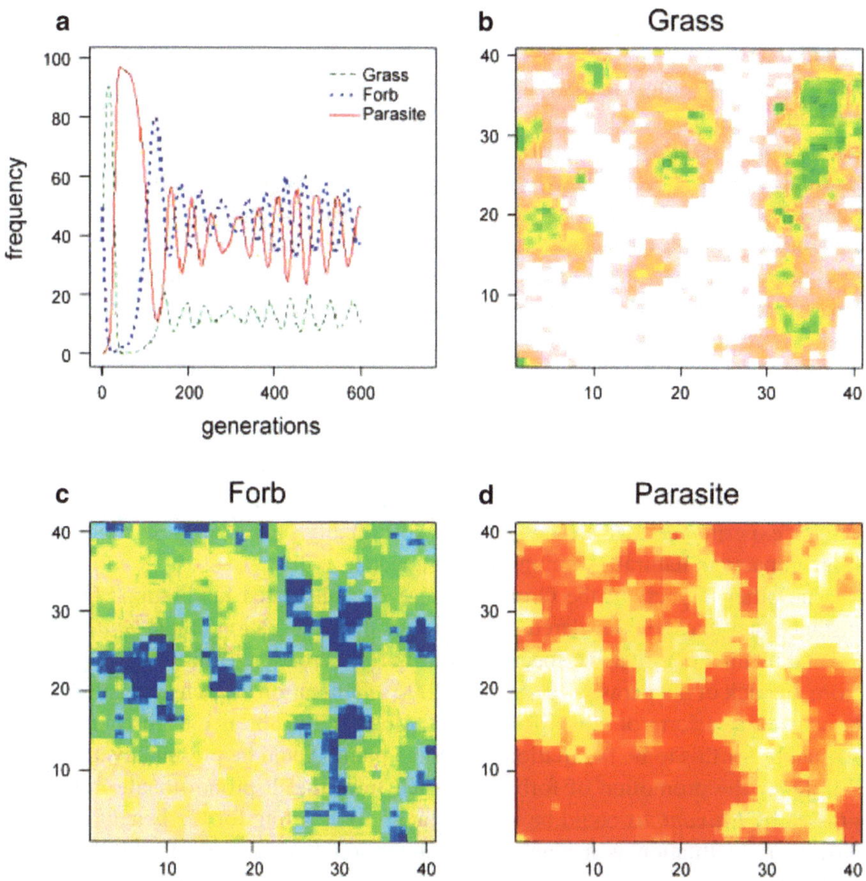

Fig. 16.2 The predictive impact of parasitic plants on the grass–forb interaction. Based on a simple difference equation model. (**a**) Temporal population dynamics of the parasite–grass–forb community. (**b–d**) Spatial structure of the parasite–grass–forb community at steady (i.e. ergodic) state after 500 generations; In this model, the parasitic plant was introduced in the low left corner (grid locations 1, 1 to 1, 20), and equal numbers of grasses and forbs were located in the rest of the grid. (**b**) *Green*—high abundance of grass. (**c**) *Blue*—high abundance of forb. (**d**) *Red*—high abundance of parasite (from Cameron et al. 2009, with permission)

and *R. alectorolophus* is reduced through either direct shading by the host plant or in communities with greater leaf area index (Matthies 1995; Joshi et al. 2000). A greater leaf area index indicates a greater carbon gain capacity of the community, and there must be a cut-off point beyond which the benefits of a potentially large carbon supply from hosts is outweighed by the reduction in the own autotrophic carbon gain of the parasite due to shading.

Facilitation has recently been proposed as a further impact of parasitic plants, in particular by invoking the concept of parasites as 'indirect facilitators' (Watson 2009). Watson (2009) illustrates the potential for Orobanchaceae to act as facilitators using the well-studied example of *Bartsia alpina* growing in sub-arctic heathland (Quested et al. 2002, 2003a, b, 2005). In that particular case, the facilitation-style impact from the decomposition of nutrient-rich litter from the decomposed parasite that promotes growth of neighbouring species and increases community biomass and species richness. Therefore, the parasite indirectly facilitates other species. These claims are in reality speculations because no evidence for these effects has been found in natural communities containing B. alpina—despite considerable effort. Whilst the concept of facilitation itself is evolving rapidly (Brooker et al. 2008), at its core remains the understanding that facilitation is a positive 'beneficial' plant-to-plant interaction. The case for facilitation by parasitic plants is perhaps less clear than proposed by Watson (2009) and certainly less so for the Orobanchaceae (Brooker and Callaway 2009). Certainly the potential is there, and pot studies have shown that nutrient-rich *B. alpina* litter can promote the growth of phytometer plants[2] (see Sect. 16.5). However, the acid test for facilitation is that it can operate in natural plant communities; thus, facilitation of Orobanchaceae from nutrient-rich litter inputs currently remains speculative. Proper experiments that are designed to specifically elucidate these nutrient-rich litter impacts in natural communities are needed.

Can other impacts of the Orobanchaceae be considered facilitation? One such instance that may be considered facilitation is the observation that *R. minor* assists in establishment of invasive species resulting from creation of safe sites in gaps left from dieback of the parasite (Joshi et al. 2000). Generation of such 'safe sites' could be viewed as a form of nurse plant effect, albeit in the absence of the nurse plant. Arguably for this to be facilitation, the nurse plant should be present, even if dead. Similarly, the promotion of forb abundance from the disproportionate impact on grasses by *Rhinanthus* could be seen as facilitation acting for the benefit of forbs. However, facilitation is a plant–plant interaction (Brooker et al. 2008) and the promotion of forb abundance partially results from their ability to *not* interact with the parasite as a result of their effective defence mechanisms. Certainly forb biomass is promoted by the reduction in competition from grasses, but then this is a forb interaction with grasses, not with the parasite. If this is facilitation, it is facilitation by proxy. Overall, 'facilitation by proxy' and 'nurse plant effects in the absence of the nurse plant' do not make a strong case for facilitation by parasitic Orobanchaceae.

[2] Model plants used in measuring physiological responses to environmental factors.

16.4 Interactions Across Multiple Trophic Levels

16.4.1 Host–Parasite–Herbivore Interactions

The parasitic Orobanchaceae can also impact other trophic levels given their considerable impact on plant community structure. Throughout the range of parasitic plants, this broad range of impacts is promoted by the fact that the parasite has both bottom-up effects (as a keystone resource) and top-down effects (as a competitor of the host) (Press and Phoenix 2005).

An example of a top-down effect is the competition for the shared resource, where snails fed less on the legume *Trifolium repens* parasitised by *Rhinanthus serotinus* than unparasitised material (Puustinen and Mutikainen (2001). The effect can work both ways, however, and partial defoliation (herbivory simulation) of the grass *Agrostis capillaris* has been shown to reduce flowering of *R. serotinus* (Salonen and Puustinen 1996).

Castilleja wightii, in contrast, provides an example of a bottom-up trophic interaction. In this case, the improved N status of the parasite when feeding on N-rich hosts improved the survival of the aphid *Nearctaphis kachena* (Marvier 1996). The fascinating tripartite interaction in this case was that *Castilleja* performed poorly on N-rich hosts when in the presence of aphids since N-rich hosts promoted a greater aphid burden on *Castilleja.*

16.4.2 Secondary Metabolites

The physiology of the host plant also has the capacity to influence higher-level trophic interactions between parasitic plants and their pollinators. Alder (2000) used an elegant approach to investigate the effects of host-borne alkaloids from *Lupinus albus* on the root hemiparasite *Castilleja indivisa. Lupinus albus* exists in two forms, a 'bitter' form that contains antifeedant compounds, alkaloids, and a 'sweet' form that only contains trace amounts of the same alkaloids. When parasitising *L. albus, C. indivisa* takes the presumably xylem-mobile alkaloids from the host, which results in a dramatic reduction in herbivore damage to the parasite (Alder 2000). Since the pollinators of *C. indivisa* actively select undamaged plants, there was a strong increase in the pollination success and hence fecundity of parasites attached to 'bitter' compared with 'sweet' individuals.

16.4.3 Parasite–Parasite Interactions

Parasitic Orobanchaceae can also interact with non-plant parasites. In the case of the host plant *Trifolium pratense*, for instance, dual parasitism by the cyst nematode *Heterodera trifolii* and *Rhinanthus serotinus* reduced *Trifolium* biomass more than the reduction caused by either parasite alone. However, since the nematode was the more aggressive parasite, which reduced *Trifolium* biomass more than *Rhinanthus*, the number and size of its cysts were not reduced by *Rhinanthus* infestation, whilst *Rhinanthus* did not gain performance when parasitising *Trifolium* if the host was already parasitised by the nematode (Puustinen et al. 2001).

16.4.4 Interactions with Soil Microbes

Given the considerable impact and drain of parasitic plants on host carbon budgets, it is to be expected that these impacts can have secondary effects on soil microbes that utilise or improve hosts carbon gain. For instance, *Melampyrum pratense* was more productive when its *Pinus sylvestris* host was colonised by ectomycorrhizal fungi (Salonen et al. 2000). The productivity benefit to *Pinus* probably enhances carbon supply to the parasite. *Melampyrum*, therefore, indirectly benefits from mycorrhizal symbioses despite probably having limited capacity itself to form such association. Similarly, *Rhinanthus minor* has greater productivity when parasitising *Lolium perenne* with arbuscular mycorrhizal symbioses, compared to non-mycorrhizal *Lolium* (Davies and Graves 1998). The mycorrhizal benefit to productivity was greater for *Rhinanthus* than it was for its host, indicating that *Rhinanthus* was a stronger competitor for carbon than the mycorrhizal fungi. This also explains why *Rhinanthus* reduced mycorrhizal infestation levels in *Lolium* (see Sect. 26.3.1).

R. minor also alters soil microbial community structure in grasslands, with reductions in soil fungal to bacterial ratios seen with increasing density of the parasite (Bardgett et al. 2006). Associated increased nitrogen mineralization rates are thought to be a consequence of the *Rhinanthus* impact on the plant community structure, altering the quality and quantity of carbon inputs to the soil (and see below in Sect. 16.5). Changes in above-ground plant community structure alter soil microbial communities belowground (Wardle et al. 2004), but direct evidence for such a mechanism operating via parasite-mediated impacts on plant community structure is yet to be found.

16.4.5 Pollination Ecology

Given the importance of parasitic plants as keystone species regulating the structure and function of both natural and agroecosystems, it is surprising that the pollination ecology of the Orobanchaceae has not been resolved in detail. No clear trend has emerged from the diverse array of pollination strategies employed by Orobanchaceae that can be associated with the evolution of the parasitic habit. The hemiparasite *Rhinanthus minor* is a large seeded hermaphrodite with both autogamous and allogamous pollination systems; allogamous pollination is largely facilitated by bumble bees (*Bombus* spp.), although floral morphology may limit the extent of outcrossing and interactions with bumble bees may even enhance autogamous pollination (Westbury 2004). Similarly, the tiny-seeded obligate hemiparasitic *Striga* species have both auto- and allogamous pollination systems to greater or lesser extents (Aigbokhan et al. 1998), with the exception of *S. aspera* and S. hermonthica—the only known two obligate self-incompatible outbreeders in the genus (Aigbokhan et al. 1998; Safa et al. 1984).

16.5 Parasitic Plant Impacts on Nutrient Cycling

The dual role that parasitic plants may have in communities, by acting as 'Dracula' ('sucking' the 'life blood' of an organism) and/or 'Robin Hood' (robbing from the rich to feed to the poor), was first proposed by Press (1998). Here, the 'Dracula' role refers to the resource extraction and negative impact on host productivity that impacts competitive interactions and changes in community structure. The comparison with 'Robin Hood' refers to the capacity of parasitic plants to extract resources from hosts and redistribute them to the rest of the community. A central pillar of this mechanism is the often nutrient-rich and readily decomposable litter of the parasite. Since little translocation of carbon and mineral nutrients occurs prior to senescence, much is returned to the soil in parasite litter, where it may be available to co-occurring plant species. This can also be considered a nutrient cycle short-circuit mechanism as well as being a mechanism for redistributing nutrients, since carbon and nutrients that would otherwise be returned in more recalcitrant host litter (or indeed stored longer in perennial hosts before senescence) are returned rapidly to the soil by the parasite (Phoenix and Press 2005).

Good examples of the potential for this effect come from the extensive work of Quested et al. (2002, 2003a, b, 2005). Using *Bartsia alpina* in sub-arctic heathland as a model system, it was shown that *B. alpina* litter can be a considerable point source of nutrients (Quested 2002; Quested et al. 2003b) and also could stimulate decomposition when mixed with other more recalcitrant litters (Quested et al. 2002, 2005). The potential for community impacts was indicated by pot studies where phytometer seedlings of *Betula nana* and *Poa alpina* grew better when supplied with *Bartsia alpina* litter compared to litter of other co-occurring species (Quested et al. 2003a). These impacts have yet to be seen in natural communities. Two lines of evidence have recently emerged suggesting such impacts could indeed occur in *Rhinanthus* species in grasslands. In the work of Bardgett et al. (2006), greater *Rhinanthus* densities increased mineralization rates. In that case, it was proposed that this impact must result from changes in plant community structure since *Rhinanthus* did not have greater tissue N concentration compared to other co-occurring species. However, that view ignores the fact that the parasite leaf litter is highly friable and decomposes more rapidly than the litter of other co-occurring species (Phoenix, unpublished data). *Rhinanthus* litter consequently provides a greater pool of resources, whether it is more nutrient enriched or not. Similarly there were greater dilutions of a ^{15}N soil tracer in *Rhinanthus minor* and *R. angustifolius* in infested Belgian meadows (Ameloot et al. 2008). This suggests that the parasite had increased the soil N pool, possibly as a result of stimulated mineralization rates. In that case, the 'mineralization stimulation' role of parasite litter was suggested as the mechanism, though impacts from changes in the co-occurring plant community could not be ruled out: reductions in meadow biomass by the parasite may have resulted in less N being removed in hay, leaving more in the soil. The effects of parasitism and the effects of the parasite litter inputs were not separated in either study, and thus, proper mechanistic links between the parasite and nutrient cycling have not been determined. Therefore, despite a number of studies showing strong evidence for the *possibility* of an impact of parasitic plant litter in natural communities, true impacts of litter in natural communities are yet to be proven. There is a clear need for experiments designed to specifically test these links with the various parasite species.

It also remains unclear as to which species in the plant community would benefit from the nutrient-rich parasite litter, so whether 'Robin Hood' will 'rob the rich to feed the poor' or 'rob the rich to feed the rich' remains unknown.

16.6 Conclusions and Future Directions

There is abundant evidence that parasitic Orobanchaceae have profound impacts on their host communities, as summarised in Fig. 16.3. By investigating the physiology of pairwise interactions in pots/microcosms coupled to the use of mesocosms, we now have a much better understanding of host–parasite interactions at the plant scale that underpin the community-level effects. However, our understanding of the ecology of Orobanchaceae is largely limited to a number of model hemiparasitic species. Community-level effects are only well resolved for certain *Rhinanthus* species. It is only in this well-studied genus that plant-level ecological interactions are understood well enough to model and scale up host–parasite interactions successfully to the community level. Mechanisms for community-level impacts have been proposed in many cases, but more rigorous testing is still needed. An understanding of the parasite–host interactions equivalent to the detail we know for *Rhinanthus* is needed to be able to understand the impacts of more Orobanchaceae at the community level. Further, for all Orobanchaceae (including *Rhinanthus*), knowledge of the indirect effects of the parasite on the plant community through, for instance, nutrient-rich litter inputs or gap generation, is much needed.

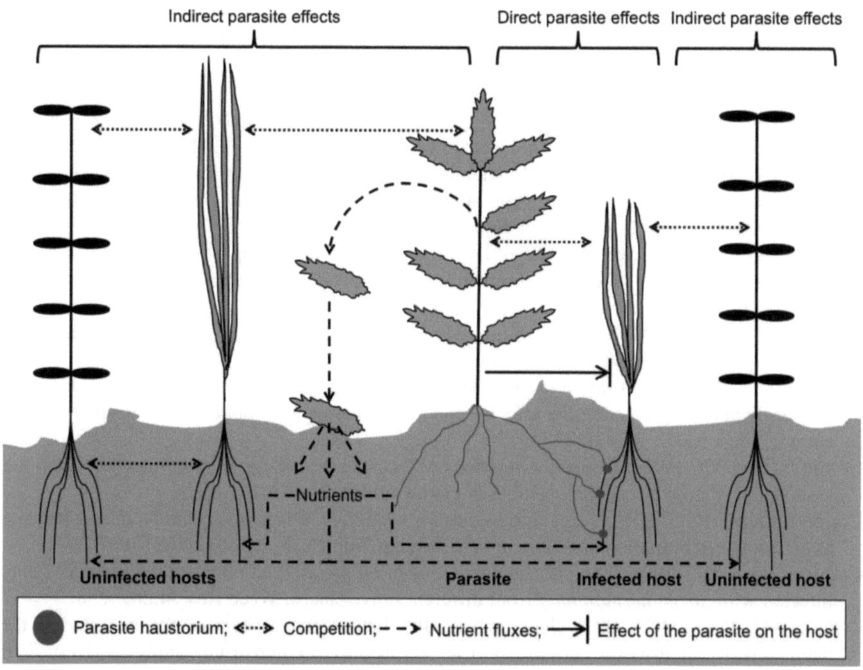

Fig. 16.3 Direct and indirect mechanisms, through which parasitic Orobanchaceae can influence their hosts and plant communities

Acknowledgements DDC thanks The Royal Society of London for the financial support provided by a University Research Fellowship (award number: UF090328).

References

Aigbokhan EI, Berner DK, Musselman LJ (1998) Reproductive ability of hybrids of *Striga aspera* and *Striga hermonthica*. Phytopathology 88:563–567

Alder LS (2000) Alkaloid uptake increases fitness in a hemiparasitic plant via reduced herbivory and increased pollination. Am Nat 156:92–99

Ameloot E, Verlinden G, Boeckx P, Verheyen K, Hermy M (2008) Impact of hemiparasitic *Rhinanthus angustifolius* and *R. minor* on nitrogen availability in grasslands. Plant Soil 311:255–268

Bardgett RD, Smith RS, Shiel RS, Peacock S, Simkin JM, Quirk H, Hobbs PJ (2006) Parasitic plants indirectly regulate below-ground properties in grassland ecosystems. Nature 439:969–972

Barker ER, Press MC, Scholes JD, Quick WP (1996) Interactions between the parasitic angiosperm *Orobanche aegyptiaca* and its tomato host: growth and biomass allocation. New Phytol 133:637–642

Brooker RW, Callaway RM (2009) Facilitation in the conceptual melting pot. J Ecol 97:1117–1120

Brooker RW, Maestre FT, Callaway RM, Lortie CL, Cavieres LA, Kunstler G, Liancourt P, Tielborger K, Travis JMJ, Anthelme F, Armas C, Coll L, Corcket E, Delzon S, Forey E, Kikvidze Z, Olofsson J, Pugnaire F, Quiroz CL (2008) Facilitation in plant communities: the past, the present and the future. J Ecol 96:18–34

Cameron DD, Seel WE (2007) Functional anatomy of haustoria formed by *Rhinanthus minor*: linking evidence from histology and isotope tracing. New Phytol 174:412–419

Cameron DD, Hwangbo J-K, Keith AM, Geniez J-M, Kraushaar D, Rowntree J, Seel WE (2005) Interactions between the hemiparasitic angiosperm *Rhinanthus minor* and its hosts: from the cell to the ecosystem. Folia Geobot 40:217–229

Cameron DD, Coats AM, Seel WE (2006) Differential resistance among host and non-host species underlies the variable success of the hemi-parasitic plant *Rhinanthus minor*. Ann Bot Lond 98:1289–1299

Cameron DD, Geniez J-M, Seel WE, Irving LJ (2008) Suppression of host photosynthesis by the parasitic plant *Rhinanthus minor*. Ann Bot Lond 101:573–578

Cameron DD, White A, Antonovics J (2009) Parasite-grass-forb interactions and rock-paper-scissor dynamics: predicting the effects of the parasitic plant *Rhinanthus minor* on host plant communities. J Ecol 97:1311–1319

Cechin I, Press MC (1993a) Nitrogen relations of the *Sorghum-Striga hermonthica* host-parasite association – growth and photosynthesis. Plant Cell Environ 16:237–247

Cechin I, Press MC (1993b) Nitrogen relations of the *Sorghum-Striga hermonthica* host-parasite association – germination, attachment and early growth. New Phytol 124:681–687

Cechin I, Press MC (1994) The influence of nitrogen on growth and photosynthesis of *Sorghum* infected with *Striga hermonthica* from different provenances. Weed Res 34:289–298

Cochrane V, Press MC (1997) Geographical distribution and aspects of the ecology of the hemiparasitic angiosperm *Striga asiatica* (L.) Kuntze: a herbarium study. J Trop Ecol 13:371–380

Dale H, Press MC (1998) Elevated atmospheric CO_2 influences the interaction between the parasitic angiosperm *Orobanche minor* and its host *Trifolium repens*. New Phytol 140:65–73

Davies DM, Graves JD (1998) Interactions between arbuscular mycorrhizal fungi and the hemiparasitic angiosperm *Rhinanthus minor* during co-infection of a host. New Phytol 139:555–563

Davies DM, Graves JD (2000) The impact of phosphorus on interactions of the hemiparasitic angiosperm *Rhinanthus minor* and its host *Lolium perenne*. Oecologia 124:100–106

Davies DM, Graves JD, Elias CO, Williams PJ (1997) The impact of *Rhinanthus* spp. on sward productivity and composition: implications for the restoration of species-rich grasslands. Biol Conserv 82:87–93

Egley GH (1971) Mineral nutrition and the host-parasite relationship of witchweed. Weed Sci 19:528–533

Farina MPW, Thomas PEL, Channon P (1985) Nitrogen, phosphorous and potassium effects on the incidence of *Striga asiatica* (L.) Kuntze in maize. Weed Res 25:443–447

Gibson CC, Watkinson AR (1989) The host range and selectivity of a parasitic plant – *Rhinanthus minor* L. Oecologia 78:401–406

Gibson CC, Watkinson AR (1991) Host selectivity and the mediation of competition by the root hemiparasite *Rhinanthus minor*. Oecologia 86:81–87

Gibson CC, Watkinson AR (1992) The role of the hemiparasitic annual *Rhinanthus minor* in determining grassland community structure. Oecologia 89:62–68

Goldwasser Y, Plakhine D, Kleifeld Y, Zamski E, Rubin B (2000) The differential susceptibility of vetch (*Vicia* spp.) to *Orobanche aegyptiaca*: anatomical studies. Ann Bot 85:257–262

Graves JD, Press MC, Stewart GR (1989) A carbon balance model of the *Sorghum-Striga hermonthica* host-parasite association. Plant Cell Environ 12:101–107

Gurney AL, Taylor A, Mbwaga A, Scholes JD, Press MC (2002) Do maize cultivars demonstrate tolerance to the parasitic weed *Striga asiatica*? Weed Res 42:299–306

Gurney AL, Grimanelli D, Kanampiu F, Hoisington D, Scholes JD, Press MC (2003) Novel sources of resistance to *Striga hermonthica* in *Tripsacum dactyloides*, a wild relative of maize. New Phytol 160:557–568
Hwangbo J (2000) Interactions between a facultative annual root hemiparasite, *Rhinanthus minor* L., and its hosts. PhD thesis, University of Aberdeen
Irving LJ, Cameron DD (2009) You are what you eat: interactions between root parasitic plants and their hosts. Adv Bot Res 50:87–138
Jiang F, Jeschke WD, Hartung W (2003) Water flows in the parasitic association *Rhinanthus minor-Hordeum vulgare*. J Exp Bot 54:1985–1993
Jiang F, Jeschke WD, Hartung W, Cameron DD (2008) Does legume nitrogen fixation underpin host quality for the parasitic plant *Rhinanthus minor*? J Exp Bot 50:917–925
Joshi J, Matthies D, Schmid B (2000) Root hemiparasites and plant diversity in experimental grassland communities. J Ecol 88:634–644
Keith AM, Cameron DD, Seel WE (2004) Spatial interactions between the hemiparasitic angiosperm *Rhinanthus minor* and its host are species-specific. Funct Ecol 18:435–442
Kelly D (1989) Demography of short-lived plants in chalk grassland. III. Population stability. J Ecol 77:785–798
Krebs CJ, Boutin S, Boonstra R, Sinclair ARE, Smith JNM, Dale MRT, Martin K, Turkington R (1995) Impact of food and predation on the snowshoe hare cycle. Science 269:1112–1115
Kuiper E, Groot A, Noordover ECM, Pieterse AH, Verkleij JAC (1998) Tropical grasses vary in their resistance to *Striga aspera*, *Striga hermonthica*, and their hybrids. Can J Bot 76:2131–2144
Last FT (1961) Direct and residual effects of *Striga* control treatments on *Sorghum* yields. Trop Agric 38:49–56
Marvier MA (1996) Parasitic plant-host interactions: plant performance and indirect effects on parasite-feeding herbivores. Ecology 77:1398–1409
Marvier MA (1998) Parasite impacts on host communities: plant parasitism in a California coastal prairie. Ecology 79:2616–2623
Matthies D (1995) Parasitic and competitive interactions between the hemi-parasites *Rhinanthus serotinus* and *Odontites rubra* and their host *Medicago sativa*. J Ecol 83:245–251
Matthies D (1996) Interactions between the root hemiparasite *Melampyrum arvense* and mixtures of host plants: heterotrophic benefit and parasite-mediated competition. Oikos 75:118–124
Matthies D, Egli P (1999) Response of a root hemiparasite to elevated CO_2 depends on host type and soil nutrients. Oecologia 120:156–161
Musselman LJ, Press MC (1995) Introduction to parasitic plants. In: Press MC, Graves JD (eds) Parasitic plants. Chapman and Hall, London, pp 1–11
Olivier A (1996) Relation between *Striga hermonthica* and its hosts – an overview. Can J Bot 74:1119–1137
Paine RT (1969) A note on trophic complexity and community stability. Am Nat 103:91–93
Parker C, Riches CR (1993) Parasitic weeds of the world: biology and control. CAB, Wallingford, UK
Patterson DT (1987) environmental factors affecting witchweed growth and development. In: Musselman LJ (ed) Parasitic weeds in agriculture. CRC, Boca Raton, pp 28–39
Pennings SC, Callaway RM (1996) Impact of a parasitic plant on the structure and dynamics of salt marsh vegetation. Ecology 77:1410–1419
Pennings SC, Callaway RM (2002) Parasitic plants: parallels and contrasts with herbivores. Oecologia 131:479–489
Phoenix GK, Press MC (2005) Linking physiological traits to impacts on community structure and function: the role of root hemiparasitic Orobanchaceae (ex-Scrophulariaceae). J Ecol 93:67–78
Press MC (1998) 'Dracula' or 'Robin Hood'? A functional role for root hemiparasites in nutrient poor ecosystems. Oikos 82:609–611
Press MC, Phoenix GK (2005) Tansley review: impacts of parasitic plants on natural communities. New Phytol 166:737–751

Press MC, Stewart GR (1987) Growth and photosynthesis in *Sorghum bicolor* infected with *Striga hermonthica*. Ann Bot 60:657–662

Press MC, Tuohy JM, Stewart GR (1987a) Gas-exchange characteristics of the *Sorghum-Striga* host- parasite association. Plant Physiol 84:814–819

Press MC, Shah N, Tuohy JM, Stewart GR (1987b) Carbon isotope ratios demonstrate carbon flux from C_4 host to C_3 parasite. Plant Physiol 85:1143–1145

Puustinen S, Mutikainen P (2001) Host-parasite-herbivore interactions: implications of host cyanogenesis. Ecology 82:2059–2071

Puustinen S, Jarvinen O, Tiilikkala K (2001) Asymmetric competition between a hemiparasitic plant and a cyst nematode on a shared host plant. Ecoscience 8:51–57

Pywell RF, Bullock JM, Walker KJ, Coulson SJ, Gregory SJ, Stevenson MJ (2004) Facilitating grassland diversification using the hemiparasitic plant *Rhinanthus minor*. J Appl Ecol 41:880–887

Quested HM (2002) Root hemiparasitic angiosperms in sub-arctic ecosystems : their potential role in ecosystem function. PhD Thesis University of Sheffield, UK

Quested HM, Press MC, Callaghan TV, Cornelissen JHC (2002) The hemiparasitic angiosperm *Bartsia alpina* has the potential to accelerate decomposition in sub-arctic communities. Oecologia 130:88–95

Quested HM, Cornelissen JHC, Press MC, Callaghan TV, Aerts R, Trosien F, Riemann P, Gwynn-Jones D, Kondratchuk A, Jonasson SE (2003a) Decomposition of sub-arctic plants with differing nitrogen economies: a functional role for hemiparasites. Ecology 84:3209–3221

Quested HM, Press MC, Callaghan TV (2003b) Litter of the hemiparasite *Bartsia alpina* enhances plant growth: evidence for a functional role in nutrient cycling. Oecologia 135:606–614

Quested HM, Callaghan TV, Cornelissen JHC, Press MC (2005) The impact of hemiparasitic plant litter on decomposition: direct, seasonal and litter mixing effects. J Ecol 93:87–98

Rümer S, Cameron DD, Wacker R, Hartung W, Jiang F (2007) An anatomical study of the haustoria of *Rhinanthus minor* attached to roots of different hosts. Flora 202:194–200

Safa SB, Jones BMG, Musselman LJ (1984) Mechanisms favouring out-breeding in *Striga hermonthica* (Scrophulariaceae). New Phytol 96:299–305

Salonen V, Puustinen S (1996) Success of a root hemiparasitic plant is influenced by soil quality and by defoliation of its host. Ecology 77:1290–1293

Salonen V, Setala H, Puustinen S (2000) The interplay between *Pinus sylvestris*, its root hemiparasite, *Melampyrum pratense*, and ectomycorrhizal fungi: influences on plant growth and reproduction. Ecoscience 7:195–200

Seel WE, Press MC (1993) Influence of the host on 3 sub-arctic annual facultative root hemiparasites 1. Growth, mineral accumulation and aboveground dry-matter partitioning. New Phytol 125:131–138

Seel WE, Press MC (1996) Effects of repeated parasitism by *Rhinanthus minor* on the growth and photosynthesis of a perennial grass, *Poa alpina*. New Phytol 134:495–502

Seel WE, Cooper RE, Press MC (1993) Growth, gas-exchange and water-use efficiency of the facultative hemi-parasite *Rhinanthus minor* associated with hosts differing in foliar nitrogen concentration. Physiol Plantarum 89:64–70

Taylor A, Martin J, Seel WE (1996) Physiology of the parasitic association between maize and witchweed (*Striga hermonthica*): is ABA involved? J Exp Bot 47:1057–1065

Thorogood CJ, Rumsey FJ, Hiscock SJ (2009) Host-specific races in the holoparasitic angiosperm *Orobanche minor*: implications for speciation in parasitic plants. Ann Bot 103:1005–1014

Wardle DA, Bardgett RD, Klironomos JN, Setälä H, van der Putten WH, Wall DH (2004) Ecological Linkages Between Aboveground and Belowground Biota. Science 304:1629–1633

Watson DM (2009) Parasitic plants as facilitators: more Dryad than 'Dracula'? J Ecol 97:1151–1159

Westbury DB (2004) Biological flora of the British Isles: *Rhinanthus minor* (L.). J Ecol 92:906–927

Westbury DB, Dunnett NP (2000) The effects of the presence of *Rhinanthus minor* on the composition and productivity of created swards on ex-arable land. Aspects Appl Biol 58:271–278

Westbury DB, Dunnett NP (2007) The impact of *Rhinanthus minor* in newly established meadows on a productive site. Appl Veg Sci 10:121–129

Westwood JH, Foy CL (1999) Influence of nitrogen on germination and early development of broomrape (*Orobanche* spp.). Weed Sci 47:2–7

Part II
The Weedy Orobanchaceae and Their Control

Chapter 17
Weedy Orobanchaceae: The Problem

Jonathan Gressel and Daniel M. Joel

Not many Orobanchaceae species are weedy and parasitize agricultural crops. Nonetheless, they have a tremendous impact on agriculture, as described in Chap. 18. The weedy species damage crops by sucking nutrients or, as in the case of *Striga* spp., also by poisoning the crops and turning them into zombies (see Sect. 7.3). In many geographical areas, they parasitize key agricultural crops, thus negatively affecting human nutrition and leading to heavy economic losses. Their common English names, 'broomrape' and 'witchweed', indicate the impact of and damage done by these parasites to crops. The following chapters specifically present the updated knowledge about these weedy Orobanchaceae and their management.

For far too long, most research efforts to manage these parasitic weeds were per se without truly understanding their biology and physiology. Millions of dollars were spent each year for decades in the USA in trying to eradicate a small infestation of *Striga asiatica*, but less than 1 % went to research on the nature of infestation. Not listening to the ancient incantation 'know your enemy' is like fighting a war without reconnaissance. The parasites seemed far more intelligent and sophisticated than the fighters, as described in detail in the first part of the book (Chaps. 2–16), which presents the various mechanisms of parasitism, the interactions of the parasitic Orobanchaceae with host and non-host plants, and the environmental, genetic and evolutionary mechanisms that may be involved in the parasitic habit.

Before discussing the various means to control the weedy parasites and manage them in agricultural areas, we present three additional basic aspects that need to be considered in any effort to control them. Chapter 18 describes the weedy species and presents a morphological key to their taxonomy, maps of their world

J. Gressel (✉)
Plant Sciences, Weizmann Institute of Science, Rehovot, Israel
e-mail: jonathan.gressel@weizmann.ac.il

D.M. Joel
Department of Weed Research, Agricultural Research Organization, Newe-Ya'ar Research Center, Ramat Yishay 30095, Israel
e-mail: dmjoel@volcani.agri.gov.il

D.M. Joel et al. (eds.), *Parasitic Orobanchaceae*,
DOI 10.1007/978-3-642-38146-1_17,

distribution and a description of their hosts and the damage they cause; Chap. 19 deals with the diversity of their populations and the factors affecting their dispersal and determining the spatial and temporal changes in their diversity, together with discussion of the various molecular methods for diversity analysis, and Chap. 20 suggests means to diagnose the parasite seedbanks in soil.

In the old days, if one wished to know which species exist in the seedbank and how many seeds abide there, one had to take soil samples and use physical methods to separate the tiny seeds from the soil and to separately identify each tiny seed under the microscope. This was complicated and therefore rarely done before planting crops, which led to surprises, e.g. cases in Spain and Israel when beans and tomatoes were attacked by dormant broomrape seeds, forgetting that there had been a heavy infestation in the same fields many years before. Testing for dormant seeds in the soil was not performed, which would have predicted the damage to the 'new' bean and tomato crops. Now there are excellent and easier to perform sampling techniques and molecular diagnostic tools for estimating who, what and how many seeds are there before planting a mistake, as outlined in Chap. 20.

Breeding crops resistant to parasitic weeds is believed to be the safest method for parasitic weed control and should be one of the key elements in the management of parasitic weeds. However, this method had first been 'trial and error' so that newly developed resistant crop lines broke down soon after their release, overcome by new strains of the parasites. The resistant genes were first only given names, as were the strains that overcame them. That was all that was really known about them until recently. This situation changed when breeders began taking a good look at the resistant lines, classifying the different modes of resistance and tolerance, and localizing the chromosomal positions of different modes of resistance. This allowed the development of physiology-based marker-assisted breeding of some crop lines with more than one mode of resistance, which are more resistant to the parasites and are far more resilient to having the parasites evolve new races that are able to parasitize them, as widely discussed in Chap. 21.

The agronomic management of parasitic weeds, which is a very basic element in any integrated pest management, also became more sophisticated as researchers better understood the 'enemy'. Far better recommendations to minimize the parasite distribution and its damage could thus be made. None of the agronomic procedures used alone provides complete or even adequate control—they must be integrated with each other, as outlined in Chap. 22, and also with the tools and technologies outlined in the other chapters to further minimize the damage.

Conventional chemical control of parasitic weeds had been problematic; to minimize damage to the crop, the parasites must be killed in the soil long before they emerge. But soil-applied herbicides seemed mostly ineffectual, requiring that the herbicide is sprayed on the crop, which posed another conundrum. Systemic herbicides, which are selective and do not kill the crop, should go from host leaves through its conductive system to the roots and then into the attached parasites. Such herbicides are mostly degraded by the crop before they reach the host–parasite junction. Thus, there can be control only for a short duration after application, necessitating repeated applications, with all the environmental and economic consequences. Low doses of some herbicides that are sublethal to the crop can also be used for parasite control, but accurate low doses are not only hard to achieve; they can also cause crop phytotoxicity and/or require many treatments throughout the season. All these issues and the ways to overcome them, including the use of crops with target-site herbicide resistances, are dealt with in Chap. 23, with a discussion on how phenological phenomena can be used in the crop/parasite systems to predict the best time to apply herbicides.

Biotechnology should bridge some gaps in the ability of conventional methods to control the parasitic weeds. Biotechnologically rendering crops resistant to herbicides, whether transgenically or by mutation or gene conversion, can achieve season-long control without crop phytotoxicity. Transgenic approaches also allow transfer of various parasite-resistant genes to sensitive crops from plant species that are naturally resistant, without need of herbicides. It may be eventually possible to design systemically moving crop-encoded RNAi that will move from the host crops to parasites, targeting and suppressing vital, parasite-specific genes. These and other options of using biotechnology for parasitic weed management are discussed in Chap. 24.

Allelopathy has long been proposed as a method for parasitic weed control, but so far only one known forage crop was found to secrete a promising parasitic plant inhibiting allelochemical, which is currently being field tested against *Striga*. More importantly, the biosynthesis pathway of the allelochemical is being elucidated such that eventually the genes for its biosynthesis can be isolated and be transformed into major crops, eliminating the 'middle man'. This issue is discussed in Chap. 25.

Biological control of the parasites by seed-eating insects is so far insufficiently effective to be considered as a biocontrol method against any parasitic weed. Nonetheless, microorganisms, especially fungi, can supply a sufficient modicum of control, as described in Chap. 26. Efforts to genetically or transgenically enhance the virulence of these biocontrol agents are believed to lower the costs and increase the benefits of biocontrol (but may be hampered by governmental regulations).

The more we elucidate and understand the basic biological aspects of the parasitic weeds, which are introduced in the first chapters of the book, the more we can come up with novel ideas and integrated strategies for their control. This basic knowledge should open new windows on how to deal with the parasites, including the use of agronomic, physiological, chemical, ecological, genetic and biotechnological tools. In the future, the preferred technologies should be integrated with those that can be put in or on the crop seed, whether bred genes or transgenic genes, or by herbicides or biocontrol agents that can be incorporated on the seed of the susceptible crop. The combination of two or more protection technologies in a single seed should be preferred when possible, in order to prevent escape of even single parasites that evolved resistance to any one methodology. This should allow seed companies to provide the elite seed, for most cost-effective crop yield, that will not only prevent parasitism but also repel or kill the parasites and reduce their productivity and dispersal to a minimum. With such in- and/or on-seed technologies, the farmers may not need other specialized technologies to deal with the parasites nor have the costs of additional inputs.

No useful agricultural weed control practice remains forever and can stand by itself; parasitic weed populations are usually highly polymorphic, and continue to evolve, particularly under the selective pressures in agricultural areas. Thus, it is imperative that growers not become overly enamoured with a single control strategy to the exclusion of others. Sustained parasite control will only prevail through integrating the practices outlined in the following chapters as well as by novel practices that should further be developed in the future.

Chapter 18
The Parasitic Weeds of the Orobanchaceae

Chris Parker

18.1 Introduction

This chapter outlines the most important members of the Orobanchaceae occurring as weeds of agriculture and horticulture worldwide, the holoparasitic broomrapes (*Orobanche* and *Phelipanche* species), the hemiparasitic *Striga* species and finally a few less important hemiparasites, including *Alectra*, *Aeginetia* and *Rhamphicarpa* species.

The distinction between weeds and non-weeds in Orobanchaceae is largely a matter of the hosts on which they thrive and the habitat in which they develop. If the host is a crop species of economic importance, then the species damaging it is certainly to be regarded as a weed. It is far from clear what distinguishes the weedy parasites from non-weedy species other than the above. However, only parasites that adapt to agricultural practices would develop a significant threat to crops (see Chap. 13). The weedy ones thrive and rapidly propagate because they are given the wonderful opportunity of large areas of their favourite host on which to grow. Schneeweiss (2007) theorises that the weedy taxa of broomrape share wide host ranges and annual life histories, whereas the majority of non-weedy broomrape species have narrow host ranges and perennial life histories. While the annual life history is indeed common to almost all known weedy parasitic Orobanchaceae, the narrow host range is not common to all. In fact, some important and widespread aggressive parasitic weeds, like *Striga hermonthica* and *Phelipanche ramosa*, have wide host ranges, while only a few species, like *O. cumana*, are limited to one principal crop.

C. Parker (✉)
Royal York Crescent, 5, Bristol BS8 4JZ, UK
e-mail: chrisparker5@compuserve.com

D.M. Joel et al. (eds.), *Parasitic Orobanchaceae*,
DOI 10.1007/978-3-642-38146-1_18, © Springer-Verlag Berlin Heidelberg 2013

18.2 The Weedy Broomrapes: *Orobanche* and *Phelipanche* Species[1]

The weedy broomrapes are separated taxonomically into the two genera: *Phelipanche* and *Orobanche*. The difference between them is based on major morphological and karyological criteria and on molecular phylogenetic tools (Schneeweiss 2007; Joel 2009). Seven broomrape species are considered below in some detail. These may be separated from each other by the following key, but there can be other species with which there could be possible confusion, and the host range may occasionally differ.

18.2.1 Identification Key for Weedy Broomrapes (Adapted from Parker and Riches 1993; Joel and Eizenberg 2002)

(A) **Stems un-branched above ground, bracteoles absent**.......................... ***Orobanche***

(B) **Stems usually branched above ground, bracteoles present**................. ***Phelipanche***

(A) **Weedy *Orobanche* species**

1. **Flowers, white with purple veins:**
   - Plants: robust, commonly up to 1 m high; **stems: strongly fragrant. Flowers: 20–30 mm long** opening out to **widely divergent pubescent lobes 5 mm long**, making total width of the flower 1.5–2 cm; **distinctly fragrant when fresh.** Stigma: white, yellow or pinkish. Stamens: inserted 2–4 mm from the base. Parasitic on legume crops, especially faba bean, pea and lentil but also carrot, parsley, potato (Fig. 18.1a–c)........ ***O. crenata***
   - Plants: up to 70 cm (occasionally 1 m); **stems: non-fragrant. Flowers: 15–20 mm long** opening out into **lobes 2–3 mm long, normally white, non-fragrant**. Stigma: usually purple. **Stamens: inserted low down in the corolla tube,**

[1] See Sect. 14.3 for phylogenetic relations.

not more than 3 mm from the base. Parasitic on a wide range of hosts, especially Fabaceae (Fig. 18.1h,i) .... ***O. minor***

2. **Flowers, deep purple:**

   - Plants: robust, up to 70 cm high. Flowers: 15–20 mm long, **deep purple**, opening out to short lobes. Stigma: yellow. Stamens: inserted 1–2 mm from the base. Parasitic on faba bean, chickpea, vetch (Fig. 18.1f).. ***O. foetida***

3. **Flowers, with lobes white to uniformly blue, without contrasting veins:**

   - Plants: **40–100 cm high.** Inflorescence: 15–30 cm long, about 3 cm diameter, **relatively lax**. **Corolla: 18–22 mm long,** moderately or sharply recurved, violet, opening to five pale blue lobes 2–3 mm long. Stigma: white. Stamens: inserted 6–8 mm from the base. **Apex of filaments: glandular-hairy**. **Anthers: usually hairy.** Parasitic on Asteraceae, especially sunflower (Fig. 18.1d,g) .... ***O. cumana***
   - Plants: **15–30 cm high**. Inflorescence: **relatively dense**, 4–5 cm diameter. **Corolla: 15–18 mm** long, tube uniformly pale, slightly downcurved, with a narrow mouth and small, 2–3 mm lobes with contrasting deep blue or purple coloration. Stigma: whitish. Stamens: inserted at least 5 mm from the base. **Apex of filaments: glabrous**. **Anthers: usually glabrous**. Parasitic on Solanaceae, especially tomato, tobacco, eggplant (Fig. 18.1e,l) .... ***O. cernua***

(B) Weedy *Phelipanche* species

- Plant: branched (except very small individuals), **usually 15–25** cm high. Flowers: **mainly uniform** pale blue/purple **outside but with** two white patches on lower lobes, tube narrowest at base and widening towards the open mouth, length usually **less than 20 mm**. **Connective between the anthers lobes glabrous or sparsely hairy.** Parasitic on a wide **range of crops but especially tomato** (Fig. 18.1j) .... ***P. ramosa***
- Plant: branched, usually 20–35 cm high, sometimes larger. Flowers: as above but longer, **always more than 20(−30) mm. Connective between anthers lobes hairy.** Parasitic on a wide range of crops, especially Solanaceae but also Brassicaceae, Cucurbitaceae and Fabaceae (Fig. 18.1k) .... ***P. aegyptiaca***

Fig. 18.1 The weedy broomrape species. (**a**) *Orobanche crenata* close-up of inflorescence; (**b**) *O. crenata* in faba bean; (**c**) *O. crenata* in carrot; (**d**) *O. cumana*; (**e**) *O. cernua*; (**f**) *O. foetida* in vetch; (**g**) *O. cumana* in sunflower; (**h**) *O. minor*; (**i**) *O. minor* in red clover; (**j**) *Phelipanche ramosa* in tomato; (**k**) *P. aegyptiaca* in carrot; (**l**) *O. cernua* in eggplant (photos: **a**, **e** from Parker and Riches (1993) and **j** from CABI (2012a) by kind permission of CAB International; **b**, **h** by Chris Parker; **i** by Hanan Eizenberg; **f** by Diego Rubiales; **c**, **d**, **g**, **j**, **k**, **l** by DM Joel)

18.2.2 Orobanche crenata *Forsk. (Fig. 18.1a–c)*

Stem: fleshy, robust, up to 12 mm in diameter and 1 m high. Scale leaves: 10–30 mm long, lanceolate, acute or acuminate. Stem: glandular and strongly clover-scented. Inflorescence: occupying 50–75 % of emerged stem, moderately dense with stem visible only in the lower half. Bracts: narrow lanceolate 4–5 × 15–20 mm, moderately to densely hairy. Calyx variable: the lateral pairs of lobes narrow, almost subulate, deeply divided to more than half way or simple, undivided, but the lateral lobes separated by sinuses to the base dorsally and ventrally. Corolla: 15–30 mm, usually 20–25 mm long, glabrous to slightly glandular-hairy outside, generally glabrous within (Fig. 18.1a). Tube: more or less cylindrical, slightly curved, opening out into pronounced crenate-edged lobes up to 5 mm long, width of mouth about 15 mm overall. Colour: generally whitish with purple veins. Fragrant. Stamens: inserted 2–4 mm from the base of the corolla tube, glabrous or hairy towards the base. Anthers: glabrous. Capsule: 10–12 mm long, splitting into two, releasing several hundred seeds about 0.3 mm long, coarsely pitted. Normally outcrossing but is facultatively autogamous (Musselman et al. 1982).

Host Range. *O. crenata* has a moderately wide host range including species mainly in Fabaceae and Apiaceae but also some in Cucurbitaceae, Solanaceae, Lamiaceae, Ranunculaceae and Asteraceae. There is no marked host specificity (Musselman and Parker 1982).

Distribution. The native distribution of *O. crenata* is predominantly around the Mediterranean including North Africa and into the Near East and Western Asia, with quite recent introductions into Sudan and Ethiopia (Fig. 18.2).

Economic Importance. *O. crenata* is of immense importance all around the Mediterranean where it infests many of the most important legume crops, particularly faba bean (Fig. 18.1b), lentil, chickpea, vetch and field peas, also carrot (Fig. 18.1c) from which it absorbs much of the root sugars (Schaffer et al. 1991). Sauerborn (1991) estimated that over 50 % of faba bean crops were infested in Spain, Portugal, Syria and Morocco, some 180,000 ha in all. Other countries in which *O. crenata* is a significant problem on legumes include Algeria, Cyprus, Iraq, Israel, Italy, Jordan, Lebanon, Malta and Tunisia, while the introductions into Sudan (Babiker et al. 2007) and Ethiopia (Reda 2006) are causing serious losses there too.

Orobanche spp. do not have the same subtle physiological effects on their hosts as *Striga* spp. (see below) but damage the host via diversion of a substantial proportion of host resources. However, where there is moisture stress, there can be greater damage to the point of total crop failure. Sauerborn (1991) estimated an average 12 % yield loss in faba bean across infested areas, but much higher losses, over 30 %, were recorded in Egypt and over 50 % in Malta and Turkey. Further surveys in 1993, 1994 and 1999 (see Parker 2009) indicated continuing or increased losses in faba bean and also increasing reports of losses in lentil and in pea,

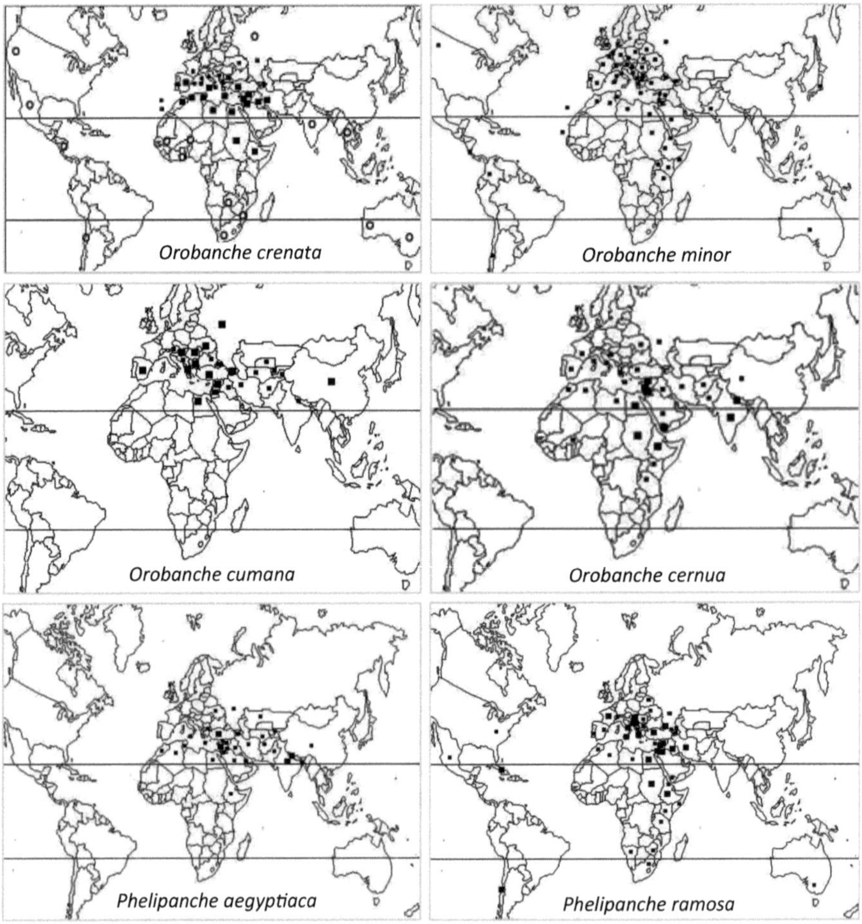

Fig. 18.2 World distributions of broomrape species. *Larger symbols* indicate countries in which it causes significant crop losses. *Smaller symbols* indicate countries in which it occurs mainly on wild hosts and/or causes lesser crop damage (reproduced from Parker (2012) by kind permission of Allen Press). *Open symbols* for *O. crenata* indicate other countries with potentially suitable climates for its growth (Grenz and Sauerborn 2007)

especially in Egypt, Morocco, Tunisia and Syria, such that farmers in several of these countries had to give up growing faba bean and the other susceptible legume crops, while Egypt had to import faba bean for the first time. In Israel, there are losses in pea, vetch, carrot and related parsley. There is a shortage of more recent surveys, but Bülbül et al. (2009) report over half of the faba bean and lentil fields infested in parts of Turkey. There is no clear evidence for any significant diminution of the problem, other than by avoidance, meaning farmers are choosing to give up growing some of their traditionally most important crops (Parker 2009).

In addition to the concern over existing infestations, there is the potential for spread to new areas, as has already occurred in Sudan and Ethiopia. Grenz and Sauerborn (2007) concluded that on the basis of climatic data, there are many parts of the world outside the existing range of *O. crenata* where it could thrive, including parts of North, Central and South America; West Africa; South and SE Asia; and Australia (Fig. 18.2).

18.2.3 Orobanche cumana *Wallr. (Fig. 18.1d, g)*

Although this species was described originally in 1825, it has until quite recently been most often treated as a subspecies of *Orobanche cernua,* i.e. *O. cernua* ssp. *cumana* (Wallr.) Soó. However, the two taxa differ quite clearly in morphology (Joel 1988; Joel et al. 2007) and in their host range, *O. cumana* being restricted to Asteraceae crops and *O. cernua* occurring as a weed almost exclusively on Solanaceae. Now, molecular evidence has been used to confirm that the two species, *O. cernua* and *O. cumana*, are quite distinct (Katzir et al. 1996). It appears that a wild form of *O. cernua* parasitizing *Artemisia* in Eastern Europe evolved in relatively recent time and attacks cultivated sunflower (*Helianthus annuus* L.) having diverged markedly in morphology and physiology from the original *O. cernua*. In Spain, Pujadas and Velasco (2000) found morphological, phenological and chemical differences between the forms attacking *Artemisia* spp. (referred to as *O. cernua* L.) and sunflower (*O. cumana*). *O. cumana* differs not only from *O. cernua* but also from most other weedy species of *Orobanche* in limited germination response to strigolactones and responsiveness to guaianolide sesquiterpene lactones such as dehydrocostus lactone (Joel et al. 2011, see Sect. 10.3.1).

The morphology of *O. cumana* is comparable to that of other members of the genus *Orobanche*, comprising unbranched stems, usually 40–65 cm high, about half of their distal length being taken up by a simple rather lax spike of flowers, each subtended by a single, undivided bract 10–12 mm long (Fig. 18.1d). The 7–9 mm long calyx is tubular at the base with acute, usually entire, teeth. The tubular corolla 19–22 mm long is almost parallel sided, 4 mm across and markedly downcurved throughout most of its length, topped by five spreading lobes, to about 10 mm across. The colour of the upper corolla varies from white to pale blue. Filaments inserted at least 4 mm from base of corolla tube. Anthers: hairy. The flowers are autogamous (Gagne et al. 1998), and each capsule, 8–10 mm long, houses several hundred elongated seeds, each about 0.3 mm long with a densely pitted surface.

Host Range. *O. cumana* is a specific parasite of sunflower.

Distribution. The distribution of *O. cumana* is centred on SE Europe, the Middle East and SW Asia and has been introduced to Spain, presumably with contaminated sunflower seed. It is also present in China (Fig. 18.2).

Economic Importance. Areas of sunflower affected (and yield losses) have been estimated at 40,000 ha (60 % losses) in Greece and 20,000 ha (20–50 % losses) in China (Parker 1994). Other countries in which it is or has been an acute problem include Spain, Hungary, Romania, Bulgaria, Russia, Ukraine, Moldova, Turkey, Syria, Israel and Egypt (Fig. 18.1g). Yakutkin and Budrevskaya (2006) describe its importance in Russia and neighbouring territories as follows: 'In the zone of low severity of the parasite, crop losses do not exceed 10 % (northern part of Central Black Soil Zone, Middle Volga region, Ural, Western and Eastern Siberia, the Far East). In the zone of moderate severity, losses are up to 30 % (Forest-steppe and Steppe Ukraine; central and southern parts of Central Black Soil Zone, central part of Volga region, Kazakhstan). In the zone of high severity, crop losses exceed 31 % (Moldova, southern part of Steppe Ukraine, Crimea; Northern Caucasus and southern part of Volga region; Transcaucasia)'.

The problem has tended to vary with time, according to the extent of use of resistant sunflower varieties on the one hand and the development of more virulent races of the parasite on the other (see Chaps. 19 and 21). There has been greater success in creating broomrape-resistant and herbicide-resistant oil-bearing sunflower cultivars, while confectionary cultivars continue to be more susceptible.

18.2.4 Orobanche cernua *Loefl. (Fig. 18.1e, l)*

Stems: fleshy, up to 35 cm high and 10 mm in diameter, usually unbranched but occasionally branched from below ground. Scales: broadly triangular/ovate, up to 10 mm long. Inflorescence: occupying about 75 % of emerged stem. Flowers: sessile, densely packed overlapping and completely hiding the stem (Fig. 18.1e). Bracts: reddish brown when dried, broadly triangular/ovate, up to 12 mm long, 3–6 mm wide. Calyx: deeply divided, to the base dorsally and ventrally, with lateral lobes narrow, acute, divided to at least half way. Stem, bracts and calyx: all hairy with some glandular hairs especially on the calyx. Corolla: almost glabrous, 15–18 mm long, slightly to markedly curved down, more or less cylindrical but constricted above the capsule, conspicuously so in fruit; lobes: small, not widely spread, 6–8 mm across. Corolla: whitish/pale yellow below but with deep blue/purple lips. The veins: not distinctly marked. The stamens: inserted at least 4–5 mm above the base of the tube; filaments and anthers: usually glabrous. Stigma: bilobed, whitish. Capsule: ovoid, 7–9 mm long, containing several hundred minute seeds, each about 0.3 mm across with coarse reticulate marking.

Host Range. The weedy *O. cernua* typically attacks Solanaceous crops, especially tomato, eggplant and tobacco (Fig. 18.1l), and, less commonly, potato. The non-weedy *O. cernua* occurs on wild Asteraceae plants.

Distribution. The natural distribution of *O. cernua* is across southern Europe, the Middle East, South Asia and Northern Africa, with possibly introduced infestations further south in Africa, in Niger, Sudan, Ethiopia, Kenya and Tanzania. It also occurs on sandy beaches of South Australia (Manen et al. 2004) (Fig. 18.2). Two other forms are recognised [e.g. by USDA/GRIN (2011)], *O. nepalensis* Reut. and *O. hansii* (A. Kern.) Beck, both with more eastern distribution.

Economic Importance. *O. cernua* has become a severe problem in large-scale plantations of tomato in South Europe and Africa and in tobacco in India.

The damaging effect of *O. cernua* on tobacco has been shown by Hibberd et al. (1998) to be proportional to the weight of the parasite, while interestingly the carbon fixation by tobacco was increased by 20 % in the infested plants (Hibberd et al. 1999). Tobacco is seriously affected in India and Pakistan and locally in Jordan, Ethiopia, Nepal and Saudi Arabia. In India about half the 40,000 ha devoted to tobacco in the state of Andhra Pradesh have been infested with losses of yield of 25–50 % and losses also in quality (Parker 1994).

In tomato, acute problems have occurred locally in Ethiopia, Israel, Jordan, Kenya and India. Lesser infestations reported in China, Iraq, Iran, Egypt, Lebanon, Saudi Arabia, Yemen, Afghanistan and Niger. Eggplant is a host in India, Israel and Ethiopia (Fig. 18.11) and potato in Jordan (Parker 1994). Interestingly, *O. cernua* has been replaced in Israel by the more competitive parasite *Phelipanche aegyptiaca* (see Sect. 18.2.8).

18.2.5 Orobanche foetida *Poir. (Fig. 18.1f)*

Stem: 20–70 cm high; scale leaves: 10–20 mm long, broadly lanceolate; inflorescence: occupying two thirds of stem, dense above, laxer below. Bracts: 15–20 mm long, narrowly lanceolate. Calyx: 8–15 mm; segments: shortly connate or all free, two-lobed. Corolla: 15–20 mm long, almost straight, glandular-pubescent or sub-glabrous, dark red, yellowish or white at the base, shining dark red inside, upper lip short, bilobed, lower lobes longer, the middle one 3 times as long as others. Filaments: inserted 1–4 mm above the base of the corolla; anthers: yellowish. Stigma: deep yellow at anthesis, dull purplish when dry (from Pujadas-Salvà et al. 2003).

Host Range. Restricted to Fabaceae, but is wide within that family, mostly wild species including *Anthyllis*, *Ononis*, *Scorpiurus*, *Trifolium*, *Medicago* and *Lotus* species but also the crops faba bean, chickpea and vetch. A degree of host preference is shown, in that populations from chickpea and faba bean in Tunisia demonstrated differential germination behaviour in response to the respective root exudates. These differences were supported by distinct genetic differentiation (Román et al. 2007). Furthermore, Vaz Patto et al. (2008) record that although *O. foetida* is common in Morocco and may occur on wild hosts in faba bean fields, the crop itself is not attacked. It is thought that the ability to attack faba bean may be the result of a recent evolutionary development. There is now evidence that vetch is being newly attacked in Morocco (Vaz Patto et al. 2008), apparently as a result of specialisation from forms attacking wild legumes. These authors caution that ‘the potential for this species to shift host ... has to be taken into account in future legume breeding'.

Distribution. The range of native *O. foetida* is limited to the Western Mediterranean; to Morocco, Tunisia, Algeria and Libya in North Africa; and to Spain, Portugal and the Balearic Islands to the north. The weedy populations occur in Tunisia and Morocco (Román et al. 2007).

Economic Importance. Although *O. foetida* occurs on a number of wild leguminous hosts, it was only first reported as a problem on faba bean in Tunisia in 1992 (Kharrat et al. 1992). It is still only in Tunisia that it poses an economic problem, mainly on faba bean and with some lesser damage on chickpea and vetch (Kharrat and Halila 1994; Abbes et al. 2007; Román et al. 2007). Damage to faba bean can be severe. Kharrat and Halila (1994) recorded increases in faba bean seed yield of over 50 % when *O. foetida* was controlled by glyphosate. Abbes et al. (2007) estimated seed yield losses of 55–93 % (depending on faba bean variety) when comparing neighbouring infested and un-infested fields.

Although as yet a relatively localised problem in Tunisia, there must be concern that this highly damaging species could spread and/or become newly virulent on crop species in other countries.

18.2.6 Orobanche minor *Sm. (Fig. 18.1h, i)*

O. minor resembles *O. crenata* in general form and colouring but is generally smaller in height and in size of flowers. Height is generally up to 50 cm though some populations in Ethiopia may exceed 1 m. Stems are up to 10 mm in diameter, and the scale leaves 10–20 mm. The inflorescence, dense above, more lax below, occupies 50–75 % of the stem, each flower subtended by a single lanceolate bract 3–5 mm wide by 10–25 m long, glandular-hairy (Fig. 18.1h). The calyx is glandular-hairy, usually with two pairs of acute, almost subulate lobes. The corolla is 10–18 mm long, rarely to 20 mm long, and the tube slightly downcurved, opening out to lobes 2–3 mm long, making total width about 10 mm at the mouth. Colour is mainly pale, whitish but with varying amounts of purple in the veins. The stems and flowers differ from those of *O. crenata* in not being fragrant. Stamens: inserted 2–4 mm from the base of the corolla tube, filaments: often hairy towards the base. Stigma: two-lobed, reddish brown, rarely yellow. Capsule: 7–10 mm long splitting longitudinally to release several hundred seeds, about 0.3 mm long with coarse reticulate marking.

Host Range. Very wide, including many species in Fabaceae (e.g. *Trifolium*, *Medicago*, *Lotus*, *Arachis*, *Vicia* spp.), in Asteraceae (*Lactuca*, *Guizotia*, *Carthamus*, *Tagetes* spp.) and in Apiaceae (*Daucus*, *Apium* spp.), Solanaceae and other families.

The hosts are usually herbaceous but can be woody, e.g. pecan (English et al. 1997). There is some evidence for host specialisation in different populations of *O. minor* (Musselman and Parker 1982; Thorogood et al. 2009), but there is rarely very strict host specificity. There have been local infestations in tobacco in New Zealand, and it occurs frequently but generally sporadically in a range of crops in Ethiopia, including faba bean, groundnut, sunflower, safflower, tobacco, niger seed and on ornamentals such as *Tagetes*, spp. and *Tropaeolum* (Parker 2006).

O. minor is strongly autogamous (Musselman et al. 1982).

Distribution. *O. minor* is very widely distributed, being native throughout most of Europe, other than the far north, Western Asia and Northern Africa, as far south as Ethiopia and Somalia, while it has also been sporadically introduced to Japan, New Zealand, Australia and several countries in North and South America (Fig. 18.2).

Economic Importance. The greatest economic damage is done in clover and lucerne crops grown for seed. This has been reported in a number of countries in Europe, and the problem has led to the abandonment of a clover seed industry in some of those countries including the UK. Eizenberg et al. (2003) found red clover (*Trifolium pratense*) very much more susceptible than white clover (*T. repens*), and currently it is threatening the red clover seed crop in NW USA, where several thousand hectares are infested (Colquhoun et al. 2006) (Fig. 18.1i). Lins et al. (2007) recorded 15–50 % reduction in total host weight, with proportionately more reduction of the host inflorescence. The problem continues to prejudice the viability of the crop as produce cannot be sold with even low contamination with parasite seed (Mallory-Smith and Colquhoun 2012). As for most other species of *Orobanche*, damage to *T. pratense* was shown to be proportional to the weight of parasite. *O. minor* population densities are rarely sufficient to cause serious harm, and this species has currently only marginal impact on agricultural crops.

18.2.7 Phelipanche ramosa *(L.) Pomel (syn.* Orobanche ramosa *L.; Fig. 18.1j)*

Stems: pale yellow/brown, 3–7 mm in diameter, non-fragrant, often branched from below, and just above, ground. Height: usually 10–30 cm, occasionally 50 cm; smaller stems: often unbranched. Scale leaves: lanceolate, acute, 5–8 mm long. Inflorescence: taking up three-quarters to 90 % or more of emerged stems and branches (Fig. 18.1j). Flowers: sessile, irregularly alternate in a sparse to moderately dense spike, overlapping 25–75 % of their length in mid-spike. Bracts: brown, ovate to lanceolate, acute, 5–10(−12) mm long, shortly hairy with simple and glandular hairs. Bracteoles: linear, 1 mm wide, 5–8 mm long, attached near base of calyx. Calyx: 5–10(−12) mm long, with 4 subequal triangular, acute lobes, the lateral pairs divided to about halfway, with a deeper sinus adjacent to the stem. Corolla: 10–20 mm long, hairy, moderately curved, cylindrical at the base, varyingly constricted above the capsule, then somewhat campanulate distally with wide-spreading lobes, up to 1 cm across, varying in colour from white at the base to pale blue or mauve to blue/purple on the lobes, but usually with distinct white patches inside the lower lobes. Stamens: inserted about 3 mm from the base of the tube. Anthers: generally glabrous about 1 mm long, acutely pointed at the base. Stigmas: obscurely lobed, whitish. Capsule: ovoid 5–8 mm long, splitting into two valves when mature, releasing several hundred minute seeds each about 0.3 mm across, coarsely reticulate.

Two taxa included here with *P. ramosa* are the subspecies *nana* (sometimes given specific status as *O. nana* Noe ex Beck) and *mutelii* (= *O. mutelii* F.J. Schultz). Musselman (1989) concluded that they represent the extremes in a continuum from the very small unbranched '*O. nana*' with corollas 10–15 mm to the more robust '*O. mutelii*' with corollas up to 22 mm long and did not deserve specific status.

Host Range. Extremely wide. *P. ramosa* attacks a range of crops in Solanaceae, especially tomato, eggplant and tobacco but also pepper and potato, and also Brassicaceae (rapeseed), Cannabaceae (hemp), Fabaceae (chickpea, clovers, groundnut, faba bean, lentil, pea), Apiaceae (carrot, celery, fennel, parsnip) and Asteraceae (lettuce, sunflower and a number of ornamental species). In addition, there are wild hosts in Chenopodiaceae, Amaranthaceae, Malvaceae, Rosaceae and many other families. It has been reported on onion but does not otherwise occur on monocots. There is evidence for some degree of host specialisation, but no marked host specificity. A population from tomato could additionally parasitize tobacco and lettuce, but one from tobacco failed to parasitize lettuce (Musselman and Parker 1982). Seed from parasite on a potato host was less virulent on crops in other plant families (Jacobsohn et al. 1991). Benharrat et al. (2005) compared samples from hemp, tobacco and rapeseed and concluded there were at least two 'pathotypes' in France, and later Brault et al. (2007) concluded that there are three different 'pathovars' of *P. ramosa*, from tobacco, from cannabis and from rapeseed. In south-west Germany, some growers began to cultivate parsley instead of tobacco, and as a result, parsley has now been infested with *P. ramosa* (Kohlschmid et al. 2011).

Distribution. The native distribution of *P. ramosa* is Europe, Middle East, West Asia and North Africa south to Ethiopia and Somalia (Fig. 18.2). New infestations are being recorded, as in Australia where *P. ramosa*, parasitizing weeds, has extended its distribution over 6,000 ha of paddocks, costing several million US$ per annum in quarantine and control operations over 200,000 ha, while there are further losses due to restrictions on sale of produce out of the infested area to prevent spread of the parasite to agricultural fields (Correll and Marvanek 2006; Warren 2006).

Economic Importance. The damaging effects of *P. ramosa* on its host may apparently exceed what is expected from the dry weight of the parasite. This may be due to dry conditions or to a disproportionate effect on the fruiting capacity of the host relative to that on the vegetative biomass. Mauromicale et al. (2008) note that a 60–70 % reduction in tomato shoot dry weight in a pot experiment was disproportionately higher than the weight of the parasite and was associated with reductions of about 50 % in both chlorophyll content and photosynthesis.

Yield losses in tomato and in tobacco are commonly reported to be 30–50 %. In Slovakia, Cagáň and Tóth (2003) measured 40–50 % yield reductions from 10 to 20 parasite shoots per plant. In Chile, Díaz et al. (2006) estimated 80 % losses in tomato. In Italy, Fracchiolla and Boari (2003) report losses of 6–20 tonne/ha in tomato and cauliflower. And in specialised tomato plantations without regular crop rotation, the problem can build up to the point that cropping has to be abandoned. In Sudan, this has resulted in the closure of the associated tomato juicing factory (Babiker et al. 1994). Other countries in which tomato and/or eggplant has been seriously affected include Italy, Greece, Lebanon, Jordan, Iraq, Iran, Syria, Turkey, Hungary, Cuba, Egypt and Ethiopia. Bülbül et al. (2009) report heavy infestation of tomato in the Eastern Mediterranean region of Turkey. Tobacco is seriously affected by *P. ramosa* in Moldova (Timus and Croitoru 2007) also in Cuba (Labrada and Perez 1988) and Italy (Zonno et al. 2000). Quality may be affected as well as yield, in both tomato and in tobacco. In France, there has been a serious increase in infestation of rapeseed; Gibot-Leclerc et al. (2003) reported that it was present in 20 of the 96 districts of the country and was affecting tobacco and hemp as well as rapeseed. Boulet et al. (2007) report over 50 % of rapeseed fields infested in the Pays de la Loire region in France. In pot experiments, Buschmann et al. (2005) recorded losses of 70–80 % biomass in this crop.

18.2.8 **Phelipanche aegyptiaca** *(Pers.) Pomel (syn.* **Orobanche aegyptiaca** *Pers.; Fig. 18.1k)*

Similar to *P. ramosa*. The two species are closely related and have been confused for some time. *P. aegyptiaca* is a more robust, taller plant than typical *P. ramosa*, but the most important differences are shown in the key above (Fig. 18.1j, k). These are the larger size of the flowers, 20–35 mm long (*P. ramosa* typically no more than 22 mm), and the hairiness of the connective between the anther lobes. The longer flowers also occur in '*P. mutelii*', but the hairiness between the anthers is then distinctive: *P. ramosa* and its subspecies may have some hairs on the filaments but not the dense hairs on the connective. The distinction between the two species has been confirmed by molecular techniques (see Sect. 14.3.1).

P. aegyptiaca is normally outcrossing but is facultatively autogamous (Musselman et al. 1982; Teryokhin 1997).

Host Range. *P. aegyptiaca* may attack most of the same crops affected by *P. ramosa*, in particular the Solanaceae tomato, potato, eggplant and tobacco, also crops in Fabaceae, Apiaceae and Asteraceae. It differs apparently in occurring on a wider range of Brassicaceae, especially various mustard species in India. It is also more important on Cucurbitaceae than is *P. ramosa* and can occasionally also occur on woody species, e.g. on olive (Eizenberg et al. 2002).

Distribution. The distribution of *P. aegyptiaca* overlaps with *P. ramosa* in South Europe, the Mediterranean and North Africa but extends much further eastwards into South Asia and China. There are no certain instances of its introduction outside these areas (Fig. 18.2).

Economic Importance. The effects of *P. aegyptiaca* on the host are not known to differ from those of *P. ramosa*. Damage can be severe, as on lentil in Turkey (Bayaa et al. 1998; Bülbül et al. 2009) and on both *Brassica juncea* (raya) and *Eruca sativa* (taramira) in India (Bedi et al. 1997). In the latter study, infestations of over 100 stems per m^2 caused 28–40 % yield reduction in *E. sativa*. Field experiments in Israel showed that infestation of 100 seeds per kg of soil caused over 30 % yield loss in carrot (Bernhard et al. 1998). Motazedi et al. (2010) estimated 71 % yield loss in potato in Iran. There are older reports of 50 % loss in water melon, 15–30 % loss in musk melon and 15 % loss in tomato in Russia (Parker and Riches 1993).

18.2.9 Other Broomrape Species

Other species mentioned by Parker and Riches (1993) as occasionally attacking crops include ***O. coerulescens* Steph. Ex Willd.**, ***O. gracilis* Sm.**, ***O. lutea* Baumg.** and ***O. solmsii* Hook.f.** There are no recent reports of these, but new instances of ***O. amethystea* Thuill.** were found on vetch, ***O. loricata* Reichenb.** on garden ornamentals and ***O. pubescens* D'Urv.** on parsley and *Tropaeolum* majus—all in Israel (Joel and Eisenberg 2002).

18.3 The Weedy Witchweeds: *Striga* Species[2]

18.3.1 Identification Key to the Main Weedy Striga *Species (Adapted from Parker and Riches 1993)*

1. **One calyx rib usually per calyx lobe:**
   - Plant: about 50 cm high (sometimes up to 1 m); spike: with many **pink** flowers, 1–2 cm across; corolla tube: 1.5–2 cm long, **bent at or just above halfway**; bracts: below each flower 2–3 mm wide, **fringed with hairs**. Parasitic on cereal crops in Africa and Arabia .... ***S. hermonthica*** (Fig. 18.3a)

[2] See Sect. 14.3 for phylogenetic relations.

- Plant: up to 50 cm high; spike of **pink** flowers about 1 cm across; corolla tube: about 1–1.5 cm long, **bent at least two thirds up** from the base; bracts: 1–2 mm wide only, mostly **without a fringe of hairs**. Parasitic on cereals in Africa only.................... ***S. aspera*** (Fig. 18.3c, d)
- Plant: 15–30 cm high; flowers: **white** in relatively dense spike, 2–6 open at a time, 0.5–1 cm across, **drying bluish**. Parasitic on cereal crops in Asia only..***S. densiflora***
- Plant: 15–30 cm high; **leaves: reduced to scales; whole plant: often pale green or purplish with little chlorophyll**. Flowers: usually **mauve/purple but sometimes white**; in spike: with only 2–6 open at a time, 0.5–1 cm across. Parasitic on cowpea, tobacco, sweet potato in Africa only... ***S. gesnerioides*** (Fig. 18.3e)

2. **Two calyx ribs usually per calyx lobe:**

- Plant: 15–30 cm high; flowers: **red, white or yellow**; in spike: with up to 6 open at a time, 0.5–1 cm across. Parasitic on cereals in Africa, Asia and the USA.. ***S. asiatica*** (Fig. 18.3b)

3. **Three calyx ribs usually per calyx lobe:**

- Plant: 15–30 cm high; leaves: narrow without conspicuous teeth; flowers: **white**, 2–6 open at a time, 0.5–1 cm across. Parasitic on cereals in Asia only... ***S. angustifolia***
- Plant: about 50 cm high; leaves: up to 1.5 cm wide, coarsely toothed; flowers: 1–2 cm across, **pale salmon-pink**, occasionally white; corolla tube: 2 cm long bent just below the corolla lobes, sparsely, coarsely hairy on the calyx rib. Parasitic on cereals in Africa only.................... .... ***S. forbesii*** (Fig. 18.3f)
- Shoots: up to 1 m high, arising from a **perennial** rootstock; leaves: up to 1 cm wide with some obscure teeth; flowers: **pink/brick red**, 2 cm across; corolla tube: 2 cm long, bent just below the corolla lobes; calyx: **covered in** short fine hairs. Parasitic on sugar cane in Africa only... ***S. latericea***

18.3.2 Striga hermonthica *(Del.) Benth.* (= S. senegalensis *Benth.) (Fig. 18.3a)*

This most damaging of the *Striga* species is also one of the largest, an erect herb up to 1 m high, especially in Eastern Africa, though usually rather shorter, about 50 cm in Western Africa (Fig. 18.3a). Larger plants may be branched. The whole plant is scabrid due to trichomes on stem and leaves. The leaves are simple, narrowly

Fig. 18.3 The important weedy witchweed species. (**a**) *Striga hermonthica* in sorghum; (**b**) *S. asiatica*; (**c**) *S. aspera*; (**d**) *Striga* flowers: left—corolla of typical West African *S. hermonthica*, centre—*S. hermonthica* form occurring in Eastern Africa, right—*S. aspera*; (**e**) *S. gesnerioides* in cowpea; (**f**) *S. forbesii* on wild grass (**a** from CABI (2012b), **e** from Parker and Riches (1993), by kind permission of CAB International)

lanceolate or elliptic, green, mainly opposite below, irregular above, 3–8 cm long and up to 1 cm wide. The upper part of the plant may be simple or branched, bearing a raceme or racemes of flowers each subtended by a single simple lanceolate or elliptical bract about 10 mm long and 2–4 mm wide with a fringe of ciliate hairs. The calyx is tubular, also about 10 m long with five distinct ribs and five short acute points 2–3 mm long. Flowers are asymmetrically campanulate, pink (very occasionally white), each about 2 cm long and 1–2 cm wide. The lower part is tubular, distinctly bent downwards, just below the midpoint in Western African populations, sometimes above the midpoint in Eastern Africa. The flower opens out to one bilobed upward-pointing and three longer lower lobes. The throat may be paler and/or streaked with purple. The inflorescence may bear up to 100 flowers. There are five stamens and a single short style. The capsules are up to 1 cm long, and each develops 500–700 minute seeds 0.3 mm long. Unlike most other weedy, *Striga* species S. *hermonthica* is strictly outcrossing (Aigbokhan et al. 1998).

Distribution. *S. hermonthica* occurs mainly in northern sub-Saharan Africa from Senegal and Gambia in the west and to Sudan, Ethiopia and Kenya in the east, with lesser incidence south of Tanzania. It occurs in the Arabian Peninsula but is otherwise restricted to Africa (Fig. 18.4).

Economic Importance. *S. hermonthica* is the most serious parasitic weed worldwide, estimated to affect many millions of hectares of crop across northern subtropical Africa (Sauerborn 1991; Parker 2009). The overall effect on the host can be devastating and lead to total crop failure. The crops affected include most of the major tropical and subtropical cereals, especially sorghum, *Pennisetum* millet and maize but also upland rice, sugar cane and finger millet (*Eleusine coracana*). In Ethiopia it is occasionally recorded on the more temperate wheat, barley and teff (*Eragrostis tef*) at altitudes over 2,000 m. Although *S. hermonthica* has functional chlorophyll, its photosynthesis is relatively inefficient (Press et al. 1987; see Sect. 6.2.2.1). But this diversion of host resources is only a small fraction of the damage caused. Even before emergence, the effects of the parasite can be obvious in stunting of the host shoot (Parker 1984) and chlorotic blotching of its foliage. Average losses of maize in Kenya are estimated at 80 % under heavy infestation (Manyong et al. 2007), while estimates of maize crop area infested vary up to 20–30 % in Ethiopia, Cameroon, Cote d'Ivoire and Guinea, 30–40 % in Togo, Mali and Nigeria and 65 % in Benin (de Groote et al. 2008). Estimates for all cereals in 1991 varied from 40 to 50 % in Ghana, Cameroon and Nigeria to over 70 % in Benin and Gambia (Sauerborn 1991). In north-east Nigeria, 85 % of cereals may be infested (Dugje et al. 2006). Across the whole of Africa, the estimates of area affected vary from 50 to 300 million hectare and the financial losses from US $300 million to more than US $1 billion. Many million farmers have been impoverished as a result.

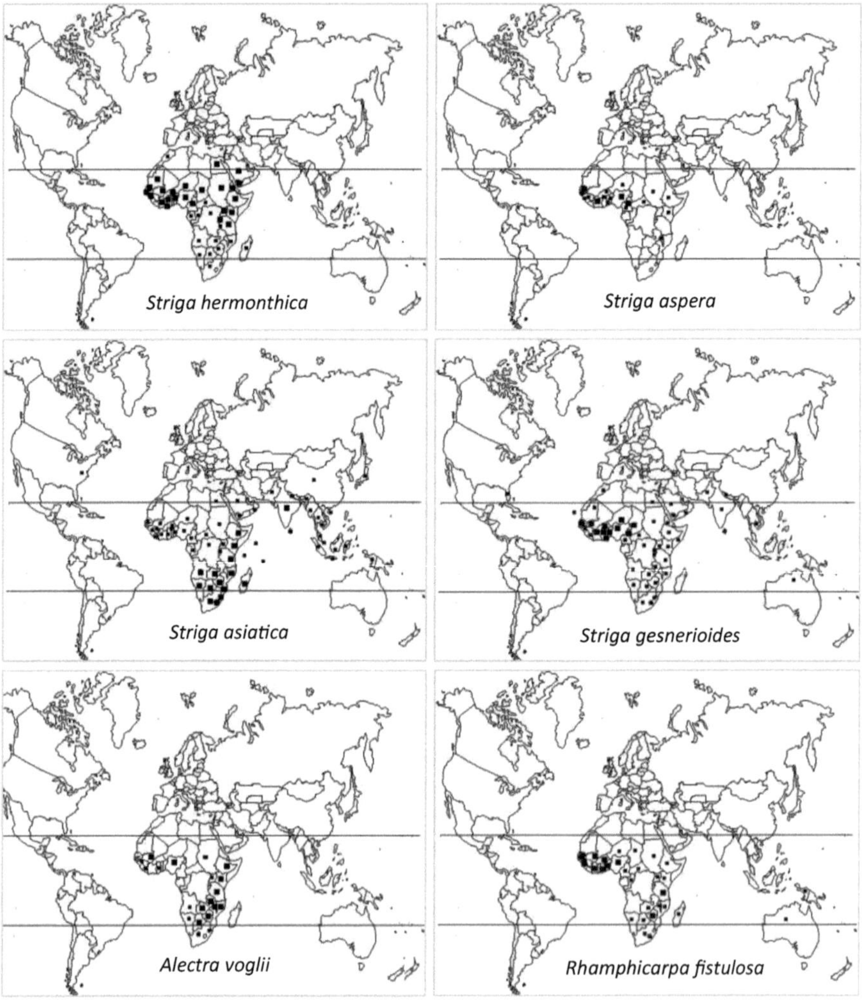

Fig. 18.4 World distributions of *Striga, Alectra and Rhamphicarpa* species. *Larger symbols* indicate countries in which it causes significant crop losses. *Smaller symbols* indicate countries in which it occurs mainly on wild hosts and/or causes lesser crop damage (maps for *Striga hermonthica*, *S. asiatica*, *S. gesnerioides* and *Alectra vogelii* reproduced from Parker (2012) by kind permission of Allen Press)

18.3.3 Striga asiatica *(L.) Kuntze (Fig. 18.3b)*

S. asiatica is much smaller than *S. hermonthica* and differs also in being autogamous. This has resulted in many distinct morphotypes with differing flower colour and host specificity. Some of these in Africa are given specific status as *S. hirsuta* Benth. or *S. lutea* Lour. (Mohamed et al. 2001), while *S. asiatica* (sensu stricto) is the usual weedy form on crops in Africa, almost always scarlet-flowered

(Fig. 18.3b) but occasionally yellow (or brick red in Ethiopia). White-flowered forms attack crops in South Asia. The weedy form is usually 15–30 cm high, covered in scabrid trichomes, usually much branched (forms occurring on wild hosts may be unbranched), with leaves 2–5 cm long about 3 mm wide. Lower bracts: leaf-like; upper bracts: smaller but always longer than the calyx, which is 7–9 mm long, tubular, with acute lobes 2–3 mm long and at least ten distinct ribs. The number of calyx lobes, basically 5, could vary up to 9. Flowers are tubular below, 10–15 mm long expanding to a broad upper lobe 3–4 mm long and narrower lower lobes 3–6 mm long. The capsule is up to 10 mm long, containing about 800 seeds not readily distinguished from those of *S. hermonthica*.

Host Range. *S. asiatica* affects all the crops attacked by *S. hermonthica*, most notably maize and sorghum. A yellow-flowered form attacks rice locally in Sumatra, Thailand and China (Parker and Riches 1993). The non-weedy forms in Africa are most often smaller and yellow flowered, while in SE Asia, there are even smaller pink- and purple-flowered forms on wild hosts. All forms are restricted to grasses.

Distribution. Figure 18.4 shows the distribution of *S. asiatica* in its widest sense, thus including forms which Mohamed et al. (2001) would refer to *S. hirsuta* and *S. lutea*. While cereal crops are occasionally attacked in West Africa, it is not fully certain whether the parasite responsible is one of these forms or the weedy *S. asiatica* (sensu stricto). Ignoring those occurrences in West Africa, it may be noted that the distribution of weedy *S. asiatica* differs markedly from that of *S. hermonthica*, being predominantly in Eastern and Southern Africa. The two species overlap in Kenya and Tanzania but rarely occur together, *S. asiatica* perhaps being commonest on light soils, though neither species is thought to be especially restricted by soil type.

The occurrence in the USA is thought to have originated from the accidental introduction of the weedy red-flowered form of *S. asiatica* from Southern Africa in the 1940s. It was a very serious infestation, affecting over 200,000 ha when first recognised in 1955 but is now quite vestigial after 50 years of eradication effort (Tasker and Westwood 2012). The occurrence of *S. asiatica* in the USA resulted in it receiving very substantial research attention, which has contributed enormously to the understanding and control of *Striga* species in Africa as well as in the USA.

Economic Importance. Where it occurs, *S. asiatica* may cause as much crop damage as *S. hermonthica*, but as an economic problem worldwide, it is perhaps an order of magnitude less serious. The physiological effects on the host are stunting, a change in host root to shoot ratio, reduction of host photosynthesis and foliage wilting even under moist conditions. Infestation can involve severe crop loss, and average losses of 10–40 % are almost certainly common. De Groote et al. (2008) estimate that 63–80 % of maize crops are affected in Malawi and substantial proportions in Angola, Swaziland and other Southern African countries, also in Madagascar (Geiger et al. 1996). There are localised infestations in sorghum, maize or rice in Kenya, Tanzania, Ethiopia and most countries of Southern Africa. Upland rice is seriously affected in Tanzania (Kayeke et al. 2007), Mozambique and Madagascar. Sadly, the problem in Africa is tending to increase rather than decrease as intensive land use and the expense, or lack, of fertilizers lead to continuing decline in soil fertility, greatly favouring its growth. In India, it is less widespread, but the white-flowered form affects sorghum and sugar cane locally (Patil and Angadi 2008).

18.3.4 Striga aspera *Willd. (Fig. 18.3c, d)*

S. aspera resembles *S. hermonthica* in general appearance and flower colour but is usually somewhat smaller, less scabrid and less robust. A useful character even in nonflowering plants is the narrowness of the leaves up to 3 mm only, bracts 1–2 mm and the lack of a fringe of hairs on the bracts. The calyx is 5-ribbed as in *S. hermonthica*. In West Africa, it is distinguished by the bend in the corolla tube occurring well above half way. The corolla is 12–15 mm long, with upper expanded lobe 3–6 mm long and lower lobes 4–8 mm (Fig. 18.3d). The capsule and seeds are comparable to those of *S. hermonthica*. The two species can hybridise, though hybrids have not been commonly reported (Aigbokhan et al. 1998).

Host Range. Includes most of the warm-climate cereals but it is less common on sorghum and pearl millet and somewhat more common on rice and sugar cane than *S. hermonthica* and is much more commonly seen on wild grasses (Parker and Riches 1993). A local infestation in Ethiopia occurs only on maize.

Distribution. *S. aspera* occurs mainly in West Africa but also eastwards to Sudan and south to Malawi (Fig. 18.4).

Economic Importance. *S. aspera* can be as damaging as *S. hermonthica*, and rice is particularly affected, seriously so in Cote d'Ivoire and Senegal (Parker and Riches 1993) and in NE Nigeria (Dugje et al. 2006). Johnson et al. (1997) recorded 50 % reduction in rice yield from 17 stems of *S. aspera* per m^2. Sugar cane can also be attacked. Maize is damaged in Ethiopia, Nigeria, Cameroon and Cote d'Ivoire, while sorghum and pearl millet are rarely if ever parasitized.

18.3.5 Other Striga *Species Affecting Cereal Crops*

Other *Striga* species affecting **cereal** crops, not considered in detail here, include the pale pink-flowered ***S. forbesii*** **Benth**. (Fig. 18.3f) with a wide distribution in East and Southern Africa and Madagascar, as well as West Africa, which is locally damaging to corn and sorghum in Zimbabwe and Tanzania and to rice in Cote d'Ivoire. The closely related perennial ***S. latericea*** **Vatke** is restricted to East Africa and is only very locally damaging to sugar cane in Ethiopia and Somalia. Two further white-flowered species attack cereals in India, ***S. densiflora*** **(Benth.) Benth.** and ***S. angustifolia*** **(Don) Saldanha,** but their current importance there, relative to *S. asiatica,* is uncertain. There have been almost no new reports on infestations of these species since the publication of Parker and Riches (1993).

18.3.6 Striga gesnerioides *(Willd.) Vatke (Figs. 14.3l and 18.3e)*

S. gesnerioides differs markedly from all the other weedy *Striga* species both morphologically and in its host range, which is **restricted to dicotyledonous (broad-leaved) hosts**. In morphology, it differs in having **very reduced foliage and low chlorophyll** (Fig. 18.3e). The stems are much branched and relatively fleshy, with scale leaves 5–10 mm long, appressed to the stem. Flowers are subtended by bracts 3–6 mm long. The calyx is 5-ribbed, 5–9 mm long including the acute lobes up to 3 mm long. The corolla tube is 10–15 mm long with upper lip 2 mm long and lower lobes 3–6 mm long. Corolla colour varies from white through mauve to purple, though it is the paler forms which predominate in the weedy forms attacking cowpea (*Vigna unguiculata* (L.) Walp.). The capsule is 5–8 mm long, containing several hundred seeds 0.33 mm long, similar to those of *S. asiatica*.

Host Range. *S. gesnerioides* is autogamous and exists in a range of races, distinct to some extent in morphology, but most particularly in host range. The main weedy races attack only cowpea, while tobacco is affected very locally in Zimbabwe (Koga et al. 2011) and sweet potato in South Africa and Ethiopia. Other races each attack a narrow range of mainly wild hosts. Mohamed et al. (2001) describe 'strains' specific to *Tephrosia*, *Indigofera*, *Euphorbia*, *Vigna*, *Ipomoea* (including sweet potato), *Merremia* and *Nicotiana*. Most of these are branched and succulent like the 'cowpea strain', but others may be unbranched and slender (on *Indigofera*, *Nicotiana*) or perennial (on *Euphorbia*). Ralston et al. (1987) describe four different forms in Botswana, including one with yellow flowers on an *Ipomoea* sp.

Distribution. *S. gesnerioides* has a much wider distribution than any other *Striga* sp., occurring mainly in Africa but also South and SE Asia (Fig. 18.4). There has also been one introduced infestation in Florida, USA, occurring on *Indigofera hirsuta*, *Jacquemontia tamnifolia* and *Alysicarpus vaginalis* (Musselman and Parker 1981).

Economic Importance. *S. gesnerioides* is extremely important on cowpea in West Africa, seriously damaging the crop from Senegal through to Chad. The physiological effects of this weed have not been intensively studied and may not be as subtle or severe as those of *S. hermonthica*, but crop loss has averaged 30 % across a range of varieties in Burkina Faso and can exceed 50 %. Currently it is considered that there are at least seven biotypes, with apparently little overlap in their distribution. Fortunately there are now varieties of cowpea with broad-range immunity to most of these biotypes (e.g. Singh et al. 2006; see Sect. 7.4.1) which should see the problem declining, though vigilance and further breeding will be needed to forestall the selection of new virulent biotypes as has occurred with sunflower and *Orobanche cumana* (see Sect. 21.3.3).

18.4 *Alectra* Species

18.4.1 Alectra vogelii *Benth (Fig. 18.5a)*

A. vogelii like the related *Striga* spp. is an obligate hemiparasite, having green foliage but a very small seed incapable of establishing without attachment to a host root. The plant is 30–45 cm high (Fig. 18.5a) and normally unbranched, the underground section of the stem bright orange. Leaves are hairy, 2–4 cm long, broadly lanceolate up to 1.5 cm wide and variably toothed. The inflorescence occupies most of the aerial stem. Flowers are subtended by a leaf-like bract. The hairy calyx is quite widely inflated, 5–8 mm across, with ten ribs, about 1 cm long including acute lobes 1–2 mm long. The corolla is tubular below, with expanded lobes about 5 mm long, all yellow, sometimes with purple streaks. The capsule is almost globose, up to 5 mm in diameter. Seeds are very distinctive having a very small kernel, 0.2 mm in diameter, suspended in a very expanded seed coat about 1 mm long, making the seed very readily transported by wind. Some populations of *Alectra* in cowpea have been identified as ***A. picta*** **(Hiern) Hemsl.,** which differs from *A. vogelii* in the hairiness of its stamen filaments (which are glabrous in *A. vogelii*), but there are doubts about the distinction between these two taxa (Parker and Riches 1993), and no attempt is made here to separate them.

Host Range. *A. vogelii* attacks various Fabaceous crops and is less restricted in its host range than *S. gesnerioides*. Cowpea is again its main host, and both parasite species can occur in the same field, but *A. vogelii* can also attack groundnut, bambara (*Vigna subterranea* (L) Verdc.), soybean and a number of other legume crops.

Distribution. *A. vogelii* occurs across much of Africa, North and South of the equator but is rarely extensive in occurrence (Fig. 18.4).

Economic Importance. Cowpea is seriously attacked in a number of West African countries, especially Nigeria and Burkina Faso, and also in Botswana, Malawi and in other countries of Southern and Eastern Africa. Damage can be severe. Emechebe et al. (1991) have reported up to 100 % yield loss.

18.4.2 Other Alectra *Species*

A. sessiliflora **(Vahl) Kuntze** is superficially very similar to *A. vogelii* but differs in having apiculate anther lobes and the calyx hairy only along the nerves. It occurs across most of Africa and Southern Asia into China and the Philippines, mainly on wild hosts in Asteraceae. Crops that are occasionally attacked include niger seed (*Guizotia abyssinica*) in Ethiopia (Parker 1988).

Fig. 18.5 Important weedy witchweed Orobanchaceae. (**a**) *Alectra vogelii* on cowpea; (**b**) *Rhamphicarpa fistulosa*; (**c**) *Aeginetia indica*; (**d**) *Buchnera hispida*; (**e**) *Odontites verna*; (**f**) *Melampyrum arvense*; (**g**) *Rhinanthus minor*; (**h**) *Seymeria cassioides* (**a**, **b**, **f**, **g** by Chris Parker; **c** by Danny Joel; **d** from Parker and Riches (1993) by permission of CAB International; **h** by Lytton Musselman)

***Alectra orobanchoides* Benth.** has reduced leaves and lower chlorophyll and has been reported on tobacco in Zambia and on sunflower in South Africa (Parker and Riches 1993). There are no recent records of occurrence on crops.

***Alectra aspera* (Cham. and Schltdl.) L.O. Williams (= *A. fluminensis* (Vell.) Stearn; *A. brasiliensis* Benth.)** is a robust South American species up to 1 m tall, which has been recorded as a parasite of sugar cane in Venezuela (Parker and Riches 1993). There are no recent reports.

18.5 *Rhamphicarpa fistulosa*

R. fistulosa is a hemiparasite, but unlike *Striga* and *Alectra*, it is a facultative parasite, due to its significantly larger seeds, 0.55 mm long, which allow it to emerge independently, even before parasitizing a host (Ouédraogo et al. 1999). It also differs from *Striga* and *Alectra* in having highly divided foliage with segments up to 10 cm long, only 1 mm wide (Figs. 18.5b and 22.1a). The flowers are carried on pedicels 10–20 mm long and have a distinctive narrow tube, 25–30 mm long, opening to five lobes, 10 mm long, usually white or cream, opening at night for pollination by moths. Calyx: 5–10 mm long with a short tube up to 2 mm long, 2–4 mm in diameter with narrow lobes 4–6 mm long. The style is up to 30 mm. Capsule: up to 10 mm including the beak, 4–7 mm wide, winged along the sutures.

Host Range. *R. fistulosa* is a facultative parasite and may occur on a wide range of hosts but is mainly seen on Poaceae and perhaps Cyperaceae.

Distribution. *R. fistulosa* is widespread across tropical Africa but is found also in New Guinea and Australia (Fig. 18.4).

Economic Importance. *R. fistulosa* is not a major problem, but it causes serious damage locally to a range of cereal crops across West, East and Southern Africa. Ouédraogo et al. (1999) recorded serious damage to maize, sorghum and rice in Mali and in Burkina Faso. Maiti and Singh (2004) also reported serious damage to pearl millet. Gworgwor et al. (2001) record infestations in rice and in sorghum in NE Nigeria. Riches and Johnson (1998) recorded serious losses in rice locally in Guinea, Benin, Tanzania and Zimbabwe and occurrence also in rice in Senegal and Ghana. They noted that severe infestations resulted in farmers abandoning their crops and failing to avoid the problem by fallowing, apparently due to the weed's persistence on wild hosts. They considered that the problem was tending to increase, and this was also suggested by Rodenburg et al. (2011) reviewing the problem in Benin, where farmers estimated rice yield losses to be at least 60 %. The reason for the apparent increase in incidence of *R. fistulosa* is not clearly explained, though the fact that it tends to be reduced by fertilizer application suggests it may be another symptom of generally declining soil fertility (see Sect. 22.3.3).

18.6 Other Orobanchaceae Occasionally Proving Weedy

A range of other members of Orobanchaceae have been recorded as weeds in the past including the following. They are noted briefly here, and recent literature reviewed. Further detail and earlier literature can be found in Parker and Riches (1993).

18.6.1 Aeginetia indica *Roxb. (Fig. 18.5c)*

An Asian holoparasitic species differing from *Orobanche* species in having no above-ground stem but solitary flowers on individual pedicels up to 30 cm high (Fig. 18.5c). Seeds: about 0.3 mm long. It ranges from India to Japan and SE Asia; *A. indica* has been recorded parasitizing sugar cane in India, Taiwan, Japan and the Philippines. It may also occur on rice and maize. The related ***Aeginetia pedunculata*** **(Roxb.) Wall.** (= ***A. acaulis*** **(Roxb.) Walp**.) has a similar distribution but larger flowers on shorter pedicels and has recently been reported damaging sugar cane in West Bengal, India (Ray and Dasgupta 2006). See damage description in Heide-Jørgensen (2008).

18.6.2 Buchnera hispida *Buch.-Ham. ex D. Don (Fig. 18.5d)*

A facultative hemiparasite, somewhat like *Striga* in form but with small, radially symmetrical, purple flowers in the axils of long bracts (Fig. 18.5d); seeds: 0.55 mm long, requiring light for germination. It is widespread in Africa and occurs also in Madagascar and India. It has a wide host range on grasses and has been recorded causing significant crop damage locally on maize, sorghum and pearl millet especially in Mali and Nigeria. See damage description in Nwoke and Okonkwo (1974).

18.6.3 Odontites verna *(Bell.) Dum.* (= Bartsia odontites *Huds.) 'Red Bartsia' (Fig. 18.5e)*

A temperate facultative hemiparasite up to 20–50 cm high with toothed lanceolate leaves and deep red flowers (Fig. 18.5e). It is a species of southern Europe, extending north to the British Isles and east into Turkey, and has been introduced to NE USA and Canada. Its host range is mainly grasses but also includes wheat, lucerne and clovers. It has been recorded as a problem on pasture grasses in Canada and on lucerne in Wisconsin, USA. It occurs on organically grown wheat in the Czech Republic (Tyšer et al. 2005) and also in wheat in a situation of very low soil fertility in the classical long-term Broadbalk Field at Rothamsted Research Station in the UK (Moss et al. 2004).

18.6.4 **Melampyrum arvense** *L. 'Cow Wheat' (Fig. 18.5f)*

Another temperate facultative hemiparasite with entire lanceolate lower leaves below; upper leaves and bracts: deeply pectinate (comb-like) (Fig. 18.5f). Flowers are yellow below but purplish towards the tip. *M. arvense* is distributed across Europe and Western Asia and has a wide host range including all the temperate cereals. It has been reported as a locally serious problem in wheat in Bulgaria in the past and more recently in cereals in Macedonia (Kostov and Pacanoski 2007) and on wheat in inner West Anatolia, Turkey (Uludag and Nemli 2009). In Poland, it has been seen as an attractive component of the roadside flora, occurring in neighbouring cereal fields to only a limited extent (Kurus and Podstawka-Chmielewska 2007).

18.6.5 **Rhinanthus minor** *L. and* **R. angustifolius** *C. Gmelin. (=* **R. serotinus** *(Schőnheit) Oborny) 'Yellow Rattle' (Fig. 18.5g)*

These two temperate facultative parasites were previously included together under *R. crista-galli* L. and may still be sometimes confused. Height: 30–50 cm; leaves and bracts: opposite, sharply toothed (Fig. 18.5g). Flowers: yellow 12–15 mm long in *R. minor*, 17–20 mm long in *R. angustifolius*. Calyx: very much inflated, enclosing an equally inflated capsule, in which the seeds, flat, 3–4 mm across, may 'rattle' when mature. Distribution includes Europe, Western Asia and North America. Host range is wide, but there can be quite marked host preference. Grasses and legumes are favoured; other herbs are less susceptible (Cameron et al. 2006). Těšitel et al. (2010) concluded that *R. minor* depended on its hosts for 50 % of its carbon metabolism. Occurrence is most commonly in pastures, where *Rhinanthus* spp. may be regarded as either damaging—Davies et al. (1997) record up to 70 % reduction in productivity, and Cameron et al. (2008) showed reduced rates of photosynthesis in a susceptible grass species—or a valuable conservation agent, contributing to wider species richness (Bullock and Pywell 2005; Westbury et al. 2006; see Sect. 16.2.2). It rarely occurs in cereal crops but has been reported damaging wheat in Bulgaria. For a comprehensive review of the biology and ecology of *R. minor*, see Westbury (2004).

18.6.6 **Seymeria cassioides** *(Walt.) Blake (Fig. 18.5h)*

Another facultative hemiparasite, known as **'senna seymeria'** in the USA, grows up to 1 m high; leaves finely divided with segments 1–3 cm long and yellow flowers 1 cm across with a purple centre (Fig. 18.5h). Seeds are 1 mm long allowing independent establishment and growth. It is native to Southern USA but occurs also in the Bahamas. It has very narrow host range, parasitizing only three species of pine, *Pinus elliottii*, *P. palustris* and *P. taeda*. The problem from this species in forestry in Florida was reviewed by Barnard and Coile (1996).

18.7 Conclusion

This chapter describes the current status of weedy Orobanchaceae across the world. It is to be hoped that the future will see a steady decline in their importance as new control methods are developed and introduced, but there is still the opposite prospect that some at least may spread or intensify as a result of careless introduction to new areas and/or of global warming. There has been no clear evidence yet that global warming will necessarily result in wider distribution and greater problems, though Hättenschwiler and Zumbrunn (2006) show improved growth of *Melampyrum pratense* and *M. sylvaticum* at higher CO_2 levels. Meanwhile, several studies have confirmed the potential for the major weedy species to thrive in areas where they are not yet present. Grenz and Sauerborn (2007) concluded that on the basis of climatic data, there are many parts of the world outside the existing range of *O. crenata* where it could thrive, including parts of North, Central and South America; West Africa; South and SE Asia; and Australia, while Mohamed et al. (2006) provide similar warnings about the potential for further spread of the major *Striga* species.

Furthermore, Schneeweiss (2012) suggests there is potential for so far non-weedy *Orobanche* species to become weedy, either because they have life traits of the weedy species (*O. picridis*, *O. canescens*, *O. haenseleri*, *O. pubescens* and *O. transcaucasica*) or because they already attack legume species, and a small change in host specificity could result in undesirable weediness (*O. densiflora*, *O. owerinii*, *O. rapum-genistae*, *O. sanguinea* and *O. variegata*).

Continued vigilance is vital.

References

Abbes Z, Kharrat M, Delavault P, Simier P, Chaïbi W (2007) Field evaluation of the resistance of some faba bean (*Vicia faba* L.) genotypes to the parasitic weed *Orobanche foetida* Poiret. Crop Prot 26:1777–1784

Aigbokhan EI, Berner DK, Musselman LJ (1998) Reproductive ability of hybrids of *Striga aspera* and *Striga hermonthica*. Phytopathology 88:563–567

Babiker AGT, Mohamed ES, El Mana ME (1994) *Orobanche* problem and management in the Sudan. In: Pieterse AH, Verkleij JAC, ter Borg SJ (eds) Proceedings of the 3rd international workshop on *Orobanche* and Related *Striga* research. Royal Tropical Institute, Amsterdam, pp 672–676

Babiker AGT, Ahme EA, Dawoud DA, Abdrella NK (2007) *Orobanche* species in Sudan: history, distribution and management. Sudan J Agric 10:107–114

Barnard EL, Coile NC (1996) Black-senna (*Seymeria cassioides* (J.F Gmel.) Blake): a root parasite of importance to forestry in Florida. Plant Pathology Circular (Gainesville) No 380, 4 pp

Bayaa B, Kumari SG, Akkaya A, Erskine W, Makkouk KM, Turk Z, Ozberk I (1998) Survey of major biotic stresses of lentil in South-East Anatolia, Turkey. Phytopathol Mediterr 37:88–95

Bedi JS, Kapur SP, Mohan C (1997) *Orobanche* – a threat to raya and taramira in Punjab. J Res (Punjab Agric Univ) 34:149–152

Benharrat H, Boule C, Theodet C, Thalouarn P (2005) Virulence diversity among branched broomrape (*O. ramosa* L.) populations in France. Agron Sustain Dev 25:123–128

Bernhard RH, Jensen J, Andreasen C (1998) Prediction of yield loss caused by *Orobanche* spp. in carrot and pea crops based on the soil seedbank. Weed Res 38:191–197

Boulet C, Pineault D, Benharrat H, Bozec D, Delavault P, Simier P (2007) Répartition de la flore adventice à l'échelle d'un paysage agricole: analyse de la diversité des pleins champs et des bordures. In: 20ème Conférence du COLUMA. Journées Internationales sur la Lutte contre les Mauvaises Herbes, Dijon, France, 11–12 décembre 2007, pp 326–345

Brault M, Betsou F, Jeune B, Tuquet C, Sallé G (2007) Variability of *Orobanche ramosa* populations in France as revealed by cross infestations and molecular markers. Environ Exp Bot 61:272–280

Bülbül F, Aksoy E, Uygur S, Uygur N (2009) Broomrape (*Orobanche* spp.) problem in the eastern Mediterranean region of Turkey. Helia 32(51):141–152

Bullock JM, Pywell RF (2005) *Rhinanthus*: a tool for restoring diverse grassland? Folia Geobot 40 (2–3):273–288

Buschmann H, Kömle S, Gonsior G, Sauerborn J (2005) Susceptibility of oilseed rape (*Brassica napus* ssp. *napus*) to branched broomrape (*Orobanche ramosa* L.). Z Pflanzenk Pflanzen 112:65–70

CABI (2012a) Crop Protection Compendium. CABI International, Wallingford, UK. http://www.cabi.org/cpc/

CABI (2012b) Invasive Species Compendium. CABI International, Wallingford, UK. http://www.cabi.org/isc/

Cagáň Ľ, Tóth P (2003) A decrease in tomato yield caused by branched broomrape (*Orobanche ramosa*) parasitization. Acta Fytotechn Zootech 6:65–68

Cameron DD, Coats AM, Seel WE (2006) Differential resistance among host and non-host species underlies the variable success of the hemiparasitic plant *Rhinanthus minor*. Ann Bot 98:1289–1299

Cameron DD, Geniez JM, Seel WE, Irving LJ (2008) Suppression of host photosynthesis by the parasitic plant *Rhinanthus minor*. Ann Bot 101(4):573–578

Colquhoun JB, Eizenberg H, Mallory-Smith CA (2006) Herbicide placement site affects small broomrape (*Orobanche minor*) control in red clover. Weed Technol 20:356–360

Correll R, Marvanek S (2006) Sampling for detection of branched broomrape. In: Preston C, Watts JH, Crossman ND (eds) 15th Australian weeds conference, Adelaide, pp 610–613

Davies DM, Graves JD, Elias CO, Williams PJ (1997) The impact of *Rhinanthus* spp. on sward productivity and composition: implications for the restoration of species-rich grasslands. Biol Conserv 82:87–93

de Groote H, Wangare L, Kanampiu F, Odendo M, Diallo A, Karaya H, Friesen D (2008) The potential of a herbicide resistant maize technology for *Striga* control in Africa. Agric Syst 97:83–94

Díaz SJ, Norambuena MH, López-Granados F (2006) Caracterizacion del holoparasitismo de *Orobanche ramosa* en tomate bajo condiciones de campo. Agricultura Técnica 66:223–234

Dugje IY, Kamara AY, Omoigui LO (2006) Infestation of crop fields by *Striga* species in the savanna zones of northeast Nigeria. Agric Syst Environ 116:251–254

Eizenberg H, Golan S, Joel DM (2002) First report of the parasitic plant *Orobanche aegyptiaca* infecting olive. Plant Dis 86:814

Eizenberg H, Colquhoun JB, Mallory-Smith CA (2003) Variation in clover response to small broomrape (Orobanche minor). Weed Sci 51:759–763

Emechebe AM, Singh BB, Leleji OI, Atokple IDK, Adu JK (1991) Cowpea-*Striga* problems and research in Nigeria. In: Kim SK (ed) Combating *Striga* in Africa. Proceedings of the international workshop, Ibadan, 1988. IITA, Ibadan, Nigeria, pp 18–28

English TJ, Norris RS, Miller AE (1997) Control of clover broomrape (*Orobanche minor* Sm.) in southwest Georgia pecan groves. Proc South Weed Sci Soc 50:81–82

Fracchiolla M, Boari A (2003) (Effect of the infestation of *Orobanche ramosa* on tomato and cauliflower yield). Informatore Fitopatologico 53(2):52–53 (in Italian)

Gagne G, Roekel-Drevett P, Grezes-Besset B, Shindrova P, Ivanon P, Grand-Ravel C, Vear F, Tourvieille de Labrouhe D, Charmet G, Nicholas P (1998) Study of the variability and evolution of *Orobanche cumana* populations infesting sunflower in different European countries. Theor Appl Genet 96:1216–1222

Geiger U, Kroschel J, Sauerborn J (1996) *Striga asiatica*, a problem in the middle west of Madagascar. In: Moreno MT, Cubero JI, Berner DK, Joel D, Musselman LJ, Parker C (eds) Advances in parasitic research. Proceedings of the 6th international parasitic weed symposium, Cordoba, Spain, pp 479–486

Gibot-Leclerc S, Brault M, Salle G (2003) L'Orobanche rameuse. La menace s'aggrave pour le colza, le chanvre et le tabac. Phytoma 561:9–12

Grenz JH, Sauerborn J (2007) Mechanisms limiting the geographical range of the parasitic weed *Orobanche crenata*. Agric Ecosyst Environ 122:275–281

Gworgwor NA, Ndahi WB, Weber H-Chr (2001) Parasitic weeds of North-eastern Nigeria: a new potential threat to crop production. In: Proceedings, British crop protection conference – weeds, pp 181–186

Hättenschwiler S, Zumbrunn T (2006) Hemiparasite abundance in an alpine treeline ecotone increases in response to atmospheric CO_2 enrichment. Oecologia 147:47–52

Heide-Jørgensen HS (2008) Parasitic flowering plants. Brill, Leiden

Hibberd JM, Quick WP, Press MC, Scholes JD (1998) Can source-sink relations explain responses of tobacco infection by the root hemiparasitic angiosperm *Orobanche cernua*? Plant Cell Environ 21:333–340

Hibberd JM, Quick WP, Press MC, Scholes JD, Jeschke WD (1999) Solute fluxes from tobacco to the parasitic angiosperm *Orobanche cernua* and the influence of infection on host carbon and nitrogen relations. Plant Cell Environ 22:937–947

Jacobsohn R, Bohlinger B, Eldar E, Agrawal VP (1991) Crop host range of *Orobanche* species in an experimental field. In: Ransom JK, Musselman, LJ, Worsham AD, Parker C (eds) Proceedings of the 5th international symposium on parasitic weeds, CIMMYT, Nairobi, pp 176–179

Joel DM (1988) *Orobanche cumana* – a new adventive weed in Israel. Phytoparasitica 16:375

Joel DM (2009) The new nomenclature of *Orobanche* and *Phelipanche*. Weed Res 49(Suppl 1):6–7

Joel DM, Eisenberg H (2002) Three *Orobanche* species newly found on crops in Israel. Phytoparasitica 30:187–190

Joel DM, Hershenhorn Y, Eizenberg H, Aly R, Ejeta G, Rich PJ, Ransom JK, Sauerborn J, Rubiales D (2007) Biology and management of weedy root parasites. In: Janick J (ed) Horticultural reviews, vol 33. Wiley, Hokobken, NJ, pp 267–349

Joel DM, Chaudhuri SK, Plakhine D, Ziadna H, Steffens JC (2011) Dehydrocostus lactone is exuded from sunflower roots and stimulates germination of the root parasite *Orobanche cumana*. Phytochemistry 72:624–634

Johnson DE, Riches CR, Diallo R, Jones MJ (1997) *Striga* on rice in West Africa; crop host range and the potential of host resistance. Crop Prot 16:153–157

Katzir N, Portnoy V, Tzuri G, Joel DM, Castejón-Muñoz M (1996) Use of random amplified polymorphic DNA (RAPD) markers in the study of the parasitic weed *Orobanche*. Theor Appl Genet 93:367–372

Kayeke J, Sibuga PK, Msaky JJ, Mbwaga A (2007) Green manure and inorganic fertiliser as management strategies for witchweed and upland rice. Afr Crop Sci J 15:161–171

Kharrat M, Halila MH (1994) *Orobanche* species on faba bean (*Vicia faba* L.) in Tunisia: problem and management. In: Pieterse AH, Verkleij JAC, ter Borg SJ (eds) Biology and management of *Orobanche*. Proceedings of the 3rd international workshop on *Orobanche* and related *Striga* research, Amsterdam, pp 639–643

Kharrat M, Halila MH, Linke KH, Haddar T (1992) First report of *Orobanche foetida* Poiret on faba bean in Tunisia. FABIS Newsl 30:46–47

Koga C, Mwenje E, Garwe D (2011) Germination stimulation of *Striga gesnerioides* seeds from tobacco plantations by hosts and non-hosts. J Appl Biosci 2011:2453–2459

Kohlschmid E, Müller-Stöver D, Sauerborn J (2011) Ausbreitung des parasitischen Unkrauts *Phelipanche ramosa* in der deutschen Landwirtschaft. Gesunde Pflanzen 63:69–74

Kostov T, Pacanoski Z (2007) Weeds with major economic impact on agriculture in Republic of Macedonia. Pak J Weed Sci Res 13:227–239

Kurus J, Podstawka-Chmielewska E (2007) Pszeniec rozowy (*Melampyrum arvense* L.) jako element krajobrazu rolniczego. Annales Universitatis Mariae Curie-Sklodowska Sectio E, Agricultura 62(2):137–144

Labrada R, Perez R (1988) Medias de lucha no quimica contra *Orobanche ramosa*. Agrotecnia de Cuba 20:35–40

Lins RD, Colquhoun JB, Mallory-Smith CA (2007) Effect of small broomrape (*Orobanche minor*) on red clover growth and dry matter partitioning. Weed Sci 55:517–520

Maiti RK, Singh VP (2004) Biotic factors affecting pearl millet (*Pennisetum glaucum* L.) R. Br.) growth and productivity – a review. Crop Res 27:30–39

Mallory-Smith C, Colquhoun, J (2012) Small broomrape (*Orobanche minor*) in Oregon and the 3 Rs: regulation, research, and reality. Weed Sci 60:277–282

Manen JF, Habashi C, Jeanmonod D, Park JM, Schneeweiss GM (2004) Phylogeny and intraspecific variability of holoparasitic *Orobanche* (Orobanchaceae) inferred from plastid rbcL sequences. Mol Phylogenet Evol 33:482–500

Manyong VM, Alene AD, Olanrewaju A, Ayedun B, Rweyendela V, Wesonga AS, Omanya G, Mignouna HD, Bokanga M (2007) Baseline study of *Striga* control using IR maize in western Kenya. Aatf/Iita *Striga* Control Project, http://aatf-africa.org/userfiles/IRmaizestudy.pdf

Mauromicale G, Lo Monaco A, Longo AMG (2008) Effect of branched broomrape (*Orobanche ramosa*) infection on the growth and photosynthesis of tomato. Weed Sci 56:574–581

Mohamed KI, Musselman LJ, Riches CR (2001) The genus *Striga* (Scrophulariaceae) in Africa. Ann Mo Bot Gard 88:60–103

Mohamed KI, Papes M, Williams R, Benz BW, Peterson AT (2006) Global invasive potential of 10 parasitic witchweeds and related Orobanchaceae. Ambio 35:281–288

Moss SR, Cussans JW, Perryman SAM, Hewitt MV (2004) The Broadbalk long-term experiment at Rothamsted: what has it told us about weeds. Weed Sci 52:864–873

Motazedi S, Jahedi A, Farnia A (2010) Integrated broomrape (*Orobanche aegyptiaca*) control by sulfosulfuron (WG 75%) herbicide with wheat mulch applied in field potato. In: Proceedings of the 3rd Iranian weed science congress, volume 2: Key papers, weed management and herbicides, Babolsar, Iran, pp 227–229

Musselman LJ (1989) *Orobanche ramosa* and *Orobanche aegyptiaca* in the Flora Palaestina. In: Wegmann K, Musselman LJ (eds) Progress in *Orobanche* research. Proceedings of the international workshop on *Orobanche* research, Obermarchtal, FRG, pp 1–5

Musselman LJ, Parker C (1981) Studies on indigo witchweed, the American strain of *Striga gesnerioides* (Scrophulariaceae). Weed Sci 29:594–596

Musselman LJ, Parker C (1982) Preliminary host ranges of some strains of economically important broomrapes (*Orobanche*). Econ Bot 36:270–273

Musselman LJ, Parker C, Dixon N (1982) Notes on autogamy and flower structure in agronomically important species of *Striga* (Scrophulariaceae) and *Orobanche* (Orobanchaceae). Beitr Biol Pflanz 56:329–343

Nwoke FIO, Okonkwo SNC (1974) Facultative hemiparasitism in *Buchnera hispida* Buch.-Ham ex D.Don. Ann Bot 38:993–1002

Ouédraogo O, Neumann U, Raynal-Roques A, Sallé G, Tuquet C, Dembélé B (1999) New insights concerning the ecology and the biology of *Rhamphicarpa fistulosa* (Scrophulariaceae). Weed Res 39:159–169

Parker C (1984) The influence of *Striga* species on sorghum under varying nitrogen fertilization. In: Parker C, Musselman LJ, Polhill RM (eds) Proceedings of the 3rd international symposium on parasitic weeds, Aleppo, pp 90–98

Parker C (1988) Parasitic plants in Ethiopia. Walia 11:21–27

Parker C (1994) The present state of the *Orobanche* problem. In: Pieterse AH, Verkleij JAC, ter Borg SJ (eds) Biology and management of *Orobanche*. Proceedings of the 3rd international workshop on *Orobanche* and related *Striga* research. Royal Tropical Institute, Amsterdam, pp 17–26

Parker C (2006) Orobanchaceae. In: Hedberg I, Edwards S, Ensermu K, Sebsebe D, Persson E (eds) Flora of Ethiopia and Eritrea. Vol 5: Gentianaceae to Cyclocheilaceae. The National Herbarium, University, Addis Ababa, Ethiopia and Department of Systematic Botany, Uppsala University, Uppsala, Sweden, pp 309–314

Parker C (2009) Observations on the current status of *Orobanche* and *Striga* problems worldwide. Pest Manag Sci 65:453–459

Parker C (2012) Parasitic weeds: a world challenge. Weed Sci 60:269–276

Parker C, Riches CR (1993) Parasitic weeds of the world: biology and control. CAB International, Wallingford, UK

Patil VL, Angadi SS (2008) Effect of management practices on *Striga* incidence, quality, yield and economics of sorghum. Plant Arch 8:185–188

Press MC, Shah N, Tuohy JM, Stewart GR (1987) Carbon isotope ratios demonstrate carbon flux from C_4 host to C_3 parasite. Plant Physiol 85:1143–1145

Pujadas A, Velasco L (2000) Comparative studies on *Orobanche cernua* L. and *O. cumana* Wallr. (Orobanchaceae) in the Iberian Peninsula. Bot J Linn Soc 134:513–527

Pujadas-Salvà AJ, Fraga i Arguinbau P, Sánchez-Gullón E, Molina-Mahedero N (2003) *Orobanche crinita* Viv. and *Orobanche foetida* Poir. (Orobanchaceae) in the West Mediterranean area. Bocconea 16:737–744

Ralston DM, Riches CR, Musselman LJ (1987) Morphology and hosts of three *Striga* species (Scrophulariaceae) in Botswana. Adansonia 2:195–215

Ray BR, Dasgupta MK (2006) Sugarcane crop loss due to wilt caused by parasitic angiosperm *Aeginetia pedunculata* (Roxb.) Wall. (Orobanchaceae). J Mycol Plant Pathol 36:31–34

Reda F (2006) *Orobanche crenata* in Ethiopia. Haustorium 49:3

Riches CR, Johnson DE (1998) *Rhamphicarpa fistulosa* on rice in Africa. Haustorium 33:2–3

Rodenburg J, Zossou-Kouderin N, Gbèhounou G, Ahanchede A, Touré A, Kyalo G, Kiepe P (2011) *Rhamphicarpa fistulosa*, a parasitic weed threatening rain-fed lowland rice production in sub-Saharan Africa – a case study from Benin. Crop Prot 30:1306–1314

Román B, Satovic Z, Alfaro C, Moreno MT, Kharrat M, Pérez-de-Luque A, Rubiales D (2007) Host differentiation in *Orobanche foetida* Poir. Flora (Jena) 202:201–208

Sauerborn J (1991) The economic importance of the phytoparasites *Orobanche* and *Striga*. In: Ransom JK, MusselmanLJ, Worsham AD, Parker C (eds) Proceedings of the fifth symposium on parasitic weeds, Nairobi, Kenya, 24–30 June 1991, pp 137–143

Schaffer AA, Jacobsohn R, Joel DM, Eliassi E, Fogelman M (1991) Effect of broomrape (*Orobanche* spp.) infection on sugar content of carrot roots. HortScience 26:892–893

Schneeweiss GM (2007) Correlated evolution of life history and host range in the nonphotosynthetic parasitic flowering plants *Orobanche* and *Phelipanche* (Orobanchaceae). J Evol Biol 20:471–478

Schneeweiss GM (2012) List of actual and potential broomrape pest species. University of Vienna, Plant Biogeography Group Available via http://www.botanik.univie.ac.at/plantchorology/documents/PestSpeciesOrobanche.pdf

Singh BB, Olufajo OO, Ishiyaku MF, Adeleke RA, Ajeigbe HA, Mohammed SG (2006) Registration of six improved germplasm lines of cowpea with combined resistance to *Striga gesnerioides* and *Alectra vogelii*. Crop Sci 46:332–333

Tasker AV, Westwood JH (2012) The U.S. witchweed eradication effort turns 50: a retrospective and look-ahead on parasitic weed management. Weed Sci 62:267–268

Teryokhin E (1997) Weed broomrapes – systematics, ontogenesis, biology and evolution. Aufstieg, Germany

Těšitel J, Plavcová L, Cameron DD (2010) Heterotrophic carbon gain by the root hemiparasites, *Rhinanthus minor* and *Euphrasia rostkoviana* (Orobanchaceae). Planta 231(5):1137–1144

Thorogood CJ, Rumsey FJ, Hiscock SJ (2009) Host-specific races in the holoparasitic angiosperm *Orobanche minor*: implications for speciation in parasitic plants. Ann Bot 103(7):1005–1014

Timus A, Croitoru N (2007) The state of tobacco culture in Republic Moldova and phytosanitary problems of tobacco production. Rasteniev'dni Nauki 44(3):209–212

Tyšer L, Hamouz P, Nováková K, Brant V (2005) Species richness and weed composition of agrophytocenoses in selected agricultural companies with conventional and organic farming systems. Herbologia 6(3):1–7

Uludag A, Nemli Y (2009) Parasitic flowering plants in Turkey. In: Rubiales D, Westwood J, Uludag A (eds) Proceedings of the 10th world congress of parasitic plants, 8–12 June 2009, Kusadasi, Turkey, p 57. Abstract 42. International Parasitic Plant Society, via http://www.parasiticplants.org/docs/IPPS_10th_Congress_Abstracts_Kusadasi_Turkey.pdf

USDA/GRIN 2011 GRIN taxonomy for plants. U.S. Department of Agriculture, Agricultural Research Service. Via http://www.ars-grin.gov/cgi-bin/npgs/html/tax_search.pl

Vaz Patto MC, Díaz-Ruiz R, Satovic Z, Román B, Pujadas-Salvá AJ, Rubiales D (2008) Genetic diversity of Moroccan populations of *Orobanche foetida*: evolving from parasitising wild hosts to crop plants. Weed Res 48:179–186

Warren P (2006) The branched broomrape eradication program in Australia. In: Preston C, Watts JH, Crossman ND (eds) 15th Australian weeds conference, Adelaide, September 2006, pp 610–613

Westbury DB (2004) *Rhinanthus minor* L. Biological Flora of the British Isles. J Ecol 92:906–927

Westbury DB, Davies A, Woodcock BA, Dunnett NP (2006) Seeds of change: the value of using *Rhinanthus minor* in grassland restoration. J Veg Sci 17:435–446

Yakutkin VI, Budrevskaya IA (2006) Area and zones of severity of sunflower broomrape. Interactive agricultural, ecological atlas of Russia and neighbouring countries. Economic plants and their diseases, pests and weeds. Via AgroAtlas http://www.agroatlas.ru/en/about/

Zonno MC, Montemurro P, Vurro M (2000) (*Orobanche ramosa*, un'infestante parassita in espansione nell'Italia meridionale). Informatore Fitopatalogico 50:13–21 (in Italian)

Chapter 19
Population Diversity and Dynamics of Parasitic Weeds

Belén Román

19.1 Introduction

Understanding population diversity within and between populations is essential if selection programmes are to target the sources of host resistance to parasite populations in different geographic areas (see Chap. 21 for the breeder's perspective). Comparative studies of the genetic diversity of parasitic biotypes in natural and agricultural habitats are also important for understanding the evolutionary path from wild parasitic plants to aggressive parasitic weeds. These facilitate in assessing the risk of the appearance of new parasite genotypes that are capable of parasitising a non-host crop.

The genetic diversity of plant populations is determined by population dynamics which include the spatial and temporal variation of population size and density. Quantitative descriptions of changes in the number of populations and in their pattern of growth or decline are useful for quantifying trends in population dynamics. Population dynamics also investigate the biological processes and physical factors causing changes in diversity that determine the genetic structure of a population. Better understanding of these causes can provide a general framework for strategies to control the parasites. Altogether, demographic studies are used to predict future infestations, to aid in decisions on strategies for parasitic weed control and to evaluate the effectiveness of long-term control measures.

B. Román (✉)
IFAPA Alameda del Obispo, Plant Breeding and Biotechnology, Apdo. 3092, 14080 Córdoba, Spain
e-mail: mariab.roman@juntadeandalucia.es

D.M. Joel et al. (eds.), *Parasitic Orobanchaceae*,
DOI 10.1007/978-3-642-38146-1_19,

19.2 Genetic Diversity and Population Dynamics

Diversity studies within and among populations of weedy parasites were first based on the parasite morphology. But parasitism has led to morphological reduction in the Orobanchaceae, so that leaves are sometimes replaced by bracts and the root system can be reduced (Musselman 1994). This results in a loss of morphological features, which allow distinguishing among species and even more so—among populations. Host range and host preference have also been used to differentiate populations of weedy parasites and races (Cubero and Moreno 1979; Radwan et al. 1988; Joel 2000; see also Sect. 19.3.3). In this way, five races of *Orobanche cumana* were identified by using a set of sunflower differentials (Vrânceanu et al. 1986; see Sect. 21.3.3). Variations in aggressiveness among *O. crenata* populations attacking faba bean (*Vicia faba*) and vetch were similarly reported (Cubero and Moreno 1979; Radwan et al. 1988; Joel et al. 2000).

However, the question whether observed variations are genetically determined or induced by environmental conditions is best answered using molecular techniques. Molecular techniques (Table 19.1) complement traditional methods for population genetic analysis, and some of them have been applied to parasitic plant populations:

- **Isozymes** were the first molecular markers used for diversity studies.
- **Random amplified polymorphic DNA** markers (RAPD). Compared to isozymes and SSRs, RAPD markers have limitations such as marker allele dominance and poor reproducibility.
- **Inter-simple sequence repeat** markers (ISSRs) are dominant markers and considered more reliable and robust than RAPDs, possibly because their primers are longer and hence polymerase chain reaction conditions are more stringent.
- **Amplified fragment length polymorphism** markers (AFLPs) show a higher capability of discriminating a large number of reproducible loci.
- The most powerful approach to characterise genetic diversity employs **simple sequence repeat** markers (SSRs, microsatellites) that are robust, reproducible, neutrally evolving and codominant markers.

The production of large numbers of minuscule seeds is an important factor affecting population dynamics of parasitic plants (see Chap. 8). This facilitates a rapid increase in parasite population density following the initial infestation. Large and durable seed banks, with seeds that often remain viable for decades in the field, provide the parasite with the potential for genetic adaptability to changes in host resistance and cultural practices. Seed persistence in soil is another fundamental character that enables weed species to survive in agroecosystems, where the soil is frequently disturbed (Roberts 1981). The seed bank reservoir is affected by seed longevity and dormancy, by variation in seed numbers at various soil depths, by soil composition and by agronomic practices (see also Sects. 19.3.4 and 22.1 for discussion of seed production, dispersal and demise).

Table 19.1 Molecular techniques used to differentiate among weedy Orobanchaceae

Techniques	Species differentiated	References
Isozymes	*Orobanche crenata*	Verkleij et al. (1986)
	Phelipanche aegyptiaca	Verkleij et al. (1986)
	O. cumana	Castejón-Muñoz et al. (1991a)
	O. cumana	Ivanov et al. (1998)
	Striga hermonthica	Bharathalakshmi et al. (1990)
RAPD	*O. crenata*	Román et al. (2001)
	O. cumana	Gagne et al. (1998)
	O. foetida	Román et al. (2007a)
	O. gracilis	Román et al. (2007b)
	P. ramosa	Brault et al. (2007)
	S. aspera	Aigbokhan et al. (2000)
	S. hermonthica	Aigbokhan et al. (2000)
ISSR	*O. crenata*	Román et al. (2002)
	O. minor	Westwood and Fagg (2004)
	P. ramosa	Benharrat et al. (2005)
	P. ramosa	Buschmann et al. (2005)
AFLP	*O. cumana*	Gagne et al. (2000)
	O. foetida	Vaz Patto et al. (2008)
	S. gesnerioides	Botanga and Timko (2005)
	S. asiatica	Botanga et al. (2002)
SSR	*S. hermonthica*	Estep et al. (2011)

19.3 Impacts of Life History on Population Demography and Genetics

Parasite species have a range of life-history strategies that affect population dynamics and, through this, affect the genetic composition and spatial structure of their populations. The following traits strongly influence the genetic structure and evolutionary trajectories of weedy parasites.

19.3.1 Mating System

Mating systems, varying from strict **inbreeding** to obligate **outcrossing**, affect the amount and partitioning of genetic diversity within and among populations. Self-pollinating species have less gene flow among populations via pollen than do mixed-mating or outcrossing species. Self-pollination promotes a more rapid differentiation among populations leading to distinct biotypes. In contrast, in mixed-mating and outcrossing species, the among-population differences are less marked (Hamrick and Godt 1989; Sweigart and Willis 2003). Consequently, outcrossing species should have higher proportions of polymorphic loci, more alleles per polymorphic locus and more genetic diversity (Hamrick and Nason 1996; Dubois et al. 2003). It is likely that the overall genetic diversity in self-pollinating species would be limited compared with that in outcrossing species, because a novel mutation arising in a population of a self-pollinating species has a lower probability of spreading to other populations than do mutations in outcrossing species, even if the allele reaches a high frequency within the population.

Molecular studies have verified the relation between genetic diversity of weedy parasites and their mating system. For example, a high genetic differentiation among populations of predominantly self-pollinating species was found by RAPD markers in *O. cumana* (Gagne et al. 1998), which has a low rate of outcrossing (Ivanov et al. 1998, and see Sect. 16.5). A high genetic differentiation among populations was similarly found with predominantly self-pollinating *P. ramosa* (Vaz Patto et al. 2009). The extremely low genetic AFLP variation among *S. gesnerioides* individuals in central Florida was attributed to inbreeding. The flowers of *S. gesnerioides* form a pollen plug, which precludes pollen dispersal and prevents a significant level of outcrossing, unlike most other *Striga* species where pollen is available to insect pollinators (Botanga and Timko 2005). In contrast, only 24 % of the total ISSR marker diversity of the outcrossing *O. crenata* was attributable to divergence between Spain and Israel, despite the 71 % of within-population diversity (Román et al. 2002).

19.3.2 Transmission and Dispersal

Understanding how parasitic plants move within and between host populations is similarly important for accurately interpreting patterns of genetic diversity. Parasites that are broadly dispersed (and thus have a high degree of gene flow) should have higher within-population genetic diversity than those with more limited dispersal. Differences in seed transmission can affect the genetic diversity and population structure. *Orobanche* and *Striga* seeds are very small, and wind and water are their major natural dispersal mechanisms (Berner et al. 1994; Ginman 2009). Patterns of dispersal in wind-dispersed species are likely to be highly stochastic and dependent on prevailing environmental conditions.

Nonetheless, dispersal associated with agricultural practices is the major relevant factor affecting populations of weedy parasites (see Sect. 22.1.1). As an example, the high variability (95 %) found among *O. crenata* individuals within a population collected from widely separated faba bean fields in the south of Spain was attributable to continuous gene migration caused by continuous dispersal of the parasite seeds by humans, machinery, animals, water and wind, as well as on host seeds (Román et al. 2001).

Geographic distance provides a barrier to gene flow when there is no commercial exchange of host seeds between regions, thus promoting genetic differentiation among regions. Whereas a low level of differentiation was found among the Spanish populations of *O. crenata* (Román et al. 2001), a clear genetic differentiation was found between the distant Spanish and Syrian populations by isozyme analysis (Verkleij et al. 1991) and between the distant Spanish and Israeli populations by ISSR markers (Román et al. 2002). It was also possible to detect differences among *P. aegyptiaca* (Joel et al. 1998) and among *S. asiatica* populations (Botanga et al. 2002) that are separated by only small distances, using RAPD and AFLP markers. In contrast, a relatively uniform low level of genetic diversity was found among 17 populations of *S. asiatica* and 24 populations of *S. hermonthica* studied in Kenya with AFLP markers, and there was no evidence of isolation by distance in any populations of the two species (Gethi et al. 2005).

Effective dispersal (and establishment) of parasites with a high degree of host specificity can only occur at any meaningful scale if there is a susceptible host in the new location. For that reason, long-distance dispersal events are likely to be more evident for generalist parasites that can infest a wide spectrum of hosts, unless there is widespread cultivation of a single crop (e.g. maize in large parts of Africa). Thus, parasites with low host specificity should have stronger patterns of spatial genetic structure and isolation by distance.

Founder events are to be expected following the introduction of infestations into new, previously uninfested areas. The founder effect is the lack of genetic variation that occurs when a new population is established by a very small number of individuals from a larger population. As a result of the decrease of genetic variation, the new population has a lower genetic diversity. Because of genetic drift, the new population may be distinctively different, both genetically and phenotypically, from the parent population from which it is derived. In extreme cases, the founder effect can lead to the speciation and subsequent evolution of new species (Provine 2004). The geographic ranges of most of the species of *Orobanche*, particularly those attacking crops, have almost certainly expanded dramatically in historical times and even more so with globalisation and climate change.

All populations outside of the centre of origin of a species would probably have less genetic variability than those in the centre of origin. The genetic uniformity of the introduced population of *S. asiatica* in North Carolina is a classic example of the founder effect in parasitic weeds (Werth et al. 1984). Similarly, a founder effect was suggested as the reason for the differences in variability within five populations of *O. cumana* in Spain (Castejón-Muñoz et al. 1991a). It was argued that the higher diversity of the parasite populations infesting confectionery sunflower, which is highly susceptible to the parasite, compared to the diversity of the other four populations could be attributed to the number of years that the susceptible sunflower crop grew in that area and consequently to the very high infestation severity. The lower variability, which was mainly manifested by the loss and fixation of some alleles in the other four populations, was due to a more recent origin from only few individuals that were transferred on sunflower achenes (Castejón-Muñoz et al. 1991b).

ISSR analysis of five *O. minor* populations in the USA revealed a low level of polymorphism, with individuals within populations having nearly all SSR fragments in common (Westwood and Fagg 2004). The reason for this low diversity may be that the populations originated from just a few founder plants. *O. minor* is native to Europe, with early introductions in the USA through ballasts of ships (Frost and Musselman 1980), or possibly in fodder or bedding for livestock or in contaminated crop seed. Two clearly different groups of populations were detected in the USA, implying that the populations originated from two separate introduction events. Similarly, there is a very low molecular differentiation among Spanish *O. crenata* populations, in contrast to *O. crenata* populations from eastern Mediterranean where this species is more widespread (Román et al. 2002). The lack of significant differences among the Spanish populations may indicate that only a single source gave rise to these populations. However, such low diversity within populations could also be due to commonly occurring apomixis in the Orobanchaceae (see Chap. 8).

Cluster analysis grouped all four *S. gesnerioides* populations from central Florida (USA) into a single group differentiating them from a separate group isolates attacking *Indigofera hirsuta* and cowpea (*Vigna unguiculata*) in West Africa. The very high level of genetic uniformity observed within and among the Florida populations suggests that there was probably a strong host-driven selection for genetic uniformity, in addition to inbreeding (Botanga and Timko 2005). Since the geographic area into which the parasite was introduced is a small area, it is likely that the uniformity is due to a single introduction.

19.3.3 Host Preference and Virulence

Host-induced selection is probably the most important selective force in parasitic weeds. There is considerable variation in host specificity among parasitic plants, and different host cultivars vary in their susceptibility to different isolates of the parasite. Genetic diversity studies within parasite species may allow the following: (a) detecting the existence of host-preference differentiation owing to host-induced selection (crop or wild host) and (b) characterisation of parasite races rather than populations, determining the relationship between their genetic variation and virulence. Molecular studies should eventually provide molecular markers for these groups of parasite populations.

19.3.3.1 Host-Induced Selection

An important aspect concerning host differentiation processes is the existence of differentiation owing to the host-induced selection and the possible adaptation of wild parasitic species to cultivated plants. Theoretically, host selection could act upon just a single gene or very small portions of the genome, while the rest of the genome is predominantly shaped by other evolutionary forces, namely, recombination (by pollen flow) and migration (by seed dispersal). The *O. foetida* populations infesting chickpea (*Cicer arietinum*) and those infecting faba bean in Tunisia have significant divergence at the molecular level (Román et al. 2007a). This parasite was only recently described from cropping lands (Kharrat et al. 1992; Rubiales et al. 2005), and the specialisation process seems to be the consequence of the strong selection pressure by the different crops. This type of differentiation has also been described in native populations of *O. minor* where ISSR markers provided preliminary evidence of host-driven divergence of the coastal clade *O. minor* ssp. *maritima* growing on sea carrot (*Daucus carota* ssp. *gummifer*) from the host-generalist lineage *O. minor* var. *minor* growing on clover (*Trifolium pratense*) (Thorogood et al. 2008, 2009). Two main causes were proposed for this host selection: (a) a distinct difference between these taxa in response to germination stimulants and (b) differential spread, size and growth rate of host root systems. A fast-growing root system is more likely to encounter parasite seeds in soil. Similarly, there is a greater RAPD marker affinity between the host species and the *P. ramosa* pathotypes, in both intensity and kinetics of infestations (Brault et al. 2007). *S. asiatica* populations in Benin, which were more adapted to maize than sorghum, were likewise distinguished by AFLP markers (Botanga et al. 2002).

19.3.3.2 Tracing the Origin of New Populations

Genetic diversity studies can also infer the origin of new parasitic populations infecting a crop host. The genetic diversity of an *O. foetida* population infecting cultivated vetch was compared, using AFLP markers, to the diversity of four populations infecting wild *Scorpiurus muricatus* and *Ornithopus sativus* in the same region. The vetch-infesting population was closer to native populations infesting *S. muricatus*, whereas the population collected on *O. sativus* was the most divergent one, suggesting that it is not a new introduction to the region and that the wild population of *O. foetida* attacking *S. muricatus* gave rise to a new population that is able to infect the crop (Vaz Patto et al. 2008).

Comparative studies between parasitic plant populations attacking wild species and those growing on crops from the same region may clarify host specialisation. Molecular diversity studies can also help to determine the risk of appearance of a new race capable of parasitising a particular crop.

19.3.3.3 Parasite Races

The existence of races in a parasitic plant species is determined by the differential aggressiveness of their populations against cultivars, landraces or breeding lines of a particular host crop. Several outbreaks of new races of parasitic plants were described: races of *O. cumana* on sunflower (Vrânceanu et al. 1980; Melero-Vara et al. 2000; Eizenberg et al. 2004), of *O. foetida* on faba bean (Kharrat et al. 1992), of *O. crenata* on vetch (Joel 2000), of *P. ramosa* on tobacco (Buschmann et al. 2005) and of *S. gesnerioides* on cowpea (Noubissie Tchiagam et al. 2010). Screening host germ plasm for resistance to broomrape lines requires characterisation of pathogen populations and genotypes as a prerequisite.

The existence of races poses an enormous challenge to breeders developing resistant cultivars. Molecular markers have been used to identify races of various parasitic weeds. Genetic diversity studies with ISSR markers allowed characterisation of two *P. ramosa* populations with different levels of pathogenicity (Buschmann et al. 2005). The genetic variability of various races of *S. gesnerioides* on cowpea was analysed using AFLP markers. It was possible to distinguish individuals within and among populations of each race and to identify specific molecular markers assisting in such differentiation. Two new races of the parasite were identified on the basis of genomic profiles and on the differential ability to parasitise specific host cultivars, suggesting that both geographic isolation and host-driven selection are critical factors defining race formation (Botanga and Timko 2006).

The evolution of new races often overcomes crop resistance to the parasite. The genetic resistance of a host crop will often be effective only as long as a new parasite race is not present. Moreover, genetic resistance bred in a host crop will often only be effective against the parasite race for which it was developed, and it can be overcome by different parasite populations in different regions (Pérez-de-Luque et al. 2009). A correlation between the level of virulence and molecular diversity of parasitic weed populations distinguishes ecotypes in the context of plant breeding programmes and can facilitate breeding multigenic resistance with a potentially longer duration of sustainability.

19.3.4 Changing Opportunities Imposed by Agriculture

The vastly different ecological constraints and changing opportunities imposed by agricultural plant communities have substantial effects on increasing genetic diversity and dynamics of parasitic weed populations compared with balance natural habitats. The persistent seed banks, forming the genetic population structure at any time, result from a long history of agricultural practices. The agricultural dispersal of parasitic weed seeds is mainly facilitated by humans and agricultural implements and by animals, water and wind (see Sect. 22.1).

Parasitic weed populations undergo regular changes in allele frequencies due to selective pressures by weed control measures, by soil cultivation and by the crops grown in the field, especially when crops are rotated (Satovic et al. 2009). This is reflected in genetic studies of parasitic weeds (see Sect. 19.3.2). Nonetheless, plants sampled in a field could be genetically variable even in self-pollinating or even apomictic species because of seed input from other populations. Since an increasing global crop-seed exchange and transport play an important role in migration of weed seeds, geographic differentiation would be difficult to discern because the population substructure may not depend solely on spatial distances or local barriers of gene flow as in natural ecosystems. This global mode of dissemination and the resulting infestation of new areas are of special relevance for seed companies and other institutions supplying crop seeds, as well as for seed transfer and quarantine regulation.

Recently developed molecular assays allow rapid and high-throughput detection of parasite seeds in crop-seed lots, e.g. by using targeted ITS sequences in qPCR assays to quantify *Phelipanche and Orobanche* seeds (Dongo et al. 2012; see discussion in Chap. 20). Such assays can be developed for all parasitic species and should be used on all crop seeds coming from infested areas.

19.4 Future Prospects

Codominant markers such as microsatellites, which have high polymorphism indices, are clearly needed for more accurate population genetic studies on weedy Orobanchaceae. Comparative studies of the genetic structures of parasite species that differ in key features of their important life-history traits will be of particular value, using similar sampling designs and genetic markers. Moreover, it would be interesting to monitor the changes in genetic diversity of populations sampled in the same field for several consecutive years, with analysis of the impact of the crops grown in the field and the control measures that are applied there.

References

Aigbokhan E, Berner D, Musselman L, Mignouna H (2000) Evaluation of variability in *Striga aspera*, *Striga hermonthica* and their hybrids using morphological characters and random amplified polymorphic DNA markers. Weed Res 40:375–386

Benharrat H, Boulet C, Theodet C, Thalouarn P (2005) Virulence diversity among branched broomrape (*Orobanche ramosa* L.) populations in France. Agron Sustain Dev 25:123–128

Berner DK, Cardwell F, Faturori BO, Ikie FO, Williams OA (1994) Relative roles of wind, crop seeds and cattle in dispersal of *Striga* spp. Plant Dis 78:402–406

Bharathalakshmi, Werth CR, Musselman LJ (1990) A study of genetic diversity among host-specific populations of the witchweed *Striga hermonthica (Scrophulariaceae)* in Africa. Plant Syst Evol 172:1–12

Botanga CJ, Timko MP (2005) Genetic structure and analysis of host and non-host interactions of *Striga gesnerioides* (witchweed) from Central Florida. Phytopathology 95:1166–1173

Botanga CJ, Timko MP (2006) Phenetic relationships among different races of *Striga gesnerioides* (Willd.) Vatke from West Africa. Genome 49:1351–1365

Botanga CJ, Kling JG, Berner DK, Timko MP (2002) Genetic variability of *Striga asiatica* (L.) Kuntz based on AFLP analysis and host-parasite interaction. Euphytica 128:375–388

Brault M, Betsou F, Jeune B, Tuquet C, Sallé G (2007) Variability of *Orobanche ramosa* populations in France as revealed by cross infestations and molecular markers. Environ Exp Bot 61:272–280

Buschmann H, Gonsior G, Sauerborn J (2005) Pathogenicity of branched broomrape (*Orobanche ramosa)* populations on tobacco cultivars. Plant Pathol 54:650–656

Castejón-Muñoz M, Suso MJ, Romero-Muñoz F, García-Torres L (1991a) Isoenzymatic study of *Orobanche* (*Orobanche cernua*) populations infesting sunflower (*Helianthus annuus*). In:

Ransom JK, Musselman LJ, Worsham AD, Parker C (eds) International symposium on parasitic weeds. CIMMYT, Nairobi, Kenya, pp 313–319

Castejón-Muñoz M, Romero-Muñoz F, García-Torres L (1991b) *Orobanche cernua* seed dispersal through sunflower achenes. Helia 14:51–54

Cubero JI, Moreno MT (1979) Agronomical control and sources of resistance in *Vicia faba* to *Orobanche crenata*. In: Bond DA, Scarascia-Mugnozza GT, Poulsen MH (eds) Some current research on *Vicia faba* in Western Europe. Commission of the European Communities, Luxemburg, pp 41–80

Dongo A, Leflon M, Simier P, Delavault P (2012) Development of a high-throughput real-time quantitative PCR method to detect and quantify contaminating seeds of *Phelipanche ramosa* and *Orobanche cumana* in crop seed lots. Weed Res 52:34–41

Dubois S, Cheptou PO, Petit C, Meerts P, Poncelet M, Vekemans X, Lefèbvre C, Escarré J (2003) Genetic structure and mating systems of metallicolous and nonmetallicolous populations of *Thlaspi caerulescens*. New Phytol 157:633–641

Eizenberg H, Plakhine D, Landa T, Achdari G, Joel DM, Hershenhorn J (2004) First report of a new race of sunflower broomrape (*Orobanche cumana*) in Israel. Plant Dis 88:1284

Estep MC, Van Mourik T, Muth P, Guindo D, Parzies HK, Koita OA, Weltzien E, Bennetzen JL (2011) Genetic diversity of a parasitic weed, *Striga hermonthica,* on sorghum and pearl millet in Mali. Trop Plant Biol 4:91–98

Frost CC, Musselman LJ (1980) Clover broomrape in the United States. Weed Sci 28:119–122

Gagne G, Roeckel-Drevet P, Grezes-Besset B, Shindrova P, Ivanov P, Grand- Ravel C, Vear F, Tourvieille de Labrouhe D, Charmet G, Nicolas P (1998) Study of the variability and evolution of *Orobanche cumana* populations infesting sunflower in different European countries. Theor Appl Genet 96:1216–1222

Gagne G, Roeckel-Drevet P, Grezes-Besset B, Shindrova P, Ivanov P, Grand-Ravel C, Vear F, Charmet G, Nicolas P (2000) Amplified fragment length polymorphism (AFLP) as suitable markers to study *Orobanche cumana* genetic diversity. J Phytopathol 148:457–459

Gethi G, Smith M, Mitchell S, Kresovich S (2005) Genetic diversity of *Striga hermonthica* and *Striga asiatica* populations in Kenya. Weed Res 45:64–73

Ginman E (2009) Dispersal biology of *Orobanche ramosa* in South Australia. MSc Thesis, The University of Adelaide

Hamrick JL, Godt MJW (1989) Allozyme diversity in plant species. In: Brown AHD, Clegg MT, Kahler AL, Weir BS (eds) Plant population genetics, breeding and genetic resources. Sinauer, Sunderland, MA, pp 43–63

Hamrick JL, Nason JD (1996) Consequences of dispersal in plants. In: Rhodes EO, Chesser RK, Smith MH (eds) Population dynamics in ecological space and time. University of Chicago Press, Chicago, IL, pp 203–236

Ivanov P, Shindrova P, Ivanova I (1998) An isoenzyme analysis of the North-East Bulgarian *Orobanche cumana* populations. In: Wegmann K, Musselman LJ, Joel DM (eds) Current problems of *Orobanche* researches. Proceedings of the 4th international workshop on *Orobanche*. 'Dubroudja', Albena, Bulgaria, pp 87–94

Joel DM (2000) A new virulent race of *Orobanche crenata*. Haustorium 37:2

Joel DM, Benharrat H, Portnoy H, Thalouarn (1998) Molecular markers for *Orobanche* species – new approaches and their potential uses. In: Wegmann K, Musselman LJ, Joel DM (eds) Current problems in *Orobanche* research. Proceedings of the 4th international *Orobanche* workshop. Albena, Bulgaria, pp 115–124

Kharrat M, Halila MH, Linke KH, Haddar T (1992) First report of *Orobanche foetida* Poiret on faba bean in Tunisia. FABIS Newsl 30:46–47

Melero-Vara JM, Domínguez J, Fernández-Martínez JM (2000) Update on sunflower broomrape situation in Spain: racial status and sunflower breeding for resistance. Helia 23:45–56

Musselman LJ (1994) Taxonomy and spread of Orobanche. In: Pieterse AH, Verkleij JAC, ter Borg SJ (eds) Biology and management of Orobanche. Proceedings of the third international workshop on *Orobanche* and related *Striga* research. Royal Tropical Institute, Amsterdam, The Netherlands, pp 27–35

Noubissie Tchiagam JB, Bell JM, Guissai Birwe S, Gonne S, Youmbi E (2010) Varietal response of cowpea (*Vigna unguiculata* (L.) Walp.) to *Striga gesnerioides* (Willd.) Vatke Race SG5 Infestation. Notulae Botanicae Horti Agrobotanici Cluj-Napoca 38(2):33–41

Pérez-de-Luque A, Fondevilla S, Pérez-Vich B, Aly R, Thoiron S, Simier P, Castillejo MA, Fernández-Martínez JM, Jorrín J, Rubiales D, Delavault P (2009) Understanding *Orobanche* and *Phelipanche*-host plant interaction and developing resistance. Weed Res 49:8–22

Provine WB (2004) Ernst Mayr: genetics and speciation. Genetics 167:1041–1046

Radwan MS, Abdalla MMF, Fischbeck G, Metwally AA, Darwish DS (1988) Variation in reaction of faba bean lines to different accessions of *Orobanche crenata* Forsk. Plant Breed 101:208–216

Roberts HA (1981) Seed banks in the soil, vol 6, Advances in applied biology. Academic, Cambridge, 55 p

Román B, Rubiales D, Torres AM, Cubero JI, Satovic Z (2001) Genetic diversity in *Orobanche crenata* populations from southern Spain. Theor Appl Genet 103:1108–1114

Román B, Satovic Z, Rubiales D, Torres AM, Cubero JI, Katzir N, Joel DM (2002) Variation among and within populations of the parasitic weed *Orobanche crenata* from Spain and Israel revealed by inter simple sequence repeat markers. Phytopathology 92:1262–1266

Román B, Satovic Z, Alfaro C, Moreno MT, Kharrat M, Perez de Luque A, Rubiales D (2007a) Host differentiation in *Orobanche foetida* Poir. Flora 202:201–208

Román B, Hernandez R, Pujadas A, Cubero JI, Rubiales D, Satovic Z (2007b) Genetic diversity in two variants of *Orobanche gracilis* Sm. [var. gracilis and var. deludens (Beck) A. Pujadas] from different regions of Spain. Electron J Biotechnol 10:221–229

Rubiales D, Sadiki M, Román B (2005) First report of *Orobanche foetida* on common vetch (*Vicia sativa*) in Morocco. Plant Dis 89:528

Satovic Z, Joel DM, Rubiales D, Cubero JI, Román B (2009) Population genetics in weedy species of *Orobanche*. Australas Plant Pathol 38:228–234

Sweigart AL, Willis JH (2003) Patterns of nucleotide diversity in two species of Mimulus are affected by mating system and asymmetric introgression. Evolution 11:2490–2506

Thorogood CJ, Rumsey FJ, Harris SA, Hiscock SJ (2008) Host-driven divergence in the parasitic plant *Orobanche minor* Sm. (Orobanchaceae). Mol Ecol 17:4289–4303

Thorogood CJ, Rumsey FJ, Harris SA, Hiscock SJ (2009) Gene flow between alien and native races of the holoparasitic angiosperm *Orobanche minor* (Orobanchaceae). Plant Syst Evol 282:31–42

Vaz Patto MC, Díaz-Ruiz R, Satovic Z, Román B, Pujadas-Salvà AJ, Rubiales D (2008) Genetic diversity of Moroccan populations of *Orobanche foetida*: from wild parasitic plants to parasitic weeds. Weed Res 48:179–186

Vaz Patto MC, Fernández-Aparicio M, Satovic Z, Rubiales D (2009) Extent and pattern of genetic differentiation within and between European populations of *Phelipanche ramosa* revealed by amplified fragment length polymorphism analysis. Weed Res 49:48–55

Verkleij JAC, Janssen J, Pieterse AH (1986) A preliminary study on isoenzyme variation in *Orobanche crenata* and *Orobanche aegyptiaca* from Syria. In: Borg SJ (ed) Proceedings of a workshop on the biology and control of *Orobanche*, Wageningen, The Netherlands, pp 154–159

Verkleij JAC, Koevoets P, López-Granados F, Egbers WS, García-Torres L, Pieterse AH (1991) Genetic variability in populations of *Orobanche crenata* from Spain. In: Ransom JK, Musselman LJ, Worsham AD, Parker C (eds) International symposium on parasitic weeds. CIMMYT, Nairobi, Kenya, pp 462–469

Vrânceanu AV, Tudor VA, Stoenescu FM, Pirvu N (1980) Virulence groups of *Orobanche cumana* Wallr. differential hosts and resistance sources and genes in sunflower. In: Proceedings of the 9th international sunflower conference, 8–13 July 1980, Torremolinos, Spain, pp 74–80

Vrânceanu, AV, Pîrvu N, Stoenescu FM, Pacureanu M (1986) Some aspects of the interaction *Helianthus annuus* L./*Orobanche cumana* Wallr. and its implications in sunflower breeding. In: ter Borg S (ed) Proceedings of a workshop on biology and control of *Orobanche* LH/VPO, Wageningen, The Netherlands, pp 181–189

Werth CR, Riopel JL, Gillespie NW (1984) Genetic uniformity in an introduced population of witchweed (*Striga asiatica*) in the United States). Weed Sci 32:645–648

Westwood JH, Fagg CM (2004) ISSR characterization of *Orobanche minor* populations in the US. In: Joel DM (ed) Proceedings of the eighth international parasitic weeds symposium. IPPS, Durban, South Africa, p 15

Chapter 20
Molecular Diagnosis of Parasite Seed Banks

Jane Prider, Kathy Ophel Keller, and Alan McKay

20.1 Introduction

Persistent soil seed banks provide the primary source for annual weed recruitment in arable land. As seed bank density determines the level of damage that parasitic plants have on their crop hosts (Bernhard et al. 1998; Linke et al. 1991; Manschadi et al. 2001; Schnell et al. 1994), seed bank depletion is an important target for parasitic weed control (see Chap. 22). An estimate of seed bank size is required to understand seed bank dynamics, to evaluate control methods and to predict potential yield losses in susceptible crops. Although the processes of seed bank depletion may be studied using seed bag burial methods (Van Mourik et al. 2005, 2011), measures of in situ seed banks are necessary for other research or management applications (see Sect. 23.6). For quantifying parasitic seed in soil, there are a number of different techniques that require physical extraction of seed from the soil; commonly by combining the soil sample with a solution of high specific gravity, so seeds and other organic matter float to the surface, followed by counting (Ashworth 1976; Kachelriess 1987; Krishnamurthy and Chandwani 1975; Linke et al. 2001; Sauerborn et al. 1991; Van Mourik 2007). These techniques generally do not require specialised equipment or expertise but require much patience. Accuracy can also vary with soil type, and all require some calibration prior to the selection of an appropriate method for a particular situation (Linke et al. 2001). These techniques are also not suitable for use where there is a need to discriminate

J. Prider (✉)
Biosecurity SA, Primary Industries and Resources South Australia, GPO Box 1671, Adelaide, SA 5001, Australia
e-mail: jane.prider@sa.gov.au

K. Ophel Keller • A. McKay
Sustainable Systems, South Australian Research and Development Institute, GPO Box 397, Adelaide, SA 5001, Australia

D.M. Joel et al. (eds.), *Parasitic Orobanchaceae*,
DOI 10.1007/978-3-642-38146-1_20, © Springer-Verlag Berlin Heidelberg 2013

amongst broomrape species and may not be appropriate where a large number of samples need to be screened.

Molecular techniques have been developed to detect and quantify a variety of plant pathogens in soil samples to assess research trials and assist farmers plan their cropping programme (Ophel-Keller et al. 2008). DNA assays can overcome identification issues for organisms with limited morphological features for discrimination, especially at different developmental stages, or limitations imposed by small size. Molecular markers to discriminate amongst broomrape species of agricultural importance have been developed which are suitable for seed isolated from soil (Joel et al. 1998; Osterbauer and Rehms 2002; Portnoy et al. 1997; Roman et al. 2007). DNA of diagnostic quality can be extracted from seeds retrieved from field soil samples (Portnoy et al. 1997) and in a variety of soil types, although the humic substances in soil can inhibit the PCR (Rehms and Osterbauer 2003).

This chapter reports on use of a DNA assay to quantify the *Phelipanche mutelii* seed bank from field samples. The technique has recently been used to detect and quantify *Phelipanche ramosa* and *Orobanche cumana* in seed lots (Dongo et al. 2012), but we report on its use in soil samples. The DNA is extracted directly from the soil sample, thereby maximising seed recovery and reducing sample processing time. The use of a high-throughput DNA extraction system and real-time PCR technology maximises the number of samples that can be processed in a single day, reducing labour costs and turnaround times for reporting. Probes can be designed to discriminate to a high level of taxonomic discrimination and enable a single soil sample to be used to detect or quantify several organisms. The sensitivity and versatility of the technique make it suitable for research and management applications.

20.2 Sample Collection

The field sampling stage is the most critical source of sampling error in the procedure for estimating seed bank size. Consideration needs to be given to the volume of the sample; the shape and size of the sampler; sampling location; and the number of samples collected.

20.2.1 Sample Volume

The ability to detect seed is constrained by the sample size that can be processed for DNA extraction and the sensitivity of the DNA assay. The DNA extraction system used in these studies can process soil samples up to 550 g dry weight; commercial kits often use samples less than 10 g (Haling et al. 2011; Ophel-Keller et al. 2008).

Soil samples are typically collected using an augur or corer. A single core can form a sample or multiple cores can be combined to form a composite sample.

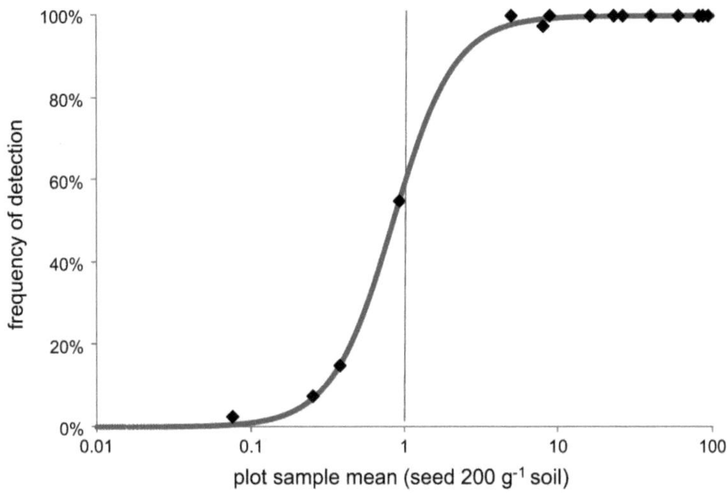

Fig. 20.1 The effect of seed density on the probability of detection in a sampled plot. Each point is the mean number of *Phelipanche mutelii* seeds detected in a 200 g soil sample versus the frequency that at least one seed was detected in each of the 20 replicates. Each sample is a composite of 20 cores

20.2.2 Probability of Detection

Low-density seed banks with a non-uniform distribution are difficult to sample with precision. In testing assay sensitivity, where the seed density approaches 10 seeds 200 g^{-1} soil, the probability that broomrape will be detected with the assay approaches one (Fig. 20.1). Where there are less than 0.8 seeds 200 g^{-1} soil, the probability of detection falls to 0.5 (i.e. at least half of the samples collected will test negative for broomrape presence). In fields where broomrape is at low density, more samples are required to increase the probability of detection.

20.2.3 Core Size

Smutny and Kren (2003) reported that larger cores can give more precise estimates of less common seeds in the soil seed bank with fewer samples. We compared two core sizes, 13 mm (composites of 25 cores) and 50 mm (composites of five cores), and a larger linear sample that was later subsampled. Each method gave different estimates of mean seed number and variance. The small cores were the most variable, whilst the larger cores gave a smaller estimate of mean seed number. The linear sampler gave the most consistent results (lowest coefficient of variation), but samples were very time consuming to process using this method. Although more cores are required, the most efficient method is a composite sample of many small cores. The small cores are easy to insert into the soil, and samples are collected over a broader area, which is important where seeds are patchy in space. As subsampling can introduce an extra source of error unless the sample is thoroughly mixed, the entire composite sample is used for the assay.

20.2.4 Sampling Location

The positioning of sampling points can affect results. Cores can be collected at random positions within each plot or field, but systematic samples may be preferable and more efficient if there is no underlying spatial pattern in seed distribution or the cores are taken far apart (Ambrosio et al. 2004). Infested areas of a field may be targeted for some applications.

The distribution of seeds within a field or plot is most likely clustered at small scales and becomes more variable at larger scales as for *Striga* seed (Van Delft et al. 1997). Our studies found that 90 % of seeds fall within 0.38 m of the source plant in a standing cereal crop and 0.76 m in a low medic pasture (Ginman 2009). With low plant densities and in the absence of cultivation, the distribution of seeds at scales larger than 1 m^2 is likely to be highly variable. For field sampling, small cores for a composite sample are collected from a 1 m grid in 5 m × 5 m plots, whereas a random method of core positioning is used for smaller research plots.

20.2.5 Number of Samples

Sample size will depend on seed density and the spatial distribution of seed in soil. Seed bank samples fit either a Poisson or a negative binomial distribution and are rarely normally distributed; therefore, a large sample size is required for precision. With compound samples, large numbers of cores are required before the sample mean approaches a normal distribution (Dessaint et al. 1992). The degree of precision required will depend on the purpose of the sampling, with a high degree of precision required for research purposes where the data are subject to statistical testing.

Consideration of the logistics of sample collection and the spatial variability of the seed bank led to the strategy of collecting a composite of many small samples to produce a single 500 g sample. In this way, a larger area of the experimental plot or field can be sampled, and seeds that are either clustered or in low densities are more likely to be included. Our research found that few seeds occurred below depths of 100 mm, so a core was adopted that was 100 mm in depth and 13 mm in diameter. To collect a 500 g soil sample, 25 such cores are collected to comprise a single composite sample used for the DNA assay.

There was a need to determine how many composite cores are required to minimise variability in seed bank estimates. A bootstrap method was used to resample from 20 samples, each a composite of 25 cores, from 11 plots that differed

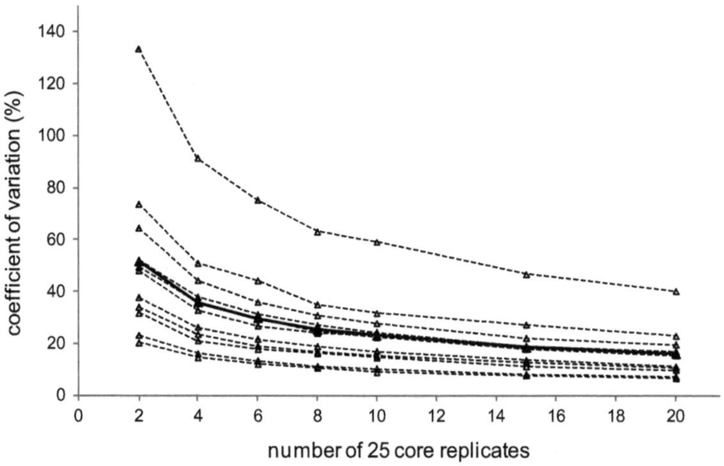

Fig. 20.2 The effect of increased sample size on variability in detected *Phelipanche mutelii* seeds. Each sample is a composite comprising 25 cores (200 g soil samples). Each *line* represents samples collected from a different plot, and the mean across all plots is shown by the bold line

in seed density. The analysis showed that the value of the coefficient of variation was 35 % with four samples per plot and this decreased to 20 % with ten composite samples per plot (Fig. 20.2). This has been used to set replicate sizes for treatment plots in research work. Smaller numbers of samples are suitable where there is likely to be a large difference between treatment plots, e.g. evaluation of soil fumigants. Where smaller treatment effects are expected, then larger numbers of replicates are required. For research, each situation demands a specific sampling protocol to be developed depending on the nature of the site, the size of the plots to be sampled and the acceptable level of variance.

20.3 Test Development

A TaqMan® MGB assay was designed to detect the broomrape infecting fields (putatively *Phelipanche mutelii*) and not *O. minor* or *O. cernua* var. *australiana* that also occur in Australia. The primers and probe designed are based on ITS1 ribosomal region sequences from local isolates and others available in GenBank (Table 20.1). The PCR conditions and production of DNA standards are described in Riley et al. (2010).

Table 20.1 Primers and probe sequences for real-time PCR assays specific for *Phelipanche mutelii*

Target	DNA sequences (5′–3′)
Forward primer	AAAAGAAGTATCTACCCCCCATTGT
Reverse primer	CGTTTTGACATTGGAGAGTGATCT
Probe	6FAM CTACCCGCAAAC MGBNFQ

20.4 Test Validation

The DNA assay is able to detect a single *P. mutelii* seed in a 200 g soil sample. Although the test can process samples up to 550 g, test sensitivity is reduced in larger soil samples and a single seed can be detected in this sample size with 30 % probability.

DNA is degraded quickly in dead organisms in moist soil (Riley et al. 2010), so the assay will not detect dead seed. Immature and non-viable seeds have very little DNA, which needs to be considered when interpreting the results.

To test assay precision, three replicate 1,300 g soil samples were spiked with known numbers of seeds, and 200 g samples were subsampled to compare seed counts with DNA results. There was reasonable congruence between the two estimates of seed numbers (Table 20.2).

20.5 Test Applications

20.5.1 Management

As a management tool, the assay can be used to detect the presence of parasitic weed seeds in a field or to assess crop infection risk relative to seed number or location. Within the area infected by *P. mutelii* in Australia, annual field inspections for emerged plants are conducted to confirm broomrape presence or absence. These surveys are important for sale of commodities to the increasing number of markets that have zero tolerance for broomrape. Repeated surveys are required as the seed bank has long-term persistence and the lack of emerged plants gives no indication of the absence of a broomrape seed bank. Using this survey method the probability of detection in a field can be very low if broomrape population density is low. Low-density, non-uniformly distributed populations are also problematic for the detection of seed in soil samples. As the aim of our programme has been to eradicate broomrape, sampling problems have precluded the use of the test to declare fields as free of broomrape. Currently, searching for emerged plants at the correct time of year is more economical than collecting the potentially large number of soil samples that would be required to evaluate seed presence to the required degree of confidence. However, the development of a protocol for sampling fields for molecular diagnosis of the broomrape seed bank would circumvent problems associated with the time limitations of the survey period. Providing quarantine

Table 20.2 Comparison of molecular and physical methods for quantifying seed banks

		Seeds in 200 g subsamples ($n = 3$)			
		Count		DNA assay	
Seeds in 1,300 g soil	Expected mean	Mean	St dev	Mean	St dev
65	10	12.66	0.58	10	1.41
390	60	53	5.29	77	20.42
3,900	600	595	3.21	579	16.82
13,000	2,000	2,006	94.88	1,913	101.04

Number of *Phelipanche mutelii* seeds estimated by counting after floating seeds from soil and by DNA assay

requirements are met, a DNA assay could be used to declare a field as broomrape-free.

The DNA assay was originally developed as a predictive tool for assessing the potential for pathogenic soil organisms to produce disease in crops (Ophel-Keller et al. 2008). Samples are collected before crops are planted to inform crop selection for a particular field. Samples are classified as low, medium or high disease risk based on estimates of the density of organisms or their propagules in soil samples. Sampling at different locations within fields can also reveal areas of higher disease risk (Heap and McKay 2009). This application is also suitable for use in broomrape-infected soils. Within the region currently infected by *P. mutelii* in South Australia, the risk of productivity loss is low so this potential application has not been fully realised. The tool could also be used to select fields with large broomrape seed banks to target for fumigation.

To assess crop risk there is a need to quantify the relationship between seed bank density and potential crop infection. For *Orobanche crenata*, a strong correlation has been demonstrated between increasing seed bank size and reductions in crop yield (Bernhard et al. 1998; Linke et al. 1991; Manschadi et al. 2001; Schnell et al. 1994). An estimate has been made of the number of seeds per kilogram of soil that results in reductions in crop dry weight or yield (Bernhard et al. 1998). This was found to differ between crops and also in pots and field situations (Linke et al. 1991). For example, moderate yield loss in carrots occurs at *O. crenata* seed densities of 200 seeds kg^{-1}, but more than 250 seeds kg^{-1} results in total pea crop failure. In another study there was no damage to faba bean with *O. crenata* seed densities up to 312 seeds kg^{-1} (Linke et al. 1991). These studies indicate that it is important to establish threshold levels for crops that are specific to broomrape species and include variation that may be expected to occur on different soil types.

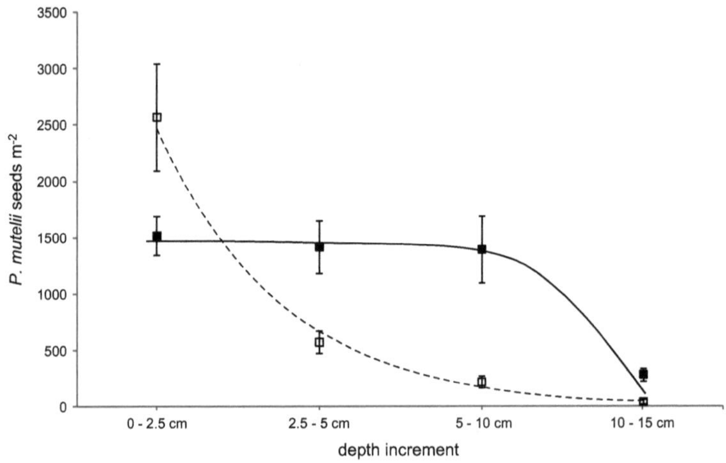

Fig. 20.3 Vertical distribution of *Phelipanche mutelii* seeds down the soil profile. *Dashed line*—fallow field; *solid line*—cultivated field. Means ± 1 SE, $n = 15$

20.5.2 Use in Research

The assay has been a useful tool for research. It provides a rapid means of assessing changes in the broomrape seed bank in relation to experimental treatments and has also allowed us to collect data to characterise the broomrape seed bank.

We have collected soil samples from different depth profiles to describe the vertical distribution of the broomrape seed bank. Samples collected from the same field in regularly cultivated areas and uncultivated areas revealed different distribution curves. Seeds were more evenly distributed in cultivated fields than in uncultivated fields, where the majority of seed occurred near the soil surface (Fig. 20.3). Very few seeds were sampled at depths of 10–15 cm.

The test has been used extensively for the evaluation of products for the destruction of broomrape seed banks (Williams et al. 2006). Soil sampling has often been an issue in these assessments. Samples cannot be collected too soon after the application of products to give the DNA of killed seeds sufficient time to decay, so samples are collected 4 weeks later. Samples must also be free of broomrape vegetative material, which can be problematic for samples collected during the growing season. Sieving of soil samples is used to ensure that this material is not included in assay samples. Spatial variability in the pretreatment seed bank has often reduced the power of statistical tests to detect treatment effects, with issues of small seed bank numbers and lack of seeds in control plots (Williams et al. 2006). As a result of limitations due to the patchiness of the in situ seed bank, we routinely use buried sachets with known numbers of seed for evaluation of seed destruction products.

Assays have also been used to monitor seed bank decline under different cropping rotations. Spatial variability in the seed bank has again reduced the usefulness of these tests in the situation where we have low population density. We have used the assay in experiments testing wind dispersal of seed (Ginman 2009). Molecular analysis of soil samples collected from the end of a wind tunnel yielded comparable seed numbers to those collected and manually counted from Bagnold traps and wind vane traps, which are used to measure sediment transport by wind. We found that even minor amounts of stubble cover reduced the distance that seeds were dispersed, especially at higher wind velocities.

20.6 Other Applications

The ability of the test to detect DNA in organic-rich substrates has increased the range of applications. The test has been used to detect and quantify broomrape seed in spiked samples of sheep egesta. As few as two seeds can be detected in a 100 g sample. The test was also able to give a reasonable estimate of seed numbers in samples spiked with up to 10 seeds 400 g^{-1} egesta. The method was used to measure the gut passage time of broomrape seeds in sheep to assess the risk sheep posed for the dispersal of broomrape seeds (see also Sect. 22.1.1). Sheep were drenched with 1×10^5 broomrape seeds and the egesta collected over the following 9 days. Samples of 100 g of egesta were assayed for broomrape seeds. The analysis revealed a classic gut retention time for broomrape seeds, with numbers in egesta peaking after 2 days and the last seeds detected at 7 days (Fig. 20.4).

The method is also being trialled for detecting broomrape and other weed contaminants in seed lots. The test will utilise seed-cleaning offal, which remains after seed has been cleaned. Current methods rely on manual examination of samples from cleaned seed lots, which are time consuming to process and have detection limitations due to the small portion of a seed lot that is sampled. Contaminants are concentrated in seed-cleaning offal, hence increasing the probability of detection of weed seeds. Dongo et al. (2012) demonstrated the use of quantitative PCR for estimating *Phelipanche ramosa* seed contamination of canola seed lots and *O. cumana* seed contamination of sunflower seed lots. They measured a detection level of 0.1 mg broomrape seed per 200 g crop seed with detection improved by assaying the residue remaining after crop seed had been filtered. These procedures could also be used for sampling broomrape contaminants in other commodities such as fodder.

20.7 Conclusions

The advancements in PCR technology that have allowed automation of procedures have substantially reduced the cost of molecular diagnosis of organisms. It is now possible to offer a commercial diagnostic testing service for the agricultural sector

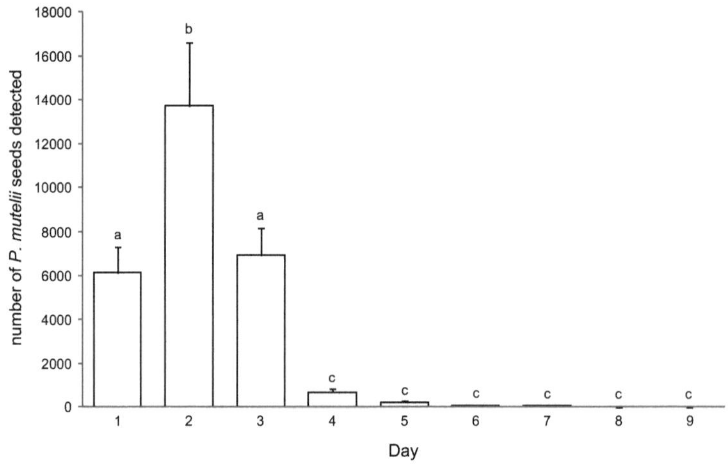

Fig. 20.4 Gut passage time for *Phelipanche mutelii* seeds through sheep. Sheep were drenched with 1.0×10^5 seeds each on day 0. Means ± 1 SE. Day 1–7, $n = 8$. Day 8 and 9, $n = 4$. *Bars* labelled with the same letter are not significantly different at $\alpha = 0.05$ (Tukey's HSD test)

that can be used for risk assessment of disease for management and quarantine applications (Ophel-Keller et al. 2008). Of particular importance for crop planning is the capability of the test to quantify pest numbers. Depending on the economic threshold, a single soil sample can be used to diagnose multiple crop diseases. With broomrape, low density seed banks are significant, so a more intensive sampling strategy will be required especially in large fields. The test may prove more useful in intensively cultivated small fields, where there is a need to assess risks associated with cultivation of high-risk or high-value crops. Correlations between seed bank density and yield loss in crops are lacking for many parasitic weed species which could be assessed using molecular diagnostic techniques.

Provided soil within a field can be sampled adequately, the method offers a means of quantifying seed numbers with at least as much precision as manual methods with the advantage that results can be obtained more quickly and species can be discriminated. With the development of a means of collecting a representative sample within a field, the assay may prove as efficient as a detection survey for emerged plants with the advantage that sampling can be done throughout the year. This has an important application in surveillance for market access purposes.

The assay has numerous applications in research, particularly in the study of seed bank dynamics and seed dispersal. Future uses of the technology target situations where seeds need to be detected and quantified within diverse substrates. The assay has proven robust to detection in organic-rich matrices such as sheep egesta, and detection in plant biomass substrates such as seed, plant litter and fodder is currently under investigation.

Acknowledgements A number of researchers have been involved in the development of the assay and its practical application. We thank Ray Correll, John Virtue, Anna Williams, Emma Ginman, Herdina, Di Hartley and John Heap for the use of unpublished analyses and data.

References

Ambrosio L, Iglesias L, Marín C et al (2004) Evaluation of sampling methods and assessment of the sample size to estimate the weed seedbank in soil, taking into account spatial variability. Weed Res 44:224–236

Ashworth LJ (1976) Quantitative detection of seed of branched broomrape in California tomato soils. Plant Dis Rep 60:380–383

Bernhard RH, Jensen JE, Andreasen C (1998) Prediction of yield loss caused by *Orobanche* spp. in carrot and pea crops based on the soil seedbank. Weed Res 38:191–197

Dessaint F, Barralis G, Beuret E et al (1992) EWRS Collaborative study of seed bank estimation 2. Estimation of the relative accuracy of the mean using the compound sample procedures. Weed Res 32:95–101

Dongo A, Leflon M, Simier P, Delavault P (2012) Development of a high-throughput real-time quantitative PCR method to detect and quantify contaminating seeds of *Phelipanche ramosa* and *Orobanche cumana* in crop seed lots. Weed Res 52(1):34–41. doi:10.1111/j 1365-1380.2011.00891x

Ginman E (2009) Dispersal biology of *Orobanche ramosa* in South Australia. Masters Thesis, University of Adelaide, Adelaide, Australia

Haling R, Simpson R, McKay A et al (2011) Direct measurement of roots in soil for single and mixed species using a quantitative DNA-based method. Plant Soil 348(1–2):123–137. doi:10.1007/s11104-011-0846-3

Heap JW, McKay AC (2009) Managing soil-borne crop diseases using precision agriculture in Australia. Crop Pasture Sci 60:824–833

Joel DM, Portnoy VH, Katzir N (1998) Use of DNA fingerprinting for soil-borne seed identification. Aspects Appl Biol 51:23–27

Kachelriess S (1987) Estimation of *Orobanche* seeds in soil. FABIS Newsl 19:17–19

Krishnamurthy GVG, Chandwani GH (1975) A method for isolation and estimation of germinated and ungerminated *Orobanche* seeds from soil. Indian Phytopathol 27:447–448

Linke KH, Sauerborn J, Saxena MC (1991) Host-parasite relationships – effect of *Orobanche crenata* seed banks on development of the parasite and yield of faba bean. Angew Bot 65:229–238

Linke K, Oswald A, Ransom J et al (2001) Methods to separate seeds from soil. In: Kroschel J (ed) A technical manual for parasitic weed research and extension. Kluwer Academic, Dordrecht, pp 45–49

Manschadi AM, Sauerborn J, Stutzel H (2001) Quantitative aspects of *Orobanche crenata* infestation in faba beans as affected by abiotic factors and parasite soil seedbank. Weed Res 41:311–324

Ophel-Keller K, McKay A, Hartley D et al (2008) Development of a routine DNA-based testing service for soilborne diseases in Australia. Australas Plant Pathol 37:243–253

Osterbauer NK, Rehms L (2002) Detecting single seeds of small broomrape (*Orobanche minor*) with a polymerase chain reaction. Plant Health Prog. doi:10.1094/PHP-2002-1111-01-RS

Portnoy VH, Katzir N, Joel DM (1997) Species identification of soil-borne *Orobanche* seeds by DNA fingerprinting. Pestic Biochem Phys 58:49–54

Rehms L, Osterbauer NK (2003) Detecting *Orobanche minor* seeds in soil using PCR. Plant Health Prog. doi:10.1094/PHP-2003-0701-01-HN

Riley IT, Wiebkin S, Hartley D et al (2010) Quantification of roots and seeds in soil with real-time PCR. Plant Soil 331:151–163

Roman B, Verdejo CIG, Satovic Z et al (2007) Detecting *Orobanche* species by using cpDNA diagnostic markers. Phytoparasitica 35:129–135

Sauerborn J, Wyrwal G, Linke KH (1991) Soil sampling for broomrape seeds in naturally infested fields. Progress in Orobanche research. Eberhard-Karls-Universitat, Tubingen, pp 35–43

Schnell H, Linke KH, Sauerborn J (1994) Trap cropping and its effect on yield and *Orobanche crenata* Forsk. infestation on following pea (*Pisum sativum* L.) crops. Trop Sci 34:306–314

Smutny V, Kren J (2003) The effect of different soil core samplers on precision of estimating weed seedbank in soil. Plant Soil Environ 49:466–472

Van Delft G, Graves JD, Fitter AH et al (1997) Spatial distribution and population dynamics of *Striga hermonthica* seeds in naturally infested farm soils. Plant Soil 195:1–15

Van Mourik TA (2007) *Striga hermonthica* seed bank dynamics: process quantification and modelling. Ph D Thesis, Wageningen University, Wageningen, The Netherlands

Van Mourik TA, Stomph TJ, Murdoch AJ (2005) Why high seed densities within buried mesh bags may overestimate depletion rates of soil seed banks. J Appl Ecol 42:299–305

Van Mourik TA, Stomph TJ, Murdoch AJ (2011) Purple witchweed (*Striga hermonthica*) germination and seedbank depletion under different crops, fallow, and bare soil. Weed Biol Manag 11:100–110

Williams AM, Virtue JG, DeDear C et al (2006) Sampling challenges in detecting branched broomrape seed bank decline. In: Preston C, Watts J, Crossman N (eds) Proceedings of the fifteenth Australian weeds conference. Weed Management Society of South Australia, Adelaide, SA, pp 622–625

Chapter 21
Marker-Assisted and Physiology-Based Breeding for Resistance to Root Parasitic Orobanchaceae

Begoña Pérez-Vich, Leonardo Velasco, Patrick J. Rich, and Gebisa Ejeta

21.1 Introduction

21.1.1 Host Plant Resistance and Tolerance to Parasitic Weeds

Under severe infestation, significant damage occurs on host plants before the root parasites emerge. This is especially true of *Striga* spp. that tend to negatively affect host photosynthesis in addition to acting as competitive sinks for water and nutrients (Watling and Press 2001). Host plant resistance is most effective at protecting yield if it acts early to counter the parasitic association (Frost et al. 1997). The brunt of the economic impact caused by parasitic weeds occurs in areas cultivated by subsistence farmers. In general, witchweeds and broomrapes are a poor farmers' problem (Ejeta 2007a). Smallholder farmers of Africa plagued by *Striga*, for instance, tend to be risk averse and slow in adopting new agricultural technologies (Ejeta 2007a). The potential success at controlling parasitic weeds in subsistence agriculture via host plant resistance is high because, provided that the improved varieties are locally adapted and acceptable, they fit within the varied agricultural practices where these weeds occur (Hearne 2009). Early efforts in improving crop performance in areas plagued by weedy root parasites identified valuable source germplasm able to produce acceptable yields in soils infested with parasitic weed seeds (Cubero 1986; Doggett 1988). These early selections were either **tolerant** (yield protected through *enduring* infestation) or **resistant**

B. Pérez-Vich (✉) • L. Velasco
Institute for Sustainable Agriculture (CSIC), Alameda del Obispo s/n, 14004 Córdoba, Spain
e-mail: bperez@cica.es

P.J. Rich • G. Ejeta
Department of Agronomy, Purdue University, 915 W. State St., West Lafayette, IN 47907, USA

D.M. Joel et al. (eds.), *Parasitic Orobanchaceae*,
DOI 10.1007/978-3-642-38146-1_21, © Springer-Verlag Berlin Heidelberg 2013

(yield protected through *reducing* infestation) to weedy root parasites. Both types of germplasm have been used in breeding improved cultivars, though often uncoupled (Hearne 2009). For most crops these characters appear to be weak. However, recent efforts have generated strong resistant sources with expansion of the search for resistance traits to the wild relatives of crop species (Wilson et al. 2000; Jan et al. 2002; Gurney et al. 2003; Labrousse et al. 2004; Rich et al. 2004; Rubiales et al. 2004; Valderrama et al. 2004; Velasco et al. 2007, 2012; Fernández-Aparicio et al. 2009a).

A practical understanding of the genetic control of tolerance and resistance traits to parasitic weeds has been slow in coming for most crops because of the series of complex biological and environmental interactions (host exudates, weed seedbank density, temperature, soil type, rainfall, soil nutrition, agronomy, etc.) influencing the degree of infestation. Resistance to parasitic weeds can be described as **vertical** (usually a strong acting character controlled by one or two genes, often through dominant alleles) or **horizontal** (usually partial resistance inherited through a series of alleles at multiple loci). Transfer of these traits into improved crop varieties through conventional approaches has been limited and slow due to crude phenotyping and the confounding environmental influences. Field screening methodologies such as artificial infestation and phenotyping by assessing damage scores and emergence counts for some crops have improved so that segregants with weed protective traits can be selected a bit more effectively (Haussmann et al. 2000; Rodenburg et al. 2005), but breeding for these traits remains a formidable challenge. Even when resistance breeding is achieved, parasitic-weed-resistant varieties often lack essential agronomic traits (disease resistance, grain quality, food and nutritional quality, and other plant characteristics). This is because the weed protective characters cannot be selectively introduced through backcrossing. When such situations occur, they tend to serve as a major obstacle to technology adoption by subsistence farmers (Hearne 2009).

21.1.2 Crop Improvement for Resistance and Tolerance to Parasitic Weeds

Part of the inherent difficulty in breeding crops for resistance to root parasitic Orobanchaceae is that initial responses to parasitism are expressed in roots where they cannot be readily monitored. Field screening criteria, at least the more common and practical ones, are usually taken on aboveground plant characteristics, such as host plant leaf verdancy, yield, and number and vigor of emerged parasite shoots. These measures are often the result of many weeks of host/parasite association and the compilation of several host and parasite genes affecting its establishment and success. Furthermore, there may truly be a paucity of host resistance to parasitic plants because of the relatively recent evolution of plant parasitism and the similarity of the foe (Pennings and Callaway 2002). Plants have had more time to evolve defenses against viruses, bacteria, fungi, insects, and herbivores, and those pests are foreign enough that protective measures are less likely to be autotoxic than those targeting plant foes. While defenses against non-plant biotic challenges might be effective against parasitic weeds, the sensory machinery to recognize other plants as intruders may not be in place to trigger the defense responses (see Sect. 7.5).

This is why significant gains can be made by a **breeding approach** informed by increased understanding of the physiology of the parasitic plant association (see Chap. 6). This physiology-based approach was first described for sorghum parasitized by *Striga* spp. (Ejeta and Butler 1993) and later extended to crops parasitized by *Orobanche* and *Phelipanche* spp. (Joel 2000). In this approach, single events of the parasitic association are examined separately by means of in vitro assays, first to characterize the particular processes involved in the establishment of a successful parasitic association (tissues and signals) and then to identify host variants that limit parasite growth and development. By this approach, the complex traits of resistance can be monitored through the simpler inheritance of individual characters. Individual resistance mechanisms can be identified and, in species with multiple mechanisms, combined into single cultivars for stronger and more durable resistance to root parasitic Orobanchaceae. This transfer into improved cultivars is further aided when **molecular markers** tightly linked to or, best of all, within the alleles encoding the resistance reactions are identified. As individual resistance mechanisms are characterized, ultimately to the DNA sequence level, the utility of these markers to crop improvement increases, not only to the target species being bred, but also to related species for which homologues may exist or be created through **targeted mutation** or **transformation**.

The path from resistance trait identification in source materials to exploitation in improved varieties is never quick or easy. The success depends first of all on the availability of **elite source traits**. The power of the **physiology-based breeding** approach is that it pushes the limit of what would traditionally be considered good source material for resistance to parasitic weeds. We might expect that a crop species sharing a common origin to the parasitic species to be the only useful source of alleles that offer resistance to the parasitic association. Sorghum, for instance, was domesticated in East Africa where *Striga hermonthica* also evolved and where several resistance (Ejeta 2007b) and tolerance (van Ast et al. 2000; Rodenburg et al. 2006) traits to *Striga* have been identified through both field and laboratory screening. But surprisingly, some *Striga* resistance has been described for maize, a crop originating in the Americas, with no strong resistance yet reported for pearl millet, a native of the same African origins as sorghum and *Striga*. Early reports of maize's tolerance to *Striga* were made serendipitously during field tests of breeding populations for other agronomic traits (Kim 1996; Oswald and Ransom 2004; Menkir et al. 2010). Increased understanding of these non-native source materials of *Striga* resistance was gained through characterization of maize and its wild progenitors grown with *Striga* in the laboratory (Gurney et al. 2003; Amusan et al. 2008; Rich and Ejeta 2008). Resistance is also described in American native species of sunflower introduced into European regions plagued by *Orobanche cumana* (Akhtouch et al. 2002; Eizenberg et al. 2004; Pérez-Vich et al. 2006).

The search for resistance or tolerance often begins by the recognition of individuals or varieties of crop plants around which fewer or less vigorous parasites are observed or the crop yield is less affected relative to other individuals or varieties growing around them in fields deliberately infested with parasitic weed seeds (so-called sick plots). An introduced crop like maize or sunflower may initially show immunity, essentially appearing to be a non-host to the local parasitic weeds. In time though, with repeated cultivation, these too eventually succumb to the weeds (Fernández-Martínez et al. 2008). Two processes are suspected to be occurring in this scenario, adaptation of the weed to parasitize the introduced crop and the non-host becoming a host. As crop breeders, we tend to focus on the latter since it is the genetics of the crop that we are able to manipulate. We have only indirect control over the genetics of the weed. So we can ask, what makes a crop plant a host to these root parasitic species? It is useful here to examine the interaction of host and parasite in the laboratory so we can observe the earliest interactions between the two. Chapters 2–5 described the normal course of development of the parasitic association and the various steps toward parasite establishment. Several co-culture methods have been described that allow these kinds of observations for both *Striga* (Hess et al. 1992; Gurney et al. 2006; Mohamed et al. 2010; Amusan et al. 2011) and *Orobanche*/*Phelipanche* (Eizenberg et al. 2005; Rubiales et al. 2006; Pérez-de-Luque et al. 2008). Ideally, through these observations we can identify the point at which the parasitic association is blocked. Does the potential host stimulate germination? Are haustoria initiated in the presence of their roots, does the parasite attach and penetrate through the epidermis, cortex, and endodermis? Is there a hypersensitive response that isolates the invading endophytic tissues? Does the parasite establish vascular connections with the host? Does the haustorium mature into a functional organ of acquisition? Does the connection support sustained growth of the parasite to the point of physiological maturity? How is the growth and yield of the host impacted?

21.1.3 Genetics of Resistance and Tolerance

Throughout the association multiple host genes are expressed that signal corresponding genes in the parasite (see Sect. 7.5 and Chap. 15). From the standpoint of crop genetics, we can consider alleles at these loci as either compatible (supporting the parasitic association) or incompatible (blocking the association). Some may be **defense genes**, functioning specifically against vital parasite processes, for example, a hypersensitive response (Mohamed et al. 2003) or the production of phytotoxins (Pérez-de-Luque et al. 2008). Others, perhaps the majority, may be **maintenance genes**, intended to serve the crop's ability to support its own growth and development, for example, exudation of strigolactones for the attraction of arbuscular mycorrhizal fungi that the parasite exploits as a germination cue (see Chap. 10).

The defense genes have received perhaps the greater attention in improving crop resistance to these plant parasites. They tend to be those vertical resistance genes that dramatically protect the crop. As useful as they are, often dominant with simple inheritance and thereby relatively easy to transfer, their effects put pressure on the parasite genome, upon which we as breeders have little control. It is the host maintenance genes that we cannot ignore, those that offer small obstacles to the parasitic association or allow a parasitized plant to maintain an acceptable yield. These are the horizontal resistance and tolerance genes that offer perhaps less dramatic but more long-term protection. The difficult challenge to breeders is to identify alleles at these loci that interfere with or protect the host from the effects of the association with the parasite and to combine them into a variety that protects yield from the parasite while allowing the crop to respond to the countless other environmental factors that contribute to its productivity. Introduction of vertical resistance genes into populations with built up horizontal resistance is powerful, as the life of the major genes can be lengthened beyond expectations since host major genes are not singly challenged and are buffered with the range of alleles in the population.

21.1.4 Breeding Informed by Biology

In dealing with the formidable challenge presented by root parasitic Orobanchaceae, an interdisciplinary research approach is essential. Crop breeders need ecologists, soil scientists, physiologists, biochemists, geneticists, and molecular biologists, as these sciences all become vital tools for trait discovery and transfer. Opportunities for resistance or tolerance are best learned through a detailed understanding of the qualities that make a particular crop a suitable host. Traits that contribute to or detract from these qualities can then be identified. It is then a matter of phenotyping breeding materials for these individual traits. This is no small task and may be impossible when the traits are combined for durable protection against these and other pests. How can one, for instance, test an entry for an incompatible response that slows or stops the haustorium from reaching the central cylinder when that entry also has a hypersensitive response? The combination of a complex horizontal trait (incompatibility) with a simple vertical resistance trait is desirable because it serves as a backup in case an individual parasite does not trigger the hypersensitive response. One cannot select for that phenotype unless one knows a particular metabolite or tissue structure to assay, which indicates the presence of the incompatible factor. The same challenge would be met when trying to select for tolerance in highly resistant entries (Rodenburg and Bastiaans 2011). This illustrates the need for detailed characterization of horizontal resistance and tolerance reactions. Defenses aside, we need to understand what in a host root supports penetration of the endophyte to what essentially becomes a compatible union. Also a better grasp of the toxic effects of *Striga* infection is needed (see Sect. 7.5).

As we define differentials of **susceptibility** and resistance, **sensitivity** and tolerance factors in host plant germplasm, and their corresponding phenotypes, we can then exercise selection by whatever breeding methods fit the target crop and its environment. For protection against parasitic root Orobanchaceae in particular, the ability to select for individual traits that confer resistance and tolerance would be greatly aided by the identification of molecular markers. With such tools, difficult (or impossible) phenotyping would not always be necessary to transfer multiple traits into locally adapted germplasm.

21.1.5 Virulence in the Parasite

It is vital to the lasting success of resistance breeding to predict and monitor **virulence** in the parasite populations (see Sect. 19.3.3). At each point in host/parasite associations where resistance characters may act, there are corresponding **avirulence** characters in the parasite that trigger the resistance reaction (e.g., cause a hypersensitive response; see Sect. 7.2). Even though, as breeders, we only directly manipulate the genetic combinations in the host plant, we cannot fully protect crops from root parasitic Orobanchaceae, even with a battery of resistance traits, as long as we do not understand the indirect **selective pressure** we may be exerting on the weed for virulence.

In some parasitic weed species, races have been defined based on differential virulence toward particular crop cultivars ("intracrop specificities"). In sunflower, where a number of single-gene resistance sources have been deployed, several virulence genes in parasites have been identified. Seven races of *Orobanche cumana* have been described based on virulence toward sunflower cultivars (Fernández-Martínez et al. 2008). These races emerged due to pressure from cultivars carrying strong (vertical) resistance which eventually broke down through the ability of certain individuals within the parasite populations to grow on these cultivars and to produce seeds. In *Striga gesnerioides* a few races on cowpea cultivars have been identified in West Africa with the aid of molecular markers (Botanga and Timko 2006).

In the *Striga* species attacking cereals (the "cereal *Striga* species"), races have not been defined, probably owing to a lack of widespread deployment of single sources of vertical resistance genes and due to the outcrossing behavior of *S. hermonthica* which makes it an extremely variable and flexible or adaptable parasite. Reports of virulence genes in cereal *Striga* species (Reda et al. 2010) lack the proper evidence in the nature of the host genes involved and the lack of control in what is being measured as evidence of susceptibility. On the other hand, host-specific strains have been reported in *S. hermonthica*. In Sudan, for instance, there are millet strains of *S. hermonthica* in regions where it is too dry for sorghum. This strain only grows on millet, which is seemingly immune when cultivated on sorghum land in moister regions. In zones where both sorghum and millet are cultivated, this host specificity is not observed, and the *Striga* from there can grow on either crop (Ali et al. 2009). A recent study in Mali found no clear distinctions in terms of diversity accessed by SSRs in *S. hermonthica* populations collected from sorghum and millet hosts in areas where both crops are cultivated (Estep et al. 2011). The emergence of virulence in parasite populations against a previously un-infested cultivar can be avoided through use of integrated parasite management practices (e.g., use of resistant cultivars with crop rotation, pulling any emerged parasites before seed set, adding fertilizer, and conserving moisture; see Sects. 22.3.1 and 22.3.3). Durable genetic protection, however, is better ensured if resistance traits are stacked in the improved cultivars such that multiple mutations would have to accumulate in the parasite population to overcome resistance genes in the host or deploying vertical resistance with horizontal resistance as described earlier.

21.2 Physiology-Based Breeding

21.2.1 Resistance to Striga *spp.*

Breaking down the complex early interactions of root parasitic Orobanchaceae into individual observable steps, through the aid of in vitro methods that allow observation of those processes normally hidden underground, has helped identify individual resistance characters in many crops. These methods can also be used to unveil the underlying physiological details of particular resistance mechanisms. This approach has been successfully used in sorghum to identify excellent sources of resistance and to characterize the various mechanisms that protect it from *Striga* spp.

The most characterized resistance mechanism in sorghum is low germination stimulant activity (see Sect. 10.4.2). Sorghum varieties that produce root exudates that do not stimulate the germination of conditioned *Striga* seeds were described over 50 years ago (Williams 1959). Identification of source "low stimulant" sorghums and selection for this trait in breeding lines improved with bioassays like the "double pot technique" (Vasudeva Rao 1985) and the agar gel assay (Hess et al. 1992). The chemical nature of the primary *Striga* germination stimulant in sorghum and other hosts of root parasitic Orobanchaceae was identified as strigolactones (Siame et al. 1993; see Sect. 10.2). The single-gene recessive character of low *Striga* germination stimulant activity in sorghum (Vogler et al. 1996) and the availability of a reliable method for phenotyping have made it relatively easy to use in *Striga* resistance improvement. Several improved *Striga*-resistant sorghum varieties contain this trait (Vaidya et al. 1988; Ejeta 2007c). One donor source of this trait, SRN39, has been extensively used, and its derivatives have shown broad *Striga* resistance over time and locations (Ejeta 2007c). This is likely due to the presence of additional resistance traits in this line beyond the germination stimulant response. SRN39 shows an incompatible response to *Striga* (Amusan et al. 2011) in addition to low germination stimulant activity (Hess and Ejeta 1992). Resistance based on low germination stimulant activity may not be due to low production of total strigolactones, but rather to the types of strigolactones exuded by the roots. SRN39 was found to exude lower amounts of 5-deoxystrigol but equal or greater amounts of certain other strigolactones relative to a susceptible cultivar (Yoneyama et al. 2010). Resistance based on low stimulant activity alone may be threatened by individual parasites within *Striga* populations able to germinate in response to other strigolactones.

Other assays developed in the Ejeta lab have allowed the identification and selection of additional resistance characters in sorghum. These include low haustorial initiation activity (Rich et al. 2004; see Sect. 4.3), a hypersensitive response (Mohamed et al. 2003), and incompatibility (Amusan et al. 2011; see Chap. 7). These mechanisms are less well characterized compared to the low germination stimulant trait, but they have been combined in sorghum varieties for improved *Striga* resistance (Ejeta 2007c).

Field resistance to *Striga* has also been reported in maize (Kim et al. 1999; Menkir 2006). Strong post-attachment resistance reactions have been described based on laboratory observations in co-culture of *Striga* with wild relatives *Tripsacum dactyloides* (Gurney et al. 2003) and *Zea diploperennis* (Lane et al. 1997) and with newly improved maize inbred lines derived from the latter species (Amusan et al. 2008). The resistance expressed in these inbred lines is manifested through less secondary branching in the root system, reduced number of parasitic attachments, failure of most attached parasites to establish vascular connections with the host, and diminished growth or eventual death of the few parasites that do achieve vascular connectivity (Amusan et al. 2008, 2011). The relatively low germination stimulant activity in *T. dactyloides* toward *S. hermonthica* (Gurney et al. 2003) and in certain maize cultivars toward *S. asiatica* (Pierce et al. 2003) has not been substantiated further, and the mode of inheritance of this character has not been determined.

Various resistance reactions in cowpea to *Striga gesnerioides* were described based on an in vitro infection system in which parasite development on host roots could be observed (Lane et al. 1993). The same system was used to differentiate the original five races of *S. gesnerioides* based on the virulence of various parasite populations, i.e., their ability to normally infect and grow on resistant cowpea varieties (Lane et al. 1996).

21.2.2 Resistance to **Orobanche/Phelipanche** *spp.*

Breeding for resistance to broomrapes (*Orobanche* and *Phelipanche* spp.) has not been closely coupled with characterization of the physiological bases underlying resistance mechanisms as in the work on *Striga* in sorghum. Selection of resistant genotypes within the host crop or its wild relatives has been largely based on pot and field experiments involving inoculation of the soil with seeds of the parasite (Rubiales et al. 2006; Fernández-Martínez et al. 2008). In these experiments, the absence or presence and, in the latter case, the intensity of parasitization symptoms (emerged broomrape stalks in most studies) determine the classification of the host genotypes as resistant or susceptible. Characterization of the resistance mechanisms usually follows the identification of resistant genotypes (Rubiales and Fernández-Aparicio 2012).

Breeding strategies for broomrape resistance largely depend upon the degree of host resistance and its genetic control. In most host species, only moderate to low levels of resistance to broomrape have been identified, with the resistance being under polygenic, non-race-specific genetic control. This is the case for a number of legumes such as faba bean (Sillero et al. 2010; Fernández-Aparicio et al. 2012), common vetch (Fernández-Aparicio et al. 2009b), and pea (Rubiales et al. 2006), as well as other crops such as tomato (Qasem and Kasrawi 1995), tobacco (Buschmann et al. 2005), rapeseed (Zehhar et al. 2003), or parsley (Goldwasser and Kleifeld 2002). Exceptionally, genotypes exhibiting a high degree of resistance have been identified in carrot (Zehhar et al. 2003) and tomato (Dor et al. 2010). In the latter case, resistance is controlled by alleles at a single locus determining a strigolactone deficiency (Koltai et al. 2010; Dor et al. 2011).

Sunflower resistance to *O. cumana* is in general qualitative, i.e., complete, race specific, and controlled by one or two loci (Fernández-Martínez et al. 2008). Transfer of such a qualitative or vertical resistance is a routine procedure for plant breeders, as it can be easily incorporated into elite cultivars, but unfortunately it has a low durability due to the ability of the parasite to overcome vertical resistance mechanisms. This has led to a need for fast replacement of varieties as old cultivars are overcome with new broomrape races in the sunflower cultivation areas of the Mediterranean basin and the Black Sea region. In these areas, novel sources of resistance to the latest local races have been rapidly overcome by the parasite (Fernández-Martínez et al. 2008). Sources of horizontal resistance to *O. cumana* have been identified in sunflower as well, and it should be possible to combine vertical and horizontal mechanisms of resistance toward the development of a more durable resistance to *O. cumana* in this crop (Pérez-Vich et al. 2006).

Characterization of host resistance mechanisms and development of assays are crucial for **pyramiding** strategies, i.e., combining multiple resistance/tolerance traits into single varieties. This requires selection at the physiological and/or molecular level. A number of studies focused on the characterization of broomrape resistance mechanisms in genotypes of crops with various levels of resistance. Similar to *Striga*, resistance to broomrape occurs at both the pre-attachment and post-attachment stages. Low production of germination stimulants and the exudation of germination inhibitors and phytoalexins are also important mechanisms that prevent germination and attaching to host roots. Once the contact has been established, different mechanisms to halt haustorial penetration can be activated at the cortex, endodermis, or even inside the central cylinder. Such mechanisms may involve reinforcement of the cell walls through protein cross-linking, suberization, or lignification, as well as production of toxic compounds such as phenolics. If the haustorium manages to establish vascular connections with the host (see Sects. 3.11 and 5.3.4), further defensive measures can include sealing of host xylem vessels and production of toxic compounds that are delivered into the parasite through the vascular system (Pérez-de-Luque et al. 2008). There are no clear indications of the effectiveness of individual mechanisms of resistance. Studies in genotypes exhibiting complete resistance, particularly in sunflower, revealed the presence of several resistance mechanisms occurring at different stages in a single-resistant genotype (Labrousse et al. 2001; Echevarría-Zomeño et al. 2006; Letousey et al. 2007).

Understanding how the presence of several complex defensive mechanisms corresponds to a simple mode of inheritance of the resistance, usually monogenic and dominant in sunflower, requires a deeper knowledge on the role of the *Or* resistance genes reported in this species and their location in the sunflower genome. Lu et al. (2000) advanced the hypothesis that one of these genes, *Or5*, might actually be a cluster of recognition-dependent resistance genes (see below).

21.3 Marker-Assisted Breeding

The key component of an efficient system for molecular breeding for resistance to root parasitic Orobanchaceae is the establishment of significant associations between genetic markers and genes determining resistance to the parasite. A number of major genes underlying simply inherited resistance have been mapped, and molecular markers linked to them have been described. Genes or chromosomal regions that control quantitative traits, called quantitative trait loci (**QTL**), which are associated to resistance to parasitic Orobanchaceae, have also been identified. Information on molecular markers development and marker-assisted selection (**MAS**) is summarized below for various host/parasite systems.

21.3.1 Resistance to Striga asiatica *and* Striga hermonthica *in Cereals*

Several mapping populations were developed in sorghum involving parents of contrasting resistance to *Striga hermonthica* and *S. asiatica* (Grenier et al. 2007). One of these was formed between a highly susceptible Chinese variety and SRN39, a highly resistant accession (Rodenburg et al. 2006). SRN39 possesses at least two resistance characters, low *Striga* germination stimulant activity (Hess and Ejeta 1992), a simply inherited recessive trait (Vogler et al. 1996), and incompatibility (Amusan et al. 2011). The recombinant inbred line (RIL) population derived from this cross was tested in the field with both *S. hermonthica* and *S. asiatica*. Six resistance QTLs were determined with four common to both species, suggesting broad protection against both parasites. The low germination stimulant trait mapped within one of the common QTLs (Grenier et al. 2007). Two other RIL populations involving *Striga* resistant × susceptible parents have been extensively field tested under *S. hermonthica* infestation (Omanya et al. 2004). The resistant parent of one population also was a low germination stimulant line, IS9830, from which this trait was mapped using an in vitro assay (Haussmann et al. 2001) and which corresponded to the most consistent and major locus from the five QTL determined common to all test environments (Haussmann et al. 2004). Five other *Striga* resistance QTL were determined in the third RIL population derived from a susceptible sorghum and N13 (Haussmann et al. 2004), a high stimulant line possessing a post-attachment resistance mechanism involving fortification of the endodermis that prevents *Striga* from penetrating the central cylinder (Maiti et al. 1984). SSR markers associated with the identified resistance QTL are currently being used and refined to introgress *Striga* resistance into farmer-preferred sorghum varieties in several African countries (Grenier et al. 2007; Kapran et al. 2007).

The locus associated with low germination stimulant activity (*lgs*) in the *Striga*-resistant sorghum SRN39 was recently fine mapped within a 5.8 cM region on chromosome 5 using 354 RILs and 367 (DArT and SSR) markers (Satish et al. 2012). The three tightest linked markers (at 0.5, 0.7, and 0.9 cM) co-segregated with germination stimulant activity toward both *S. asiatica* and *S. hermonthica* in an in vitro assay among 23 diverse accessions used to validate the markers. The only dissention arose in one wild sorghum accession with low germination stimulant activity that did not share the SRN39 type (mutant) alleles at *lgs*, suggesting perhaps the presence of other major loci with control on *Striga* germination stimulant activity in sorghum.

QTL associated with resistance to *S. hermonthica* in rice have also been reported (Gurney et al. 2006; see Sect. 7.4.1). In a population derived from Nipponbare, a rice variety possessing an incompatible reaction to *Striga* infection (post-attachment resistance, see Sect. 7.3), with a susceptible rice cultivar, four QTL with major effects on resistance to *S. hermonthica* were identified. In order to find candidate resistance genes in these intervals, expression profiling was used and three candidate genes coding for uncharacterized proteins were identified within one of the major QTLs associated with *Striga* resistance (Swarbrick et al. 2008). QTL with major effects on tolerance to *S. hermonthica* have also been reported in rice (Kaewchumnong and Price 2008).

A recent study from the University of Sheffield (Huang et al. 2012) underscores the need for breeders to pay attention to the genetics of the root parasitic Orobanchaceae populations in the target areas of their improved varieties. They conducted a diversity study comparing individuals from a common seed source of *S. hermonthica* collected from a maize field in Kenya grown in the laboratory on three rice accessions of varying resistance to *Striga*. One rice entry was susceptible, the second partially resistant, and the third, Nipponbare, highly resistant to *S. hermonthica* at the early post-attachment stage of the parasitic association (see Sect. 7.3). They sampled individual parasites growing on each of the three rice accessions and assessed the diversity among individuals of the three subpopulations (one on the susceptible, one on the moderately resistant, and one on the highly resistant rice hosts) by means of AFLP markers. In pairwise comparisons between subpopulations, they found that the one able to grow on the highly resistant rice was less diverse, with respect to 24 of 191 AFLPs assayed, than that growing on the susceptible rice, and a quarter of those also distinguished between the diversity levels of the subpopulations growing on the moderately resistant rice and on the susceptible rice. This is a powerful demonstration of what we too often ignore, that a resistant variety selects for individuals in parasite populations with respect to loci affecting host specificity. The hope in this is that such loci in the *Striga* genome can be marked, and the effects of specific host characters on these can be studied, for improved durability of resistance.

21.3.2 Resistance to **Striga gesnerioides** *in Cowpea*

Although MAS is not yet practiced in breeding cowpeas for resistance to *S. gesnerioides*, there have been molecular markers, nearly a dozen AFLPs and two SCARs, found to be associated with resistance to specific races of the parasite (Timko and Singh 2008). Some of these have been placed on the cowpea genetic map (Ouedraogo et al. 2002). An SSR marker was also found to be associated with vertical resistance to race SG3. The resistance involved a strong hypersensitive response that stopped *Striga* parasitism after attachment. This marker was determined to be within the gene causing the reaction to SG3 which allowed it to be cloned and characterized. The gene, named *RSG3-301*, codes for a transcription factor that is triggered by an avirulence factor in SG3 in a classical gene-for-gene resistance (Li and Timko 2009; see Sect. 7.4.1).

21.3.3 Resistance to **Orobanche cumana** *in Sunflower*

Initial studies on the sunflower/*O. cumana*, association by Vrânceanu et al. (1980) indicated that it fitted the gene-for-gene model (see Sect. 7.2). Using a set of differential lines, they found that resistance to sunflower broomrape races A through E was determined by dominant alleles at single genes named *Or1* through *Or5*, respectively. This was later confirmed by other authors (Ish-Shalom-Gordon et al. 1993; Sukno et al. 1999). However, the nature of *Or1* through *Or5* genes involved in this interaction has not been determined. Molecular mapping studies revealed that the *Or5* gene conferring resistance to race E is located on a telomeric region of linkage group (LG) 3 of the sunflower genetic map (Lu et al. 2000; Tang et al. 2003; Pérez-Vich et al. 2004). Molecular markers linked to *Or5* were identified on one side of the gene (Tang et al. 2003; Pérez-Vich et al. 2004). It has been hypothesized that *Or5* might be a cluster of recognition-dependent resistance genes encoding proteins characterized by the presence of leucine-rich repeat (LRR) motifs and a nucleotide binding site (NBS) N-terminal to the LRR domain (Lu et al. 2000), similar to those NBS-LRR clusters conferring dominant resistance to different races of downy mildew in sunflower (Bouzidi et al. 2002; Radwan et al. 2003). This is supported by the mapping of three NBS-LRR loci to the upper segment of LG3, closely linked to *Or5* (Radwan et al. 2008), two of them derived from *Helianthus tuberosus*, which has been extensively used as a source of broomrape resistance genes (Fernández-Martínez et al. 2008). Our current understanding of the linkage arrangement of clusters of resistance genes in the sunflower genome is weak (Radwan et al. 2008). They are often located in regions of high recombination such as telomeric regions, as it is the case of *Or5*, or in gaps in the genetic map, which makes the identification of tightly linked flanking markers more problematic. For a marker-assisted breeding program, it is desirable to have markers flanking the gene closely on both sites, as selection based on a single marker is not satisfactory in most cases (Hospital 2003). Márquez-Lema et al. (2008) used sunflower target region amplification polymorphism (TRAP) markers developed from *Arabidopsis* telomeric sequences, previously mapped to linkage group ends of the sunflower linkage map (Hu 2006), to saturate the *Or5* region, and to find markers flanking the gene. They found that *Or5* was probably located at the TRC27133 to ZVG406/CRT392c marker interval, being TRC27133 a telomere-associated TRAP marker defining LG3 upper end (Hu 2006), ZVG406 the uppermost RFLP marker on LG3 (Berry et al. 1996), and CRT392c the uppermost SSR marker on LG3 described to date (Yu et al. 2003).

In addition to *Or5*, resistance to *O. cumana* race E in sunflower has been reported to have a quantitative component. In a QTL study, Pérez-Vich et al. (2004) determined that resistance to this race was the result of the major gene *Or5* in combination with a quantitative component for which at least four QTL were identified. Such QTL had a minor effect, in some cases non-race specific, and determined mainly the number of broomrape shoots per plant. Studies based on the determination of resistance stages, histological observations, and gene expression demonstrated the existence of several different resistance mechanisms to race E (Labrousse et al. 2001; de Zélicourt et al. 2007; Letousey et al. 2007). These studies support the existence of a polygenic component in race E resistance and provide a valuable source of candidate genes for understanding the defensive mechanisms and biochemical pathways involved.

Resistance to *O. cumana* race F in germplasm derived from cultivated sunflower (P-96 and KI-534) has been found to be controlled by recessive alleles at two loci (Rodríguez-Ojeda et al. 2001; Akhtouch et al. 2002). Pérez-Vich et al. (2004) determined that recessive resistance to race F in the sunflower line P-96 was determined by six QTL with a small to moderate effect in decreasing the number of broomrapes per plant. Some of these QTL were non-race specific and stable over environments and jointly explained more than half of the phenotypic variation for the resistance traits evaluated (Pérez-Vich et al. 2004). Cytological and cytochemical observations in compatible and incompatible reactions in resistant and susceptible genotypes have identified several defense mechanisms to race F of broomrape, e.g., a physical barrier to prevent parasite intrusion or a chemical response by secreting toxic compounds (Echevarría-Zomeño et al. 2006).

The rapid evolution of new virulent *O. cumana* races is probably the result of the use of sunflower hybrids that are nearly exclusively based on resistance determined by single race-specific dominant genes (Fernández-Martínez et al. 2008; Molinero-Ruiz et al. 2008). New breeding strategies such as pyramiding of major genes or combining vertical and horizontal resistance mechanisms are required for the development of more durable resistant cultivars. The implementation of MAS programs or physiology-based breeding supported by good bioassays will be essential for these strategies to be effective in deploying durable cultivars. Molecular markers will also be useful for pyramiding tightly linked resistance genes within the same resistance gene cluster.

21.3.4 Resistance to **Orobanche** *and* **Phelipanche** *spp. in Legumes*

Unlike sunflower, qualitative or vertical resistance to broomrape (*Orobanche* and *Phelipanche* spp.) conferred by a reduced number of major genes has not been identified in legume species. Broomrape resistance in legumes is generally polygenic and non-race specific (Rubiales and Fernández-Aparicio 2012). Several genetic studies on resistance to *O. crenata* in faba bean and common vetch (*Vicia sativa* L.) concluded that the genetic system controlling this trait was quantitative, with a very strong additive component (Gil et al. 1987; Cubero and Hernández 1991). Dominance, if present, was generally partial, and in the direction of susceptibility, although fully dominant resistance has been reported in some faba bean germplasm (Cubero and Hernández 1991), and in common vetch (Gil et al. 1987). This is consistent with studies on pathogenic variation in *O. crenata* parasitizing faba bean, where no clear evidence supporting the existence of races was found (Cubero and Moreno 1979; Radwan et al. 1988), and with molecular studies carried out in *O. crenata* populations from Spain (Román et al. 2001), Egypt (Zeid et al. 1997), and Israel (Paran et al. 1997) attacking faba bean or pea, which detected low differentiation among parasite populations and a considerable variation at the intrapopulation level (see Chap. 19).

Genes or chromosomal regions that control the quantitative *O. crenata* resistance have been identified in faba bean and pea. In faba bean, Román et al. (2002) identified three resistance QTLs (*Oc1*, *Oc2*, and *Oc3*) in a RAPD-seed protein gene-isozyme-map developed from an F_2 population derived from the cross between a susceptible (Vf6) and a resistant (Vf136) parent. The three QTL explained a high percentage (74 %) of the phenotypic variation for resistance to *O. crenata*, with the major QTL *Oc1* explaining more than 35 % of the phenotypic variance. By using recombinant inbred lines (RILs) and a linkage map with a higher marker density, Díaz-Ruiz et al. (2010) validated the *Oc2* and *Oc3* QTL in two further environments and refined their position. Additionally, two other environmentally sensitive QTL (*Oc4* and *Oc5*) were identified (Díaz-Ruiz et al. 2010). However, the total variance explained by all QTL detected in the RIL population (16–28 % depending on the environment) was lower than that observed in the F_2 population. The RIL population was also evaluated for resistance to *O. foetida*. Two QTL (*Of1* and *Of2*) for resistance to this parasite were found, though they were unstable across environments and explained little of the phenotypic variation (Díaz-Ruiz et al. 2009).

In pea, Valderrama et al. (2004) identified two QTL conferring resistance to *O. crenata* in a RAPD-STS genetic linkage map from an F_2 population evaluated under field conditions, obtained by crossing a resistant wild pea (*Pisum sativum* subsp. *syriacum*) strain and a susceptible pea variety. The QTL explained only a moderate portion of the observed variation.

The precision of QTL studies largely depends on the accuracy of phenotypic evaluations, which is generally low under field conditions in cases in which quantitative resistance to broomrape is involved, mainly because of the influence of environmental factors and of heterogeneity of soil seedbanks. Additionally, field evaluations are based on counting emerged broomrape shoots, which does not consider escape factors and/or resistance mechanisms acting in previous phases of the infection process (Valderrama et al. 2004; Sillero et al. 2010). For example, broomrape emergence in extremely susceptible lines is hindered because of the establishment of many tubercles per plant, with the subsequent competition for nutrients at an early stage of host development (Sillero et al. 1996). In these cases, extreme susceptibility could wrongly be interpreted as resistance if evaluation is only based on emerged broomrape shoots (Rubiales et al. 2006).

In order to overcome these disadvantages, Fondevilla et al. (2010) carried out a QTL study on a RIL population derived from the same cross of Valderrama et al. (2004), but instead of scoring resistance only as the final number of emerged parasite, they phenotyped the populations at various different phases of the parasite cycle under controlled conditions in the lab and also under field conditions. Using an improved RAPD-STS map, the authors were able to identify four genomic regions controlling field resistance, explaining also a low proportion of the phenotypic variation, but also additional QTL governing-specific mechanisms of resistance such as low induction of *O. crenata* seed germination, low number of established tubercles per host root length unit, and slow tubercle development, which altogether explained a higher proportion of the observed variation. This study also revealed the influence of host vigor in the level of broomrape attack under field conditions. Thus, three out of the four QTL identified for resistance in the field evaluations co-localized with QTL controlling plant vigor traits, and small plants, although they might be susceptible, were not able to harbor a high number of broomrapes. The authors concluded the importance of selection for those QTL governing resistance mechanisms but not those linked to low host biomass.

Molecular markers linked to broomrape resistance QTL developed thus far in legumes have a limited value in practical breeding due to relatively large distances between the flanking markers and the QTL (Pérez-de-Luque et al. 2009). This can be solved by developing markers based on the sequence of the genes underlying resistance QTL. Molecular approaches such as candidate gene and expression analyses can be helpful for the identification of these genes. In faba bean, candidate resistance genes belonging to the NBS gene family have been isolated and (Palomino et al. 2006) and are currently being mapped in RIL populations segregating for broomrape resistance (Torres et al. 2010).

Gene expression studies also provide a valuable source of candidate genes for broomrape resistance. Die et al. (2007, 2009) analyzed the transcriptional profile during infection by *O. crenata* of a *Medicago truncatula* genotype showing incomplete late resistance by means of a suppression subtractive hybridization cDNA library (SSH) analysis and real-time quantitative reverse transcription PCR. These authors identified a collection of *Medicago truncatula* expressed sequence tags (ESTs) corresponding to genes regulated in this process. It was concluded that resistance to *O. crenata* in this model legume species comprised induction of general defense-related mechanisms as well as more specific responses (Die et al. 2009). The transcriptional profile of *M. truncatula* during infection by *O. crenata* in two resistant genotypes, with complete early and incomplete late resistances, was also analyzed using a microarray approach (Dita et al. 2009). Several hundred genes upregulated in this process were identified, concluding that resistance mechanisms activated in both genotypes were temporally and spatially different and resembled those associated with plant resistance to microbial pathogens (see Sect. 7.5).

Proteomic studies are starting to provide a deeper knowledge of different aspects of plant–parasite interactions and can serve also as a source of candidate genes. This approach has been used to study the plant response to *O. crenata* infection. In pea the presence of higher levels of defense- and stress-related proteins induced upon infection was reported (Castillejo et al. 2004). Among these, several proteins were identified with protease activity which could play a role in preventing the penetration and connection to the vascular system of the parasite (Castillejo et al. 2012). In *M. truncatula,* the changes in the proteome of resistant plants parasitized by *O. crenata* corresponded to a general increase in the amounts of defense-related proteins, such as proteinase inhibitors, pathogenesis-related (PR) proteins, cell wall modifying proteins, reactive oxygen species (ROS) detoxifying enzymes, and enzymes involved in the synthesis of secondary metabolites (Castillejo et al. 2009).

In the absence of major genes conferring high levels of resistance to broomrape in legume crops, the development of cultivars with an adequate level of quantitative resistance under field conditions requires stacking of several genes with minor or moderate effects into individual cultivars. Because of the difficulty of pyramiding minor resistance genes on the basis of phenotypic evaluations, MAS emerges as an indispensable strategy for developing germplasm with strong and durable resistance. However, the development of appropriate tools for accurate MAS for quantitative broomrape resistance is a complex task. It requires the identification of QTL associated with resistance traits, the development of closely linked markers, or preferably markers based on the gene sequences, and the identification of the genes and their role in the resistance mechanisms. Important advances toward these objectives have already been made, but the routine use of MAS in most breeding programs is not yet practiced or documented. Recent studies in different host species on QTL identification, QTL validation across genetic backgrounds and environments, marker refinement, analyses of candidate genes, profiling of gene expression, and accurate phenotypic evaluation—they all represent valuable advances toward the development of effective MAS strategies for broomrape resistance in legume crops.

References

Akhtouch B, Muñoz-Ruz J, Melero-Vara JM, Fernández-Martínez JM, Domínguez J (2002) Inheritance of resistance to race F of broomrape (*Orobanche cumana* Wallr.) in sunflower lines of different origin. Plant Breed 121:266–269

Ali RAMA, El-Hussein AA, Mohamed KI, Babiker AGT (2009) Specificity and genetic relatedness among *Striga hermonthica* strains in Sudan. Life Sci Int J 3:1159–1166

Amusan IO, Rich PJ, Menkir A, Housley T, Ejeta G (2008) Resistance to *Striga hermonthica* in a maize inbred line derived from *Zea diploperennis*. New Phytol 178:157–166

Amusan I, Rich PJ, Housley T, Ejeta G (2011) An *in vitro* method for identifying post-attachment *Striga* resistance in maize and sorghum. Agron J 103:1472–1478

Berry ST, Leon AJ, Challis P, Livini C, Jones R, Hanfrey CC, Griffiths S, Roberts A (1996) Construction of a high density, composite RFLP linkage map for cultivated sunflower (*Helianthus annuus* L.). In: Proceedings of the 14th international sunflower conference, 12–20 June 1996, Beijing, China, pp 1155–1160

Botanga CJ, Timko MP (2006) Phenetic relationships among different races of *Striga gesnerioides* (Willd.) Vatke from West Africa. Genome 49:1351–1365

Bouzidi MF, Badaoui S, Cambon F, Vear F, Tourvieille de Labrouhe D, Nicolas P, Mouzeyar S (2002) Molecular analysis of a major locus for resistance to downy mildew in sunflower with specific PCR-based markers. Theor Appl Genet 104:592–600

Buschmann H, Gonsior G, Sauerborn J (2005) Pathogenicity of branched broomrape (*Orobanche ramosa*) populations on tobacco cultivars. Plant Pathol 54:650–656

Castillejo MA, Amiour N, Dumas-Gaudot E, Rubiales D, Jorrín JV (2004) A proteomic approach to studying plant response to crenate broomrape (*Orobanche crenata*) in pea (*Pisum sativum*). Phytochemistry 65:1817–1828

Castillejo MA, Maldonado AM, Dumas-Gaudot E, Fernández-Aparicio M, Susín R, Diego R, Jorrín JV (2009) Differential expression proteomics to investigate responses and resistance to *Orobanche crenata* in *Medicago truncatula*. BMC Genomics 10:294

Castillejo MA, Fernández-Aparicio M, Diego R (2012) Proteomic analysis by two-dimensional differential in gel electrophoresis (2D DIGE) of the early response of *Pisum sativum* to *Orobanche crenata*. J Exp Bot 63:107–119

Cubero JI (1986) Breeding for resistance to *Orobanche* and *Striga*: a review. In: ter Borg SJ (ed) Biology and control of *Orobanche*. LH/VPO, Wageningen, pp 127–139

Cubero JI, Hernández L (1991) Breeding faba bean (*Vicia faba* L.) for resistance to *Orobanche crenata* Forsk. In: JI Cubero and MC Saxena (eds) Present status and future prospects of faba bean production and improvement in the Mediterranean countries. Options Méditerranéennes, Serie A: Seminaires Méditerranéennes 10. Zaragoza, Spain. CIHEAM/ECC/ICARDA, pp 51–57

Cubero JI, Moreno MT (1979) Agronomical control and sources of resistance in *Vicia faba* to *Orobanche crenata*. In: Bond DA, Scarascia- Mugnozza GT, Poulsen MH (eds) Some current research on *Vicia faba* in Western Europe. Commission of the European Communities, Luxemburg, pp 41–80

de Zélicourt A, Letousey P, Thoiron S, Campion C, Simoneau P, Elmorjani K, Marion D, Simier P, Delavault P (2007) Ha-DEF1, a sunflower defensin, induces cell death in *Orobanche* parasitic plants. Planta 226:591–600

Díaz-Ruiz R, Torres A, Gutierrez MV, Rubiales D, Cubero JI, Kharrat M, Satovic Z, Román B (2009) Mapping of quantitative trait loci controlling *Orobanche foetida* Poir. resistance in faba bean (*Vicia faba* L.). Afr J Biotechnol 8:2718–2724

Díaz-Ruiz R, Torres AM, Satovic Z, Gutierrez MV, Cubero JI, Román B (2010) Validation of QTLs for *Orobanche crenata* resistance in faba bean (*Vicia faba* L.) across environments and generations. Theor Appl Genet 120:909–919

Die JV, Dita MA, Krajinski F, González-Verdejo CI, Rubiales D, Moreno MT, Román B (2007) Identification by suppression subtractive hybridization and expression analysis of *Medicago truncatula* putative defence genes in response to *Orobanche crenata* parasitization. Physiol Mol Plant Pathol 70:49–59

Die JV, González-Verdejo CI, Dita MA, Nadal S, Román B (2009) Gene expression analysis of molecular mechanisms of defence induced in *Medicago truncatula* parasitized by *Orobanche crenata*. Plant Physiol Biochem 47:635–641

Dita MA, Die JV, Román B, Krajinski F, Küster H, Moreno MT, Cubero JI, Rubiales D (2009) Gene expression profiling of *Medicago truncatula* roots in response to the parasitic plant *Orobanche crenata*. Weed Res 49(Suppl 1):66–80

Doggett H (1988) Coordination of *Striga* research. In: Kim SK (ed) Combating *Striga* in Africa. IITA, Ibadan, pp 126–133

Dor E, Alperin B, Wininger S, Ben-dor B, Somvanshi VS, Koltai H, Kapulnik Y, Hershenhorn J (2010) Characterization of a novel tomato mutant resistant to the weedy parasites *Orobanche* and *Phelipanche* spp. Euphytica 171:371–380

Dor E, Yoneyama K, Wininger S, Kapulnik Y, Yoneyama K, Koltai H, Xie X, Hershenhorn J (2011) Strigolactone deficiency confers resistance in tomato line SL-ORT1 to the parasitic weeds *Phelipanche* and *Orobanche* spp. Phytopathology 101:213–222

Echevarría-Zomeño S, Pérez-de-Luque A, Jorrín J, Maldonado A (2006) Pre-haustorial resistance to broomrape (*Orobanche cumana*) in sunflower (*Helianthus annuus*): cytochemical studies. J Exp Bot 57:4189–4200

Eizenberg H, Plakhine D, Hershenhorn J, Kleifeld Y, Rubin B (2004) Variation in responses of sunflower cultivars to the parasitic weed broomrape. Plant Dis 88:479–484

Eizenberg H, Shtienberg D, Silberbush M, Ephrath JE (2005) A new method for *in-situ* monitoring of the underground development of *Orobanche cumana* in sunflower (*Helianthus annuus*) with a mini-rhizotron. Ann Bot 96:1137–1140

Ejeta G (2007a) The *Striga* scourge in Africa: a growing pandemic. In: Ejeta G, Gressel J (eds) Integrating new technologies for *Striga* control: towards ending the witch-hunt. World Scientific, Singapore, pp 3–16

Ejeta G (2007b) Dissecting the complex trait of *Striga* resistance to simpler components for effective breeding of sorghums with high level of resistance to *S. hermonthica*. In: Ejeta G, Gressel J (eds) Integrating new technologies for *Striga* control: towards ending the witch-hunt. World Scientific, Singapore, pp 87–98

Ejeta G (2007c) Breeding for *Striga* resistance in sorghum: exploitation of an intricate host-parasite biology. Crop Sci 47:S216–S227

Ejeta G, Butler LG (1993) Host plant resistance to *Striga*. In: Buxton DR et al (eds) International crop science I. Crop Science Society of America, Madison, WI, pp 561–569

Estep M, Van Mourik T, Muth P, Guindo D, Parzies H, Koita O, Weltzien E, Bennetzen J (2011) Genetic diversity of a parasitic weed, *Striga hermonthica*, on sorghum and pearl millet in Mali. Trop Plant Biol 4:91–98

Fernández-Aparicio M, Sillero JC, Rubiales D (2009a) Resistance to broomrape in wild lentils (*Lens* spp.). Plant Breed 128:266–270

Fernández-Aparicio M, Sillero JC, Rubiales D (2009b) Resistance to broomrape species (*Orobanche* spp.) in common vetch (*Vicia sativa* L.). Crop Prot 28:7–12

Fernández-Aparicio M, Moral A, Kharrat M, Rubiales D (2012) Resistance against broomrapes (*Orobanche* and *Phelipanche* spp.) in faba bean (*Vicia faba*) based in low induction of broomrape seed germination. Euphytica 186:897–905

Fernández-Martínez JM, Domínguez J, Pérez-Vich B, Velasco L (2008) Update on breeding for resistance to sunflower broomrape. Helia 31:73–84

Fondevilla S, Fernández-Aparicio M, Satovic Z, Emeran AA, Torres AM, Moreno MT, Rubiales D (2010) Identification of quantitative trait loci for specific mechanisms of resistance to *Orobanche crenata* in pea (*Pisum sativum* L.). Mol Breed 25:259–272

Frost DL, Gurney AL, Press MC, Scholes JD (1997) *Striga hermonthica* reduces photosynthesis in sorghum: the importance of stomatal limitations and a potential role for ABA? Plant Cell Environ 20:483–492

Gil J, Martín LM, Cubero JI (1987) Genetic of resistance in *Vicia sativa* L to *Orobanche crenata* Forks. Plant Breed 99:134–143

Goldwasser Y, Kleifeld Y (2002) Tolerance of parsley varieties to *Orobanche*. Crop Prot 21:1101–1107

Grenier C, Haussmann BIG, Kiambi D, Ejeta G (2007) Marker-assisted selection for *Striga* resistance in sorghum. In: Ejeta G, Gressel J (eds) Integrating new technologies for *Striga* control: towards ending the witch-hunt. World Scientific, Singapore, pp 159–172

Gurney AL, Grimanelli D, Kanampiu F, Hoisington D, Scholes JD, Press MC (2003) Novel sources of resistance to *Striga hermonthica* in *Tripsacum dactyloides*, a wild relative of maize. New Phytol 160:557–568

Gurney AL, Slate J, Press MC, Scholes JD (2006) A novel form of resistance in rice to the angiosperm parasite *Striga hermonthica*. New Phytol 169:199–208

Haussmann BIG, Hess DE, Welz HG, Geiger HH (2000) Improved methodologies for breeding *Striga*-resistant sorghums. Field Crops Res 66:195–201

Haussmann BIG, Hess DE, Omanya GO, Reddy BVS, Welz HG, Geiger HH (2001) Major and minor genes for stimulation of *Striga hermonthica* seed germination in sorghum, and interaction with different *Striga* populations. Crop Sci 41:1507–1512

Haussmann BIG, Hess DE, Omanya GO, Folkertsma RT, Reddy BV, Kayentao M, Welz HG, Geiger HH (2004) Genomic regions influencing resistance to the parasitic weed *Striga* hermonthica in two recombinant inbred populations of sorghum. Theor Appl Genet 109:1005–1016

Hearne SJ (2009) Control – the *Striga* conundrum. Pest Manag Sci 65:603–614

Hess DE, Ejeta G (1992) Inheritance of resistance to *Striga* in sorghum genotype SRN39. Plant Breed 109:233–241

Hess DE, Ejeta G, Butler LG (1992) Selecting sorghum genotypes expressing a quantitative biosynthetic trait that confers resistance to *Striga*. Phytochemistry 31:493–497

Hospital F (2003) Marker-assisted breeding. In: Newbury HJ (ed) Plant molecular breeding. Blackwell, London, pp 30–56

Hu J (2006) Defining the sunflower (*Helianthus annuus* L.) linkage group ends with the *Arabidopsis*-type telomere sequence repeat-derived markers. Chromosome Res 14:535–548

Huang K, Whitlock R, Press MC, Scholes JD (2012) Variation for host range within and among populations of the parasitic plant *Striga hermonthica*. Heredity 108:96–104

Ish-Shalom-Gordon N, Jacobsohn R, Cohen Y (1993) Inheritance of resistance to *Orobanche cumana* in sunflower. Phytopathology 83:1250–1252

Jan CC, Fernández-Martínez JM, Ruso J, Muñoz-Ruz J (2002) Registration of four sunflower germplasms with resistance to *Orobanche cumana* race F. Crop Sci 42:2217–2218

Joel DM (2000) The long-term approach to parasitic weeds control: manipulation of specific developmental mechanisms of the parasite. Crop Prot 19:753–758

Kaewchumnong K, Price AH (2008) A study on the susceptibility of rice cultivars to *Striga hermonthica* and mapping of *Striga* tolerance quantitative trait loci in rice. New Phytol 180:206–216

Kapran I, Grenier C, Ejeta G (2007) Introgression of *Striga* resistance into African sorghum landraces. In: Ejeta G, Gressel J (eds) Integrating new technologies for *Striga* control: towards ending the witch-hunt. World Scientific, Singapore, pp 129–142

Kim S-K (1996) Horizontal resistance: core to a research breakthrough to combat *Striga* in Africa. Integr Pest Manag Rev 1:229–249

Kim SK, Akintunde AY, Walker P (1999) Responses of maize inbreds during development of *Striga hermonthica* infestation. Maydica 44:333–339

Koltai H, LekKala SP, Bahattacharya C, Mayzlish-Gati E, Resnick N, Wininger S, Dor E, Yoneyama K, Yoneyama K, Hershenhorn J, Joel DM, Kapulnik Y (2010) A tomato strigolactone-impaired mutant displays aberrant shoot morphology and plant interactions. J Exp Bot 61:1739–1749

Labrousse P, Arnaud M-C, Serieys H, Bervillé A, Thalouarn P (2001) Several mechanisms are involved in resistance of *Helianthus* to *Orobanche cumana* Wallr. Ann Bot 88:859–868

Labrousse P, Arnaud M-C, Griveau Y, Fer A, Thalouarn P (2004) Analysis of resistance criteria of sunflower recombined inbred lines against *Orobanche cumana* Wallr. Crop Prot 23:407–413

Lane JA, Bailey JA, Butler RC, Terry PJ (1993) Resistance of cowpea to *Striga gesnerioides* Vatke, a parasitic angiosperm. New Phytol 125:405–412

Lane JA, Moore THM, Child DV, Cardwell KF (1996) Characterization of virulence and geographic distribution of *Striga gesnerioides* on cowpea in West Africa. Plant Dis 80:299–301

Lane JA, Child DV, Moore THM, Arnold GM, Bailey JA (1997) Phenotypic characterisation of resistance in *Zea diploperennis* to *Striga hermonthica*. Maydica 42:45–51

Letousey P, de Zélicourt A, Vieira Dos Santos C, Thoiron S, Monteau F, Simier P, Thalouarn P, Delavault P (2007) Molecular analysis of resistance mechanisms to *Orobanche cumana* in sunflower. Plant Pathol 56:536–546

Li J, Timko MP (2009) Gene-for-gene resistance in *Striga*-cowpea associations. Science 325:1094

Lu YH, Melero-Vara JM, García-Tejada JA, Blanchard P (2000) Development of SCAR markers linked to the gene *Or5* conferring resistance to broomrape (*Orobanche cumana* Wallr.) in sunflower. Theor Appl Genet 100:625–632

Maiti RK, Ramaiah KV, Bisen SS, Chidley VL (1984) A comparative study of the haustorial development of *Striga asiatica* (L.) Kuntze on sorghum cultivars. Ann Bot 54:447–457

Márquez-Lema A, Delavault P, Letousey P, Hu J, Pérez-Vich B (2008) Candidate gene analysis and identification of TRAP and SSR markers linked to the *Or5* gene conferring sunflower resistance to race E of broomrape (*Orobanche cumana* Wallr.). In: Proceedings of the 17th international sunflower conference, 8–12 June 2008, Córdoba, Spain, pp 661–666

Menkir A (2006) Assessment of reactions of diverse maize inbred lines to *Striga hermonthica* (Del.) Benth. Plant Breed 125:131–139

Menkir A, Adetimirin VO, Yallou CG, Gedil M (2010) Relationship of genetic diversity of inbred lines with different reactions to *Striga hermonthica* (Del.) Benth and the performance of their crosses. Crop Sci 50:602–611

Mohamed A, Ellicott A, Housley TL, Ejeta G (2003) Hypersensitive response to *Striga* infection in *Sorghum*. Crop Sci 43:1320–1324

Mohamed AH, Housley TL, Ejeta G (2010) An *in vitro* technique for studying specific *Striga* resistance mechanisms in sorghum. Afr J Agric Res 5:1868–1875

Molinero-Ruiz ML, Pérez-Vich B, Pineda-Martos R, Melero-Vara JM (2008) Indigenous highly virulent accessions of the sunflower root parasitic weed *Orobanche cumana*. Weed Res 48:169–178

Omanya GO, Haussmann BIG, Hess DE, Reddy BVS, Kayentao M, Welz HG, Geiger HH (2004) Utility of indirect and direct selection traits for improving *Striga* resistance in two sorghum recombinant inbred populations. Field Crops Res 89:237–252

Oswald A, Ransom JK (2004) Response of maize varieties to *Striga* infestation. Crop Prot 23:89–94

Ouedraogo JT, Tignegre JB, Timko MP, Beizile FJ (2002) AFLP markers linked to resistance against *Striga gesnerioides* race 1 in cowpea (*Vigna unguiculata*). Genome 45:787–793

Palomino C, Satovic Z, Cubero JI, Torres AM (2006) Identification and characterization of NBS-LRR class resistance gene analogs in faba bean (*Vicia faba* L.) and chickpea (*Cicer arietinum* L.). Genome 49:1227–1237

Paran I, Gidoni D, Jacobsohn R (1997) Variation between and within orobanche (*Orobanche*) species revealed by RAPD markers. Heredity 78:68–74

Pennings SC, Callaway RM (2002) Parasitic plants: parallels and contrasts with herbivores. Oecologia 131:479–489

Pérez-de-Luque A, Moreno MT, Rubiales D (2008) Host plant resistance against broomrapes (*Orobanche* spp.): defence reactions and mechanisms of resistance. Ann Appl Biol 152:131–141

Pérez-de-Luque A, Fondevilla S, Pérez-Vich B, Aly R, Thoiron S, Simier P, Castillejo MA, Fernández JM, Jorrín J, Rubiales D, Delavault P (2009) Understanding *Orobanche* and *Phelipanche*-host plant interaction and developing resistance. Weed Res 49(S1):8–22

Pérez-Vich B, Akhtouch B, Knapp SJ, Leon AJ, Velasco L, Fernández-Martínez JM, Berry ST (2004) Quantitative trait loci for broomrape (*Orobanche cumana* Wallr.) resistance in sunflower. Theor Appl Genet 109:92–102

Pérez-Vich B, Velasco L, Muñoz-Ruz J, Domínguez J, Fernández-Martínez JM (2006) Registration of three sunflower germplasms with quantitative resistance to race F of broomrape. Crop Sci 46:1406–1407

Pierce S, Mbwaga AM, Press MC, Scholes JD (2003) Xenognosin production and tolerance to *Striga asiatica* infection of high-yielding maize cultivars. Weed Res 43:139–145

Qasem JR, Kasrawi MA (1995) Variation of resistance to broomrape (*Orobanche ramosa*) in tomatoes. Euphytica 81:109–114

Radwan MS, Abdalla MMF, Fischbeck G, Metwally AA, Darwish DS (1988) Variation in reaction of faba bean lines to different accessions of *Orobanche crenata* Forsk. Plant Breed 101:208–216

Radwan O, Bouzidi MF, Vear F, Philippon J, Tourvieille de Labrouhe D, Nicolas P, Mouzeyar S (2003) Identification of non-TIR-NBS-LRR markers linked to *PL5*/*PL8* locus for resistance to downy mildew in sunflower. Theor Appl Genet 106:1438–1446

Radwan O, Gandhi S, Heesacker A, Whitaker B, Taylor C, Plocik A, Kesseli R, Kozik A, Michelmore RW, Knapp SJ (2008) Genetic diversity and genomic distribution of homologs encoding NBS-LRR disease resistance proteins in sunflower. Mol Genet Genomics 280:111–125

Reda F, Dierick A, Verkleij JAC (2010) Virulence study of *Striga hermonthica* populations from Tigray region (northern Ethiopia). World J Agric Sci 6:676–682

Rich PJ, Ejeta G (2008) Towards effective resistance to *Striga* in African maize. Plant Signal Behav 3:618–621

Rich PJ, Grenier C, Ejeta G (2004) *Striga* resistance in the wild relatives of sorghum. Crop Sci 44:2221–2229

Rodenburg J, Bastiaans L (2011) Host-plant defence against *Striga* spp.: reconsidering the role of tolerance. Weed Res 51:438–441

Rodenburg J, Bastiaans L, Weltzien E, Hess DE (2005) How can field selection for *Striga* resistance and tolerance in sorghum be improved? Field Crops Res 93:34–50

Rodenburg J, Bastiaans L, Kropff M (2006) Characterization of host tolerance to *Striga hermonthica*. Euphytica 147:353–365

Rodríguez-Ojeda MI, Fernández-Escobar J, Alonso LC (2001) Sunflower inbred line (KI-374) carrying two recessive genes for resistance against a highly virulent Spanish population of *Orobanche cernua* Loefl./*O. cumana* Wallr. race "F". In: Proceedings of the 7th international parasitic weed symposium, 5–8 June 2001, Nantes, France, pp 208–211

Román B, Rubiales D, Torres AM, Cubero JI, Satovic Z (2001) Genetic diversity in *Orobanche crenata* populations from southern Spain. Theor Appl Genet 103:1108–1114

Román B, Torres AM, Rubiales D, Cubero JI, Satovic Z (2002) Mapping of quantitative trait loci controlling broomrape (*Orobanche crenata* Forks.) resistance in faba bean (*Vicia faba* L.). Genome 45:1057–1063

Rubiales D, Fernández-Aparicio M (2012) Innovations in parasitic weeds management in legume crops. A review. Agron Sustain Dev 32:433–449

Rubiales D, Alcantara C, Sillero JC (2004) Variation in resistance to *Orobanche crenata* in species of Cicer. Weed Res 44:27–32

Rubiales D, Pérez-de-Luque A, Férnández-Aparicio M, Sillero JC, Román B, Kharrat M, Khalil S, Joel DM, Riches C (2006) Screening techniques and sources of resistance against parasitic weeds in grain legumes. Euphytica 147:187–199

Satish K, Gutema Z, Grenier C, Rich P, Ejeta G (2012) Molecular tagging and validation of microsatellite markers linked to the low germination stimulant gene (*lgs*) for *Striga* resistance in sorghum [*Sorghum bicolor* (L.) Moench]. Theor Appl Genet 124:989–1003

Siame BA, Weerasuriya Y, Wood K, Ejeta G, Butler LG (1993) Isolation of strigol, a germination stimulant for *Striga asiatica*, from host plants. J Agric Food Chem 41:1486–1491

Sillero JC, Rubiales D, Cubero JI (1996) Risks of *Orobanche* screenings based only on final number of emerged shoots per plant. In: Moreno MT, Cubero JI, Berner D, Joel D, Musselman LJ, Parker C (eds) Advances in parasitic plant research. Junta de Andalucía, Sevilla, pp 652–657

Sillero JC, Villegas-Fernández AM, Thomas J, Rojas-Molina MM, Emeran AA, Fernández-Aparicio M, Rubiales D (2010) Faba bean breeding for disease resistance. Field Crops Res 115:297–307

Sukno S, Melero-Vara JM, Fernández-Martínez JM (1999) Inheritance of resistance to *Orobanche cernua* Loefl. in six sunflower lines. Crop Sci 39:674–678

Swarbrick PJ, Huang K, Liu G, Slate J, Press MC, Scholes JD (2008) Global patterns of gene expression in rice cultivars undergoing a susceptible or resistant interaction with the parasitic plant *Striga hermonthica*. New Phytol 179:515–529

Tang S, Heesacher A, Kishore VK, Fernández A, Sadik ES, Cole G, Knapp SJ (2003) Genetic mapping of the *Or5* gene for resistance to *Orobanche* race E in sunflower. Crop Sci 43:1021–1028

Timko MP, Singh BB (2008) Cowpea, a multifunctional legume. In: Moore PH, Ming R (eds) Genomics of tropical crop plants, 1st edn. Springer, New York, pp 227–258

Torres AM, Avila CM, Gutierrez N, Palomino C, Moreno MT, Cubero JI (2010) Marker-assisted selection in faba bean (*Vicia faba* L.). Field Crops Res 115:243–252

Vaidya PK, Raghavender B, Mukuru SZ (1988) Progress in breeding resistance to *Striga asiatica* in sorghum at the ICRISAT center. In: Kim SK (ed) Combating *Striga* in Africa. IITA, Ibadan, pp 81–89

Valderrama MR, Román B, Satovic Z, Rubiales D, Cubero JI, Torres AM (2004) Locating quantitative trait loci associated with *Orobanche crenata* resistance in pea. Weed Res 44:1–6

van Ast A, Bastiaans L, Kropff MJ (2000) A comparative study on *Striga hermonthica* interaction with a sensitive and a tolerant sorghum cultivar. Weed Res 40:479–493

Vasudeva Rao MJ (1985) Techniques for screening sorghums for resistance to *Striga*. In: Musselman LJ (ed) Parasitic weeds in agriculture. CRC, Boca Raton, FL, pp 281–304

Velasco L, Pérez-Vich B, Jan CC, Fernández-Martínez JM (2007) Inheritance of resistance to broomrape (*Orobanche cumana* Wallr.) race F in a sunflower derived from wild sunflower species. Plant Breed 126:67–71

Velasco L, Pérez-Vich B, Yassein AAM, Jan CC, Fernández-Martínez JM (2012) Inheritance of resistance to sunflower broomrape (*Orobanche cumana* Wallr.) in an interspecific cross between *Helianthus annuus* and *Helianthus debilis* subsp. *tardiflorus*. Plant Breed 131:220–221

Vogler RK, Ejeta G, Butler LG (1996) Inheritance of low production of *Striga* germination stimulant in sorghum. Crop Sci 36:1185–1191

Vrânceanu AV, Tudor VA, Stoenescu FM, Pirvu N (1980) Virulence groups of *Orobanche cumana* Wallr. differential hosts and resistance sources and genes in sunflower. In:

Proceedings of the 9th international sunflower conference, 8–13 July 1980, Torremolinos, Spain, pp 74–80

Watling JR, Press MC (2001) Impacts of infection by parasitic angiosperms on host photosynthesis. Plant Biol 3:244–250

Williams CN (1959) Resistance of sorghum to witchweed. Nature 184:1511–1512

Wilson JP, Hess D, Hanna WW (2000) Resistance to *Striga hermonthica* in wild accessions of the primary gene pool of *Pennisetum glaucum*. Phytopathology 90:1169–1172

Yoneyama K, Awad AA, Xie X, Yoneyama K, Takeuchi Y (2010) Strigolactones as germination stimulants for root parasitic plants. Plant Cell Physiol 51:1095–1103

Yu JK, Tang S, Slabaugh MB, Heesacker A, Cole G, Herring M, Soper J, Han F, Chu WC, Webb DM, Thompson L, Edwards KJ, Berry ST, Leon A, Olungu C, Maes N, Knapp SJ (2003) Towards a saturated molecular genetic linkage map for cultivated sunflower. Crop Sci 43:367–387

Zehhar N, Labrousse P, Arnaud MC, Boulet C, Bouya D, Fer A (2003) Study of resistance to *Orobanche ramosa* in host (oilseed rape and carrot) and non-host (maize) plants. Eur J Plant Pathol 109:75–82

Zeid M, Madkour M, Koraiem Y, Nawar A, Soliman M, Zaitoun F (1997) Molecular studies on *Orobanche*. J Phytopathol 145:351–355

Chapter 22
Integrated Agronomic Management of Parasitic Weed Seed Banks

Yaakov Goldwasser and Jonne Rodenburg

22.1 Introduction

The main difficulties in controlling parasitic weeds of the Orobanchaceae arise from the distinctive properties of their seeds: their immense number, their minute size, their extreme longevity and the ease of their dispersal (Bebawi et al. 1984, and see Chap. 8). These cause rapid increase in the parasite soil seed banks, even when the original infestation area is very limited. Containment of infested areas and prevention of seed distribution should therefore be major objectives of parasitic weed management strategies, in addition to direct control interventions against the parasites (Parker 1991; Ramaiah 1987; Rubiales et al. 2009).

22.1.1 Seed Dispersal

The distribution of seeds to near and afar fields is possible due to various factors. As with non-parasitic invasive weeds, human practices are responsible to a large extent for field infestation by parasitic weeds. Parasitic weed seeds are transported to other fields through contaminated soil and water (by run-off) and because parasitic weed seeds adhere the fur of grazing animals, farming implements like ploughs or boots and clothing. Seed dispersal is also caused by local, national and international trade

Y. Goldwasser (✉)
The Robert H. Smith Institute of Plant Sciences & Genetics in Agriculture, The Robert H. Smith Faculty of Agriculture, Food & Environment, The Hebrew University of Jerusalem, Rehovot 76100, Israel
e-mail: gold@agri.huji.ac.il

J. Rodenburg
Africa Rice Center (AfricaRice), East and Southern Africa, P.O. Box 33581, Dar es Salaam, Tanzania
e-mail: j.rodenburg@cgiar.org

D.M. Joel et al. (eds.), *Parasitic Orobanchaceae*,
DOI 10.1007/978-3-642-38146-1_22,

and by farmer-to-farmer seed exchange of contaminated crop seeds. Seed traders on local markets were found to be a more important vector for witchweed seeds than wind or water (Berner et al. 1994).

Animal manure is another important source for infestation. Parasite seeds remain viable after passing through animal's digestive system (Jacobsohn et al. 1987; Berner et al. 1994). *Phelipanche aegyptiaca* seeds lose viability if incubated in a cow stomach for 72 h or if submerged in a cow manure slurry pit for at least 1 month. In reality, viable seeds can be released through direct droppings in the field, because of the much shorter durations of such unfavourable conditions within the digestive system (Goldwasser et al. 2011) (see Chap. 20.6 for quantification of parasite seeds in animal egesta).

Climate changes may also favour parasitic weeds to extend their range (Rodenburg et al. 2011a). Species of weedy Orobanchaceae have the ability to adapt to environmental changes (Kroschel 1998) because of their rather broad climatic tolerance (Mohamed et al. 2006) and high genetic variation within field populations (Satovic et al. 2009, see Sect. 19.3.2). So far the spread of *Striga* spp. is limited mainly to tropical regions, but species that thrive under conditions of low and variable rainfall or temporary submergence (e.g. *Rhamphicarpa fistulosa*) may benefit from extreme or erratic climates (Rodenburg et al. 2010).

22.1.2 Seed Bank Development

Even a small seed bank rapidly increases in subsequent seasons when suitable host plants (crops or weeds) grow in the field (Lopez-Granados and Garcia-Torres 1993). The seed production of only 2–3 *Striga* plants per m^2 would be high enough to replenish the seasonal seed bank losses (van Delft et al. 1997). Westerman et al. (2007) conservatively estimated success rates from a viable *S. hermonthica* seed to production of viable seeds. The rate of seed survival in the seed bank was estimated at 40 %, the rate of host-stimulated germination at 10 % (and germination caused by a non-host root at 20 %), the proportion of successful attachments at 20 %, successful establishment at 50 %, the chance of emergence at 30 % and the proportion of those plants reaching maturity at 34 % of which only 76 % would produce seeds, with 7,500 seeds per plant and 70 % seed viability. Hence, the chance that an individual seed in the seed bank develops into a seed-producing plant is very low, but even with these conservative estimates, the seasonal multiplication rate would still be over 150 %, meaning a fivefold increase of the seed bank every four seasons. Similarly, only 0.003 % of the *Orobanche crenata* seed bank successfully attaches to host roots, 9 % of these emerge from the soil, and about 43 % of the subsequently produced seeds are lost through various natural processes (dispersion, predation, degradation), and hence only 57 % of the total number of produced seeds is incorporated back in the soil. The annual soil seed bank still multiplies more than threefold in eight years due to the high seed production from each successful *O. crenata* plant (Lopez-Granados and Garcia-Torres 1993).

22.1.3 Seed Bank Control

Completely eliminating the parasite seed bank in the soil is practically impossible. Near complete elimination (from 174,824 infested ha in 1957 to 809 infested ha in 2011) has been achieved in *Striga asiatica* infestations in North and South Carolina (USA), thanks to massive financial investments, by the use of advanced equipment and expensive control agents which were integrated with strict phytosanitary regulations that were enforced by the Federal Government (Eplee 1981, 1992; Tasker and Westwood 2012). However, even this approach is no guarantee for complete eradication. Despite the implementation of an expensive control programme and strict quarantine measures for the eradication of *Phelipanche ramosa* in California (USA), there have been several reoccurring outbreaks of infestations, including in recent years (R. Smith, personal communication).

The South Australian branched broomrape (*P. ramosa*) eradication programme has been launched in 2001. A *P. ramosa* infestation discovered at first in 1911 was reported as extinct within a few years; but in 1991 an additional infestation was detected in Glenelg 90 km east of the 1911 infestation and was eradicated by methyl bromide; however, within the next 7 years an additional 22 infestations were detected within 15 km, leading to the launching of the 2001 eradication programme (Panetta and Lawes 2007). This programme crosses national and state boundaries, includes all three levels of government and is supported by the community, government, industry and research organizations with the common objective to eradicate the parasite through quarantine protocols, land management practices, fumigation and herbicide treatments (Government of South Australia 2011). Recent progress reports of this programme reveal that human-mediated dispersal, mainly by machinery, is the most important parasite seed vector. A high overall effectiveness of containment after 2000 is reported, and though there was an increase in the number of infested paddocks, only few of them were beyond the boundary of the quarantined area (Panetta 2012).

Only integrated measures are expected to lead to successful containment of the parasitic weeds problem (Rubiales and Fernandez-Aparicio 2012). Such measures should be targeted at (1) reduction of existing seed banks, (2) prevention of further seed production and (3) avoiding seed dissemination. These objectives are mutually dependent. The seed bank can only be reduced when new seed inputs are smaller than the output caused by unsuccessful germination, pathogens, seed predation or natural death of the seed (Lonsdale 1993; Westerman et al. 2003; Grundy et al. 2003; van Mourik et al. 2005, 2011). Reduction of the parasite reproduction and increase of seed bank demise while preventing seed dispersal to other places can all be achieved by cultural measures. In this chapter, we discuss the use of agronomic practices with proven or putative effectiveness for the control of parasitic weeds and suggest options for integrated parasitic weed management.

22.2 Phytosanitary Measures

A number of measures need to be instituted to prevent new infestations: (1) Movement of seeds out of an infested field must be avoided. Therefore, vehicles, farm machinery and planting material must not carry infested soil when moving from one field to another. (2) Farm animals should be prevented from becoming vectors for parasitic weeds by (a) limiting their movements between fields, (b) cleaning them to avoid transfer of infested soil or seeds on their body and (c) avoiding use of fodder originating from infested fields. (3) Plant material should come only from certified sources and be free from parasite seeds. (4) Irrigation or flooding should not come from ponds, canals or reservoirs that are located adjacent to infested fields or contain seed-contaminated water or sewage, without fine filtering of the water. (5) Erosion of infested soil, by wind or water, should be prevented (Jacobsohn 1986; Parker 1991; Government of South Australia 2011).

22.3 Reduction of Parasite Seed Production and Crop Damage

A range of agronomic measures is available to contain or reduce the extent of parasite presence in infested fields. These measures are discussed in more detail below.

22.3.1 Hand Weeding

Hand weeding is the most obvious and therefore most common control method against parasitic weeds. Early weeding is presently the only way subsistence farmers can control *Rhamphicarpa fistulosa* in their lowland rice fields (Fig. 22.1a, b). Unlike *Striga*, *Rhamphicarpa* is a facultative parasite that will only start to parasitize after its emergence (Ouédraogo et al. 1999). Weeding *Rhamphicarpa* is therefore effective both for preventing damage to the host and for avoiding production and dispersal of the parasite seeds.

However, weeding emerging or flowering obligate parasites like *Striga*, *Phelipanche* and *Orobanche* (Fig. 22.1c) can be conducted only after spotting the appearance of the parasite above ground, hence when the host plant has already suffered from the parasite. Nevertheless, hand weeding is an important measure to reduce future infestations if it is carried out prior to seed production and release (Parker and Riches 1993; Goldwasser and Kleifeld 2004; Rodenburg et al. 2006). It is essential to remove the weeded plants from the field and to effectively destroy them because seeds can ripen even when the plants are disconnected from the host (Fig. 22.1d).

Fig. 22.1 Parasite seed dispersal. (**a**, **b**) Hand weeding *Rhamphicarpa fistulosa* in rain-fed lowland rice in central Benin. (**c**) Flowering shoots of *Phelipanche aegyptiaca* left on the soil in a pepper greenhouse in Israel, leading to maturation of seeds and further infestation. (**d**) *P. aegyptiaca* inflorescences left to rot in plastic bags in the sun after being weeded in a tomato field in Israel (photos: **a**, **b** J. Rodenburg; **c**, **d** Y. Goldwasser)

22.3.2 Transplanting and Deep Sowing

Transplanting reduces infestation by parasitic weeds. It gives the crop a time advantage and thereby renders it more competitive with parasitic weeds such as *Striga* (Oswald et al. 2001). Transplanting is often used in lowland rice production as a general measure to avoid weed competition (Rodenburg and Johnson 2009) and to facilitate weeding operations (in row-planted crops). For these reasons, transplanting was suggested as a good practice in lowland rice fields infested by *Rhamphicarpa fistulosa* (Rodenburg et al. 2010). Transplanting has also proven an effective way to reduce *Striga* infestations in sorghum, compared to seed-sown sorghum (van Ast et al. 2005).

For the same reason, deep sowing can also contribute to reduced parasitism. A study conducted in sorghum and maize showed that after using the standard sowing depths (at 2.5 cm) under high *Striga* infestation levels, the seed bank increased by 1,397 and 1,876 seeds dm^3, depending on crop variety, whereas following deep sowing (at 15–20 cm, in cone-shaped holes) seed bank increases were reduced on average by 55 % (van Delft et al. 2000). Under moderate *Striga* infestation levels (836 and 970 seeds dm^3), the increase in seed banks was reduced by 25 %.

When deep planting was combined with no-tillage, *Striga* seed production was even completely suppressed. Practices like these require, however, more labour (e.g. transplanting compared to direct sowing, digging planting holes compared to standard sowing practices) and also have clear trade-offs with other management operations such as land preparation and mechanical weeding.

22.3.3 Enhancing Chemical Soil Fertility

Soil fertility plays a key role in *Striga* and *Rhamphicarpa* management. Phosphate deficiency, and in some cases nitrogen deficiency, has shown to play an important role in the biosynthesis of strigolactones, the parasite germination stimulants (see Sect. 10.2). Hence, increasing the level of N and P in the soil by fertilizer application could reduce germination and consequently infestation rates of parasitic weeds (Lopez-Raez et al. 2009; Jamil et al. 2011, 2012). Fertilizer application can reduce the population density of *Rhamphicarpa* (Fig. 22.2; Rodenburg et al. 2011b). Enhancing soil fertility by applying urea 3 weeks after sowing helped in reducing the number of emerged *S. asiatica* plants in rice (Riches et al. 2005) and helped delaying and reducing parasite infestation while raising rice yields in *S. hermonthica*-infested fields (Adagba et al. 2002). Although some studies reported insignificant effects of nitrogen fertilization (e.g. Kamara et al. 2007; Sinebo and Drennan 2001; Tesso and Ejeta 2011), *Striga* is generally considered to be closely associated with poor soil fertility (Emechebe et al. 2004; Parker 2009), and increasing the soil fertility by fertilizer applications seems an effective way to reduce infestation (Bebawi 1981; Ransom 2000; Showemimo et al. 2002; Jamil et al. 2012). However, increased soil fertility does not necessarily reduce *Phelipanche* and *Orobanche* parasitism, and crops can be heavily infested by broomrapes despite high applications of mineral and organic fertilizers as in the case of Israeli agriculture (Y. Goldwasser, personal observations). See further discussion on strigolactones and soil fertility in Sect. 10.4.2.

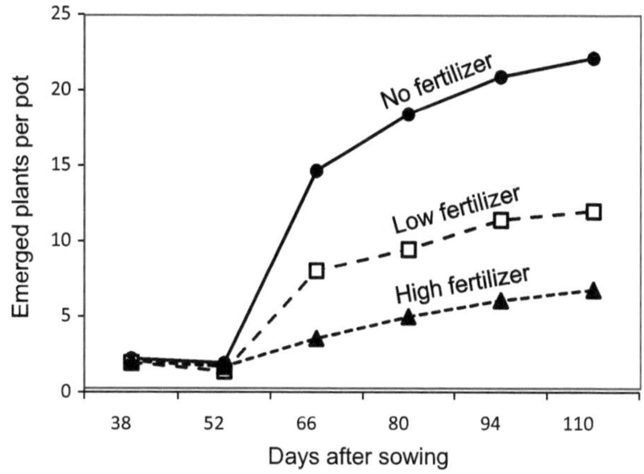

Fig. 22.2 Reduction of *Rhamphicarpa fistulosa* germination by fertilizer. *Filled circles*: no fertilizer; *open squares*: 32.5 kg N, 15 kg P and 15 kg K ha^{-1}; *filled triangles*: 65 kg N, 30 kg P and 30 kg K ha^{-1}. Averaged over seven rice cultivars (from Rodenburg et al. 2011b, with permission)

22.4 Methods to Reduce Existing Seed Banks

22.4.1 Soil Fumigation and Ethylene

Various cultural and chemical control measures are available to kill parasitic weed seeds in the soil. The most effective are non-selective soil applied fumigation pesticides that are also used for the control of various soilborne pests and pathogens. All soil fumigation agents are expensive, extremely toxic to humans and can cause air and water contamination. They thus require special expensive application equipment. Effective soil fumigation requires adequate soil preparation, soil moisture control, precise and uniform application and means to avoid evaporation of the fumigant. Fumigants are toxic to crops thus requiring ventilation and a waiting period prior to crop planting (Goldwasser and Kleifeld 2004). Precise performance of all aspects of application is essential for ensuring fumigation success.

Methyl bromide is an effective soil fumigant to reduce the soil seed bank of broomrape (Wilhelm et al. 1958) and witchweed (Gurney et al. 1995). This fumigant was widely used for broomrape control mainly in cash crops and in eradication programmes such as the national eradication project of *P. ramosa* in introduced infestations in California (Wilhelm et al. 1959; Goldwasser and Kleifeld 2004). In 2005 world health and agricultural authorities banned the use of methyl bromide due to its negative environmental effects, igniting the search for an alternative cost-effective fumigant. This search is still ongoing.

Fumigants that release the toxic compound methyl isothiocyanate have been used for *Orobanche* and *Phelipanche* eradication. Metham sodium, a liquid injected directly into the soil or by chemigation (i.e. the process of applying pesticides via irrigation water), and dazomet, a powder incorporated mechanically into the soil followed by irrigation, were found to be effective for *Orobanche* and *Phelipanche* control. The latter fumigant has been successfully used in the South Australia

Branched Broomrape Eradication Program. These compounds can be effective in deeper soil layers but show more erratic results in the upper soil layer, due to their rapid evaporation (Goldwasser et al. 1995). Control failures were also attributed to the build-up of microbial biodegradation in the soil.

The nematocidal fumigant 1,3-dichloropropene is applied into the soil followed by sealing of the soil with a roller or by sprinkler irrigation. The method was very effective against *O. crenata* but less so against *Phelipanche aegyptiaca* (Jacobsohn et al. 1991). The difficulties in application of this fumigant and its narrow pest control range resulted in only limited use in small-scale intensive farming.

Chloropicrin is an old fumigation agent currently re-evaluated as a methyl bromide substitute for *P. aegyptiaca* control in Israel. Two formulations, one for soil injection and the second for application via drip irrigation, were applied under polyethylene strip mulching in greenhouse trials. Results show a significant reduction in broomrape infestation during the subsequent crop, but the efficiency of this product depends on the number of drip lines and the volume of water applied to the treated area (G. Sagi, personal communication).

A new soil fumigation agent, dimethyl disulfide, available for soil injection or for application through drip irrigation, is evaluated for *P. aegyptiaca* control mainly in greenhouses under polyethylene mulching. Results of these studies reveal good control of the parasite (G. Sagi, personal communication 2012).

In a number of laboratory and field trials, methyl iodide was shown to be as effective as methyl bromide for control of many soilborne pests. Significant decrease in *P. aegyptiaca* emergence was achieved in trials in Israel in which the fumigant was delivered via drip irrigation systems in small plots (in greenhouses) under polyethylene mulching (O. Naot, personal communication 2012).

In a current soil fumigation research programme in Israel, initial results of experiments under controlled conditions and in small field plots show that metham sodium, chloropicrin, dimethyl disulfide, methyl iodide and mixtures of these fumigants at reduced rates efficiently control *P. aegyptiaca* seeds (A. Gamliel, personal communication 2012). The success of all these fumigants in this research programme is probably due to the sandy soil conditions and very precise application procedures. Further studies in different soils and conditions are in progress.

Striga control can be achieved using ethylene gas (Rodenburg et al. 2005). Ethylene can be injected in the soil using a backpack ethylene applicator as described by Bebawi et al. (1985) or by tractor-mounted injectors (Eplee 1992). The gas should be injected at regular intervals in the soil at reasonable depth (10–20 cm) at least 1 week prior to sowing of the crop and in the absence of any other vegetation. These gas injections provoke preconditioned *Striga* seeds to germinate and die in the absence of a suitable host (Logan and Stewart 1991). Ethylene application has been successfully employed as part of the *Striga* eradication project in the USA (Eplee 1981, 1992; Tasker and Westwood 2012). However, ethylene injections do not guarantee total eradication of *Striga*, and the practice is very expensive (Rodenburg et al. 2005). Both ethylene and the injector are not readily available in remote places, rendering this technology less suitable for subsistence farmers living in rural Africa.

22.4.2 Soil Solarization

Solarization is a method that facilitates trapping of solar radiation in moist soil under transparent plastic sheets to kill soilborne insects, fungi and weed seeds (Katan et al. 1976). Solarization can also effectively be used to control broomrape seed banks (Jacobsohn et al. 1980; Sauerborn et al. 1989; Abu-Irmaileh 1991; Mauromicale et al. 2005). It can be applied in regions where the field can be exposed to direct sunlight for a long period. Broomrape seed control is achieved in the upper soil levels if the soil is covered for 5–6 weeks, allowing the peak soil temperatures to reach 50 °C and higher during the day. However, in heavy clay soils, such lethal high temperatures cannot be reached below the upper 15–20 cm soil layer, allowing broomrape seeds to escape from control (Kleifeld et al. 1999; Joel et al. 2007). Therefore, solarization in heavy clay soils is sufficient only for crops that develop shallow root systems, such as carrot. Only imbibed or preconditioned parasite seeds do not survive high temperatures. Under lab conditions, a 100 % loss of germinability in preconditioned *P. aegyptiaca* seeds was achieved by 4 h heat treatment at 55 °C (Fig. 22.3). It is important to retain high soil moisture under the transparent plastic sheets for effective solarization and to prevent drying during the treatment. This requires thick plastic sheets, a good gluing system for the seams, and protection of the treated areas from wind and roaming animals.

Although solarization would also be an effective control strategy for *Striga* spp. and *Rhamphicarpa fistulosa*, the common weak financial situation of the farming systems in sub-Saharan Africa would prohibit the use of this method on a large scale. The same is true for control of broomrape in developing countries, where it would only be viable for high-value export crops.

22.4.3 Flooding and Irrigation

Flooding of broomrape-infested fields causes decay of parasitic weed seeds, leading to a decrease in infestation. Long-period flooding significantly reduced *O. crenata* infestation in subsequent host crops, but the effect of shorter flooding periods was not consistent (Parker and Riches 1993; Linke 1999). In a recent lab study, *P. aegyptiaca* seeds placed in flooded soil containers completely lost their viability after 9 days of submergence (Y. Goldwasser, unpublished). Based on the preferred

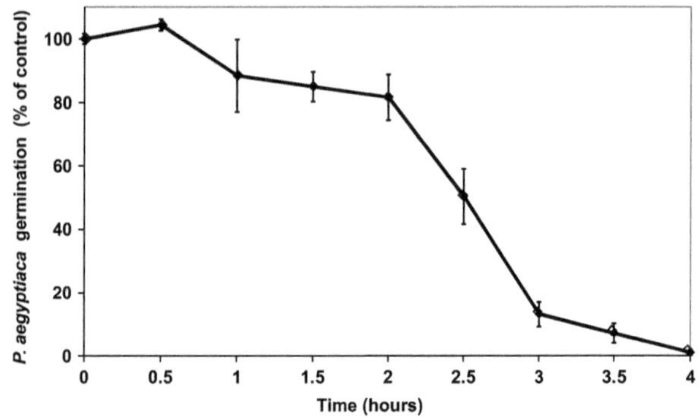

Fig. 22.3 Influence of heat treatments on parasite seed germination. Germination of preconditioned *Phelipanche aegyptiaca* seeds following different durations of heat treatment in a 55 °C incubator (Y. Goldwasser, unpublished)

habitat of *R. fistulosa*, which is in temporarily flooded fields (Hansen 1975), improved water management, e.g. controlled continuous flooding, would also be an effective measure to control this parasite in lowland rice (Rodenburg et al. 2010), a hypothesis that has recently been supported by results of a pot trial (Fig. 22.4). Following the ecological preference of *Striga* spp., most of the weedy species of this genus could be controlled by continuous flooding as well. However, apart from rice, none of the important crops that are susceptible to *Striga* attack can be grown under such flooded conditions, while in rice systems where *Striga* is prevalent, i.e. rain-fed uplands (Rodenburg et al. 2010), flood irrigation would mostly not be feasible due to the free-draining characteristic of the predominant soils in these agroecosystems and due to a common lack of available water or lack of irrigation infrastructure.

Irrigation of parasitized crops can result in increased yields as demonstrated by the case of *Orobanche cumana* on confectionary sunflower under dry-land farming in Israel; heavy infestations with *O. cumana* under dry-land farming resulted in total loss of crop yield, while the same field under irrigation regime resulted in some commercial yield (Y. Goldwasser, personal observations).

22.4.4 Enhancing Biological Soil Fertility

An increase of soil organic matter accelerates seed decomposition and facilitates seed predation by respectively encouraging the development of the soil microflora and soil fauna (Ahonsi et al. 2002; van Mourik et al. 2003; Ayongwa et al. 2011b). A negative correlation was found between *Striga* soil infestation and soil fertility (Sauerborn et al. 2003). The quantity and quality of organic matter, in particular the N-release characteristic determined by the C:N ratio, are important in this respect, but large amounts (>5 tonnes/ha) of organic matter are usually needed per cropping season in order to be effective (Ayongwa et al. 2011a).

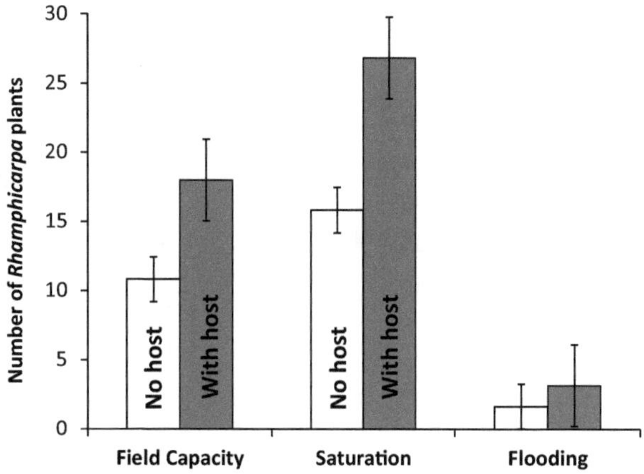

Fig. 22.4 Effect of flooding and drainage on *Rhamphicarpa fistulosa*. Results of a pot experiment indicating that the development of *R. fistulosa* can be controlled by improved water management either by effective drainage, i.e. reducing soil water levels from 'saturation' to 'field capacity', or by permanent flooding. Average numbers of *R. fistulosa* plants per pot at 89 days after sowing (J. Rodenburg, unpublished)

22.4.5 Cropping Systems

Intercropping and relay intercropping can help in reducing parasitic weed infestations and seed banks and increasing crop yields in fields infested by parasitic weed. This is particularly effective against the obligate witchweeds and broomrapes. Adding a non-host crop in the field can contribute to parasitic weed control in three ways: (1) The intercrop can act as a cover crop, thereby providing shade, reducing soil temperature and suppressing weeds and parasitic weeds in particular *Striga* spp.; (2) the intercrop can improve the soil fertility through nitrogen fixation or through the production of organic matter followed by decomposition causing the release of nutrients and increase of biological activity and the production of ethylene gas (see Sect. 22.4.1); (3) the intercrop can produce root exudates that cause suicidal germination in case it is not a suitable host for the parasite (Parker and Riches 1993).

Trap crops ('false hosts') are non-host crops that can stimulate parasite germination (see Sect. 10.4.1), but further development of the parasite is impeded as no viable connection of the parasite haustorium with the host root is established (Parker and Riches 1993; Goldwasser et al. 1997).

Striga may be controlled by rotating or intercropping the cereal crop with groundnut (*Arachis hypogea*) (e.g. Carson 1989), cowpea (*Vigna unguiculata*) (e.g. Carsky et al. 1994), soybean (*Glycine max*) (e.g. Robinson and Dowler 1966; Carsky et al. 2000), pigeon pea (*Cajanus cajan*) (e.g. Oswald and Ransom 2001), cotton (*Gossypium* spp.; e.g. Murdoch and Kunjo 2003) or yellow gram (*Cicer arietinum)* (Oswald et al. 2002). Intercropping with sesame (*Sesamum*

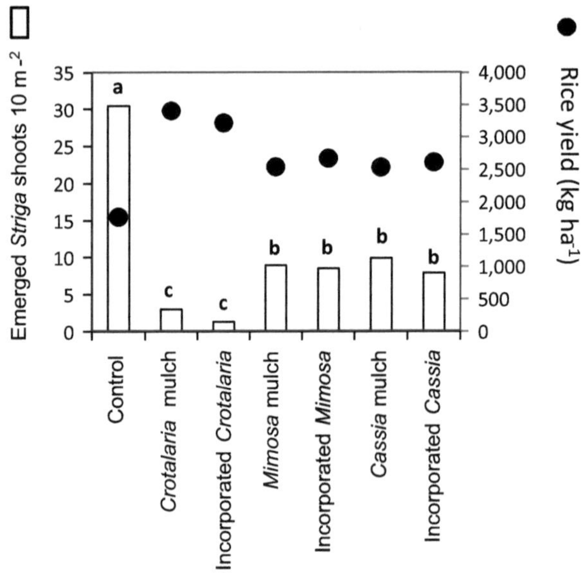

Fig. 22.5 Effect of green manure mulching or incorporation on *Striga asiatica* and rice. *S. asiatica* emergence in rice and rice yields, following mulching with various green manure species, or their incorporation into the soil (based on data from Kayeke 2004, with permission)

indicum) or cowpea reduces the *Striga hermonthica* seed bank in millet fields in the long run, as was shown by van Mourik et al. (2008). Rotations with the green manure crops *Cajanus cajan* or *Crotalaria ochroleuca* improved upland rice yields in *Striga asiatica*-infested fields in Tanzania, mainly because of the enhanced soil fertility and because the root exudates provoke suicidal *Striga* germination (Riches et al. 2005). Hence, in addition to positive results on rice yields, reduction in the number of emerged *S. asiatica* was obtained when *C. ochroleuca* was grown for 2 months, followed by mulching or soil incorporation of *C. ochroleuca* biomass, prior to growing rice (Fig. 22.5) (Kayeke 2004). Other well-known intercrop species with proven success in *Striga* suppression can be found in the genus *Desmodium* (e.g. Pickett et al. 2010, see Chap. 25). Growing these forage legumes improves the soil fertility and, most importantly, causes suicidal germination and inhibition of parasite attachments to the host roots, because of the flavonoid compounds in their root exudates (e.g. Khan et al. 2010; see Sect. 25.2). Through these mechanisms it is considered an effective intercrop to reduce the witchweed seed bank.

Important crops that were reported in reducing broomrape seed banks are sorghum (*Sorghum bicolor*), barley (*Hordeum vulgare*), maize (*Zea mays*), vetches (*Vicia villosa* and *V. atropurpurea*), clover (*Trifolium* spp.), flax (*Linum usitatissimum*), coriander (*Coriandrum sativum*), pepper (*Capsicum annuum*), cowpea (*Vigna unguiculata*), hemp (*Cannabis sativa*), mung bean (*Phaseolus aureus*), snap bean (*Phaseolus vulgare*), alfalfa (lucerne) (*Medicago sativa*), soybean (*Glycine max*) and chickpea (*Cicer arietinum*) (Krisnamurthy and Rao 1976; Abu-Irmaileh 1984; Sauerborn and Saxena 1986; Al-Menoufy 1991; Saxena et al. 1994; Kleifeld et al. 1994; Abebe et al. 2005).

Most of these crops were effective in pot experiments but not under heavily infested field conditions.

Catch crops are host plants that support normal parasitism but are removed from the field after the parasite seeds germinated and before the flowering and seed dispersal stages of the parasite. By this method, the parasite seed bank is reduced in a manner similar to that of trap crops. Important crops reported as potential catch crops for broomrape control are faba bean (*Vicia faba*), field mustard (*Brassica campestris*), white mustard (*Sinapis alba*), lentil (*Lens culinaris*), berseem clover (*Trifolium alexandrinum*) and fenugreek (*Trigonella foenum-graecum*) (Parker and Riches 1993; Sauerborn and Saxena 1986; Dhanapal et al. 1996; Acharya et al. 2002; Fernandez-Aparicio et al. 2008, 2010).

Theoretically the activity of trap and catch crops by inducing 'suicidal germination' of the parasite seeds in soil is an effective measure to reduce the parasite soil seed bank, as viable parasitic weed seeds germinate in vain and are not replenished by newly produced seeds. Numerous potential trap crops have been suggested (Parker and Riches 1993), but most of them were examined only in vitro or in small pots and not in the field. In many cases there are contradicting reports regarding the effectiveness of certain crops under field conditions (Goldwasser and Kleifeld 2004), which may result from environmental differences and from the use of different cultivars. Repeated cropping with trap and catch crops is required for effective reduction of the parasite seed bank in the soil, but in most cases it is not economical or a good agronomical practice because of depletion of soil nutrients and build-up of specific harmful crop diseases and pests, including the crop becoming a volunteer weed in subsequent crops. Thus, this measure should only be used as part of an integrated management scheme (Goldwasser and Kleifeld 2004; see Sect. 22.5).

Recently, van Mourik et al. (2011) compared *Striga hermonthica* seed bank depletion rates, attained under bare or weedy fallow and mono- or intercropping with a non-host crop, to the rates attained under cereal (sorghum or millet) mono-cropping. They found that seasonal depletion rates of the soil seed bank under continuous mono-cropping of the host crop (75–82 % under sorghum depending on variety and 74 % under millet) were higher than when intercropped with a non-host crop (49–66 %), which in turn depleted the soil seed bank more than when the non-host crop was grown as mono-crop (35–43 %) or when the soil was under a weedy fallow (47 %) or left bare (28–43 %). These results suggest that if parasitic weed seed production can be prevented, e.g. by timely hand weeding, growing a suitable host can actually be more effective in reducing the weed seed bank than growing a non-host crop as mono- or intercrop or leaving the field fallow

Crop rotations can be an effective way of reducing the *Striga* seed bank, in particular when the rotation crop can act as trap crop for the parasite (Oswald and Ransom 2001; see *trap crops* above). It is effective in reducing the seed bank primarily because it interrupts the seasonal production of parasitic seed weeds, it improves the *Striga* suppressive capacity of the soil (Parkinson et al. 1987) and it can cause suicidal germination as described above. Although the above-described measures can potentially provide partially sustainable solutions to parasitic weed problems, the farmer needs to carefully select the rotation or intercrop species to avoid growing crops that can themselves act as hosts for parasitic weeds. In order to avoid competition with the subsequent crop, fallow plant residues should be either burnt, removed, incorporated in the soil or mulched prior to a new cropping season. The intercrop, rotation or improved-fallow species should also be rotated regularly to avoid that populations of other pests and disease develop into economically harmful proportions (Teasdale 2003). While these kinds of crop management technologies (e.g. rotations, intercropping) may be technically sound, adoption rates by farmers are often low (e.g. Tarawali et al. 1999), which is caused by land tenure insecurities, additional costs and work load and difficult or unsatisfactory establishment of the intercrop or catch crop (Faulkner 1934; Langyintuo and Dogbe 2005). Moreover, farmers might find many of these cropping system alternatives impractical or not sufficiently profitable. Suitable legumes should therefore at least combine parasitic weed control characteristics with an additional economic benefit to increase the likelihood to be acceptable to farmers (Becker and Johnson 1999; Ransom 2000), and they should possess good environmental adaptation.

Resistant crop varieties (see Chap. 21) are similarly able to reduce parasitic weed infestation rates because they act as trap crops. **Tolerant crop** varieties are able to reduce the negative effects of parasitic weed infestation on crop yields but do not prevent seed production by the parasite (e.g. Rodenburg and Bastiaans 2011). Various crop varieties are resistant to parasitic weeds. Resistant varieties, while able to reduce immediate damage to the crop (in particular when combined with tolerance), are unlikely to significantly reduce the seed bank simultaneously in a similar way as trap crops. When the natural seasonal seed bank depletion is 46 %, the production of only 8 seed capsules per m^2 for *S. hermonthica* would fully replenish a low-density seed bank of 30,000 seeds per m^2 during one season. In a multi-variety study on *Striga*-infested sorghum, only with the very resistant sorghum variety N13, and only when it was grown under low infestation levels, this threshold was not reached. In all other cases, the estimated production of *Striga* seed greatly surpassed replenishment of the seasonal losses and hence increased the soil seed bank (e.g. Rodenburg et al. 2006).

22.5 Integrating Agronomic Management Practices

No single control practice described above will accomplish full- and long-term control of parasitic weeds as a stand-alone measure. For effective and durable control, measures need to be combined (e.g. Perez-de-Luque et al. 2010). The real challenge is to integrate practices that obtain optimum efficiency in terms of reduction of existing seed banks, prevention of seed production and avoidance of seed dissemination, with input rates affordable to individual and often resource-poor farmers.

The impact of integrated control measures on broomrape seed bank has been studied and modelled by several researchers. Grenz et al. (2005) assessed strategies for *Orobanche* spp. control using a combined seed bank and competition model approach. They quantified the effects of environment, rotation, tillage, hand pulling and combined strategies on parasite seed bank dynamics and bean yields and concluded that only by combining several management approaches, such as delayed sowing, no-till and hand pulling, can parasite populations be contained. A computer simulation model was developed to predict broomrape species infestation levels in a dynamic and deterministic way based on published data (Kebreab and Murdoch 2001; see Chap. 11). The simulations also stress the importance of integrating control measures and preventing new seeds from entering the soil seed bank.

The *best-bet* integrated parasitic weed management approach should follow three main principles: (1) Different control measures should not conflict with one another. (2) The individual components should be adapted to the local conditions such as the access to inputs, the presence of other important pests or diseases, climate and soil characteristics and the specific cropping systems, and preferably address other production constraints. (3) Any integrated management strategy should contribute to the following three objectives:

(a) Reducing the seed bank
(b) Preventing or reducing parasitic weed seed production
(c) Avoiding further seed dissemination

The main obstacle in the long-term management of infested fields is the near indestructible seed bank. There is an urgent need to implement novel integrated parasitic weed management programmes to overcome this obstacle. These should be based on new findings in molecular biology and physiology of plant-pathogen interactions and the use of monitoring and decision support systems enabling precision agriculture and site-specific farming technologies (Rubiales et al. 2009, see Sect. 23.4).

A recent example of successful integrated *Striga* management is the combination of resistant varieties, tied-ridge tillage and N-fertilizer as shown in sorghum systems in Ethiopia (Tesso and Ejeta 2011). This example seems to satisfy most, if not all, of the above criteria. The measures are adapted to local circumstances and also address other production constraints, such as low soil fertility and drought. The measures are also synergistic or at least additive in the sense that they do not seem to conflict each other. Resistant crop varieties, for instance, are generally considered a useful component of an integrated approach (Haussmann et al. 2000) that is usually easy to combine with other measures such as soil fertility amendments, land preparation or soil tillage. Finally the approach tested by Tesso and Ejeta (2011) could in theory help to achieve the three objectives of containment, reduction and prevention. By increasing the soil moisture content and the chemical soil fertility, the tied-ridges and N-fertilizer could attract more biological activity and through that enhance *Striga* suppression ability of the soil and thereby reduce the seed bank (objective a). By reducing the number of successful parasites, the resistant varieties could theoretically reduce the *Striga* seed production (objective b). By limiting or slowing down soil run-off, the tied ridges could reduce or arrest dissemination to neighbouring fields (objective c). Unfortunately, Tesso and Ejeta (2011) only provided data on sorghum yield and *Striga* infestation levels (emergence counts and vigour scores) and not on *Striga* reproduction, seed bank and dispersal rates. Consequently there is no proof that the three objectives can indeed be met using this approach.

As integrated management practices need to be locally adjusted, it presumably requires multiple seasons of farmer-participatory on-farm testing and fine-tuning before the management strategy is fully optimized and tailored to the local bio-physical and socioeconomic conditions, implying intensive interactions between researchers, extension and farmers (e.g. Abang et al. 2007; Ransom 2000). It must be emphasized that all it takes is a single season of neglecting a small existing parasite population to initiate a gradual increase in parasitic weed seed bank densities and associated crop infestations leading to progressively declining crop yields.

22.6 Conclusions

Difficulties in containment and elimination of the parasitic weed seed bank in agricultural situations are primarily due to the specialized traits of the parasite that ensure mass production of seeds, their vast dispersal, longevity in the soil and germination in close association with host plants. Cultural strategies that reduce the parasitic weed seed bank, in one way or another, include phytosanitary measures, hand weeding, alternative planting methods, flooding, soil solarization, crop rotations or intercropping with catch and trap crops. These measures by themselves are not sufficiently effective to completely eliminate the seed bank of parasitic weeds but can impede or reduce seed production and dispersal. Reduction of existing infestations can be achieved by integrating these measures with additional measures, such as the use of resistant host varieties, soil fumigation and herbicide application, that are discussed in other chapters.

As with many other pests, prevention and early containment of parasitic weed seed banks is essential to avoid new infestations and seed bank build-up in already infested fields. Once the parasitic weed seeds are in the soil, it is practically impossible to completely eradicate them. An approach whereby effective and locally available and affordable measures are used in an integrated manner, preferably with different technologies targeting the various objectives of long-term reduction, containment and prevention of parasitic weed infestations, is most likely the only way the problem of parasitic weeds can be effectively and sustainably managed. Very few locally adapted on-site integrated parasitic weed management strategies have been developed, implemented and tested. There is a great need for quantitative assessment of the effectiveness of these strategies in terms of seed bank reduction as well as agronomic and economic feasibility. Developing and fine-tuning of a *best-fit* strategy requires intensive research, farmer-participatory work and a long-term commitment.

References

Abang MM, Bayaa B, Abu-Irmaileh B, Yahyaui A (2007) A participatory farming system approach for sustainable broomrape (*Orobanche* spp.) management in the Near East and North Africa. Crop Prot 26:1723–1732

Abebe GG, Sahile A, Al-Tawaha ARM (2005) Evaluation of potential trap crops on *Orobanche* soil seed bank and tomato yield in the central rift valley of Ethiopia. World J Agric Sci 1:148–151

Abu-Irmaileh BE (1984) Effect of planting flax on subsequent infestation of tomato by *Orobanche ramosa*. In: Parker C, Musselman LJ, Polhill RM, Wilson AK (eds) Proceedings of the third international symposium on parasitic weeds. ICARDA, Aleppo, Syria, pp 250–255

Abu-Irmaileh BE (1991) Soil solarization controls broomrapes (*Orobanche* spp.) in host vegetable crops in the Jordan valley. Weed Technol 5:575–581

Acharya BD, Khattri GB, Chettri MK, Srivastava SC (2002) Effect of *Brassica campestris* var. toria as a catch crop on *Orobanche aegyptiaca* seed bank. Crop Prot 21:533–537

Adagba MA, Lagoke STO, Imolehin ED (2002) Nitrogen effect on the incidence of *Striga hermonthica* (Del.) Benth in upland rice. Acta Agron Hung 50:145–150

Ahonsi MO, Berner DK, Emechebe AM, Lagoke ST (2002) Effects of soil pasteurization and soil N status on severity of *Striga hermonthica* (Del.) Benth. in maize. Soil Biol Biochem 34:1675–1681

Al-Menoufy OA (1991) Crop rotation as a control measure of *Orobanche crenata* in *Vicia faba* fields. In: Wegmann K, Musselman LJ (eds) Proceedings of the international workshop in *Orobanche* research. Eberhard-Karls-Univesitat, FRG, Tubingen, pp 241–247

Ayongwa GC, Stomph TJ, Beder P, Leffelaar PA, Kuyper TW (2011a) Organic matter and seed survival of *Striga hermonthica* – mechanisms for seed depletion in the soil. Crop Prot 30:1594–1600

Ayongwa GC, Stomph TJ, Kuyper TW (2011b) Host-parasite dynamics of *Sorghum bicolor* and *Striga hermonthica* – the influence of soil organic matter amendments of different C:N ratio. Crop Prot 30:1613–1622

Bebawi FF (1981) Response of *Sorghum bicolor* cultivars and *Striga hermonthica* population to nitrogen fertilization. Plant Soil 59:261–268

Bebawi FF, Eplee RE, Harris CE, Norris RS (1984) Longevity of witchweed (*Striga asiatica*) seed. Weed Sci 32:494–497

Bebawi FF, Eplee RE, Harris CE, Norris RS (1985) The dispersion of backpack-applied ethylene in soil. Weed Sci 33:74–77

Becker M, Johnson DE (1999) The role of legume fallows in intensified upland rice-based systems of West Africa. Nutr Cycl Agrosyst 53:71–81

Berner DK, Cardwell KF, Faturoti BO, Ikie FO, Williams OA (1994) Relative roles of wind, crop seeds, and cattle in dispersal of *Striga* spp. Plant Dis 78:402–406

Carsky RJ, Singh L, Ndikawa R (1994) Suppression of *Striga hermonthica* on sorghum using a cowpea intercrop. Exp Agric 30:349–358

Carsky RJ, Berner DK, Oyewole BD, Dashiell K, Schulz S (2000) Reduction of *Striga hermonthica* parasitism on maize using soybean rotation. Int J Pest Manag 46:115–120

Carson AG (1989) Effect of intercropping sorghum and groundnuts on density of *Striga hermonthica* in the Gambia. Trop Pest Manag 35:130–132

Dhanapal GN, Struik PC, Udayakumar M, Timmermans P (1996) Management of broomrape (*Orobanche* spp.) – a review. J Agron Crop Sci 176:335–359

Emechebe AM, Ellis Jones J, Schulz S, Chikoye D, Douthwaite B, Kureh I, Tarawali G, Hussaini MA, Kormawa P, Sanni A (2004) Farmers' perception of the *Striga* problem and its control in Northern Nigeria. Exp Agric 40:215–232

Eplee RE (1981) *Striga*'s status as a plant parasite in the United States. Plant Dis 65:951–954

Eplee RE (1992) Witchweed (*Striga asiatica*): an overview of management strategies in the USA. Crop Prot 11:3–7

Faulkner OT (1934) Some experiments with leguminous crops at Ibadan, Southern Nigeria, 1925–1933. Empire J Exp Agric 2:93–102

Fernandez-Aparicio M, Andolfi A, Evidente A, Perez-De-Luque A, Rubiales D (2008) Fenugreek root exudates show species-specific stimulation of *Orobanche* seed germination. Weed Res 48:163–168

Fernandez-Aparicio M, Emeran AA, Rubiales D (2010) Inter-cropping with berseem clover (*Trifolium alexandrinum*) reduces infestation by *Orobanche crenata* in legumes. Crop Prot 29:867–871

Goldwasser Y, Kleifeld Y (2004) Recent approaches to *Orobanche* management – a review. In Inderjit (ed) Weed biology and management. Kluwer Academic, Dordrecht, pp 439–466

Goldwasser Y, Kleifeld Y, Golan S, Barguti A, Rubin B (1995) Dissipation of metham-sodium from soil and its effect on the control of *Orobanche aegyptiaca*. Weed Res 35:445–452

Goldwasser Y, Kleifeld Y, Plakhine D, Rubin B (1997) Variation in vetch (*Vicia* spp.) response to *Orobanche aegyptiaca*. Weed Sci 45:756–762

Goldwasser Y, Meron J, Sibony M, Sorek P, Joseph E, Rubin B (2011) Spread of weed seeds by cattle manure – is there a danger of infesting farmers' fields with weeds? In: 3rd International conference on novel and sustainable weed management in arid and semi-arid ecosystems, Huesca, Spain, p 5

Government of South Australia (2011) Branched broomrape eradication program – 2010/2011 Annual Report. 48 pp

Grenz JH, Manschadi AM, DeVoil P, Meinke H, Sauerborn J (2005) Assessing strategies for *Orobanche* sp. control using a combined seed bank and competition model. Agron J 97:1551–1559

Grundy AC, Mead A, Burston S (2003) Modelling the emergence response of weed seeds to burial depth: interactions with seed density, weight and shape. J Appl Ecol 40:757–770

Gurney AL, Ransom JK, Press MC (1995) The parasitic angiosperm *Striga hermonthica* can reduce photosynthesis of its sorghum and maize hosts in the field. J Exp Bot 46:1817–1823

Hansen OJ (1975) The genus *Rhamphicarpa* Benth. emend. Engl. (Scrophulariaceae). A taxonomic revision. Botanisk Tidsskrift 70:103–125

Haussmann BIG, Hess DE, Welz HG, Geiger HH (2000) Improved methodologies for breeding *Striga*-resistant sorghum. Field Crop Res 66:195–211

Jacobsohn R (1986) Broomrape avoidance and control: agronomic problems and available methods. In: ter Borg S (ed) Biology and control of *Orobanche*. Proceedings of the workshop on biology and control of *Orobanche*, L/HVPO, Wageningen, The Netherlands, pp 18–24

Jacobsohn R, Greenberger A, Katan J, Levi M, Alon H (1980) Control of Egyptian broomrape (*Orobanche aegyptiaca*) and other weeds by means of solar heating of soil by polyethylene mulching. Weed Sci 28:312–316

Jacobsohn R, Ben-Gedalia D, Marton K (1987) Effect of animal's digestive system on the effectivity of *Orobanche* seeds. Weed Res 27:87–90

Jacobsohn R, Kleifeld Y, Agrawal VP, Jha P, Marton K (1991) Soil fumigation with Telone II for broomrape (*Orobanche* spp.) control. In: Wegmann K, Musselman LJ (eds) Proceedings of the international workshop on *Orobanche* research. Eberhard-Karls-Universität, Obermarchtal, Tübingen, pp 185–190

Jamil M, Charnikhova T, Cardoso C, Jamil T, Ueno K, Verstappen F, Asami T, Bouwmeester HJ (2011) Quantification of the relationship between strigolactones and *Striga hermonthica* infection in rice under varying levels of nitrogen and phosphorus. Weed Res 51:373–385

Jamil M, Kanampiu FK, Karaya H, Charnikhova T, Bouwmeester HJ (2012) *Striga hermonthica* parasitism in maize in response to N and P fertilisers. Field Crop Res 134:1–10

Joel DM, Hershenhorn Y, Eizenberg H, Aly R, Ejeta G, Rich PJ, Ransom JK, Sauerborn J, Rubiales D (2007) Biology and management of weedy root parasites. In: Janick J (ed) Horticultural reviews, vol 33. Wiley, Hokobken, NJ, pp 267–349

Kamara AY, Menkir A, Chikoye D, Omoigui LO, Ekeleme F (2007) Cultivar and nitrogen fertilization effects on *Striga* infestation and grain yield of early maturing tropical maize. Maydica 52:415–423

Katan J, Greenberger A, Alon H, Grinstein A (1976) Solar heating by polyethylene mulching for control of diseases caused by soil-borne pathogens. Phytopathology 66:683–688

Kayeke J (2004) Evaluation of the potential of green manure and plant extracts for the control of *Striga asiatica* (L.) Kuntze on upland rice in Kyela, Tanzania. Ph.D thesis, Sokoine University of Agriculture, Morogoro, Tanzania. 162 pages

Kebreab E, Murdoch AJ (2001) Simulation of integrated control strategies for *Orobanche* spp. based on a life cycle model. Exp Agric 37:37–51

Khan ZR, Midega CAO, Bruce TJA, Hooper AM, Pickett JA (2010) Exploiting phytochemicals for developing a 'push-pull' crop protection strategy for cereal farmers in Africa. J Exp Bot 61:4185–4196

Kleifeld Y, Goldwasser Y, Herzlinger G, Joel DM, Golan S, Kahana D (1994) The effects of flax (*Linum usitatissimum*) and other crops as trap and catch crops for control of Egyptian broomrape (*Orobanche aegyptiaca* Pers.). Weed Res 34:37–44

Kleifeld Y, Goldwasser Y, Plakhine D, Eizenberg H, Herzlinger G, Golan S (1999) Selective control of *Orobanche* spp. in various crops with sulfonylurea and imidazolinone herbicides. In: Joint action to control *Orobanche* in the WANA-region: experiences from Morocco. Proceedings of the regional workshop, CTZ GmbH, Rabat, Morocco, pp 251–256

Krisnamurthy S, Rao UM (1976) Control of *Orobanche* through crop rotation. Indian Farming 25:23

Kroschel J (1998) *Striga* – how will it affect African agriculture in the future? – an ecological perspective. In: Martin K, Muther J, Auffarth A (eds) Agro-ecology, plant protection and the human environment: views and concepts. Margraf, Weikersheim, pp 137–158

Langyintuo AS, Dogbe W (2005) Characterizing the constraints for the adoption of a *Callopogonium mucunoides* improved fallow in rice production systems in northern Ghana. Agric Ecosyst Environ 110:78–90

Linke KH (1999) Status quo of *Orobanche* management: preventive, cultural and physical control. In: Joint action to control *Orobanche* in the WANA-region: experiences from Morocco. Proceedings of the regional workshop, CTZ GmbH, Rabat, Morocco, pp 107–133

Logan DC, Stewart GR (1991) Role of ethylene in the germination of the hemiparasite *Striga hermonthica*. Plant Physiol 97:1435–1438

Lonsdale WM (1993) Losses from the seed bank of *Mimosa pigra* – soil microorganisms vs temperature fluctuations. J Appl Ecol 30:654–660

Lopez-Granados F, Garcia-Torres L (1993) Population dynamics of crenate broomrape (*Orobanche crenata*) in faba bean (*Vicia faba*). Weed Sci 41:563–567

Lopez-Raez JA, Matusova R, Cardoso C, Jamil M, Charnikhova T, Kohlen W, Ruyter-Spira C, Verstappen F, Bouwmeester H (2009) Strigolactones: ecological significance and use as a target for parasitic plant control. Pest Manag Sci 65:471–477

Mauromicale G, Lo Monaco A, Longo AMG, Restuccia A (2005) Soil solarization, a nonchemical method to control branched broomrape (*Orobanche ramosa*) and improve the yield of greenhouse tomato. Weed Sci 53:877–883

Mohamed KI, Papes M, Williams R, Benz BW, Peterson TA (2006) Global invasive potential of 10 parasitic witchweeds and related Orobanchaceae. Ambio 35:281–288

Murdoch AJ, Kunjo RM (2003) Depletion of natural soil seed banks of *Striga hermonthica* in West Africa under different integrated management regimes. Aspect Appl Biol 69:261–268

Oswald A, Ransom JK (2001) *Striga* control and improved farm productivity using crop rotation. Crop Prot 20:113–120

Oswald A, Ransom JK, Kroschel J, Sauerborn J (2001) Transplanting maize and sorghum reduces *Striga hermonthica* damage. Weed Sci 49:346–353

Oswald A, Ransom JK, Kroschel J, Sauerborn J (2002) Intercropping controls *Striga* in maize based farming systems. Crop Prot 21:367–374

Ouédraogo O, Neumann U, Raynal Roques A, Sallé G, Tuquet C, Dembélé B (1999) New insights concerning the ecology and the biology of *Rhamphicarpa fistulosa* (Scrophulariaceae). Weed Res 39:159–169

Panetta FD (2012) Evaluating the performance of weed containment programs. Divers Distrib 18:1024–1032

Panetta FD, Lawes R (2007) Evaluation of the Australian branched broomrape (*Orobanche ramosa*) eradication program. Weed Sci 55:644–651

Parker C (1991) Protection of crops against parasitic weeds. Crop Prot 10:6–22

Parker C (2009) Observations on the current status of *Orobanche* and *Striga* problems worldwide. Pest Manag Sci 65:453–459

Parker C, Riches CR (1993) Parasitic weeds of the world: biology and control. CAB International, Wallingford

Parkinson V, Efron Y, Bello L, Dashiell K (1987) Trap crops as a cultural measure in *Striga* control in Africa. FAO Plant Protect Bull 35:51–54

Perez-de-Luque A, Eizenberg H, Grenz JH, Sillero JC, Avila C, Sauerborn J, Rubiales D (2010) Broomrape management in faba bean. Field Crop Res 115:319–328

Pickett JA, Hamilton ML, Hooper AM, Khan ZR, Midega CAO (2010) Companion cropping to manage parasitic plants. In: VanAlfen NK, Bruening G, Leach JE (eds) Annual Review of Phytopathology, vol 48, pp 161–177

Ramaiah KV (1987) Control of *Striga* and *Orobanche* species – a review. In: Weber HC, Forstreuter W (eds) Proceedings of the 4th ISPFP, Marburg, Germany, pp 637–664

Ransom JK (2000) Long-term approaches for the control of *Striga* in cereals: field management options. Crop Prot 19:759–763

Riches CR, Mbwaga AM, Mbapila J, Ahmed GJU (2005) Improved weed management delivers increased productivity and farm incomes from rice in Bangladesh and Tanzania. Aspect Appl Biol 75:127–138

Robinson EL, Dowler CC (1966) Investigations of catch and trap crops to eradicate witchweed (*Striga asiatica*). Weeds 14:275–276

Rodenburg J, Bastiaans L (2011) Host-plant defence against *Striga* spp.: reconsidering the role of tolerance. Weed Res 51:438–441

Rodenburg J, Johnson DE (2009) Weed management in rice-based cropping systems in Africa. Adv Agron 103:149–218

Rodenburg J, Bastiaans L, Weltzien E, Hess DE (2005) How can field selection for *Striga* resistance and tolerance in sorghum be improved? Field Crop Res 93:34–50

Rodenburg J, Bastiaans L, Kropff MJ, van Ast A (2006) Effects of host plant genotype and seed bank density on *Striga* reproduction. Weed Res 46:251–263

Rodenburg J, Riches CR, Kayeke JM (2010) Addressing current and future problems of parasitic weeds in rice. Crop Prot 29:210–221

Rodenburg J, Meinke H, Johnson DE (2011a) Challenges for weed management in African rice systems in a changing climate. J Agric Sci 149:427–435

Rodenburg J, Zossou-Kouderin N, Gbèhounou G, Ahanchede A, Touré A, Kyalo G, Kiepe P (2011b) *Rhamphicarpa fistulosa*, a parasitic weed threatening rain-fed lowland rice production in sub-Saharan Africa – a case study from Benin. Crop Prot 30:1306–1314

Rubiales D, Fernandez-Aparicio M (2012) Innovations in parasitic weeds management in legume crops. Agron Sustain Dev 32:433–449

Rubiales D, Fernandez-Aparicio M, Wegmann K, Joel DM (2009) Revisiting strategies for reducing the seed bank of *Orobanche* and *Phelipanche* spp. Weed Res 49:23–33

Satovic Z, Joel DM, Rubiales D, Cubero JI, Román B (2009) Population genetics in weedy species of *Orobanche*. Australas Plant Pathol 38:228–234

Sauerborn J, Saxena MC (1986) A review on agronomy in relation to *Orobanche* problems in faba bean (*Vicia faba* L.). In: ter Borg S (ed) Biology and control of *Orobanche*. Proceedings of the workshop on biology and control of *Orobanche*, L/HVPO, Wageningen, The Netherlands, pp 160–165

Sauerborn J, Linke KH, Saxena MC, Koch W (1989) Solarization, a physical control method for weeds and parasitic plants (*Orobanche* spp.) in Mediterranean agriculture. Weed Res 29:391–397

Sauerborn J, Kranz B, Mercer Quarshie H (2003) Organic amendments mitigate heterotrophic weed infestation in savannah agriculture. Appl Soil Ecol 23:181–186

Saxena MC, Linke KH, Sauerborn J (1994) Integrated control of *Orobanche* in cool-season food legumes. In: Pieterse AH, Verkleij JAC, Borg SJ (eds) Biology and management of *Orobanche*. Royal Tropical Institute, Amsterdam, pp 419–431

Showemimo FA, Kimbeng CA, Alabi SO (2002) Genotypic response of sorghum cultivars to nitrogen fertilization in the control of *Striga hermonthica*. Crop Prot 21:867–870

Sinebo W, Drennan DSH (2001) Vegetative growth of sorghum and *Striga hermonthica* in response to nitrogen and the degree of host root infection. Eur J Plant Pathol 107:849–860

Tarawali G, Manyong VM, Carsky RJ, Vissoh PV, Osei-Bonsu P, Galiba M (1999) Adoption of improved fallows in West Africa: lessons from mucuna and stylo case studies. Agroforest Syst 47:93–122

Tasker AV, Westwood JH (2012) The U.S. witchweed eradication effort turns 50: a retrospective and look-ahead on parasitic weed management. Weed Sci 60:267–268

Teasdale JR (2003) Principles and practices of using cover crops in weed management systems. In: Labrada R (ed) Weed management for developing countries (Addendum 1). FAO, Rome, pp 109–143

Tesso TT, Ejeta G (2011) Integrating multiple control options enhances *Striga* management and sorghum yield on heavily infested soils. Agron J 103:1464–1471

Van Ast A, Bastiaans L, Katile S (2005) Cultural control measures to diminish sorghum yield loss and parasite success under *Striga hermonthica* infestation. Crop Prot 24:1023–1034

van Delft GJ, Graves JD, Fitter AH, Pruiksma MA (1997) Spatial distribution and population dynamics of *Striga hermonthica* seeds in naturally infested farm soils. Plant Soil 195:1–15

van Delft GJ, Graves JD, Fitter AH, van Ast A (2000) *Striga* seed avoidance by deep planting and no-tillage in sorghum and maize. Int J Pest Manag 46:251–256

Van Mourik TA, Stomph TJ, Westerman PR (2003) Estimating *Striga hermonthica* seed mortality under field conditions. Aspects Appl Biol 69:187–194

Van Mourik TA, Stomph TJ, Murdoch AJ (2005) Why high seed densities within buried mesh bags may overestimate depletion rates of soil seed banks. J Appl Ecol 42:299–305

Van Mourik TA, Bianchi FJJA, Van Der Werf W, Stomph TJ (2008) Long-term management of *Striga hermonthica*: strategy evaluation with a spatio-temporal population model. Weed Res 48:329–339

Van Mourik TA, Stomph TJ, Murdoch AJ (2011) Purple witchweed (*Striga hermonthica*) germination and seed bank depletion under different crops, fallow, and bare soil. Weed Biol Manag 11:100–110

Westerman PR, Wes JS, Van der Kropff MJ, Werf W (2003) Annual losses of weed seeds due to predation in organic cereal fields. J Appl Ecol 40:824–836

Westerman PR, van Ast A, Stomph TJ, van der Werf W (2007) Long-term management of the parasitic weed *Striga hermonthica*: strategy evaluation with a population model. Crop Prot 26:219–227

Wilhelm S, Benson RC, Sagen J (1958) Studies on the control of broomrape on tomatoes: soil fumigation by methyl bromide is promising control. Plant Dis Rep 42:645–651

Wilhelm S, Srokan RCE, Benson RC, Sagen JE, Carpenter T (1959) Large scale soil fumigation against broomrape. Phytopathology 49:530–532

Chapter 23
Chemical Control

Hanan Eizenberg, Joseph Hershenhorn, Jhonathan H. Ephrath, and Fred Kanampiu

23.1 Introduction: The Complexity of Chemical Control of Parasitic Weeds

Herbicides may be applied before weed emergence (PRE) or following weed emergence (POST).Whereas for nonparasitic weeds the herbicide rates are adjusted according to the observed phenological stage of the weed and its density aboveground, this option is usually not applicable for root-parasitic weeds, because their crucial stages of development occur underground. While hemiparasitic weeds, such as *Striga* (witchweed) and *Alectra*, develop foliage and can be treated after emergence with some herbicides (e.g., auxin herbicides when attacking cereals), weedy holoparasites, such as the broomrapes *Phelipanche* and *Orobanche*, must be treated during their underground stages of development because they emerge above the ground only during flowering when most of the damage to the host has already been done (Eizenberg et al. 2006). In both cases treatment after emergence is usually too late to prevent yield losses, and herbicides applied on emerged parasites mainly help in limiting parasite seed dispersal.

Some herbicides may control the parasites only at a specific phenological stage. Thus, knowledge of their underground phenology is essential in any attempt to

H. Eizenberg (✉) • J. Hershenhorn
Department of Weed Research, Agricultural Research Organization, Newe Ya'ar Research Center, P.O. Box 1021, Ramat Yishay, Israel
e-mail: eizenber@volcani.agri.gov.il

J.I.H. Ephrath
Wyler Dept. of Dryland Agriculture, French Associates Institute for Agriculture and Biotechnology of Drylands, Jacob Blaustein Institutes for Desert Research, Ben-Gurion University of the Negev, Sede Boqer Campus, Sede Boqer 84990, Israel

F. Kanampiu
International Maize and Wheat Improvement Center (CIMMYT), P.O. Box 1041-00621, Nairobi, Kenya

D.M. Joel et al. (eds.), *Parasitic Orobanchaceae*, DOI 10.1007/978-3-642-38146-1_23, © Springer-Verlag Berlin Heidelberg 2013

effectively control them in infested fields. Specific methodologies and technologies have been developed to address this problem (see Sect. 23.4).

As broomrapes are non-photosynthetic, and *Striga* and *Alectra* lack chlorophyll during their underground development, the number of herbicides that can be considered for root-parasite control is limited, e.g., to those that do not target photosynthesis-related processes. When searching effective herbicides for the control of root-parasitic weeds, one should also consider that the conductive tissues of the parasites are directly connected to those of the host (see Sect. 3.9), allowing systemic herbicides to be translocated from host to parasite and vice versa (see Sect. 6.3). Therefore, in order to allow safe control of the parasite, the host plant should be selective to the applied herbicide either by metabolic or by target-site resistance. In case of metabolic resistance, the metabolites must be toxic to the parasite (see Sect. 24.2).

Several approaches that have been proposed for parasitic weed control are discussed in this chapter. In most cases this is done separately for broomrapes and witchweeds because they differ not only in phenology and in their photosynthetic abilities but also in their agricultural environments and predominant hosts (see Chap. 18).

23.2 Herbicides

23.2.1 Potential Herbicides

Several systemic herbicides have so far been proposed for broomrape control in vegetables and field crops (Table 23.1; Foy et al. 1989; Hershenhorn et al. 1998b, c, 2009; Qasem 1998; Eizenberg et al. 2004a, 2006, 2009a). These herbicides include the aromatic amino acid synthesis inhibitor glyphosate, which targets the enzyme 5-enolpyruvylshikimate-3-phosphate synthase (EPSPS), and the branched-chain amino acid synthesis inhibitors imidazolinones, sulfonylureas, and pyrithiobac-sodium, which target the enzyme acetolactate synthase (ALS) (Schloss 1990).

Whereas glyphosate is commonly absorbed by foliage, some ALS-inhibiting herbicides can be absorbed by the roots, others by the shoots, and in this latter case they are rapidly transferred to the roots or to meristematic tissues that act as sinks. Shoot-applied imidazolinones may be exuded from the plant roots (Little and Shaner 1991; Pester et al. 2001; Kanampiu et al. 2002; Colquhoun et al. 2006), which is relevant to the development of control strategies for ALS-sensitive root-parasitic weeds. The leaching potential of any given ALS inhibitor, as well as the rates of its degradation in the soil and the crop, may influence the level of control and the period of control activity.

Various herbicides were used in *Striga* control (Table 23.2), depending on the target developmental stages. Systemic herbicides such as the hormonal herbicides 2,4-D and dicamba have effectively controlled *Striga* (Eplee and Norris 1987;

Table 23.1 Chemical control of broomrapes under field conditions

Host		Broomrape		Rate	Number of	
Family	Crop	species	Herbicide	(g a.i/ha)	applications	Reference
Solanaceae	Tomato	*P. aegyptiaca*	Sulfosulfuron	37.5–75.0	1–3[a]	Eizenberg et al. (2012a)
			Imazapic	4.8	1–2[d]	Eizenberg et al. (2012a)
			Chlorsulfuron	2.5	3[b] + 2[c]	Hershenhorn et al. (1998a)
			Triasulfuron	7.5	3[b] + 2[c]	Hershenhorn et al. (1998a)
			Rimsulfuron	37.5–50.0	3–4[b]	Eizenberg et al. (2003)
	Potato	*P. aegyptiaca*	Rimsulfuron	37.5–50.0	3[a]	Goldwasser et al. (2001)
	Tobacco	*P. ramosa*	Glyphosate	90	2[d]	Lolas (1994)
Apiaceae	Carrot	*O. crenata*	Imazapic	2.5	3[d]	Jacobsohn et al. (1996)
		P. aegyptiaca *O. crenata*	Glyphosate	36–72	2–3[d]	Jacobsohn and Kelman (1980)
	Parsley	*P. aegyptiaca* *O. crenata*	Glyphosate	36–72	3–4[d]	Goldwasser et al. (2003)
		O. crenata	Imazapic	2.5–5.0	3–4[d]	Goldwasser et al. (2003)
Fabaceae	Pea	*O. crenata*	Imazethapyr	20–60	1[e]	Jurado-Expósito et al. (1996)
		O. crenata	Glyphosate	18	2–3[d]	Jacobsohn and Kelman (1980)
	Faba bean	*O. crenata*	Imazethapyr	20–40	1[e]	Jurado-Expósito et al. (1997)
		O. crenata	Imazapyr	5	1[e]	Jurado-Expósito et al. (1997)
		O. crenata	Imazethapyr	2.5–5.0	3[d]	Garcia-Torres and Lopez-Granados (1991)
		O. crenata	Glyphosate	54	2–3[d]	Jacobsohn and Kelman (1980)
		O. crenata	Glyphosate	60	1[d]	Mesa-García and García-Torres (1985)

(continued)

Table 23.1 (continued)

Host		Broomrape		Rate	Number of	
Family	Crop	species	Herbicide	(g a.i/ha)	applications	Reference
	Lentils	*O. crenata*	Imazethapyr	20–40	1[e]	Jurado-Expósito et al. (1997)
		O. crenata	Imazapyr	5	1[e]	Jurado-Expósito et al. (1997)
	Red clover	*O. minor*	Imazamox	10–40	1–2[d]	Lins et al. (2005)
Asteraceae	Sunflower	*O. cumana*	Imazapic	2.4	3[d]	Aly et al. (2001), Eizenberg et al. (2012b)
			Imazapyr	10	1–2[d]	Garcia-Torres et al. (1995)
			Pronamide	2K	1[e]	Diaz-Sanchez et al. (2003)

[a]Foliar applications followed by overhead irrigation at 300 m^3 ha^{-1}
[b]Herbigation—herbicides delivered through 300 m^3 ha^{-1} sprinkler irrigation
[c]Herbigation—herbicides delivered through 300 m^3 ha^{-1} drip irrigation
[d]Foliar application
[e]Seed coating

Awad et al. 1991). Dicamba was thought to be the most promising nonselective post attachments herbicide for effective *Striga* control when applied soon after attachment to reduce yield losses (Odhiambo and Ransom 1993, 1997).

23.2.2 Site of Herbicide Action

The mode of herbicide uptake by root-parasitic Orobanchaceae is often not known. Reports on successful parasite control by herbicides usually do not mention whether the herbicide reached the parasite systemically through the conductive tissues of the host with or through the soil. However, Kanampiu et al. (2002) have shown that imazapyr and pyrithiobac that are applied on maize leaf whorls of imidazolinone-resistant maize were systemically transferred out of maize roots and inhibited germination of *S. hermonthica* in the soil.

Once we know how the herbicide reaches the host, a decision can be made whether it should be applied to host foliage or directly to the soil. A simple method to determine whether a herbicide reaches the parasite through the soil or translocated from the foliage employs surface-activated charcoal as a soil cover that prevents leakage of herbicide from the plant foliage to the soil. Using this method it was shown that sulfonylurea herbicides are active against broomrapes

Table 23.2 Chemical control of *Striga* under field conditions

Host						
Family	Crop	*Striga* spp.	Herbicide	Rate (g a.i/ha)	Number of applications	Reference
Poaceae	Maize	*S. asiatica*	Imazapyr	15–45	1[a]	Kanampiu et al. (2003), Kabambe et al. (2007)
		S. hermonthica	Imazapyr	27	1[c]	Abayo et al. (1998)
			Imazamox	71	1[c]	Abayo et al. (1998)
			Imazethapyr	71	1[c]	Abayo et al. (1998)
			Sulfometuron	50	1[c]	Abayo et al. (1998)
			Rimsulfuron	50	1[c]	Abayo et al. (1998)
			Metsulfuron	50	1[c]	Abayo et al. (1998)
			Halosulfuron	120	1[c]	Abayo et al. (1998)
		S. hermonthica	Dicamba	280–750	1[d]	Odhiambo and Ransom (1993)
		S. hermonthica	Imazapyr	45	1[a]	Kanampiu et al. (2002)
			Pyrithiobac	21	1[a]	Kanampiu et al. (2002)
		S. hermonthica	Imazapyr	30–45	1[a]	Kanampiu et al. (2003)
			Pyrithiobac	11–21	1[a]	Kanampiu et al. (2003)
	Sorghum	*S. hermonthica*	Brine	0.5–2.0 M	1[b]	Gworgwor et al. (2002)
			2,4-D	0.5 % w/v	1[b]	Dembélé et al. (2005)
		S. asiatica	Imazapyr	19–75 μg a.i/seed	1[b]	Tuinstra et al. (2009)
			Metsulfuron methyl	3–12 μg a.i/seed	1[b]	
Fabaceae	Cowpea	*S. hermonthica*	Imazaquin	1.8–7.2 mg a.i/ml	1[a]	Berner et al. (1994)

[a]Seed coating
[b]Seed priming
[c]Drenching
[d]Foliar application

mainly through the soil solution (Fig. 23.1a, b) whereas the activity of the imidazolinones is highly effective via translocation through the host plant to the attached parasites (Plakhine et al. 2001; Eizenberg et al. 2004a; Colquhoun et al. 2006).

Another important aspect to be considered when planning the use of herbicides against root-parasitic weeds is the depth distribution of the parasite seeds in soil, because the various herbicides behave differently in soil (Eizenberg et al. 2007). Optimizing soil applications of each herbicide should correspond not only to the crop root architecture but also to the location of parasite seeds that can potentially parasitize the crop.

23.3 The Use of Herbicides and Fumigants

Chemical control of parasitic weeds has extensively been explored since the 1970s (Kasasian 1973). Four main approaches for chemical control of broomrapes and *Striga* are currently used in agriculture: soil fumigation, foliar application of herbicides, soil application of herbicides, and herbicide application on crops that are resistant to herbicides. These are discussed in the following sections (and see Chap. 24 for further discussion of the use of herbicide resistances).

23.3.1 Soil Fumigation

Soil disinfection by fumigation is used for the control of parasite seedbanks. This method requires considerably long preparations and is usually conducted a few months before growing of host crops. Fumigation is successful when targeting broomrape seeds in the field before winter crops such as carrot and potato and in the greenhouse before various vegetables (Jacobsohn et al. 1988). So far only few fumigants have been successfully applied against broomrape, mainly methyl bromide, dazomet, metham sodium, and 1,3-dichloropropene (Foy et al. 1989; Goldwasser et al. 1995). The soil needs to be well prepared and well watered to allow optimal delivery of the fumigant to the relevant soil depths (usually 0–30 cm below the surface). The fumigants can either be injected by needles directly into the soil or applied through drip irrigation systems.

Soil fumigation with methyl bromide has so far been the most effective fumigant, being highly successful in killing broomrape seeds in the soil. Nevertheless, it may no longer be used due to its harmful environmental effects. Other fumigants are rarely used because of their high cost, complicated application procedures, and limited efficacy. Application of metham sodium for soil fumigation under transparent polyethylene mulching can increase broomrape control efficacy if applied in the summer, due to the solarization effect that is produced by this mulching (Goldwasser et al. 1995, and Sect. 22.4).

Fig. 23.1 Parasitic weeds control by herbicides. (**a**) *Phelipanche aegyptiaca* control by sulfosulfuron applied to tomato rhizosphere in polyethylene bag; (**b**) nontreated control. (**c**) *Striga hermonthica* control on imidazolinone-resistant (IR) maize (behind the farmers) grown in Kenya from seeds that were coated with a low dose of imazapyr; the control maize (in the front of the farmers), raised from a nontreated local maize hybrid, was highly infested by *Striga*. (**d**) Sequential photographs of *Orobanche cumana* tubercles parasitizing sunflower roots 20 cm under soil subsurface, captured by video camera of a minirhizotron (Eizenberg et al. 2005b) at four different growing degree days (GDD); vital *O. cumana* tubercles were observed in the image series of the non-treated control, while *Orobanche* tubercle degeneration was seen in images taken following the foliar applications of imazapic at 720 GDD. *T* parasite tubercle, *DT* dead tubercle

Economic aspects concerning the application of soil fumigants considerably differ in the open field and in the greenhouse. Soil fumigation is often used annually in the greenhouse for the control of nematodes and/or pathogens, which may cause broomrape seed demise without further expense. For economic reasons fumigation in the field is usually considered only for cash crops, such as carrots, potatoes, or tomatoes, and is therefore recommended to be included in 4–5 years of crop rotation that includes at least 2–3 non-host crops.

Ethylene gas has been widely used to induce suicidal germination of *Striga* seeds (Egley et al. 1990; see Sect. 22.4.1). This procedure was a basic element in the *Striga* eradication program in the USA (Langston and English 1990).

23.3.2 Foliar Application of Herbicides

Parasites can infest host plants throughout its growing season, and herbicides need to be available throughout the earliest susceptible stages of parasite development to avoid crop damage and during late infection stages to prevent the parasites from producing seeds and replenishing the seedbank. This can be done either by one application of higher herbicide doses that kill the crop or by repeating the application of low doses several times. When the herbicide is only partly selective to the host and no other herbicides are available to effectively control the parasite, cautious application of sequential low doses of the herbicide may be applied to reduce infestation, as demonstrated with imazapic in carrot (Jacobsohn et al. 1996). While this treatment reduces carrot quality, it still allows achieving a yield. This is not the case with foliar applications of imazapic on potato to control of *P. aegyptiaca* and *P. ramosa*, which cause severe deformation of the potato tubers (Goldwasser et al. 2001), and with imazapic application on sunflower, during early flowering which causes head deformations containing only few seeds (Aly et al. 2001).

Low rates of the systemic herbicide glyphosate, applied up to three times, were effective for broomrape control only on a few hosts that are less susceptible to the herbicide, i.e., members of the Apiaceae (carrot, celery, and parsley), Fabaceae (faba bean, vetch, pea, *Vicia narbonensis*, *V. sativa*), and various Brassicaceae (Kasasian 1973; Jacobsohn and Kelman 1980; Foy et al. 1989; Nadal et al. 2008; Nandula et al. 1999). Glyphosate controls broomrape in the above crops, but it may at the same time also reduce host resistance to pathogenic microorganisms, presumably because it blocks the synthesis of phytoalexins that are induced during pathogen attack (Lévesque and Rahe 1992).

Herbicide movement along the soil profile may potentially harm susceptible intercrops; this is particularly relevant for Africa, where cereal-legume systems are a common practice. Nevertheless, when intercropping the herbicide-sensitive cowpea and yellow gram (*Vigna radiata*) with herbicide-resistant maize, the legumes are not affected by imazapyr and pyrithiobac that are applied on the cereals, in spite of being sensitive to the herbicides, provided that the sensitive crops are planted at minimal distance of 15 cm from the treated crop (Kanampiu et al. 2002; see Sect. 24.2.3).

23.3.3 Herbicide Application to Crop Seeds

Herbicides can be applied to crop seeds before sowing, which lowers the amount of herbicide that is applied compared to what would be sprayed in the field, and preclude the need of spraying that relies on specialized application equipment that is not available to many farmers in developing countries (Jurado-Expósito et al. 1996).

Two main methods are employed for the application of herbicides to crop seeds: seed coating, in which the herbicides are incorporated to the surface of crop seeds, and seed priming in which the herbicide is incorporated into seed tissues by imbibing seeds in a herbicide solution. The latter is only possible in crops with target-site herbicide resistance.

The use of coated or primed seeds enables a more efficient, economic, and environmentally friendly control of root parasites in infested fields, targeting only the near vicinity of the host roots rather than the whole field. In this way the herbicide reaches the parasite either directly through the soil when it moves from the seed coat with the water column or indirectly from the emerging host once the parasite attaches to host roots or when the herbicide leaches from host roots. Herbicides that can be applied to the seeds are chosen according to the level of their selectivity and their ability to control the parasite. The efficacy of seed treatments depends on having a sufficient margin of selectivity for both the crop and the parasite, allowing a significant control of the parasite while avoiding crop damage and yield loss (Jurado-Expósito et al. 1996, 1997; Diaz-Sanchez et al. 2003; Dembélé et al. 2005).

Parasitic weed control can be achieved with herbicides if the crop has target-site resistance and does not metabolize the herbicide (Gressel 2009). In order to allow the use of ALS-inhibiting herbicides against *Striga,* efforts have therefore been invested in delivering such herbicides to seeds of ALS-resistant maize genotypes. Application of pyrithiobac and imazapyr in seed coating or priming provided effective *Striga* control in the field (Kanampiu et al. 2001; Kabambe et al. 2007, 2008a) while improving maize yields nearly threefold (Abayo et al. 1996, 1998; Berner et al. 1997; Kanampiu et al. 2001). The best *Striga* control in maize (Fig. 23.1c) was achieved by seed dressing with imazapyr or pyrithiobac, and such dressed seeds are now widely commercialized in Western Kenya (Ransom et al. 2012). In these treatments almost all *Striga* seeds were killed in the upper soil layers, and approximately 80 % were killed at 30 cm depth (Kanampiu et al. 2002; De Groote et al. 2007).

23.3.4 Application of Herbicides Through the Soil

Applying herbicides through the soil for the management of root-parasitic weeds targets the parasite seedlings and its young attachments. The success of this mode of herbicide application depends on the availability of herbicide in the soil layer where the host roots are parasitized.

Herbicides can be delivered to the target area using mechanical incorporation or by rainfall before host sowing or planting. They can also be delivered through the irrigation system or through the host plant after attachment to the host. Application by herbigation, i.e., through irrigation systems, requires sufficient amounts of water that will reach the target areas. Herbigation can be done by sprinklers or by drippers. Sprinkler herbigation spreads the herbicides homogenously only when applied under optimal environmental conditions (e.g., no wind), while dripper herbigation, which is not affected by wind, is more complicated technically and does not provide an even distribution in soil. In both cases low doses must be carefully calibrated so as not to kill the crop and to meet the approval of the regulatory authorities. The advantage of herbigation was demonstrated in tomatoes, where chlorsulfuron and triasulfuron successfully controlled *P. aegyptiaca*, by integrating overhead sprinkler irrigation and drip irrigation (Hershenhorn et al. 1998a). This method may fail when sprinkler irrigation is applied under wind conditions or when sprinklers or drippers are not adjusted to optimal application. The use of low-discharge drippers significantly increases control efficacy due to the higher uniformity of herbicide delivery in the soil. Foliar application of sulfosulfuron or rimsulfuron and incorporating it into the soil with the aid of water is considerably safer than herbigation, because it requires only a short term of optimal wind conditions, only during spraying the herbicide. The use of sulfosulfuron has recently been commercialized in Israel for *P. aegyptiaca* control in tomato (Eizenberg et al. 2012a; Hershenhorn et al. 2009; Eizenberg et al. 2004a, 2006; Goldwasser et al. 2001).

One should however note that imidazolinone herbicides can successfully be translocated through the plant to the parasite without being metabolized by the host, and therefore these herbicides do not need to be incorporated into the soil. Sulfonylurea herbicides are also translocated, but much is metabolized before reaching the parasites through the roots.

Shallow incorporation of the preemergence herbicide trifluralin, which inhibits root development by interrupting mitosis, can control *Striga* in the soil when maize or sorghum seeds are planted in a furrow below the trifluralin-treated soil. Care should however be taken to ensure that no treated soil lies directly over the crop seeds (Langston et al. 1991).

Many soil-surface applied PRE herbicides are easy to use in *Striga* and *Alectra* control. Low vapor pressure herbicides like metolachlor (Kabambe et al. 2008b) and pendimethalin reduced and/or delayed *Striga* development when rainfall incorporated the herbicides into the soil and controlled *Striga* before its emergence (Langston et al. 1991). Similarly, *Alectra vogelii* can be effectively controlled when

PRE herbicide mixtures containing metazachlor and antidote are applied, followed by postemergence application of imazaquin or pendimethalin or both herbicides in tank mix (Magani and Lagoke 2008). Nonetheless, adapting these technologies to the low-input systems in Africa has so far been difficult.

23.4 Models for Optimizing Herbicide Application

The optimal timing for successful control of root-parasitic weeds, in terms of efficacy and damage, is achieved when treating the parasite during its initial stages of parasitism. However, the fact that these stages take place underground is a key obstacle in determining the timing. Introducing the minirhizotron video camera and its adaptation for in situ monitoring of broomrape development in the soil (Fig. 23.1d) (Eizenberg et al. 2005b; and see Sect. 23.5) allowed the study of the parasite phenology under field conditions. The parasite develops in a stepwise manner which is strongly influenced by the accumulation of specific quantities of heat in the soil (Grenz et al. 2005; Eizenberg et al. 2004b; 2005a; 2009b; 2012a; Ephrath and Eizenberg 2010; Ephrath et al. 2012). Based on this knowledge, models could be developed that predict the timing of the most sensitive stages of broomrape development in particular crops. Once heat accumulation is followed in the relevant soil layers, the timing of the various developmental stages of the parasite, i.e., germination, attachment, and tubercle development, can be predicted (Eizenberg et al. 2012a). A measure of heat accumulation in the soil is GDD (growing degree days, also known as thermal time, see Sect. 11.6.2) that can be computed by various equations but must be fitted to individual crops and parasites. GDD has already been used to predict the development of *O. minor* in red clover (Mallory-Smith and Colquhoun 2012), of *O. cumana* in sunflower (Eizenberg et al. 2009a; Ephrath and Eizenberg 2010; Eizenberg et al. 2012b), and of *P. aegyptiaca* in field tomato (Eizenberg et al. 2012a; Ephrath et al. 2012).

The germination of parasite seeds is only possible when both the host and the parasite reached their base temperature (see Sect. 11.6.2), because the host needs to be able to supply the stimulants for parasite germination and the parasite needs to be able to germinate. Therefore, the base temperature T_{base} that needs to be used for predicting parasite germination in the presence of a host (e.g., in the field) is that of the partner with the higher base temperature. This is particularly important when the parasite T_{base} is significantly different from host T_{base}, which is the case when tomato is grown in a *P. aegyptiaca*-infested field; although the base temperature of the parasite is 4.9 °C (see Sect. 11.6.2), it will not be able to germinate under 10 °C (Ephrath et al. 2012), which is the base temperature of tomato development.

Once this information is available, it can be combined with other factors that are important for optimal herbicide application in the field, and a decision support system (DSS) can be developed for management of the parasite in the field by using a modeling approach (Eizenberg et al. 2012a; Grenz et al. 2008). Five years of field validation confirmed the advantage of the modeling approach for successful

O. cumana control in sunflower (Eizenberg et al. 2012b). A decision support system, named "PICKIT" (Eizenberg et al. 2012a), was developed for optimal prophylactic soil treatments with sulfosulfuron and post-attachment treatment with imazapic for *P. aegyptiaca* management in tomato. PICKIT determines the timing and number of herbicide applications needed for optimal control of the parasite, based on GDD accumulation data starting at tomato planting, and on risk assessment based on geographical information system (GIS) of previously known infestations and/or updated infestation mapping. This provides the farmer with a package that allows growing tomato under various infestation levels without yield losses.

23.5 Broomrape Control by Herbicide-Resistant Crops

Herbicides used for parasitic plant control are in most cases not adequately selective to the crops or are metabolized into nontoxic compounds within the tissues of the host plant. The crop plant should carry target-site resistance to herbicides that are effective against the parasites to effectively control root-parasitic weeds (Joel et al. 1995; Surov et al. 1998). These requirements are not met by the majority of crops that are affected by parasitic weeds, and they are therefore the main limitation for successful implementation of chemical broomrape and witchweed control in these crops. Several technologies were applied to overcome this problem: classical breeding, mutagenesis, and genetic engineering (see Chap. 24). So far only the first two methods have reached practical application in parasitic weed control, though genetic engineering is already used in various crops for the control of nonparasitic weeds.

Imidazolinone-resistant sunflowers were developed by crossing a weedy imazethapyr-resistant sunflower population with inbreds used for hybrid sunflower production and back-crossing them for many generations. The resistant trait was introduced into cultivated sunflower lines and commercialized (Tan et al. 2005). In this way imidazolinone-resistant sunflower varieties occupy about 25 % of Turkey's sunflower growing area, which is severely affected by *O. cumana*, and are so far effective in controlling all available parasite races (Süzer and Büyük 2010). *O. cumana* control with imazamox on resistant sunflower varieties is also being used on a large scale in Russia and Serbia. Successful broomrape control with imazapyr was achieved with a mutagenized rapeseed (*Brassica napus*) seeds bearing ALS target-site resistance that is highly tolerant to imazethapyr (Tan et al. 2005; Gressel 2009, also see Chap. 24).

23.6 New and Future Approaches

Future approaches for chemical control of root-parasitic weeds may fit the trend of "global environmentally clean tech." Precision agriculture and in particular site-specific weed management (SSWM) are based on recording parasite heterogeneity within fields, analyzing and defining the sources of heterogeneity (see Chap. 19 for population analysis) and, as a result, applying the optimal herbicide rates only at the infested locations in the field. Knowledge of the spatial distribution of the parasite seedbank in the field must therefore be available in advance (see Chap. 20 for seedbank analysis). Field history and patch mapping of infestations in previous years can be used for herbicide applications under management zones. Field history documentation and GIS technologies should be among the most promising means for increasing the precision of parasitic weeds control under SSWM, as it combines infestation mapping and field history data through growing seasons.

Technologies of low flow drip irrigation systems that deliver the water uniformly along the bed at low flow rate (e.g., 0.6 l h^{-1}) promise development of new herbigation methodologies. Hyper-spectral cameras and specifically near-infrared reflectance spectroscopy may assist in early detection of broomrape-infested patches in the field based on the transpiration rates, which affect leaf temperature, may vary between infested and noninfested hosts (see Sect. 6.2.1). The rapid development of Internet applications for farmers and virtual communities allows data transfer between farmers and the extension, as well as between regions and countries. This data may include resistance to herbicides, meteorological data for modeling, and decision support systems.

Further research should be invested in breeding crops resistant to herbicides, either genetically modified or by mutagenesis or by selection. Despite the fact that the introduction of genetically modified crops is currently limited, herbicide-producing companies are highly motivated to release Clearfield cultivars also for nonparasitic weed control, which makes the production of these crops more economic. There is no doubt that crop resistance to herbicides represents the next generation in chemical control of parasitic weeds. However, the use of herbicide-resistant crops should be carefully managed in order to reduce the risk of evolving herbicide resistances in the parasites (Beckie 2006; see Chap. 24).

In addition, massive research into the mechanism of herbicide action in host–parasite systems may open the way for new herbicide families for the control of parasitic weeds. This should reduce the use of ALS-inhibiting herbicides as well as the risk of the evolution of ALS resistance in the parasitic weeds, which have not yet been found.

23.7 Conclusions

Our knowledge of parasitic weeds has broadened in the last three decades, in particular regarding host–parasite relationships and parasitism dynamics. This knowledge allows focusing on specific developmental stages at which the parasites are most susceptible to herbicides. Farmers have already obtained effective tools to chemically control broomrapes and witchweeds in some crops. The solutions should fit the target environments, taking into account the limitations of herbicide application. Where technology is limited, for example, absence of precision sprayers or irrigation facilities (Parker 2012), the technology should be transferred to the target region ready for use (e.g., supply treated seeds) as reported for *Striga* control in Kenya (Kanampiu et al. 2001, 2002; Ransom et al. 2012). Conversely, whenever the use of chemical control is possible, it should consider precision agriculture techniques for environmental friendly herbicide application at the optimal timing and in the exact location. Numerous field experiments that were conducted over a whole decade resulted in a feasible chemical approach for *P. aegyptiaca* control in processing tomato in Israel, which is successfully implemented by farmers, but needs adaptation when applied in other agricultural systems and different soil conditions.

Being one of the only groups of herbicides that have so far been found as selective to host crops and effective in broomrape and witchweed control, ALS-inhibiting herbicides are so far the best available option for chemical control of parasitic weeds. The application of ALS-inhibiting herbicides may however lead to the evolution of herbicide resistances in the parasites. In order to reduce this risk, the use of herbicides should be reduced to minimum, and research should develop protocols to allow using herbicides with diverse modes of action.

References

Abayo GO, Ransom JK, Gressel J, Odhiambo GD (1996) *Striga hermonthica* control with acetolactate synthase inhibiting herbicides seed-dressed to corn with target site resistance. In: Moreno MT, Cubero JI, Berner DK, Joel D, Musselman LJ, Parker C (eds) Advances in parasitic weed research. Junta de Andalucia, Cordoba, Spain, pp 762–768

Abayo GO, English T, Eplee RE, Kanampiu FK, Ransom JK, Gressel J (1998) Control of parasitic witchweeds (*Striga* spp.) on corn (*Zea mays*) resistant to acetolactate synthase inhibitors. Weed Sci 46:459–466

Aly R, Goldwasser Y, Eizenberg H, Hershenhorn J, Golan S, Kleifeld Y (2001) Broomrape (*Orobanche cumana*) control in sunflower (*Helianthus annuus*) with imazapic. Weed Technol 15:306–309

Awad AE, Worsham AD, Corbin FT, Eplee RE (1991) Absorption, translocation and metabolism of foliarly applied 14-C dicamba in sorghum (*Sorghum bicolor*) and corn (*Zea mays*) parasitized with witchweed (*Striga asiatica*). In: Ransom JK, Musselman LJ, Worsham AD, Parker C (eds) Proceedings of the 5th international symposium of parasitic weeds. CIMMYT, Nairobi, pp 535–536

Beckie HJ (2006) Herbicide-resistant weeds: management tactics and practices. Weed Technol 20:793–814

Berner DK, Awad AE, Aigbokhan EI (1994) Potential of imazaquin seed treatment for control of *Striga gesnerioides* and *Alectra vogelii* in cowpea (*Vigna unguiculata*). Plant Dis 78:18–23

Berner DK, Ikie FO, Green JM (1997) ALS-inhibiting herbicide seed treatments control *Striga hermonthica* in ALS-modified corn (*Zea mays*). Weed Technol 11:704–707

Colquhoun JB, Eizenberg H, Mallory Smith CA (2006) Herbicide placement site affects small broomrape (*Orobanche minor*) control in red clover (*Trifolium pratense*). Weed Technol 20:356–360

De Groote H, Wangare L, Kanampiu F (2007) Evaluating the use of herbicide-coated imidazolinone-resistant (IR) maize seeds to control *Striga* in farmers' fields in Kenya. Crop Prot 26:1496–1506

Dembélé B, Dembélé D, Westwood JH (2005) Herbicide seed treatments for control of purple witchweed (*Striga hermonthica*) in sorghum and millet. Weed Technol 19:629–635

Diaz-Sanchez J, Jurado-Exposito M, Lopez-Granados F, Castejon-Munoz M, Garcia-Torres L (2003) Pronamide applied to sunflower seeds for *Orobanche cumana* control. Weed Technol 17:314–319

Egley GH, Eplee RE, Norris RS (1990) Discovery and development of ethylene as a witchweed seed germination stimulant. In: Sand PF, Eplee RE, Westbrooks RG (eds) Witchweed research and control in the United States. WSSA, Champaign, IL, pp 46–55

Eizenberg H, Hershenhorn J, Graph S, Manor H (2003) *Orobanche aegyptiaca* control in tomato with sulfonylurea herbicides. ACTA Hortic 613:205–208

Eizenberg H, Goldwasser Y, Golan S, Plakhine D, Hershenhorn J (2004a) Egyptian broomrape (*Orobanche aegyptiaca*) control in tomato with sulfonylurea herbicides greenhouse studies. Weed Technol 18:490–496

Eizenberg H, Colquhoun JB, Mallory-Smith CA (2004b) The relationship between temperature and small broomrape (*Orobanche minor*) parasitism in red clover. Weed Sci 52:735–741

Eizenberg H, Colquhoun JB, Mallory-Smith CA (2005a) A predictive degree-days model for small broomrape (*Orobanche minor*) parasitism in red clover in Oregon. Weed Sci 53:37–40

Eizenberg H, Shtienberg D, Silberbush M, Ephrath JE (2005b) A new method for monitoring early stages of *Orobanche cumana* development in sunflower (*Helianthus annuus*) with minirhizotron. Ann Bot 96:137–140

Eizenberg H, Colquhoun JB, Mallory-Smith CA (2006) Imazamox application timing for small broomrape (*Orobanche minor*) control in red clover. Weed Sci 54:923–927

Eizenberg H, Lande T, Achdari G, Roichman A, Hershenhorn J (2007) Effect of Egyptian broomrape (*Orobanche aegyptiaca*) seed-burial depth on parasitism dynamics and chemical control in tomato. Weed Sci 55:152–156

Eizenberg H, Lande T, Achdari G, Smirnov E, Hershenhorn J (2009a) PICKIT – a decision support system for rational control of *Phelipanche aegyptiaca* in tomato. In: Proceedings of the 10th world congress on parasitic plants, Kusadasi, Turkey, p 79

Eizenberg H, Hershenhorn J, Ephrath JE (2009b) Factors affecting the efficacy of *Orobanche cumana* chemical control in sunflower. Weed Res 49:308–315

Eizenberg H, Aly R, Cohen Y (2012a) Technologies for smart chemical control of broomrape (*Orobanche* and *Phelipanche* spp.). Weed Sci 60:316–323

Eizenberg H, Hershenhorn J, Achdari G, Ephrath J (2012b) A thermal time model for predicting parasitism of *Orobanche cumana* in irrigated sunflower Field validation. Field Crops Res 137:49–55

Ephrath JE, Eizenberg H (2010) Quantification of the dynamics of *Orobanche cumana* and *Phelipanche aegyptiaca* parasitism in confectionery sunflower. Weed Res 50:140–152

Ephrath JE, Hershenhorn J, Achdari G, Bringer S, Eizenberg H (2012) The use of sigmoid equations for the detection of the initial parasitism phase of *Phelipanche aegyptiaca* in tomato. Weed Sci 60:57–63

Eplee RE, Norris RS (1987) Chemical control of *Striga*. In: Musselman LJ (ed) Parasitic weeds in agriculture, vol I: *Striga*. CRC, Boca Raton, FL, pp 173–182
Foy CL, Jain R, Jacobsohn R (1989) Recent approaches for chemical control of broomrape (*Orobanche* spp.). In: Foy CL (ed) Reviews of weed science, vol 4. Weed Science Society of America, Champaign, IL, pp 123–152
Garcia-Torres L, Lopez-Granados F (1991) Control of broomrape *Orobanche crenata* Forsk.) in broad bean (*Vicia faba* L.) with imidazolinones and other herbicides. Weed Res 31:227–235
Garcia-Torres L, Castejon-Muños M, Lopez-Granados F, Jurado-Exposito M (1995) Imazapyr applied postemergence in sunflower (*Helianthus annuus*) for broomrape (*Orobanche cernua*) control. Weed Technol 9:819–824
Goldwasser Y, Kleifeld Y, Golan S, Bargutti A, Rubin B (1995) Dissipation of metham-sodium from soil and its effect on the control of *Orobanche aegyptiaca*. Weed Res 35:445–452
Goldwasser Y, Eizenberg H, Golan S, Hershenhorn J, Kleifeld Y (2001) *Orobanche aegyptiaca* control in potato. Crop Prot 20:403–410
Goldwasser Y, Eizenberg H, Golan S, Kleifeld Y (2003) Control of *Orobanche crenata* and *O. aegyptiaca* in parsley. Crop Prot 22:295–305
Grenz JH, Manschadi AM, Uygur FN, Sauerborn J (2005) Effects of environment and sowing date on the competition between faba bean (*Vicia faba*) and the parasitic weed *Orobanche crenata*. Field Crop Res 93:300–313
Grenz JH, Istoc VA, Manschadi AM, Sauerborn J (2008) Interactions of sunflower (*Helianthus annuus*) and sunflower broomrape (*Orobanche cumana*) as affected by sowing date, resource supply and infestation level. Field Crops Res 107:170–179
Gressel J (2009) Crops with target-site herbicide resistance for *Orobanche* and *Striga* control. Pest Manag Sci 65:560–565
Gworgwor NA, Hudu AI, Joshua AD (2002) Seed treatment of sorghum varieties with brine (NaCl) solution for control of *Striga hermonthica* in sorghum. Crop Prot 21:1015–1021
Hershenhorn J, Goldwasser Y, Plakhine D, Lavan Y, Blumenfeld T, Bucsbaum H, Herzlinger G, Golan S, Chilf T, Herzlinger H, Kleifeld Y (1998a) *Orobanche aegyptiaca* control in tomato fields with sulfonylurea herbicides. Weed Res 38:343–349
Hershenhorn J, Goldwasser Y, Plakhine D, Lavan Y, Herzlinger G, Golan S, Chilf T, Kleifeld Y (1998b) Effect of sulfonylurea herbicides on Egyptian broomrape (*Orobanche aegyptiaca*) in tomato (*Lycopersicon esculentum*) under greenhouse conditions. Weed Technol 12:115–120
Hershenhorn J, Plakhine D, Goldwasser Y, Westwood JH, Foy CL, Kleifeld Y (1998c) Effect of sulfonylurea herbicides on early development of Egyptian broomrape (*Orobanche aegyptiaca*) in tomato (*Lycopersicon esculentum*). Weed Technol 12:108–114
Hershenhorn J, Eizenberg H, Dor E, Kapulnik Y, Goldwasser Y (2009) *Phelipanche aegyptiaca* management in tomato. Weed Res 49:34–47
Jacobsohn R, Kelman Y (1980) Effectiveness of glyphosate in broomrape (*Orobanche* spp.) control of four crops. Weed Sci 28:692–699
Jacobsohn R, Kelman Y, Shared R, Klein L (1988) Broomrape (*Orobanche* spp.) control with ethylene dibromide and chloropicrin. Weed Res 28:151–157
Jacobsohn R, Tanaami Z, Eizenberg H (1996) Selective control of broomrape in carrot and vetch with foliar applied imidazolinone herbicides. Phytoparasitica 24:207
Joel DM, Kleifeld Y, Losner-Goshen D, Herzlinger G, Gressel J (1995) Transgenic crops to fight parasitic weeds. Nature 374:220–221
Jurado-Expósito M, Castejón-Muñoz M, García-Torres L (1996) Broomrape (*Orobanche crenata*) control with imazethapyr applied to pea (*Pisum sativum*) seeds. Weed Technol 10:774–780
Jurado-Expósito M, García-Torres L, Castejón-Muñoz M (1997) Broad bean and lentil seed treatments with imidazolinones for the control of broomrape (*Orobanche crenata*). J Agric Sci 129:307–314
Kabambe VH, Kanampiu F, Nambuzi SC, Kauwa A (2007) Evaluation of the use of herbicide (Imazapyr) and fertilizer application in integrated management of *Striga asiatica* in maize in Malawi. Afr J Agric Res 2:687–691

Kabambe VH, Kanampiu F, Ngwira A (2008a) Imazapyr (herbicide) seed dressing increases yield, suppresses *Striga asiatica* and has seed depletion role in maize (*Zea mays* L.) in Malawi. Afr J Biotechnol 17:3293–3292

Kabambe VH, Kauwa AE, Nambuzi SC (2008b) Role of herbicide (metolachlor) and fertilizer application in integrated management of *Striga asiatica* in maize in Malawi. Afr J Agric Res 3:140–146

Kanampiu FK, Ransom JK, Gressel J (2001) Imazapyr seed dressings for *Striga* control on acetolactate synthase target-site resistant maize. Crop Prot 20:885–895

Kanampiu FK, Ransom JK, Friesen D, Gressel J (2002) Imazapyr and pyrithiobac movement in soil and from maize seed coats controls *Striga* while allowing legume intercropping. Crop Prot 21:611–619

Kanampiu FK, Kabambe V, Massawe C, Jasi L, Friesen D, Ransom JK, Gressel J (2003) Multi-site, multi-season field tests demonstrate that herbicide seed-coating herbicide-resistance maize controls *Striga* spp. and increases yields in several African countries. Crop Prot 22:697–706

Kasasian L (1973) Control of *Orobanche*. Proc Natl Acad Sci USA 19:368–371

Langston MA, English TJ (1990) Vegetative control of witchweed and herbicide evaluation of techniques. In: Sand PF, Eplee RE, Westbrooks RG (eds) Witchweed research and control in the United States. WSSA, Champaign, IL, pp 108–113

Langston M, English T, Eplee R (1991) Herbicides for control of *Striga asiatica* in the USA. In: Ransom JK, Musselman LJ, Worsham AD, Parker C (eds) Proceedings of the 5th international symposium on parasitic weeds. CIMMYT, Nairobi, pp 400–406

Lévesque CA, Rahe JE (1992) Herbicide interactions with fungal root pathogens, with special reference to glyphosate. Annu Rev Phytopathol 30:579–602

Lins R, Colquhoun JB, Cole CM, Mallory-Smith CA (2005) Postemergence herbicide options for small broomrape (*Orobanche minor*) control in red clover (*Trifolium pratense*). Weed Technol 19:411–415

Little DL, Shaner DL (1991) Absorption and translocation of the imidazolinone herbicides. In: Shaner DL, O'Connor SL (eds) The imidazolinone herbicides. CRC, Boca Raton, FL, pp 53–69

Lolas PC (1994) Herbicides for control of broomrape (*Orobanche ramosa* L.) in tobacco (*Nicotiana tabacum* L.). Weed Res 34:205–209

Magani IH, Lagoke STO (2008) Pre-emergence and post-emergence herbicide control of the parasitic plant *Alectra vogelii* in cowpea. J Anim Plant Sci 1:26–32

Mallory-Smith C, Colquhoun J (2012) Small Broomrape (*Orobanche minor*) in Oregon and the 3 Rs: Regulation, Research, and Reality. Weed Sci 277–282

Mesa-García J, García-Torres L (1985) *Orobanche crenata* (Forsk) control in *Vicia faba* (L.) with glyphosate as affected by herbicide rates and parasite growth stages. Weed Res 25:129–134

Nadal S, Moreno MT, Roman B (2008) Control of *Orobanche crenata* in *Vicia narbonensis* by glyphosate. Crop Prot 27:873–876

Nandula VK, Foy CL, Orcutt DM (1999) Glyphosate for *Orobanche aegyptiaca* control in *Vicia sativa* and *Brassica napus*. Weed Sci 47:486–491

Odhiambo GD, Ransom JK (1993) Effect of Dicamba on the control of *Striga hermonthica* in maize in western Kenya. Afr Crop Sci J 1:105–110

Odhiambo GD, Ransom JK (1997) On-farm evaluation of an integrated approach to *Striga* control in western Kenya. Proc Afr Crop Sci Conf 3:887–893

Parker C (2012) Parasitic weeds: a world challenge. Weed Sci 60:269–276

Pester TA, Nissen SJ, Westra P (2001) Absorption, translocation, and metabolism of imazamox in jointed goat grass and feral rye. Weed Sci 49:607–612

Plakhine D, Eizenberg H, Hershenhorn J, Goldwasser Y, Kleifeld Y (2001) Control of *Orobanche aegyptiaca* with sulfonylurea herbicides in tomato: polyethylene bag studies. In: Fer A, Thalouran P, Joel DM, Musselman LJ, Parker C, Verkleij JAC (eds) Proceedings of the 7th international parasitic weed symposium, Nantes, France, pp 294–295

Qasem JR (1998) Chemical control of branched broomrape (*Orobanche ramosa*) in glasshouse grown tomato. Crop Prot 17:625–630

Ransom JK, Kanampiu F, Gressel J, Groote HD, Burnet M, Odhiambo G (2012) Herbicide applied to imidazolinone resistant-maize seed as a *Striga* control option for small-scale African farmers. Weed Sci 60:283–289

Schloss JV (1990) Acetolactate synthase, mechanism of action and its herbicide binding site. Pestic Sci 29:283–292

Surov T, Aviv D, Aly R, Joel DM, Goldman-Guez T, Gressel J (1998) Generation of transgenic asulam-resistant potatoes to facilitate eradication of parasitic broomrapes (*Orobanche* spp.), with the *sul* gene as the selectable marker. Theor Appl Genet 96:132–137

Süzer S, Büyük H (2010) Residual effects of spraying imidazolinone-family herbicides on Clearfield® sunflower production from the point of view of crop rotation. Helia 33:25–36

Tan S, Evans RR, Dahmer ML, Singh BK, Shaner D (2005) Imidazolinone-tolerant crops: history, current status and future. Pest Manag Sci 61:246–257

Tuinstra RM, Soumana S, Al-Khatib K, Kapran I, Toure A, van Ast A, Bastiaans L, Ochanda NW, Salami I, Kayentao M, Dembele S (2009) Efficacy of herbicide seed treatments for controlling *Striga* infestation of sorghum. Crop Sci 49:923–929

Chapter 24
Biotechnologies for Directly Generating Crops Resistant to Parasites

Jonathan Gressel

24.1 Introduction

Breeding (Chap. 21) has been moderately and transiently successful as a delaying tactic with broomrape species, e.g. *Orobanche cumana* attacking sunflower quickly evolved resistance to each new gene when breeding was performed one resistance gene at a time. It appears that with *Striga* success may remain longer due to the polygenic breeding of a confluence of genes, and time will tell how long these varieties with multiple recessive genes painstakingly crossed into local varieties will remain immune to evolutionary forces. Theoretically, the more genes introduced, the longer they should remain effective. The chemical strategies described in Chap. 23 are limited to selective herbicides degraded by the crop, such that they have a very short window to effectively control the parasites. Some crop plants are naturally resistant to the parasites, but they have limited geographic distribution (such as *Desmodium*) (Khan et al. 2007) and/or are not the species the farmer wishes to cultivate. Biotechnology has the potential to combine the best of all of these technologies by taking genes for resistance from wherever they may exist, or using genes for herbicide resistance that do not degrade the herbicide, as well as add some tricks of its own from knowing the genome of the parasite. Biotechnology is only the first step. After getting genes that confer resistance, they must be bred into locally adapted varieties. Basically, biotechnology has drastically expanded the scope of genes the breeder may use to attain a crop not devastated by the parasites. Thus, biotechnological techniques are being rapidly adopted by a new generation of breeders who want to use all available breeding tools and not to be constrained by the limited gene pool of a given crop and its wild interbreeding relatives.

J. Gressel (✉)
Plant Sciences, Weizmann Institute of Science, Rehovot 76100, Israel
e-mail: jonathan.gressel@weizmann.ac.il

D.M. Joel et al. (eds.), *Parasitic Orobanchaceae*,

DOI 10.1007/978-3-642-38146-1_24,

It is not enough to increase crop yield by a parasite control strategy. For long-term sustainability, it is necessary to vastly reduce the seed bank levels of parasite seeds (see Chap. 22). The densities of these seed banks have skyrocketed due to monoculture, low fertility and lack of cost-effective control strategies that were acceptable to farmers. Various population dynamic models for seed bank reduction of *Striga* (Smith and Webb 1996) and *Orobanche* (Kebreab and Murdoch 2001) come to similar conclusions that any treatment that does not reduce seed bank input by more than 95 % would hardly reduce parasite infestation in sites with large seed banks. Of the biotechnological strategies discussed below, only herbicide seed treatments in short season maize can presently attain and exceed this goal. Dealing only with crop yield is short-sighted, and the treatments that only reduce parasite infestation without decreasing seed bank size may be selecting for their own failure by leaving behind large populations that can evolve (see Chap. 19). The parasite plants that do survive in strategies where yield is enhanced but the parasites are not killed were probably heavily stressed. There is ample evidence that mutation frequencies increase and the rate of evolution is quickened in stressed organisms (Gressel 2011), including evolution of resistance to whatever strategy stressed the parasites but did not kill them. This should be kept in mind in reading the following sections. Most of the strategies described below, engineering the crop not to make germination stimulant, or to make allelochemicals toxic to the parasite, have the crop make RNAi's or micro RNAs that are toxic to the parasite, disseminate genes through multi-copy transposons that harm the parasite, or engineer in genes that confer resistance for physiologically unclear reasons, are presently at best with promising preliminary data or just plain science fiction. Indeed there are more than an order of magnitude more papers (or chapters such as this) extolling and reviewing the promise of biotechnology for dealing with the parasites than there are publications describing concrete results. We need successes with the promising technologies, as the one strategy that does work—herbicide resistance, is only a temporary stopgap, as the parasites are bound to evolve resistance to the herbicides. Evolution is inevitable in biology, and replacement strategies will be needed.

24.2 Target Site Herbicide Resistances[1]

Selective herbicides have been tested for nearly half a century for control of parasitic weeds, as discussed in an extensive 1989 review of the historical literature by Foy et al. (1989) and updated in Chap. 23, to which the reader is referred for this information. The only mode of herbicide selectivity initially available to agriculture in general and to control parasitic weed in crops in particular was metabolic; the

[1] Section 24.2 is condensed and updated from Gressel J, Crops with target site herbicide resistance for *Orobanche* and *Striga* control. Pest Management Science 65: 560–565, © 2009 Society of Chemical Industry and published by John Wiley & Sons Ltd., with their permission.

crop was able to protect itself by catabolising the herbicide to inactive compounds. Such foliar applied herbicides initially could be used only after the weed has emerged. If soil-applied, they had to quickly kill the parasite before the crop catabolised the herbicide taken up. Systemic non-selective herbicides have also been used at doses sublethal to the crop. Such doses are tricky to attain, especially when backpack sprayers with non-uniform applications are used. Sublethal doses to the crop are easier to achieve and safer to use when there is a heavy parasite infestation; the parasites act as a sink, sucking the herbicides (such as glyphosate) with the photosynthate. The larger the sink, the quicker the herbicide is removed from the crop. An additional problem has occurred with glyphosate. It is well known that sublethal doses of glyphosate suppress phytoalexin biosynthesis in legumes (Lévesque and Rahe 1992; Sharon et al. 1992), so that when *Orobanche* was controlled on faba beans, chocolate spot disease could more readily manifest itself because crop immunity was compromised (Foy et al. 1989). Methyl bromide was widely used to kill *Orobanche* seeds prior to planting high-value vegetable crops, but that expensive option is hardly available due to an international ban (see Sect. 22.4).

Crops bearing target site herbicide resistances have the target enzyme normally affected by the herbicide modified such that the enzyme still binds its normal substrate, but no longer binds the herbicide, leaving the herbicide un-metabolised in the crop. Not all target site resistances will control root-parasitic weeds. The first such target site resistance bred into a crop would not be effective. Resistance to photosystem 2-inhibiting herbicides (e.g. atrazine) was bred into oilseed rape (*Brassica napus*) from a related weed that evolved resistance (Souza-Machado 1982), and is still cultivated, especially in Australia. This resistance is inappropriate for parasite control for two reasons: (a) this group of herbicides is not systemic and the herbicides will not be translocated through the hosts to the parasites; (b) the herbicides target photosynthesis, which is non-existent in *Orobanche* and *Phelipanche, and* occurs only late in life in *Striga* and *Alectra.*

Similarly the target site resistance that has evolved to protox (protoporphyrinogen oxidase)-inhibiting herbicides (Patzoldt et al. 2006), or generated by site-directed mutagenesis (Volrath et al. 1999), or in other ways (Li and Nicholl 2005) might be useful for general weed control but not for root parasites. These herbicides actively kill plants by photodynamic action, and the light required will not penetrate to the soil depths where the parasites must be killed.

24.2.1 Systemic Target Site Resistances

The advent of three target site resistances, generated by transgenic and mutagenic techniques, gave rise to suggestions that they might have utility in controlling parasitic weeds (Foy et al. 1989; Gressel 1992). This was contrary to conventional wisdom at the time, as it was thought that the parasites derived their organic nitrogen from the host crop (Press 1995). These herbicides suppress either acetolactate synthase (ALS) leading to branched chain amino acids (e.g. chlorsulfuron and imazapyr) or 5-enolpyruvylshikimate-3-phosphate synthase (EPSPS) suppressing the shikimate pathway leading to aromatic amino acid and other biosyntheses (glyphosate) or inhibit dihydropteroate synthase, the biosynthesis of the vitamin folic acid (asulam). It was posited (Gressel 1992) that conventional wisdom about parasites receiving their organic nitrogen from hosts was not accurate vis-à-vis these organic nitrogen products, because herbicides that specifically affect these pathways kill the parasites (Foy et al. 1989) and because the parasites can be cultured on tissue culture media devoid of organic nitrogen (Ben-Hod et al. 1991).

The approach has been quite successful in the lab and greenhouse in controlling root-parasitic weeds (Table 24.1), but only the mutant target site resistances have been commercialised so far.

24.2.2 Target Site Resistances for **Orobanche** *Control*

Target site resistant tobacco strains, one with asulam resistance and one with ALS resistance, and oilseed rape with glyphosate resistance were obtained with some difficulty from the developers. A single foliar application of asulam, chlorsulfuron and glyphosate, respectively, controlled *Phelipanche aegyptiaca* (Joel et al. 1995). Transgenic metabolic resistance to glufosinate was used as negative control. Indeed, *P. aegyptiaca* could not be controlled in transgenic glufosinate-resistant tobacco with glufosinate. This was repeated with various Orobanchaceae, resistant crops and herbicides (Table 24.1a) with the same result.

Other crops that are parasitised by Orobanchaceae have been genetically engineered with target site herbicide resistances (see: http://www.isb.vt.edu/search-release-data.aspx—accessed August 2011) that should allow parasite control in these crops, but there have been no reports of experiments ascertaining whether this can be useful (Table 24.1b). *Orobanche minor*, which has become a problem in alfalfa in the USA (Ross et al. 2004), should be easy to control on transgenic glyphosate-resistant alfalfa, where the resistance is due to a modified EPSPS (Monsanto and Forage Genetics), but not the alfalfa with metabolic resistance conferred by the *GAT* gene (Pioneer). Soybeans can be attacked by Orobanchaceae (yet are rarely cultivated in parasite-infested areas), which could easily be controlled on the EPSPS glyphosate-resistant soybean varieties. Potatoes and beets with target site resistance to glyphosate and tomatoes with ALS resistance that have been field-tested should also be useful for *Orobanche* control. Glyphosate-resistant lettuce, tomatoes, peas and carrots that have been field-tested by Seminis Vegetable Seeds may or may not be useful for both broomrape and *Striga* control, as which resistance genes are being used is not stated by those performing the field tests.

There are oilseed rape lines in Canada that have both transgenic target site glyphosate resistance and non-target site glufosinate resistance, and

Table 24.1 Herbicide control of root-parasitic weeds in crops with target site resistance

Representative herbicide	Target[a]	Crop	Mode of generation	Parasite	References
(a) Demonstrated to be effective against root-parasitic weeds					
Asulam	DHPTA	Tobacco	Transgenic	*P. aegyptiaca*	Joel et al. (1995)
Asulam	DHPTA	Potato	Transgenic	*P. aegyptiaca*	Surov et al. (1998)
Glyphosate	EPSPS	Oilseed rape	Transgenic	*P. aegyptiaca*	Joel et al. (1995)
Glyphosate	EPSPS	Tomato	Transgenic	*P. ramosa*	Kotoula-Syka (2003)
Chlorsulfuron	ALS	Tobacco	Transgenic	*P. aegyptiaca* *P. ramosa*	Joel et al. (1995) Slavov et al. (2005)
Imazapyr	ALS	Carrot	Transgenic	*P. aegyptiaca*	Aviv et al. (2002)
Imazapyr	ALS	Oilseed rape	Mutant	*P. aegyptiaca*	J. Gressel, unpub.
Imazethapyr	ALS	Sunflower	Mutant from weedy sunflower	*O. cernua* *O. cumana*	Alonso et al. (1998) Malidza et al. (2004)
Imazamox	ALS	Tomato	EMS mutant	'Broomrape'	Dor et al. (2011)
Imazapyr	ALS	Maize	Mutant	*S. hermonthica* *S. asiatica*	Kanampiu et al. (2007) Kabambe et al. (2008)
Imazapyr	ALS	Sorghum	Mutant from shattercane[b]	*S. hermonthica*	Tuinstra et al. (2009)
(b) Potentially effective against root-parasitic weeds, but not reported					References to resistant material
Glyphosate	EPSPS	Maize	Transgenic	*S. hermonthica* *S. asiatica*	James (2011)
Glyphosate	EPSPS	Soybeans	Transgenic	*S. gesnerioides*	James (2011)
Imazapyr	ALS	Rice	Mutant[b]	*S. hermonthica* *R. fistulosa*	cf. Gressel and Valverde (2009)
Glyphosate		Rice	Transgenic[b]	*S. hermonthica* *R. fistulosa*	APHIS[c]
Glyphosate	EPSPS	Sugarcane	Transgenic	*S. hermonthica*	APHIS[c]
Glyphosate	EPSPS	Alfalfa	Transgenic	*O. minor*	APHIS[c]
Glyphosate	EPSPS	Potatoes	Transgenic	*Orobanche* spp.	APHIS[c]
Glyphosate	EPSPS	Beets	Transgenic[b]	*Orobanche* spp.	APHIS[c]
Not stated	ALS	Tomatoes	Transgenic	*Orobanche* spp.	APHIS[c]

[a]*DHPTS* dihydropteroate synthase (in pathway to folic acid), *EPSPS* 5-enolpyruvylshikimate-3-phosphate synthase

[b]Because of gene flow issues to pernicious weedy relatives, this should best be done transgenically with mitigation genes

[c]Has been field-tested in the USA according to the official website for APHIS field releases: http://www.isb.vt.edu/CFDOCS/fieldtests3.cfm (accessed August 2011)

non-transgenic target site ALS resistance. These were derived from natural crosses between varieties planted in proximity (Hall et al. 2000). They would be useful as a crop where the ranges of *Phelipanche* and oilseed rape overlap. In warmer areas this could be used as a killer-catch crop (see Sect. 22.4.5), with seed treatments of imazapyr (the most persistent imidazolinone ALS-inhibiting herbicide) to control early germinating *Orobanche* and with a mid- or late season treatment with glyphosate to control parasites that germinated after imazapyr dissipation or that evolved resistance to imazapyr. This should significantly reduce the *Orobanche* seed bank, especially if the rapeseed is densely planted in narrow rows or broadcast to assure an even and extensive crop root distribution, which will stimulate extensive *Orobanche* germination. Still, the cultivation of present oilseed rape varieties is contrary to the spirit of the world-wide ban on methyl bromide. This crop emits about 10,000 tons of methyl bromide per year into the atmosphere, as oilseed rape naturally methylates the trace amounts of bromine in the soil (extrapolated from a 1998 estimate) (Gan et al. 1998). This could be rectified transgenically by suppressing the known halide methylation gene (Rhew et al. 2003).

Despite the intractable problems with Orobanchaceae, none of the transgenic technologies described above have been commercialised. Attempts should be made to obtain regulatory approval in spite of the regulatory costs and the Luddite approach to acceptance of transgenics in the target market areas.

24.2.3 Target Site Resistances for Striga *Control*

Two types of target site resistance were being developed in the early 1990s that could be considered for *Striga hermonthica* and *S. asiatica* control: maize with transgenic target site resistance to glyphosate and mutation-derived ALS-resistant maize with target site resistance to imidazolinone herbicides (Table 24.1b). ALS gene sulfometuron methyl-resistant sugarcane, field-tested by Louisiana State University, should be appropriate for *Striga* control (see: http://www.isb.vt.edu/search-release-data.aspx—accessed August 2011). Similarly, various companies are field-testing herbicide-resistant rice, but do not state the gene being used, so it is not clear whether such rice/herbicide resistance combinations would be appropriate for controlling *Striga* in upland rice and the related *Rhamphicarpa fistulosa*, which has become important in lowland rice (Rodenburg et al. 2011). The metabolic glufosinate-resistant rice (Song et al. 2011) would not be appropriate. Additionally, as soybeans are parasitised by *S. gesnerioides*, the glyphosate-resistant soybeans could be considered for use against this weed. More than 80 % of the soybeans cultivated in the world contain this transgene (James 2011), but none in Africa, where *S. gesnerioides* is a problem. At the time these crops were being developed, the developers would not make seed available for testing with *Striga*, as they did not consider *Striga* control in Africa a viable market. After ALS mutant imidazolinone-resistant (IR) maize was commercially released, it was purchased and first tested for *S. asiatica* control in the USA, using preemergence field treatments (Abayo et al. 1998).

This led to brainstorming on how such treatments might be made viable in the economic conditions of subsistence farming in Africa. It was posited that seed dressing of the herbicides might be possible, precluding the need for sprayers, and would potentially lower herbicide costs because the herbicide would not be spread over the fields; the herbicide would remain in the rhizosphere just where needed, under the crop seed. This seed treatment concept was first tested outside Africa, and *Phelipanche aegyptiaca* was used as a model with transgenic glyphosate-resistant oilseed rape (Gressel and Joel 1997) as well as with mutant imidazolinone-resistant oilseed rape. It was even possible to coat tiny transgenic asulam-resistant tobacco seed with herbicide by producing pelleted seed (Joel et al. 2000). In all these cases *Phelipanche* was successfully controlled by the herbicide on the crop seed.

A seed drench of commercial, detergent-formulated imazapyr was initially used with US (non-tropical) IR maize, because non-formulated material was not available. The success with this material (Abayo et al. 1996, 1998) as well as similar material coated with other ALS-inhibiting herbicides (Berner et al. 1997) led to testing various imidazolinone and sulfonylurea herbicides, and imazapyr gave the longest season control with the least signs of phytotoxicity to the crop. Thus, imazapyr was then purified from detergent-formulated imazapyr, and various salts were synthesised and screened (Kanampiu et al. 2001). In the meantime, the ALS-resistant gene was backcrossed into tropical east and southern Africa elite material and testing continued in experiment stations and in farmers' fields (Kanampiu et al. 2002, 2007; De Groote et al. 2008; Kabambe et al. 2008; Menkir et al. 2010).

It is probable that the crop plants both take up and exude the systemic ALS-inhibiting herbicides from their roots. This is the best explanation for the duration of weed control achieved by these herbicides, which would be washed away from the root zone if they were not taken up and 'recycled'. There is evidence from laboratory experiments that the herbicides are exuded from roots. When imazapyr was applied to maize leaves, *Striga* was killed as it approached maize roots before it attached (Kanampiu et al. 2002).

One important issue to be addressed was whether the seed treatment technology would be 'appropriate' for African conditions, where farmers often interplant an edible legume between the hills of maize. Would the herbicide injure the legume intercrop? If various legumes were 15 cm or more apart from the maize plants, then there was no effect of the imazapyr on the legume (Kanampiu et al. 2002). The legumes are typically planted 30 cm away from the maize, so this issue was settled, and intercropping can be practised without fear (see Chap. 22).

The imidazolinone-resistant (IR) maize has been commercialised in western Kenya and farmer acceptance has been outstanding, as measured by studies on declared willingness to purchase (De Groote et al. 2008) as well as the actual purchases of unsubsidized seed by farmers. This is both because of the *Striga* control and also because the gene was put in a superior background and the varieties outproduce local hybrids, even when there is no *Striga* infestation. It is estimated that there is an annual market for more than 2,500 T of IR maize seed for planting in western Kenya and 150,000 T for the *Striga*-infested areas throughout Africa. The technology is only useful after the IR gene is backcrossed into elite material locally adapted for each agro-ecological zone and for the maize types local populations prefer (De Groote et al. 2008). The present material appropriate for western Kenya is being tested in a similar agro-ecological zone in Uganda and Tanzania, with testing planned in 2008 for 10,000 farmers' field trials (Anonymous 2008).

Transgenic target site glyphosate-resistant maize has been introduced to South Africa for general weed control and has been a considerable success among subsistence farmers. This is being cultivated so far in areas free of *Striga*, but experience with *Orobanche* control in glyphosate-resistant crops suggests that glyphosate-resistant maize should be invaluable for controlling *Striga* as well.

Imidazolinone-(ALS) resistant sorghum is being developed both by mutation breeding (David Ndungu, pers comm.) as well as by transfer of resistance genes from feral sorghum (shattercane) that evolved ALS resistance in the USA (Tuinstra et al. 2009). This non-transgenic material should be relatively simple to register throughout Africa. The fact that the IR gene could quickly introgress into feral weedy sorghum in Africa is not of worry to African scientists, as herbicides are not commonly used for weed control in sorghum (A.-G. Babiker, pers. comm.). This view may be short-sighted, as Africa must clearly develop, and eventually herbicides will be used in agriculture. Intermediate forms between wild and cultivated sorghum have been found in sorghum-growing areas of Kenya (Mutegi et al. 2010), so there may well be a gene flow problem when herbicide usage becomes more widespread. Mitigation strategies are available for transgenic sorghum that could prevent the establishment of feral sorghum that introgressed a herbicide-resistant transgene, e.g. coupling the herbicide resistance gene in tandem with a gene preventing seed shattering (Gressel 2012). No major gene-encoding seed non-shattering is known in sorghum as yet, but it is clear that the non-shattering genes differ among closely related cereals (Li and Gill 2006). Such strategies will not work with the ALS mutant sorghum, but would work with the same gene inserted transgenically.

The existing glyphosate target site resistances in sugarcane and rice (http://www.isb.vt.edu/CFDOCS/fieldtests3.cfm) should allow *Striga* control in these crops, despite no reports ascertaining whether they will be useful. It should be possible to similarly control *Striga* in upland rice and the related *Rhamphicarpa fistulosa* in lowland rice (Rodenburg et al. 2011) using mutant ALS-resistant 'Clearfield^{TM}' rice, but the same problem exists as in sorghum; the resistance gene will rapidly introgress into weedy feral rice strains, as it had in many parts of the world. Again, there are methods to mitigate gene flow from rice to weedy rice strains that could be instituted (Gressel and Valverde 2009), and in the case of rice, a major non-shattering gene has been sequenced (Konishi et al. 2005) and could be used in a mitigation package.

It is unfortunate that those who have genetically engineered cowpeas and other African legumes for insect and disease resistance have used antibiotic resistance as a selectable marker (Popelka et al. 2006; Solleti et al. 2008). Had they used the

EPSPS or *ALS* genes conferring herbicide resistance as selectable markers, they would have facilitated *S. gesnerioides* control as well as dealing with insects.

24.3 When Will the Parasites Evolve Herbicide Resistance?

The evolution of herbicide resistance is an exceedingly common phenomenon (Gressel 2002; Heap 2011), especially to ALS-inhibiting herbicides (Corbett and Tardif 2006). One in a million weed seeds is considered to be resistant before herbicide use. Thus, when IR maize was beginning to be developed for Africa, modelling was performed to predict how quickly herbicide resistance would evolve and how quickly resistance would spread to a point where the technology would be useless (Gressel et al. 1996). Many assumptions were inserted into the equations: seed production; the 10^{-6} mutation frequency to resistance; almost near normal fitness; and a slow spread in fields, as mechanical corn pickers and combine harvesters, the main source of weed seed spread in the developed world, are not used in the parts of Africa where *Striga* is prevalent.

The model suggested that five new resistant *Striga* plants should be expected per cropping season per hectare, and the expanding coverage with resistant *Striga* would render the herbicide resistance technology useless in about eight seasons if unchecked (Gressel et al. 1996). The model also predicted that if farmers could find and remove four of every five *Striga* plants that appeared, before they set seed, the technology would last over 20 years. Of course if all emerging stalks were removed before seed set, the technology would last forever. What is more surprising is that thousands of hectares have been planted to IR maize and no resistant individuals have been reported. Why does nature not follow logical mathematical models? It rapidly became apparent that a major logical assumption was incorrect for the situation with IR maize—the mutation frequency of 10^{-6}. This frequency is correct for field rates of the herbicide and non-parasitic weeds, but much higher doses surround the seed. During the backcrossing of the IR gene from temperate to tropical maize, far lower doses of herbicide had to be used to select between susceptible and heterozygous-resistant individuals than were later used with the recessively homozygous-resistant material finally bred. The dose used with the homozygous recessive-resistant maize would still be toxic to *Striga* plants with only a single mutant allele. Thus, the mutation frequency put in the models should not have been 10^{-6}; it should have been $10^{-6} \times 10^{-6} = 10^{-12}$. Instead of five resistant plants per hectare, one should expect five plants per million hectares (Gressel 2005). With one small mistaken assumption in modelling, there can be an error of a factor of a million, but not quite. The herbicide dissipates from the rhizosphere during the season, and late in the season the level might be low enough that the one in a million heterozygotes can survive and then interbreed with each other, yielding some highly resistant homozygous parasites. With the short season maize varieties used in the equatorial areas of Africa with two rainy seasons, there are rarely later season attachments that give rise to flowering plants that set seed, so there is little worry that resistance will evolve, except when there are early season heavy rainfalls that leach much of the herbicide to lower soil layers where it is ineffective, and allow heterozygously resistant individuals to reproduce. Thus, vigilance is still required and roguing is a necessity when this happens.

There is a much greater worry about the sustainability of the technology with IR maize varieties bred for longer seasons (Menkir et al. 2010) as well as the ALS-resistant sorghum lines being bred (Tuinstra et al. 2009). In both cases there is a considerable amount of late-emerging *Striga*, which hardly affects crop yield. One critical bit of information is missing from those published studies—whether the herbicides completely suppressed *Striga* seed production. This is important for long-term sustainability of the technology. The mid-/late season attachments occur after the herbicides have somewhat dissipated, probably to low enough herbicide levels that heterozygote-resistant individuals can survive. If there is enough time to flower and cross pollinate, you will have 25 % homozygote resistant individuals, and the technology is lost. The late-emerging *Striga* did not have time to set seed in the short season maize varieties (Kanampiu et al. 2001, 2002, 2007). Similar problems must be considered if the ALS-resistant tomato mutant (Dor et al. 2011) is to be commercialised.

With slow release herbicide formulations the amount available can also be less, and heterozygote selection may occur (section 24.3.2). Thus, resistance management is still of utmost importance, especially warning farmers to scout their fields and pull easy to see flowering *Striga*. Other resistance management strategies are described in the following sections.

24.3.1 Resistance Management Using Sequential Herbicide Treatments and Stacked Genes

One strategy that would considerably delay the evolution of herbicide resistance in these root parasites and would add many side benefits would be the sequential use of resistance to the two most effective herbicide groups. The ALS inhibitors are substantially more expensive than glyphosate. The very low dose per hectare (but high in the vicinity of the seed) used in seed treatments alleviates this cost problem. The seed treatment has an additional benefit; other non-parasitic weeds germinating close to the treated crop seed are also controlled. Their closeness to the crop renders them hardest to weed manually, yet they are the weeds that compete most with crops. Glyphosate can be applied to the seeds bearing the respective target site resistance (Gressel and Joel 1997), but as observed in South Africa, it is cheap enough for use by subsistence farmers on glyphosate-resistant maize. This mid-season treatment for general weed control provided a huge financial advantage accruing from rapid weed control at the correct mid-season timing, which cannot be achieved manually.

Thus, especially on the longer season crops, where heterozygous-resistant parasite individuals may set seed, a stacking of ALS and glyphosate resistances may provide excellent parasite and general weed control. The use of seed treatments with ALS inhibitors will provide the early season weed control of parasites as well as weeds germinating near the crop, and a mid-season foliar glyphosate treatment would kill late attaching parasites and alleviate the hand weeding typical in subsistence agriculture. The use of the ALS will kill any glyphosate-resistant individuals that may evolve and the use of glyphosate will control any ALS-resistant parasites that may evolve. A modicum of cooperation by the competing manufacturers of herbicides is required to attain this goal that is advantageous to all parties.

24.3.2 The Next Generation: Slow Release Formulations of Herbicides

The seed treatments for *Striga* control are not without drawbacks. There can be crop phytotoxicity when rains are sparse after maize planting. This is manifested by seedling emergence from herbicide-dressed seed a day or so later than from untreated seed and also by a reduction in crop stand. This does not usually result in yield loss because the farmers typically plant far too densely. The plasticity of the crop precludes yield loss. The treated hybrid seed used is of much higher quality than much of what farmers had been used to, and presumably they could save considerably on seed costs by planting less seed, if they did not have to worry about crop phytotoxicity in dry seasons.

When rains are exceedingly heavy, the herbicide can be washed beyond the root zone too quickly, precluding season long control. Season long control is not necessary from a yield point of view. It is the early infesting *Striga* that damages the crop. Still, full season control precludes replenishing the *Striga* seed bank in the soil. Diminishing seed bank so that *Striga* sensitive crops can be cultivated in rotation is a desirable goal.

The present IR technology provides season long control in equatorial tropical Africa where there are two plantings per year of 12–14-week maturing maize under normal rainfall conditions. The single seed treatment will probably not give season long control where there is only one 18–22-week maize crop planted in other parts of *Striga*-stricken Africa, i.e. with a single, long cropping season per year. Slow release formulation technologies were developed to prevent early (dry) season crop phytotoxicity and extend control further into the season. This would mean less herbicide available early season, preventing early season phytotoxicity to the crop and possibly extending the duration that the herbicide is not leached out of the crop rhizosphere. Such formulations were initially achieved by ionically binding the negatively charged free acid form of imazapyr to various anion exchangers. The ones tested so far have prevented early season phytotoxicity (Kanampiu et al.

2009). It was not possible to ascertain whether the duration of complete control without *Striga* seed set can be extended by these slow release seed treatments. The low rainfall conditions in those seasons prevented getting this information. Newer controlled-release formulations have been prepared, but not yet tested in the field (Burnet et al. 2010). These can be used as seed coat as well as a herbicide pellet to be planted with the seed into the hill. The pellet would allow a greater degree of dose control, e.g. to use smaller or fewer pellets for the short rainy season in equatorial maize, and larger or more pellets for the long rainy season, or with long season maize or sorghum (Ransom et al. 2012).

24.3.3 Integrating Target Site Resistance with Other Parasite Management Technologies

Because parasitic weeds can set tens of millions of seeds per hectare, it is easy for natural selection to select for resistance to any technology with such large populations from which to choose. Thus, it is imperative to integrate the most desirable target site resistances with other technologies (or with other herbicide resistances, as outlined above).

Herbicide resistance would integrate well and augment the biocontrol strategies being developed for *Striga* and for *Orobanche* (see Chap. 26), as it is envisaged that the biocontrol agents too would be disseminated on the crop seed. If the biocontrol fungi are not resistant to the target site herbicide used, the pathogens will either have to be transformed or mutagenized and selected for herbicide resistance.

Classical breeding has typically been for resistance to parasitic weeds (which is often incomplete and inadequate). One attempt has been made to augment polygenic breeding for *Striga* resistance, with a single gene for herbicide resistance (Menkir et al. 2010). Thus, if the parasite evolves resistance to the breeding strategy, as it has so many times in breeding sunflower resistant to *Orobanche* (Labrousse et al. 2001; see Chap. 21), or if the parasite evolves resistance to the herbicide, the alternative mechanism still controls the parasite.

Herbicide resistance is needed at least initially for intercrops such as *Desmodium* (Khan et al. 2007; see Chap. 25). *Desmodium* intercrops take a few years to establish sufficiently to control *Striga*, and *Striga* control is needed during this interim period. It is hard to predict whether *Striga* will evolve resistance to the allelochemical(s) exuded by *Desmodium*, so another precautionary mechanism may be continually needed even after the intercrop has established. Alternatively, if the crop could make the allelochemicals, the 'middleman' *Desmodium* might be excluded, as discussed in the following sections.

24.4 Biotechnologically Directly Conferring Crop Resistance to the Parasites

There are far more reviews on what could be done (such as the rest of this chapter) than papers articles on what has been done and what might actually be successful. Still, what was dreamed about a decade ago may become a reality. For a long period the authorities that funded DNA sequencing work would only consider 'economic' plants (read 'crop'), and it required a considerable amount of explaining that plants with considerable negative economic value (read 'weeds') should also be sequenced (Stewart 2009), including the root-parasitic Orobanchaceae. This allows researchers to use the differences between crop and weed genes to design molecular strategies for weed control that do not require herbicides. It could also allow ascertaining the genetic differences between the wild progenitors of the parasitic weeds that do not attack crops with the vengeance of the highly virulent weedy parasites (see Sect. 14.3.1). For example *S. hermonthica* is partially interfertile with *S. aspera*, which is considered its progenitor (Aigbokhan et al. 1998). *S. hermonthica* is an agricultural pest having rare or no native hosts, and the presumed progenitor *S. aspera* is found on native grasses and hardly damages crops (see Sect. 18.3.4). Knowing the genetic differences could facilitate some of the strategies discussed below. Additionally, expression profiling mRNA generated after parasite attack of resistant plants vs. their susceptible relatives allows pinpointing genes that might confer resistance. Thus, there is reason to be optimistic that molecular biotechnology will contribute to providing solutions to the parasite problems.

24.4.1 Moving Parasite-Resistant Genes from Crop to Crop

Various species are not affected by these root parasites. Why they are resistant is not clear. If the genes were known, they could be used to confer resistance. An easier approach is to take genes from crops that have resistant and susceptible lines, such as the sorghum or the *Zea diploperennis* lines described used for breeding (see Sect. 21.2.1). The strategy used mostly is gene mapping and using the information for marker-assisted breeding. Marker-assisted breeding is rendered increasingly effective as the resolution of the map increases. Differential expression profiling pinpoints genes that are expressed in one line, but not the other, and can be picked out. Differential expression profiling could also be used more as a tool to increase mapping resolution as well as to verify which up- or downregulated genes are the ones responsible for resistance, and together with marker-assisted breeding and sequencing, the technologies would facilitate gene isolation. Too often breeders are interested only in their own crop, but resistance gene isolation could assist molecular breeders of other crops. When the sorghum genes separately conferring a lack of germination stimulant, poor attachment, poor penetration and poor establishment are isolated, they could be put in a single construct for engineering into any parasite sensitive crop; the closest to 'one size fits all' you can get in biology. If the resistance genes are in a single construct, they will be inherited as a single dominant trait, and backcrossing to multiple varieties will be much simpler than backcrossing four recessive genes, each on a different chromosome. Even though they would be inherited as a single gene, such a cluster will act as a polygenic trait vis-à-vis high resilience against parasite evolution of resistance. Thus gene isolation could help the breeder of the crop that isolated them to more quickly disseminate the parasite resistance traits to many locally adapted varieties, as well as breeders of other crops. Isolated resistance genes from a number of species can be stacked (mixed) into a single construct that would be inherited as a single dominant gene and can be used in a number of crops. In some cases where antisense or RNAi is used to suppress crop genes, or transcription factors or other controlling elements are used to turn on genes, it will be necessary to re-engineer the multigene construct with the precise sequences of each species. Where novel peptides are part of the multigene construct, it may not be necessary to change the codon usages unless very high levels of expression are needed.

The sequencing of crops and the new chip technologies are rendering it easier to isolate the resistance genes, even if their metabolic function is unknown. Thus, a quantitative trait locus (QTL) in rice conferring resistance has rapidly been whittled down to a 1.5 Mbase pair segment of rice chromosome 4 (Swarbrick et al. 2009). Three other levels of resistance, each inherited on separate genes, have also been identified in rice (Yoshida and Shirasu 2009). Now that sorghum and rice have been sequenced, it is hoped that the four QTLs conferring resistance can soon be isolated as genes for transformation into other species. One of the resistance genes has been isolated and cloned from cowpeas (Li et al. 2009). Suppressing the gene in resistant cowpea renders it susceptible to *Striga*. It is now important to demonstrate that adding the gene confers resistance, hopefully in other species as well. This approach led to finding genes encoding the NAD(P)H reduction of quinones to compounds inducing haustorial development of the hemiparasite *Triphysaria*. Silencing one of these genes resulted 'in a dramatic decrease in the number of haustoria produced' (Bandaranayake et al. 2011), suggesting that it would be an appropriate gene to be silenced in other species.

A recessive mutant in tomato was reported that confers resistance to *Orobanche* and *Phelipanche* spp. (Dor et al. 2010). It produces less strigolactone, but also has aberrant shoot morphology (Koltai et al. 2010). Since the tomato genome is published, it should not be too hard to isolate the gene, and if re-engineered back into tomato in antisense form or RNAi should confer dominant resistance. The gene should have many consensus sequences to the same gene in other species that would allow in silico isolation of a gene with similar function from sequenced crop species and by PCR from species that have not been fully sequenced.

A lesser discussed source of genes for resistance should be incompatible species that induce parasite germination but are immune to their effects, especially those that are attacked and fend off the attack at various stages of development. They are just beginning to become studied (see Chap. 7; Yoshida and Shirasu 2009), and some of those species have been fully sequenced, e.g. *Arabidopsis* that is attacked by *Striga*, but does not support the parasite.

24.4.2 Suppressing Parasites with Crop RNAi and Antisense Translocated from the Crop

Various macromolecules can traverse from the host to the root parasite, including peptides (which could be toxic) and polynucleotides (cf. Tomilov et al. 2008; Westwood et al. 2009; Aly et al. 2011). At present there is more information on the use of translocated RNAi than peptides to suppress parasites. Interference RNA (RNAi) is a technology where long pieces of a gene to be suppressed are put in a special construct that gets chopped into shorter pieces of RNAi. These in turn basically suppress that gene from being expressed. This is best known as a method for suppressing viruses. Similarly, antisense technology, where a piece of a gene is put in reverse direction in a high expression cassette, suppresses the sense form of the mRNA when transformed into plants.

Antisense was used in what may retrospectively be considered an ill-advised manner as far as parasite control is concerned. It was used to suppress the carotenoid cleavage dioxygenase gene in tomato encoding a critical enzyme in the pathway to strigolactone germination stimulants. It achieved 90 % reduction in *P. ramosa* germination in a laboratory experiment, which would be useful possibly in mildly infested soils, but of little use in a highly infested field (Vogel et al. 2010). Furthermore, the lack of this branch-inhibiting hormone caused the tomato plants to become far too bushy to be of economic interest (see further discussion in Sect. 10.4.2).

Short pieces of RNAi mRNA can be transported short distances (Kragler 2010). For example, plants have been transformed to make an RNAi that targets a nematode gene. The nematodes attacking the plants are severely inhibited after ingesting the RNAi (Huang et al. 2006). RNAi can traverse the junction between host and parasite (Tomilov et al. 2008; Westwood et al. 2009). Thus it was posited that an RNAi that specifically affects a metabolic pathway in root parasite, which has gene sequences not occurring in the crop, should have no effect on the crop, but should kill parasite. The first trials with *S. asiatica* and maize targeting the same genes that herbicides target, but with parasite-specific sequences were ineffectual (de Framond et al. 2007). This approach should have worked because a herbicide need not completely suppress an enzyme to kill a plant. Various groups are still trying this approach, hoping to hit on the right gene (Yoder et al. 2009). The same approach has been used to target mannose-6-phosphate reductase, an enzyme found in Orobanchaceae (see Sect. 6.2.2.2) and not in most crops (Robert et al. 1999). While the results were promising (Aly et al. 2009), the level of control was far too low to be of value to farmers.

Expression profiling is being used to ascertain early post attachment parasite-specific genes that could be expressed as RNAi. In one case three putative invertase genes were found with similar activity levels in *P. ramosa* (Draie et al. 2011). Targeting one gene would be useless, as there easily could be compensation by the others. Thus the RNAi would have to target a consensus sequence of all three genes or be a stacked construct with all three RNAi's.

Perhaps there is a mistaken paradigm with this approach. Even when RNAi was used with nematodes and pathogens, it was unacceptable as an agricultural tool due to insufficient levels of control. A construct containing RNAi's to various parasite-specific genes may be additive or even synergistic, but there are yet no reports that this has been tried. It could easily be done by genetically combining the maize lines with RNAi's for the various *Striga* targets (de Framond et al. 2007) that are already available.

Much research is going on in this exciting area, and it is hoped that results that are more than promising will soon be forthcoming. A possible new 'packaging agent' to convey the RNAi has possibly not been considered. Broomrapes can be infected by cucumber mosaic virus and possibly other viruses from their hosts (Gal-on et al. 2009). Perhaps non-destructive 'attenuated' forms of these viruses could be used to carry parasite-specific RNAi's from pre-infected crop to the parasites.

24.4.3 Do It Yourself Herbicides: Host-Generated Allelochemicals

There have been various efforts to genetically engineer crop roots to exude parasite-toxic chemicals into the rhizosphere. One effort has been partially successful, engineering a gene for production of sarcotoxin into tomato (Aly et al. 2006). The cecropin (insect haemolymph-derived antimicrobial peptides) sarcotoxin is typically phytotoxic to all plants, but at the level expressed under a host root-specific promoter, it significantly reduced broomrape attachment and enhanced crop yield, but not enough for a farmer. Perhaps constructs should be made to obtain higher expression, but under a wound inducible promoter, so that only attacked roots would be harmed in a manner akin to hypersensitive responses of leaves, where necrotic areas surrounding a pathogen stop its advance. As other cecropins are known to kill nematodes (Chalk et al. 1995), the transgenic tomatoes should be checked for nematode resistance, as dual effectiveness would surely increase the value of such a product.

Another approach is to generate transgenic crops secreting allelochemicals known to control parasites in nature, e.g. from *Desmodium* (see Chap. 25). The effectiveness of the *Desmodium* allelochemical may go beyond *Striga* control. *Desmodium* species, planted between tomatoes, reduced *P. ramosa* emergence and increased tomato yield significantly in one of two seasons (Idris 2009). It will not necessarily be easy to obtain transgenic crops emitting this allelochemical. It took over a decade of intensive research to isolate and characterise the major allelochemical, isoschaftoside, a glycosylflavonoid from *Desmodium* (Hooper et al. 2010; see Chap. 25). Many other inactive isoflavones were isolated from this species before this compound was found. It is yet to be determined how many and what genes are needed to synthesise isoschaftoside from a primary metabolite(s) and then if there are sufficient amounts of the primary metabolite(s) in other hosts to support the transgenic biosynthesis of the allelochemical at the levels required to control root parasites. One could also contemplate genetically engineering *Desmodium* to produce less of the inactive compounds by using antisense or RNAi or micro RNA technologies to close the branch pathways leading to the inactive compounds, resulting in a channelling of more secondary metabolism to isoschaftoside.

Various fungi produce toxins that are selectively toxic to the parasites. For example, the AAL toxin of *Alternaria* is 100 times more potent in killing *Orobanche* than killing tomato, one of the more sensitive crops to this toxin (de Zelicourt et al. 2007). A gene cluster on a conditionally dispensable *Alternaria* chromosome controlling the genes encoding their biosynthesis has been isolated (Kodama et al. 2008). Such pathways are often clustered in fungi (Walton 2000), and if the genes are indeed in a cluster, they could be transferred to crops in a human-facilitated horizontal gene transfer, similar to that occurring among unrelated microorganisms in nature. For efficiency, the gene cluster will have to be resynthesised with crop codon usage and promoters. In many cases, the genes would have to be under a highly specific root promoter, as many of the fungal toxins that inhibit parasites (e.g. fumonisins) are also carcinogenic, and must not enter the edible portions of the crop. Even when all this is known, and plants are transformed, there may be too high a metabolic cost in synthesising complex allelochemicals, such that crop yields are unduly lowered due to the metabolic load. In performing such engineering, one must always keep in mind the ecological paradigm that plants either use a strategy of growth or one of defence, and defence has a high cost at the expense of growth (Herms and Mattson 1992). Part of the expense can be alleviated by using parasite attack-specific promoters, such that the allelochemicals or toxins are only produced when there is a need. More often than not, agrochemical treatments 'cost' less than the yield penalty of producing allelochemicals. A notable exception is transgenic Bt production, which is far cheaper than insecticides.

A simpler approach is to engineer peptides encoded by single genes as allelochemicals. The *Ha-Def*1 gene encoding a 28 amino acid defensin peptide toxic to *O. cumana* is the resistant determinant in a strain of sunflowers (de Zelicourt et al. 2009). Presumably the gene could be transformed into other crops for broomrape resistance, but this defensin is not toxic to *Striga* (de Zelicourt et al. 2007). Many antimicrobial peptides including some that have been rationally designed have been successfully transformed into plants to control phytopathogens, including broomrape susceptible crops such as potato, tomato, eggplant and rice (Montesinos 2007; Marcos et al. 2008), but there is no evidence that they were screened for conferring resistance to parasitic weeds. It is clear that it is worthwhile to both screen the existing transformed plants as well as screen the multitudes of isolated peptides in an in vitro germination test with parasite seeds or with parasite cell cultures to ascertain which of these peptides might be effective in staving off parasite infestation when the genes are transformed into crops.

24.4.3.1 Needed: Easy Transformation Screens for Transgenic Resistance to Orobanchaceae

After in vitro tests for control of the root parasites, one needs easily transformable systems to assay whether parasites are controlled when the genes are expressed in plants. Both tobacco and tomato are readily transformable and easy to assay with *Orobanche* and *Phelipanche* species. The situation for *Striga* is less simple. While maize, sorghum, millet and rice have been transformed, it is a more tedious process, and their regeneration is much harder and time and labour consuming than tobacco or tomato. An alternative model plant might be *Digitaria sanguinalis*, which is very easy to transform grass. One can rapidly obtain *Digitaria* plantlets that are easy to propagate by cutting, precluding the need to wait for seeds to perform experiments (Chen et al. 1998). It is probably susceptible to attack by *S. aspera,* as the closely related crop *D. exilis* (fonio), which is highly susceptible (cf. Robson and Broad 1989). It would be well worthwhile to ascertain whether either *Digitaria* species can be developed into good rapid model assay for genes conferring resistance to *Striga*.

24.4.4 Making Outcrossing Parasites Double Cross Themselves

In a series of innovative papers, it was proposed to rapidly distribute conditionally lethal genes via multi-copy transposons throughout a pest population (see Grigliatti et al. 2007) using their TAC-TIC model: 'Transposons with Armed Cassettes for Targeted Insect Control'. They proposed to use a chemically induced promoter to activate genes that would prevent feeding and mating or otherwise kill the insect, i.e. chemically assisted-suicide genes, termed by this author as '*kev*' (Kevorkian) genes. Not many transgenic pests would be needed if the transgenes are transmitted in multi-copy transposons because all the offspring of any cross with one parent carrying the multi-copy transposon will also carry the transposon, as will all their future progeny.

This approach was modified for *Striga hermonthica* and other outcrossing weeds (Gressel and Levy 2000; Gressel 2002). The *kev* genes can be introduced into a transposon cassette, preferably under the control of a promoter that is turned on by an indigenous molecule in one of the host species of the parasite. That host would only be planted after the transposons had disseminated throughout the population. Alternatively, it could be a chemical applied to the crop that would translocate systemically to the parasite and activate the *kev* gene. The *kev* gene could be a chemically induced gene that will cause pollen sterility a generation hence, as had been proposed for protecting crop varieties, i.e. the so-called terminator genes of the popular press (see Oliver and Li 2012). Herbicidally lethal *kev* genes could be used, i.e. antisense or RNAi constructs of any herbicide target gene. These constructs would work in the same manner as the herbicide.

The TAC-TIC concept was taken one step further by Rector (2009). He suggested using a sterile-female technique where the *kev* gene causes female sterility. This could be *barnase* under an ovary specific promoter. It would be disseminated via a multi-copy transposon if introduced into the parasite genome. Male fertility would be maintained. In his words, the female sterility would "act without induction and spread through a weed population via pollen from female-sterile target plants to conspecific wild-type target plants, which would serve as the female parents." "The female-sterility construct would replicate during meiosis and be sexually transmitted in pollen." He concludes that "with successive post-release generations, female-sterile seed" would "make up an increasing proportion of the seed bank, as pollen from female-sterile plants competes with wild-type target pollen to fertilise decreasing numbers of wild-type pistils." "The parasite population would (eventually) crash as female-sterile seed germinate but do not reproduce" (condensed from Rector 2009).

The TAC-TIC technologies could only work with *S. hermonthica*, the only outcrosser among the major *Striga* pests (Mohamed et al. 1996; see Sect. 18.3.2). Unfortunately, the technology may not work as well with the weedy broomrapes as both weedy *Orobanche* and *Phelipanche* species are known to be obligate outcrossers and most have a mechanism that ensures selfing if not pollinated by insects (Teryokhin 1997; see Chap. 8).

24.5 Other Biotechnological Approaches

Any approach that reduces seed bank offers the prospect of cultivating root-parasite susceptible crops or those that have a modicum of tolerance to lower level soil infestations. Thus having rotational catch and trap crops is often suggested. Such crops can be biotechnologically enhanced to lower the seed bank to a greater extent. One such system with two target site herbicide resistances in oilseed rape is discussed in Sect. 24.2.2.

24.5.1 Have Non-hosts Overproduce Stimulant

As more information accumulates on the biosynthesis pathways of germination stimulants, the better the possibility of generating plants that overproduce stimulant, such that the stimulant will permeate more of the rhizosphere and cause suicidal germination (see Sect. 10.4). Leads on the genes are beginning to come in. T-DNA activation tagging of *Arabidopsis* led to three lines with less stimulant (Denev et al. 2007). This could assist in finding the non-mutant genes in the wild type and, if over expressed, could have a higher level of stimulation. Many of the germination stimulants are strigolactones. As discussed in an earlier section, under-expressing stimulants did lead to less attachments to a susceptible crop but also led to too much branching. If over-expressed strigolactones get to the shoot, there may be too few branches, and engineered gene constructs would have to target over-expressed crop root excretion of the stimulants. It appears that not all stimulants are strigolactones; e.g. the sunflower stimulant for *O. cumana* is synthesised on a different (yet unknown) pathway (Joel et al. 2011). It will be interesting if over-expressing this stimulant will have other effects, especially on branching, as commercial sunflowers are unbranched, unlike their wild, ornamental and weedy relatives. Similarly, *Brassica napus* stimulates the germination of *P. ramosa* via rhizosphere metabolites of root secreted glucosinolates (Auger et al. 2011) and not strigolactones.

24.6 Conclusions

Target site herbicide resistance is an ‘instant gratification’ mechanism for the farmer providing immediate control of parasitic weeds, even at very high seed bank densities, where breeding often breaks down. In herbicide-treated short season imidazolinone-resistant maize, no *Striga* flowers set seed, depleting the *Striga* seed bank. Herbicide resistance has evolved in other weeds and can evolve in the parasites. Thus, care is needed to manage herbicide-resistant crops, and additional technologies are needed.

World food insecurity and rising agricultural commodity prices may convince Africa and Europe to join the rest of the world and realise that transgenics are a safe and inexpensive way to deal with parasitic weeds. Most root-parasitising weeds are concentrated in Africa and Europe, where transgenic herbicide resistance would be more useful. The non-transgenic ALS herbicide target site resistance is available in a smaller number of crops. Weeds have evolved resistance faster to ALS-inhibiting herbicides than any other herbicide group, and one should be especially wary about their use in whole field spray applications (such as being used in sunflowers for *O. cumana* control).

Until better technologies become available, integration and good management are imperative for target site resistance to the more than an excellent stopgap technology. Otherwise, it will have been nice to have target site-resistant crops, while they lasted. Industry has almost completely stopped search for new modes of action of herbicides, so new systemic herbicides, where new target site-resistant crops can be developed, are unlikely to be available in the conceivable or even distant future.

There is an unfortunate trend in the donor community that results in inefficiencies when dealing with the molecular tools available to attain parasite control. Donor support is typically available for dealing with a given parasite in a particular major crop, and not with the parasites on a multitude of crops. The specific crop focused approach to parasite control was logical when it was not possible to move genes into unrelated species, but those days are over. Since the results of molecular strategies may be far more universal, initial model plants that are easier to deal with than many more crops would surely require less investment and would get more rapid results, including eventually to the farmer in the field. This is especially the case with the crops susceptible to *Striga*; none of those used at present can give rapid results. For this reason, *Digitaria* was suggested as a possible model that could supply fast information on some genes that may have utility with crops. The faster useful findings with model species can be extrapolated to real crops, the better, as the transgenic solutions are needed to most of the intractable parasitic weed problems. Such solutions are imperative for both local and global food security.

Acknowledgements The author deeply acknowledges Okesie Akobundu (1991) for introducing him to *Striga* and Daniel Joel for introducing him to *Orobanche* two decades ago. Useful input and information for this chapter was graciously supplied by D.M. Joel and S.O. Duke. The author's work on *Orobanche* was funded by grant HNE-0158-G-00-3060-00 from the trilateral US-Israel-Egypt US Agency for International Development (AID) program and a DGF trilateral Israel-German-Palestinian project and on *Striga* by the Rockefeller Foundation.

References

Abayo GO, Ransom JK, Gressel J, Odhiambo GD (1996) *Striga hermonthica* control with acetolactate synthase inhibiting herbicides seed-dressed to corn with target site resistance. In: Moreno MT, Cubero JI, Berner DK, Joel DM, Musselman LJ, Parker C (eds) Advances in parasitic weed research. Junta de Andalucia, Cordoba, Spain, pp 762–768

Abayo GO, English T, Eplee RE, Kanampiu FK, Ransom JK, Gressel J (1998) Control of parasitic witchweeds (*Striga* spp.) on corn (*Zea mays*) resistant to acetolactate synthase inhibitors. Weed Sci 46:459–466

Aigbokhan EI, Berner DK, Musselman LJ (1998) Reproductive ability of hybrids of *Striga aspera* and *Striga hermonthica*. Phytopathology 88:563–567

Akobundu IO (1991) Weeds in human affairs is sub-Saharan Africa. Weed Technol 5:680–690

Alonso LC, Rodriguez-Ojeda MI, Fernandez-Escobar J, Lopez-Ruiz-Calero G (1998) Chemical control of broomrape (*Orobanche cernua* Loefl.) in sunflower (*Helianthus annuus* L.) resistant to imazethapyr herbicide. Helia 21:45–54

Aly R, Plakhin D, Achdari G (2006) Expression of sarcotoxin IA gene via a root-specific tob promoter enhanced host resistance against parasitic weeds in tomato plants. Plant Cell Rep 25:297–303

Aly R, Cholakh H, Joel DM, Leibman D, Steinitz B, Zelcer A, Naglis A, Yarden O, Gal-On A (2009) Gene silencing of mannose 6-phosphate reductase in the parasitic weed *Orobanche aegyptiaca* through the production of homologous dsRNA sequences in the host plant. Plant Biotechnol J 7:487–498

Aly R, Hamamouch N, Abu-Nassar J, Wolf S, Joel DM, Eizenberg H, Kaisler E, Cramer C, Gal-On A, Westwood JH (2011) Movement of protein and macromolecules between host plants and the parasitic weed *Phelipanche aegyptiaca* Pers. Plant Cell Rep 30:2233–2241. doi:10.1007/s00299-011-1128-5

Anonymous (2008) More STRIGAWAY maize demons in Uganda. In: Strides in *Striga* Management. African Agricultural Technology Foundation, Nairobi. http://www.aatf-africa.org/UserFiles/File/Strides_Issue-2_Mar08.pdf

Auger B, Pouvreu J-B, Monteau F, Pouponneau K, Gauthier M, Prat E, Berges H, Delavault P, Simier P (2011) Stimulants of *Phelipanche ramosa* germination from oilseed rape roots. In: Proceedings of the 11th world congress on parasitic plants, Martina Franca, Italy, p 61

Aviv D, Amsellem Z, Gressel J (2002) Transformation of carrots with mutant acetolactate synthase for *Orobanche* (broomrape) control. Pest Manage Sci 58:1187–1193

Bandaranayake PCG, Filappova T, Tomilov A, Tomilov N, Yoder JI (2011) Parasitic plant genes necessary for host development. In: Proceedings of the 11th world congress on parasitic plants, Martina Franca, Italy, p 15

Ben-Hod G, Losner D, Joel DM, Mayer AM (1991) In vitro culture of *Orobanche aegyptiaca*. Ann Bot 68:413–416

Berner DK, Ikie FO, Green JM (1997) ALS-inhibiting herbicide seed treatments control *Striga hermonthica* in ALS-modified corn (*Zea mays*). Weed Technol 11:704–707

Burnet M, Guse JH, Reisser M (2010) Novel pesticide formulations. US patent application # 2010 11/476,479

Chalk R, Townson H, Ham PJ (1995) *Brugia pahangi* – the effects of cecropins on microfilariae in-vitro and in *Aedes aegypti*. Exp Parasitol 80:401–406

Chen WS, Lennox SJ, Palmer KE, Thomson JA (1998) Transformation of *Digitaria sanguinalis*: a model system for testing maize streak virus resistance in Poaceae. Euphytica 104:25–31

Corbett CAL, Tardif FJ (2006) Detection of resistance to acetolactate synthase inhibitors in weeds with emphasis on DNA-based techniques: a review. Pest Manage Sci 62:584–597

de Framond ARP, McMillan J, Ejeta G (2007) Effects on *Striga* parasitism of transgenic maize armed with RNAi constructs targeting essential *S. asiatica* genes. In: Ejeta G, Gressel J (eds) Integrating new technologies for *Striga* control – towards ending the witch hunt. World Scientific, Singapore, pp 185–196

De Groote H, Wangare L, Kanampiu F, Odendo M, Diallo A, Karaya H, Friesen D (2008) The potential of a herbicide resistant maize technology for *Striga* control in Africa. Agric Syst 97:83–94

de Zelicourt A, Letousey P, Thoiron S, Campion C, Simoneau P, Elmorjani K, Marion D, Simier P, Delavault P (2007) Ha-DEF1, a sunflower defensin, induces cell death in *Orobanche* parasitic plants. Planta 226:591–600

de Zelicourt A, Montiel G, Pouvreau JB, Thoiron S, Delgrange S, Simier P, Delavault P (2009) Susceptibility of *Phelipanche* and *Orobanche* species to AAL-toxin. Planta 230:1047–1055

Denev I, Deneva B, Batchvarova R, Westwood J (2007) Use of T-DNA activation tag *Arabidopsis* mutants in studying formation of germination stimulants for broomrapes (*Orobanche* spp.). Biotechnol Biotec Equip 21:403–407

Dor E, Alperin B, Wininger S, Ben-Dor B, Somvanshi VS, Koltai H, Kapulnik Y, Hershenhorn J (2010) Characterization of a novel tomato mutant resistant to the weedy parasites *Orobanche* and *Phelipanche* spp. Euphytica 171:371–380

Dor E, Smirnov E, Achdary G, Hershenhorn Y (2011) Characterisation of a novel tomato mutant HRT-1 resistant to acetolactate synthase inhibiting herbicides for broomrape management. In: Proceedings of the 11th world congress on parasitic plants, Martina Franca, Italy, p 101

Draie R, Peron T, Pouvreau JB, Veronesi C, Jegou S, Delavault P, Thoiron S, Simier P (2011) Invertases involved in the development of the parasitic plant *Phelipanche ramosa*: characterization of the dominant soluble acid isoform, PrSAI1. Mol Plant Pathol 12:638–652

Foy CL, Jain R, Jacobsohn R (1989) Recent approaches for chemical control of broomrape (*Orobanche* spp.). Rev Weed Sci 4:123–152

Gal-On A, Naglis A, Leibman D, Ziadna H, Kathiravan K, Papayiannis L, Holdengreber V, Guenoune-Gelbert D, Lapidot M, Aly R (2009) Broomrape can acquire viruses from its hosts. Phytopathology 99:1321–1329

Gan J, Yates SR, Ohr HD, Sims JJ (1998) Production of methyl bromide by terrestrial higher plants. Geophys Res Lett 25:3595–3598

Gressel J (1992) The needs for new herbicide-resistant crops. In: Denholm I, Devonshire AL, Hollomon DW (eds) Achievements and developments in combating pesticide resistance. Elsevier, London, pp 283–294

Gressel J (2002) Molecular biology of weed control. Taylor & Francis, London, 520 pp

Gressel J (2005) Problems in qualifying and quantifying assumptions in plant protection models: resultant simulations can be mistaken by a factor of million. Crop Prot 24:1007–1015

Gressel J (2011) Low pesticide rates may hasten the evolution of resistance by increasing mutation frequency. Pest Manage Sci 67:253–257

Gressel J (2012) Containing and mitigating transgene flow from crops to weeds, to wild species, and to crops, Chapter 32. In: Altman A, Hasegawa PM (eds) Plant biotechnology and agriculture: prospects for the 21st century. Elsevier, Amsterdam, pp 495–509

Gressel J, Joel DM (1997) Use of glyphosate salts in seed dressing herbicidal compositions. US Patent 6,096,686

Gressel J, Levy A (2000) Giving *Striga hermonthica* the DT's. In: Haussmann BIG, Hess DE, Koyama ML, Grivet L, Rattunde HFW, Geiger HH (eds) Breeding for striga resistance in cereals. Margraf Verlag, Weikersheim, pp 207–224

Gressel J, Valverde BE (2009) A strategy to provide long-term control of weedy rice while mitigating herbicide resistance transgene flow, and its potential use for other crops with related weeds. Pest Manage Sci 65:723–731

Gressel J, Segel LE, Ransom JK (1996) Managing the delay of evolution of herbicide resistance in parasitic weed. Int J Pest Manage 42:113–129

Grigliatti TA, Meister G, Pfeifer TA (2007) TAC-TICS: transposon-based biological pest management systems. In: Vurro M, Gressel J (eds) Novel biotechnologies for biocontrol agent enhancement and management. Springer, Dordrecht, pp 327–351

Hall LM, Topinka K, Huffman J, Davis L, Good A (2000) Pollen flow between herbicide-resistant *Brassica napus* is the cause of multiple resistant *B. napus* volunteers. Weed Sci 48:688–694

Heap IM (2011) International survey of herbicide-resistant weeds. http://www.weedscience.org. Accessed Aug 2011

Herms DA, Mattson WJ (1992) The dilemma of plants to grow or to defend. Quart Rev Biol 67:283–335

Hooper AM, Tsanuo MK, Chamberlain K, Tittcomb K, Scholes J, Hassanali A, Khan ZR, Pickett JA (2010) Isoschaftoside, a C-glycosylflavonoid from *Desmodium uncinatum* root exudate, is an allelochemical against the development of *Striga*. Phytochemistry 71:904–908

Huang GZ, Allen R, Davis EL, Baum TJ, Hussey RS (2006) Engineering broad root-knot resistance in transgenic plants by RNAi silencing of a conserved and essential root-knot nematode parasitism gene. Proc Natl Acad Sci USA 103:14302–14306

Idris KI (2009) Evaluation of tick-trefoil (*Desmodium* species) for drought tolerance and control of witchweed (*Striga hermonthica* (Del.) Benth.) in grain sorghum and branched broomrape (*Orobanche ramosa* l.) in tomato. PhD Thesis, University of Gezira, Wad Medani (Sudan),

141 pp. http://agris.fao.org/agris-search/search/display.do?f=2010/SD/SD1010.xml; SD2010000215. Accessed Aug 2011

James C (2011) Global status of commercialized biotech/GM crops: 2011. ISAAA, Ithaca, NY

Joel DM, Kleifeld Y, Losner-Goshen D, Herzlinger G, Gressel J (1995) Transgenic crops against parasites. Nature 374:220–221

Joel DM, Plakhine D, Creange P, Dupuis JM, Kamoda J, Gressel J (2000) Broomrape (*Orobanche aegyptiaca*) control using asulam-resistant seeds coated with asulam. Weed Sci Soc Am Abstr 40:137

Joel DM, Chaudhuri SK, Plakhine D, Ziadna H, Steffens JC (2011) The natural germination stimulant of *Orobanche cumana* is not a strigolactone. In: Proceedings of the 11th world congress on parasitic plants, Martina Franca, Italy, p 65

Kabambe VH, Kanampiu F, Ngwira A (2008) Imazapyr (herbicide) seed dressing increases yield, suppresses *Striga asiatica* and has seed depletion role in maize (*Zea mays* L.) in Malawi. Afr J Biotechnol 7:3293–3298

Kanampiu FK, Ransom JK, Gressel J (2001) Imazapyr seed dressings for *Striga* control on acetolactate synthase target-site resistant maize. Crop Prot 20:885–895

Kanampiu FK, Ransom JK, Friesen D, Gressel J (2002) Imazapyr and pyrithiobac movement in soil and from maize seed coats controls *Striga* in legume intercropping. Crop Prot 21:611–619

Kanampiu F, Diallo A, Burnet M, Karaya H, Gressel J (2007) Success with the low biotech of seed-coated imidazolinone-resistant maize. In: Ejeta G, Gressel J (eds) Integrating new technologies for *Striga* control: ending the witch-hunt. World Scientific, Singapore, pp 145–158

Kanampiu F, Karaya H, Burnet M, Gressel J (2009) Needs for and effectiveness of slow release herbicide seed treatment *Striga* control formulations for protection against early season crop phytotoxicity. Crop Prot 28:845–853

Kebreab E, Murdoch AJ (2001) Simulation of integrated control strategies for *Orobanche* spp. based on a life cycle model. Exp Agric 37:37–51

Khan ZR, Midega CAO, Hassanali A, Pickett JA (2007) Field developments on *Striga* control by *Desmodium* in a "push-pull" strategy. In: Ejeta G, Gressel J (eds) Integrating new technologies for Striga control: ending the witch-hunt. World Scientific, Singapore, pp 241–251

Kodama M, Akagi Y, Akamatsu H, Otani H, Yamamoto M, Tsuge T (2008) A gene cluster on a conditionally dispensable chromosome controlling AAL-toxin biosynthesis and pathogenicity in the tomato pathotype of *Alternaria alternata*. Biol Plant-Microbe Interact 6:18/1–18/4 (citation and information supplied by Dr. S. O. Duke)

Koltai H, Lekkala SP, Bahattacharya C, Mayzlish-Gati E, Resnick N, Wininger S, Dor E, Ka Y, Ko Y, Hershenhorn J, Joel DM, Kapulnik Y (2010) A tomato strigolactone-impaired mutant displays aberrant shoot morphology and plant interactions. J Exp Bot 61:1739–1749

Konishi S, Lin SY, Ebana K, Fukuta Y, Izawa T, Sasaki T, Yano M (2005) Molecular cloning of a major QTL, *QSH-1*, controlling seed shattering habit in rice. Plant Cell Physiol 46:S198–S198

Kotoula-Syka E (2003) *Orobanche ramosa* control in tomato. In: Biology and control of broomrape. COST, Athens, p 13. http://cost849.ba.cnr.it/athens%20abstract.PDF. Accessed 15 Aug 2011

Kragler F (2010) RNA in the phloem: a crisis or a return on investment? Plant Sci 178:99–104

Labrousse P, Arnaud M, Serieys H, Berville A, Thalouarn P (2001) Several mechanisms are involved in resistance of *Helianthus* to *Orobanche cumana* Wallr. Ann Bot 88:859–868

Lévesque CA, Rahe JE (1992) Herbicide interactions with fungal root pathogens, with special reference to glyphosate. Annu Rev Phytopathol 30:579–602

Li W, Gill BS (2006) Multiple genetic pathways for seed shattering in the grasses. Funct Integr Genomics 6:300–309

Li XG, Nicholl D (2005) Development of PPO inhibitor-resistant cultures and crops. Pest Manage Sci 61:277–285

Li JX, Lis KE, Timko MP (2009) Molecular genetics of race-specific resistance of cowpea to *Striga gesnerioides* (Wilid.). Pest Manage Sci 65:520–527

Malidza G, Jocić S, Škorić DO, Orbovic B (2004) Clearfield production system for sunflower. Zbornik radova 40:279–290 (in Serbian with English abstract—Source: FAO AGRIS record)
Marcos JF, Munoz A, Perez-Paya E, Misra S, Lopez-Garcia B (2008) Identification and rational design of novel antimicrobial peptides for plant protection. Annu Rev Phytopathol 46:273–301
Menkir A, Chikoye D, Lum F (2010) Incorporating an herbicide resistance gene into tropical maize with inherent polygenic resistance to control *Striga hermonthica* (Del.) Benth. Plant Breed 129:385–392
Mohamed KI, Musselman LJ, Aigbokhan EI, Berner DK (1996) Evolution and taxonomy of agronomically-important *Striga* species. In: Moreno MT, Cubero JI, Berner D, Joel D, Musselman LJ, Parker C (eds) Advances in parasitic plant research. Junta de Andalucia, Cordoba, Spain, pp 53–73
Montesinos E (2007) Antimicrobial peptides and plant disease control. FEMS Microbiol Lett 270:1–11
Mutegi E, Sagnard F, Muraya M, Kanyenji B, Rono B, Mwongera C, Marangu C, Kamau J, Parzies H, de Villiers S, Semagn K, Traoré P, Labuschagne M (2010) Ecogeographical distribution of wild, weedy and cultivated *Sorghum bicolor* (L.) Moench in Kenya: implications for conservation and crop-to-wild gene flow. Genet Resour Crop Evol 57:243–253
Oliver MJ, Li Y (eds) (2012) Gene containment. Wiley, Ames
Patzoldt WL, Hager A, McCormick JS, Tranel PJ (2006) A codon deletion confers resistance to herbicides inhibiting protoporphyrinogen oxidase. Proc Natl Acad Sci USA 103:12329–12334
Popelka J, Gollasch S, Moore A, Molvig L, Higgins TJV (2006) Genetic transformation of cowpea (*Vigna unguiculata* L.) and stable transmission of the transgenes to progeny. Plant Cell Rep 25:304–312
Press MC (1995) Carbon and nitrogen relations. In: Press MC, Graves JD (eds) Parasitic plants. Chapman and Hall, London, pp 103–124
Ransom J, Kanampiu F, Gressel J, De Groote H, Burnet M, Odhiambo GD (2012) Herbicides applied to the seed of IR-maize as a *Striga* control option for small-scale African farmers. Weed Sci 60:283–289
Rector BG (2009) A sterile-female technique proposed for control of *Striga hermonthica* and other intractable weeds: advantages, shortcomings and risk management. Pest Manage Sci 65:596–602
Rhew RC, Ostergaard L, Saltzman ES, Yanofsky MF (2003) Genetic control of methyl halide production in Arabidopsis. Curr Biol 13:1809–1813
Robert S, Simier P, Fer A (1999) Purification and characterization of mannose 6-phosphate reductase, a potential target for the control of *Striga hermonthica* and *Orobanche ramosa* Austral. J Plant Physiol 26:233–237
Robson TO, Broad HR (eds) (1989) *Striga* – improved management in Africa. FAO, Rome, pp 206. http://books.google.com/books?hl=en&lr=&id=x0WMSFETzS4C&oi=fnd&pg=PA190&dq=Striga+–+Improved+management+in+Africa. Accessed Aug 2011
Rodenburg J, Zossou-Kouderin N, Gbèhounou G, Ahanchede A, Touré A, Kyalo G, Kiepe P (2011) *Rhamphicarpa fistulosa*, a parasitic weed threatening rain-fed lowland rice production in sub-Saharan Africa a case study from Benin. Crop Prot 30:1306–1314
Ross KC, Colquhoun JB, Mallory-Smith CA (2004) Small broomrape (*Orobanche minor*) germination and early development in response to plant species. Weed Sci 52:260–266
Sharon A, Amsellem Z, Gressel J (1992) Glyphosate suppression of an elicited defense response. Plant Physiol 98:654–659
Slavov S, Valkov V, Batchvarova R, Atanassova S, Alexandrova M, Atanassov A (2005) Chlorsulfuron resistant transgenic tobacco as a tool for broomrape control. Transgenic Res 14:273–278
Smith MC, Webb M (1996) Estimation of the seed bank of *Striga* spp. (Scrophulariaceae) in Malian fields and the implications for a model of biocontrol of *S. hermonthica*. Weed Res 36:85–92

Solleti SK, Bakshi S, Purkayastha J, Panda SK, Sahoo L (2008) Transgenic cowpea (*Vigna unguiculata*) seeds expressing a bean alpha-amylase inhibitor 1 confer resistance to storage pests, bruchid beetles. Plant Cell Rep 27:1841–1850

Song X, Wang Z, Qiang S (2011) Agronomic performance of F1, F2 and F3 hybrids between weedy rice and transgenic glufosinate-resistant rice. Pest Manage Sci 67:921–931

Souza-Machado V (1982) Inheritance and breeding potential of triazine tolerance and resistance in plants. In: LeBaron HM, Gressel J (eds) Herbicide resistance in plants. Wiley, New York, pp 257–273

Stewart CN Jr (ed) (2009) Weedy and invasive plant genomics. Wiley-Blackwell, Ames, 253 pp

Surov T, Aviv D, Aly R, Joel DM, Goldman-Guez T, Gressel J (1998) Generation of transgenic asulam-resistant potatoes to facilitate eradication of parasitic broomrapes (*Orobanche spp.*) with the *sul* gene as the selectable marker. Theor Appl Genet 96:132–137

Swarbrick PJ, Scholes JD, Press MC, Slate J (2009) A major QTL for resistance of rice to the parasitic plant *Striga hermonthica* is not dependent on genetic background. Pest Manage Sci 65:528–532

Teryokhin ES (1997) Weed broomrapes: systematics ontogenesis biology evolution. Aufstieg-Verlag, Landshut (information and citation kindly provided by DM Joel)

Tomilov AA, Tomilova NB, Wroblewski T, Michelmore R, Yoder JI (2008) Trans-specific gene silencing between host and parasitic plants. Plant J 56:389–397

Tuinstra MR, Soumana S, Al-Khatib K, Kapran I, Toure A, van Ast A, Bastiaans L, Ochanda NW, Salami I, Kayentao M, Dembele S (2009) Efficacy of herbicide seed treatments for controlling *Striga* infestation of sorghum. Crop Sci 49:923–929

Vogel JT, Walter MH, Giavalisco P, Lytovchenko A, Kohlen W, Charnikhova T, Simkin AJ, Goulet C, Strack D, Bouwmeester HJ, Fernie AR, Klee HJ (2010) SlCCD7 controls strigolactone biosynthesis, shoot branching and mycorrhiza-induced apocarotenoid formation in tomato. Plant J 61:300–311

Volrath SL, Johnson MA, Ward JER, Heifetz PB (1999) DNA molecules encoding plant protoporphyrinogen oxidase and inhibitor-resistant mutants thereof. US Patent 5,939,602

Walton JD (2000) Horizontal gene transfer and the evolution of secondary metabolite gene clusters in fungi: an hypothesis. Fungal Genet Biol 30:167–171

Westwood JH, Roney JK, Khatibi PA, Stromberg VK (2009) RNA translocation between parasitic plants and their hosts. Pest Manage Sci 65:533–539

Yoder JI, Gunathilake P, Wu B, Tomilova N, Tomilov AA (2009) Engineering host resistance against parasitic weeds with RNA interference. Pest Manage Sci 65:460–466

Yoshida S, Shirasu K (2009) Multiple layers of incompatibility to the parasitic witchweed, *Striga hermonthica*. New Phytol 183:180–189

Chapter 25
Allelopathy

John A. Pickett, Antony M. Hooper, Charles A.O. Midega, and Zeyaur R. Khan

25.1 Introduction

Allelopathy refers to an effect on an organism by another and was originally confined to deleterious but now includes beneficial and symbiotic effects (Belz 2007; Macias et al. 2007). The various mechanisms by which allelopathy can control parasitic weeds have been reviewed in consideration of plant defence against parasitic plants by the hosts (see Chap. 7) and for exploiting this using the form of companion cropping termed "intercropping" (Pickett et al. 2010). In intercropping, plants are grown between the rows of the main crop stand and can thereby interfere with weed development and particularly that of parasitic weeds. However, although potentially exploitable by intercropping, the creation of physical barriers that prevent access to the host by the parasitic plant and provision of broadly phytotoxic materials all offer less valuable mechanisms than the specific generation of allelopathic secondary metabolites that can selectively interfere with parasite development, particularly at the early stage. Specifically for the control of *Striga*, e.g. *S. hermonthica*, a number of legume companion crops have previously been recommended, including sesame (*Sesamum indicum*) (Hess and Dodo 2004) and groundnuts (*Arachis hypogaea*) (Tenebe and Kamara 2002), with the most effective being cowpea (*Vigna unguiculata*) (Carsky et al. 1994), but the mechanisms involved appear more related to nutrition and ground cover than strictly allelopathic processes involving secondary metabolism. However, the discovery that forage legumes, *Desmodium* spp., particularly silverleaf (*D. uncinatum*), would give dramatic control of *S. hermonthica* when planted as one-to-one intercrops with cereals such as maize and sorghum, came as a

J.A. Pickett (✉) • A.M. Hooper
Rothamsted Research, Harpenden, Hertfordshire AL5 2JQ, UK
e-mail: john.pickett@rothamsted.ac.uk

C.A.O. Midega • Z.R. Khan
International Centre of Insect Physiology and Ecology (icipe), Nairobi, Kenya

D.M. Joel et al. (eds.), *Parasitic Orobanchaceae*,
DOI 10.1007/978-3-642-38146-1_25, © Springer-Verlag Berlin Heidelberg 2013

breakthrough in providing a suitable technology for resource-poor farmers (Amudavi et al. 2009a, b; Hassanali et al. 2008; Khan et al. 2008a–c, 2009, 2010a).

25.2 Allelopathic Mechanism by Which *Desmodium* Controls *Striga* in Maize

Originally, the discovery that *D. uncinatum* could control *S. hermonthica* was made in the field. Indeed, this discovery was made by chance when using *Desmodium* spp. as an intercrop for repelling stem borer moths (Lepidoptera) and was found not only to be highly effective in repelling the ovipositing females but also in attracting parasitic wasps to attack any ensuing larvae from the limited number of eggs that were laid (Khan et al. 1997; Khan et al. 2000; Midega et al. 2009). Initially, non-allelopathic mechanisms such as ground cover and improved nitrogen availability were investigated, and although both are contributed by intercropping with *D. uncinatum*, a very strong allelopathic mechanism was determined when the intercrop was used as opposed to providing alternative ground cover and nitrogen fertiliser (Khan et al. 2002). Furthermore, water percolating through *D. uncinatum* roots growing in sterile soil, when then allowed to run through soil containing maize and *S. hermonthica* seeds, gave almost complete protection against the parasite, whereas water percolating through just the soil or other legume roots did not (Khan et al. 2002). Bioassay-guided fractionation of root exudates and root extracts, using high-performance liquid chromatography (HPLC), followed by nuclear magnetic resonance (NMR) spectroscopy and mass spectrometry (MS), identified prenylated novel isoflavonones, named uncinanone A and B as germination stimulants and uncinanone C as an inhibitor of radicle elongation, together with a series of isoflavones and pterocarpans, which were not active in the bioassay systems that recorded both the positive and negative aspects of early parasite development (Tsanuo et al. 2003; Guchu et al. 2007). At this stage, it was realised that, combined with the effect of the host plant, the germination stimulant effect and the radicle elongation inhibitory effect combined to give a valuable suicidal germination of the seed. This in turn reduces the parasitic seed bank (Khan et al. 2008d; Vanlauwe et al. 2008). However, inhibition of radicle development alone was used as an assay to investigate further more polar fractions when it was subsequently found that this physicochemical property gave maximal inhibitory effects. Thus, at very low concentrations of *ca* 1 ppm, an active component was identified as the di-*C*-glycosylflavone 6-C-α-L-arabinopyranosyl-8-*C*-β-D-glucopyranosylapigenin, also known as isoschaftoside (**1**, Fig. 25.1) (Hooper et al. 2010), by microprobe NMR spectroscopy, electrospray MS and electron impact MS after permethylation. Although other *C*-glycosylated flavones have since been discovered, the main effort has been with the *C*-arabinosylated and *C*-glucosylated apigenin, and this is the target for further development.

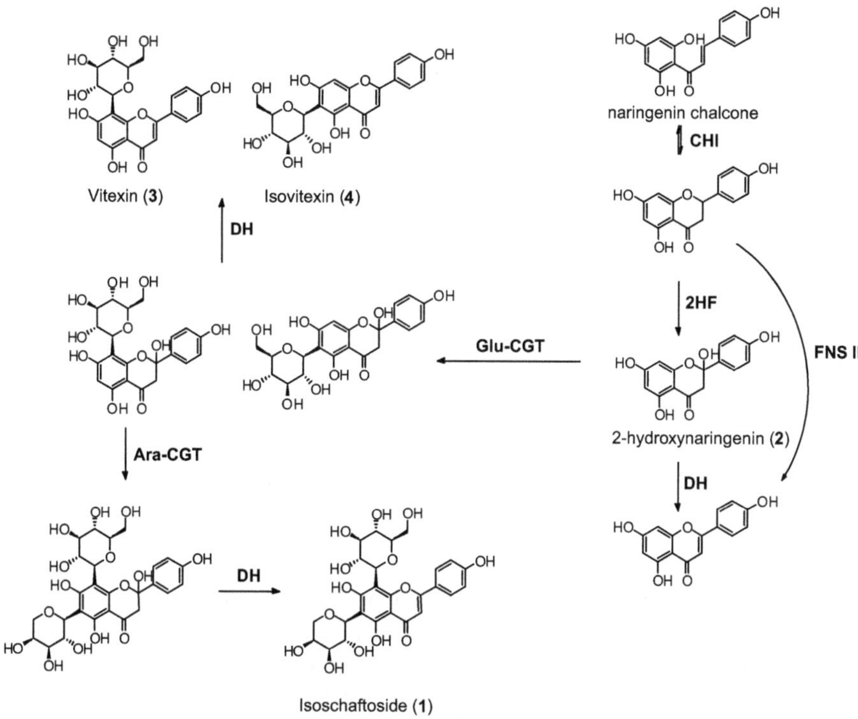

Fig. 25.1 Biosynthesis of *C*-glucosylflavonoids *in planta*. *CHI* isomerise, *FNSII* flavone synthase, *2HF* 2-hydroxyflavanone synthase, *DH* dehydratase, *CGT C*-glucosyltransferase

25.2.1 *Biosynthesis of 6-C-α-L-Arabinopyranosyl-8-C-β-D-Glucopyranosylapigenin*

25.2.1.1 Overall Mechanism

It might be expected that 6-C-α-L-arabinopyranosyl-8-*C*-β-D-glucopyranosylapigenin would be formed by *C*-glycosylation of apigenin and that two *C*-glycosyltransferases would be involved, one to create the *C*-glucosyl moiety, the other for the *C*-arabinosyl. However, previous biosynthesis studies had shown (Kersher and Franz 1987) that an earlier biosynthetic intermediate, the flavonone, 2-hydroxynaringenin (**2**, Fig. 25.2), is the target of *C*-glycosylation, at least for the specific case of 8*C*- or 6*C*-β-D-glucopyranosylapigenin (vitexin (**3**) or isovitexin (**4**), respectively), in seedlings of Buckwheat (*Fagopyrum esculentum* M.; Kersher and Franz 1988) (Fig. 25.1).

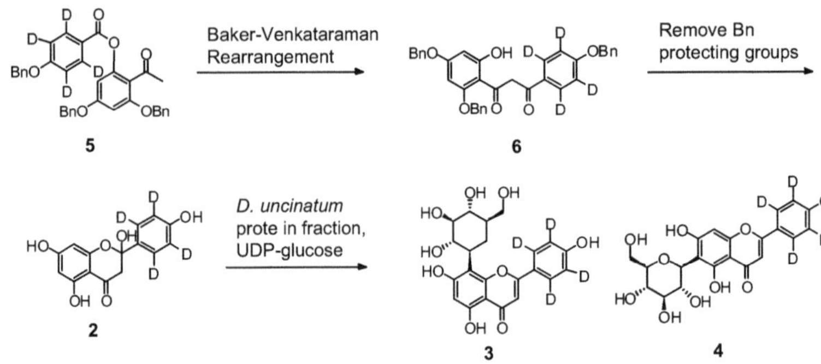

Fig. 25.2 Preparation of 2-hydroxynaringenin (2) via a Baker–Venkataraman rearrangement and incorporation of 2 by a *D. uncinatum* CGT into vitexin and isovitexin

25.2.1.2 Testing 2-Hydroxynaringenin as the Substrate for the C-Glucosyltransferases in *D. uncinatum*

A successful synthesis of 2-hydroxynaringenin was achieved with the key step being the Baker–Venkataraman rearrangement of a 6-acyloxy-3,5-dibenzyloxyacetophenone (**5**, Fig. 25.2), which gives the open chain, protected precursor of 2-hydroxynaringenin (**6**). This route, by incorporating deuterium into the ester moiety through esterification with (2,3,5,6-$^2H_4$)-4-benzyloxybenzoic acid, gave the corresponding tetradeuteriated (2′,3′,5′,6′-$^2H_4$)-2-hydroxynaringenin (Hamilton et al. 2009). This substrate, when added to protein extracts of *D. uncinatum*, gave 6*C*- and 8*C*-β-D-glucopyranosylapigenin and, where labelled, with the expected deuterium replacement as recorded by MS and allowed further purification of the protein fraction responsible (Fig. 25.2). Candidate purified proteins are now the subject of mass spectrometric protein sequencing and N-terminal sequencing prior to isolation of the corresponding gene through standard molecular biology techniques. Although highly purified and active *C*-glucosyltransferase protein extracts have been obtained, these have not yet yielded full sequence data (Hamilton et al. 2012). The complete biosynthesis of isoschaftoside also requires introduction of an arabinosyl moiety. In both *D. intortum* and *D. uncinatum*, we have shown that the two *C*-glycosylations are sequential. First, a *C*-glucosyltransferase adds glucose from UDP-glucose and this step is required for subsequent *C*-arabinosylation which is performed by a separate enzyme that utilises UDP-arabinose to form *C*-glucosyl-*C*-arabinosyl-2-hydroxynaringenin (Hamilton et al. 2012). It is likely that the dehydration of this doubly glycosylated hydroxyflavanone, to remove water and generate the flavone 2-3-double bond, fixing the sugar regiochemistry of isoschaftoside and its regioisomer schaftoside is enzyme mediated in the plant. Although spontaneous dehydration does occur in our assays, the ratio of regioisomers in the plant does not match that of non-enzymic chemical dehydration supporting the presence of a separate dehydratase enzyme activity not associated with the *C*-glycosyltransferases (Fig. 25.1).

25.3 Long-Term Needs

25.3.1 Overall Possibilities

The technology involving intercropping of cereals with *Desmodium* spp. against *S. hermonthica* and *S. asiatica* has been designed primarily for resource-poor farmers. These farmers are unable to purchase seasonal inputs such as fertiliser, pesticides and seed, and so a perennial intervention is most appropriate for their needs (Khan et al. 2010a, b). Already, over 50,000 farmers in western Kenya are using the technology, each designated by GPS registration, and a major extension to the millions more farmers who would benefit is now being attempted not only in Kenya but also in more arid regions of East Africa (Kenya, Ethiopia, Tanzania) and West Africa (Nigeria) for which alternative *Desmodium* species with suitable agronomic tolerance and correct secondary metabolism are required. The technology is suitable for other cereal crops, including sorghum (Khan et al. 2006a, 2007) and millets, although taxonomically diverse, including finger millet, *Eleusine coracana* (Midega et al. 2010), and upland rice (Khan et al. 2010a; Pickett et al. 2010). Upland rice is developing rapidly as a crop in Africa because of it having no requirement for irrigation, but the NERICA (NEw RICe for Africa) cultivars suffer badly from *S. hermonthica* to which they are completely unadapted. However, with *D. uncinatum* as an intercrop, the parasite is controlled with high yield benefits (Fig. 25.3; Khan et al. 2010a; Pickett et al. 2010).

It has now proved possible to include edible beans in the main cropping system deploying *D. uncinatum* to control *S. hermonthica* in maize (Khan et al. 2009). However, the farmers have also made it known, through the extension services and at their meetings known locally as barazas, that they would like the option of just an edible bean intercrop, and so to transfer the *Desmodium* trait that controls *Striga* is being attempted (Khan et al. 2010a). In addition to the direct control of *Striga* in cereals by new crops having this trait, there are possibilities of the trait being useful against other parasitic weeds and even beyond the Orobanchaceae, to include non-parasitic weeds. To this end, transferring the trait to rhizosphere occupying organisms could also be considered. To date, the chemistry shown to be associated with the *Striga*-controlling trait of *Desmodium* has only been found in other *Desmodium* spp. forage legumes and not in edible bean legumes (Khan et al. 2006b). Therefore, for creating new edible beans with the *Striga*-controlling trait, and certainly for cereal crop plants with this trait, genetic modification offers the best approach.

Fig. 25.3 *Striga* control in rice intercropped with *Desmodium*. Rice field planted with Nerica4 rice variety. On the *left-hand side* is a rice plot intercropped with *Desmodium uncinatum*, and on the *right-hand side* is a rice monocrop field heavily infested with *Striga hermonthica*

25.3.2 Application of the Desmodium *Allelopathic Activity in Other Crops*

Although the genetics by which *D. uncinatum* and other species create the *C*-glycosylated flavonones still remains to be fully elucidated, the enzymology is showing great promise to date. *C*-Glycosylflavonoids have been isolated from the tissues of many other plants including major cereal crops and have had biological activities attributed to them both *in planta* and as dietary components for animals. An enzyme (OsCGT) catalysing the UDP-glucose-dependent *C*-glucosylation of 2-hydroxyflavanone precursors has been identified and cloned from rice (*Oryza sativa* ssp. *indica*), with a similar protein characterised in wheat (*Triticum aestivum*) (Brazier-Hicks et al. 2009). Currently (in collaboration with Professor Mike Timko, University of Virginia) we have transferred this known OsCGT into cowpea, using transformation technology already developed that has demonstrated gene-for-gene resistance in the *Striga*/cowpea association (Li and Timko 2009; see Chap. 7). This will allow exploration for the presence of the *C*-glycosylation substrate (2-hydroxynaringenin) which is an intermediate in flavone biosynthesis and may already be sufficient in the cowpea. Otherwise, transformation constructs have been prepared so that this substrate may be prepared *in vivo* from naringenin using the additional 2-hydroxyflavanone synthase gene that has been characterised from rice (designated *CYP93G2*) (Du et al. 2010) that can be introduced alongside the OsCGT. In the same way, the genes for specific *C*-glycosyltransferases and the desaturase responsible for later processing of the initial *C*-glycosyl adducts will be available for the precise reconstruction of this pathway in edible legumes and eventually in cereal crop plants.

25.4 Conclusions

The work describing the discovery and development potential, specifically for the allelopathic mechanism by which *Desmodium* spp. control the parasitic *Striga* spp. in cereals, particularly maize, creates a new example for this approach that will underpin development of allelopathy in host plants themselves against parasitic plants. Other allelopathic situations referred to above and in the review Pickett et al. (2010) and other systems yet to be discovered should also be investigated. In addition, the evolutionary nature of the adverse effects of allelopathy on parasitic plants needs further study, not least to establish why a cattle forage legume can protect a cereal crop plant against a parasite in the event that, by understanding the origin of this interaction, further valuable examples may be identified.

Acknowledgements This work was funded by the Gatsby Charitable Foundation (UK), Kilimo Trust (East Africa), the Rockefeller Foundation and the Biovision Foundation (Switzerland). Rothamsted Research receives grant-aided support from the Biotechnology and Biological Sciences Research Council (BBSRC) and was funded through the BBSRC/DFID SARID initiative. The authors also acknowledge the assistance provided by *icipe* field staff, Ministry of Agriculture extension staff and the farmers.

References

Amudavi DM, Khan ZR, Wanyama JM, Midega CAO, Pittchar J, Hassanali A, Pickett JA (2009a) Evaluation of farmers' field days as a dissemination tool for push-pull technology in Western Kenya. Crop Prot 28:225–235

Amudavi DM, Khan ZR, Wanyama JM, Midega CAO, Pittchar J, Nyangau IM, Hassanali A, Pickett JA (2009b) Assessment of technical efficiency of farmer teachers in the uptake and dissemination of push-pull technology in Western Kenya. Crop Prot 28:987–996

Belz RG (2007) Allelopathy in crop/weed interactions – an update. Pest Manag Sci 63:308–326

Brazier-Hicks M, Evans KM, Gershater MC, Puschmann H, Steel PG, Edwards R (2009) The *C*-glycosylation of flavonoids in cereals. J Biol Chem 284:17926–17934

Carsky RJ, Singh L, Ndikawa R (1994) Suppression of *Striga hermonthica* on sorghum using a cowpea intercrop. Exp Agric 30:349–358

Du Y, Chu H, Chu IK, Lo C (2010) CYP93G2 is a flavanone 2-hydroxylase required for *C*-glycosylflavone biosynthesis in rice. Plant Physiol 154:324–333

Guchu SM, Yenesew A, Tsanuo MK, Gikonyo NK, Pickett JA, Hooper AM, Hassanali A (2007) *C*-methylated and *C*-prenylated isoflavonoids from root extract of *Desmodium uncinatum*. Phytochemistry 68:646–651

Hamilton ML, Caulfield JC, Pickett JA, Hooper AM (2009) *C*-Glucosylflavonoid biosynthesis from 2-hydroxynaringenin by *Desmodium uncinatum* (Jacq.) (Fabaceae). Tetrahedron Lett 50:5656–5659

Hamilton ML, Kuate SP, Brazier-Hicks M, Caulfield JC, Rose R, Edwards R, Torto B, Pickett JA, Hooper AM (2012) Elucidation of the biosynthesis of the di-*C*-glycosylflavone isoschaftoside, an allelopathic component from *Desmodium* spp. that inhibits *Striga* spp. development. Phytochemistry 84:169–176

Hassanali A, Herren H, Khan ZR, Pickett JA, Woodcock CM (2008) Integrated pest management: the push-pull approach for controlling insect pests and weeds of cereals, and its potential for other agricultural systems including animal husbandry. Philos Trans R Soc B 363:611–621

Hess DE, Dodo H (2004) Potential for sesame to contribute to integrated control of *Striga hermonthica* in the West African Sahel. Crop Prot 23:515–522

Hooper AM, Tsanuo MK, Chamberlain K, Tittcomb K, Scholes J, Hassanali A, Khan ZR, Pickett JA (2010) Isoschaftoside, a *C*-glycosylflavonoid from *Desmodium uncinatum* root exudates, is an allelochemical against the development of *Striga*. Phytochemistry 71:904–908

Kersher F, Franz G (1987) Biosynthesis of vitexin and isovitexin: enzymatic synthesis of the *C*-glucosylflavones vitexin and isovitexin with an enzyme preparation from *Fagopyrum esculentum* M. seedlings. Z Naturforsch 42c:519–524

Kersher F, Franz G (1988) Isolation and some properties of an UDP-Glucose: 2-hydroxyflavanone-6(or 8)-*C*-glucosyltransferase from *Fagopyrum esculentum* M. cotyledons. J Plant Physiol 132:110–115

Khan ZR, Ampong-Nyarko K, Chiliswa P, Hassanali A, Kimani S, Lwande W, Overholt WA, Pickett JA, Smart LE, Wadhams LJ, Woodcock CM (1997) Intercropping increases parasitism of pests. Nature 388:631–632

Khan ZR, Pickett JA, van den Berg J, Wadhams LJ, Woodcock CM (2000) Exploiting chemical ecology and species diversity: stem borer and *Striga* control for maize and sorghum in Africa. Pest Manag Sci 56:957–962

Khan ZR, Hassanali A, Overholt W, Khamis TM, Hooper AM, Pickett JA, Wadhams LJ, Woodcock CM (2002) Control of witchweed *Striga hermonthica* by intercropping with *Desmodium* spp., and the mechanism defined as allelopathic. J Chem Ecol 28:1871–1885

Khan ZR, Midega CAO, Hassanali A, Pickett JA, Wadhams LJ, Wanjoya A (2006a) Management of witchweed, *Striga hermonthica,* and stemborers in sorghum, *Sorghum bicolor*, through intercropping with greenleaf desmodium, *Desmodium intortum.* Int J Pest Manag 52:297–302

Khan ZR, Pickett JA, Wadhams LJ, Hassanali A, Midega CAO (2006b) Combined control of *Striga* and stemborers by maize-*Desmodium* spp. intercrops. Crop Prot 25:989–995

Khan ZR, Midega CAO, Hassanali A, Pickett JA, Wadhams LJ (2007) Assessment of different legumes for the control of *Striga hermonthica* in maize and sorghum. Crop Sci 47:728–734

Khan ZR, Amudavi DM, Midega CAO, Wanyama JM, Pickett JA (2008a) Farmers' perceptions of a 'push-pull' technology for control of cereal stemborers and *Striga* weed in western Kenya. Crop Prot 27:976–987

Khan ZR, Midega CAO, Amudavi DM, Hassanali A, Pickett JA (2008b) On-farm evaluation of the 'push-pull' technology for the control of stemborers and *Striga* weed on maize in western Kenya. Field Crops Res 106:224–233

Khan ZR, Midega CAO, Njuguna EM, Amudavi DM, Wanyama JW, Pickett JA (2008c) Economic performance of push-pull' technology for stemborer and *Striga* control in smallholder farming systems in western Kenya. Crop Prot 27:1084–1097

Khan ZR, Pickett JA, Hassanali A, Hooper AM, Midega CAO (2008d) *Desmodium* species and associated biochemical traits for controlling *Striga* species: present and future prospects. Weed Res 48:302–306

Khan ZR, Midega CAO, Wanyama JM, Amudavi DM, Hassanali A, Pittchar J, Pickett JA (2009) Integration of edible beans (*Phaseolus vulgaris L.*) into the push-pull technology developed for stemborer and *Striga* control in maize-based cropping systems. Crop Prot 28:997–1006

Khan ZR, Midega CAO, Bruce TJA, Hooper AM, Pickett JA (2010a) Exploiting phytochemicals for developing a 'push-pull' crop protection strategy for cereal farmers in Africa. J Exp Bot 61:4185–4196

Khan ZR, Pickett JA, Hamilton ML, Hassanali A, Hooper AM, Kuate SP, Midega CAO, Pittchar J, Torto B (2010b) Control of stem borers and *Striga* in African cereals: a low input push-pull approach with rapidly expanding impact. Aspects Appl Biol 56:145–151

Li J, Timko MP (2009) Gene-for-gene resistance in *Striga*-cowpea associations. Science 325:1094

Macias FA, Molinillo JMG, Varela RM, Galindo JCG (2007) Allelopathy – a natural alternative for weed control. Pest Manag Sci 63:327–348

Midega CAO, Khan ZR, van den Berg J, Ogol CKPO, Bruce TJ, Pickett JA (2009) Non-target effects of the 'push–pull' habitat management strategy: parasitoid activity and soil fauna abundance. Crop Prot 28:1045–1051

Midega CAO, Khan ZR, Amudavi DM, Pittchar J, Pickett JA (2010) Integrated management of *Striga hermonthica* and cereal stemborers in finger millet (*Eleusine coracana* (L.) Gaertn.) through intercropping with *Desmodium intortum*. Int J Pest Manag 56:145–151

Pickett JA, Hamilton ML, Hooper AM, Khan ZR, Midega CAO (2010) Companion cropping to manage parasitic plants. Annu Rev Phytopathol 48:161–177

Tenebe VA, Kamara HM (2002) Effect of *Striga hermonthica* on the growth characteristics of sorghum intercropped with groundnut varieties. J Agron Crop Sci 188:376–381

Tsanuo MK, Hassanali A, Hooper AM, Khan ZR, Kaberia F, Pickett JA, Wadhams LJ (2003) Isoflavanones from the allelopathic aqueous root exudates of *Desmodium uncinatum*. Phytochemistry 64:265–273

Vanlauwe B, Kanampiu F, Odhiambo G, De Groote H, Wadhams LJ, Khan ZR (2008) Integrated management of *Striga hermonthica*, stemborers, and declining soil fertility in western Kenya. Field Crops Res 107:102–115

Chapter 26
Biocontrol

Alan K. Watson

26.1 Introduction

Biological control of parasitic weeds involves the use of living organisms that reduce the density of the parasite and reduce crop damage. There are two foremost methods available to implement biocontrol, the classical and the inundative methods. Classical biological control is defined as: '*The intentional introduction of an exotic, usually co-evolved, biological control agent for permanent establishment and long-term pest control*' (Eilenberg et al. 2001). In classical biological weed control, the origins of the weed and the biotic agent are exotic. The classical method is an ecological approach and can only be implemented where parasitic *weeds* have been unintentionally introduced. Most *Orobanche* and *Phelipanche* spp. are native to the temperate northern hemisphere but now are distributed worldwide. *Striga* species are endemic to Africa; thus little opportunity exists for classical biological control via the introduction of natural enemies from areas outside of Africa.

Inundative biological control is defined as: '*The use of living organisms to control pests when control is achieved exclusively by the released organisms themselves*' (Eilenberg et al. 2001). In inundative biological weed control, indigenous herbivores and pathogens are managed by mass rearing and periodically releasing these organisms on the target weed population. Endemic phytophagous insects or pathogens capable of rapidly attacking and severely damaging or killing the target weed are applied in an inundative fashion. Initial collection and screening of potential biocontrol agents focuses on finding aggressive, damaging natural enemies of the target weed pest (Bailey 2010). The host range of selected biocontrol agents is determined to ensure their safety to non-target organisms and the environment. Once developed, these microbial and arthropod biocontrol agents are used in

A.K. Watson (✉)
McGill University, 21,111 Lakeshore Rd., Ste-Anne-de-Bellevue, QC, Canada H9X 3V9
e-mail: alan.watson@mcgill.ca

D.M. Joel et al. (eds.), *Parasitic Orobanchaceae*,
DOI 10.1007/978-3-642-38146-1_26,

a manner similar to chemical herbicides. Weed infestations are treated with an inundative application of the bioherbicidal agent that aims to produce multiple infestations of the agent resulting in weed control. Typically, the inundative approach does not produce sustainable multiyear control of the weed, but requires annual applications of the biocontrol agent which provides the commercial incentive for the development of these products. The inundative bioherbicide method with microbial pathogens has received increased interest and support as a feasible method for controlling weeds (Barton 2005; Yandoc-Ables et al. 2006). In contrast to the ecological approach of classical biocontrol, the inundative approach is a technological approach requiring repeated interventions.

The goal of biocontrol of parasitic weeds is to minimize the negative impacts on crop plants. This goal may be realized and success evaluated through preventing immediate damage caused by existing infestations as expressed in increased crop yield, restricting seed dissemination, and/or preventing future damage by reducing existing seed bank populations. Little is known about seed banks of parasitic weeds except that they are thought to be exceedingly high. The persistent seed bank emphasizes the need for rigorous prevention of seed production for the successful control of parasitic weeds. Germination, attachment, survival to maturity, and fecundity are the most sensitive growth parameters that can be manipulated by control strategies (van Mourik et al. 2008). Various natural enemies with biocontrol potential have been recorded on *Orobanche*, *Phelipanche*, and *Striga* spp. This chapter will review the developments and opportunities in the biological control of broomrape and witchweed.

26.2 Insects Attacking Broomrapes and Witchweeds

Although various insects have been reported to occur on *Orobanche*, *Phelipanche*, and *Striga* spp., most are polyphagous with broad host ranges and cause minimal damage to these parasitic plants (Greathead 1984; Bashir 1987; Klein and Kroschel 2002; Kroschel et al. 1995). In biological weed control, monophagous and oligophagous insects must not attack economic crops and other non-target plant species. Two parasite-restricted, pre-dispersal seed predatory insects were the centre of the biological control research efforts for several decades. The seed-head fly, *Phytomyza orobanchia* (Kaltenbach) (Diptera: Agromyzidae), is host specific, attacking only *Orobanche* or *Phelipanche* spp.; similarly, the seed gall-forming weevils, *Smicronyx (Sm.)* species (Coleoptera: Curculionidae), are specialized on *Striga* spp. These insects limit seed production through the development of larvae inside the seed capsules of their target hosts and may contribute to the reduction of the reproductive capacity and spread of these noxious parasitic weeds when infestations are very low.

What good is a biocontrol agent after the crop is ruined? When these seed-feeding insects are parasitizing the flowers and seed capsules, the devastation to the host crop has already occurred. Pre-dispersal seed predators have been a common choice for biological weed control agents, but they have generally had a low rate of success in controlling invasive plants (McFadyen 1998; Garren and Strauss 2009). Seed predators can have large impacts on seed production, but compensation for seed loss through plant plasticity and density-dependent processes are explanations for the high percentage of failed biological control agents (Crawley 2000). Other sources of mortality are needed to reduce population sizes to very low plant densities before pre-dispersal agents can become effective tools in weed control (Garren and Strauss 2009).

26.2.1 Biocontrol of Broomrapes with Phytomyza

Ph. orobanchia is a monophagous agromyzid fly that oviposites into the buds, flowers, or stems of 20 of the 140 described *Orobanche* and *Phelipanche* species (Klein and Kroschel 2002). *Phytomyza orobanchia* has coevolved with *Orobanche* and *Phelipanche* spp. and they coexist throughout their indigenous distribution ranges (Fig. 26.1a). Some larvae directly consume the seeds, while other larvae mine the shoots leading to the deterioration and rotting of fungi-infected capsules. Seed predation can range from 11 to 79 % (Klein and Kroschel 2002) and infestation levels of seed capsules can be as high as 95 % (Linke et al. 1990). Outwardly, it appears that *Ph. orobanchia* has great potential to reduce broomrape seed production, but even at over 90 % seed reduction, natural populations of the fly alone are unable to effectively reduce population levels of the parasite (see Smith et al. 1993).

Natural populations of *Ph. orobanchia* are reduced by their own natural enemies and cropping practices including soil cultivation and pesticide application (Klein and Kroschel 2002). Mass rearing and timed release of more flies might help avoid natural enemies and detrimental cropping practices. *Ph. orobanchia* was successfully mass reared and applied on a large scale in the former Soviet Union and in some East European countries to control *O. cumana* and *O. cernua* (Parker and Riches 1993; Klein and Kroschel 2002). The inundative approach was attempted in northern Morocco by collecting and distributing infested *Orobanche* shoots in an infested faba bean field when *O. crenata* was expected to emerge. The release of *Ph. orobanchia* reduced seed viability up to 90 % (Klein and Kroschel 2002). Similar results were obtained when adult flies were released in an *O. crenata*-infested faba bean field in Egypt (Shalaby et al. 2004). Of note, these and earlier reports do not mention the crucial effect on crop yield or infestation levels the following year. It can be assumed that these levels of seed predation did not result in increased crop yield as the damage to the crop had already occurred. Effective control of *Orobanche* with *Ph. orobanchia* would require relentless inundative applications over many years to reduce the broomrape seed bank to levels that would reduce the incidence and severity of the parasite that could be expressed in increased crop yield. Cost-effectiveness of a mass release approach would be indeed doubtful.

Fig. 26.1 Biological control of weedy Orobanchaceae. (**a**) Adult female *Phytomyza orobanchia* Kaltenbach (from Singh 1989, with permission). (**b**) Galls (*arrow*) on *Striga hermonthica* caused by *Smicronyx umbrinus* (*source*: Chris Parker/CAB International, Crop Protection Compendium, CABI (http://www.cabi.org/cpc, with permission). (**c**) Numerous healthy *Striga* seedlings (SL) attached via haustorium (ha) to sorghum roots (SR) in the control treatment 3 weeks after sowing in root chamber; s, *Striga* seed (from Elzein et al. 2010, with permission). (**d**) Diseased/dead *Striga* seedlings (SL) on Foxy 2-treated sorghum root (SR) 3 weeks after sowing in root chamber (from Elzein et al. 2010, with permission). (**e**) The effect of *Fusarium oxysporum* f. sp. *strigae* isolate M12-4A on *Striga* emergence and sorghum health. Left—NO M12-4A. Right—PLUS M12-4A (photo by Marie Ciotola, McGill University)

There has been one attempt at classical biocontrol of Orobanchaceae in Chile against *P. ramosa* and *O. minor* that were unintentionally introduced into Chile and became pests in tomato and other crops. *Ph. orobanchia* was released in 1998 and the fly population became established (Norambuena 2003). Follow-up has been limited and the present status of this classical biocontrol release is unknown, but assumed to have failed.

Alone, the fly will not be sufficient to achieve effective control, but releases of the fly may serve as a component in an integrated approach when combined with other control and management strategies (Amsellem et al. 2001a; Klein and Kroschel 2002). In an integrated approach, *Ph. orobanchia* would help reduce the seed bank and help prevent further infestation and dissemination. *Ph. orobanchia* has been used and promoted for biological control of broomrapes in the past, but there is no information to indicate that there are currently any deliberate efforts to exploit this organism for the control of *Orobanche* or *Phelipanche*.

26.2.2 Biological Control of Striga *with Insects*

Striga has many insect pests, but most also damage major crop plants (Greathead 1984; Bashir 1987). These natural enemies of *Striga* include defoliators, gall formers, shoot, stem and root borers, miners, and inflorescence and fruit feeders. Most of these insects rarely cause significant damage or control in their indigenous areas, as their populations are regulated by their own parasites (Parker and Riches 1993). The exceptions are the gall-forming weevils, *Smicronyx* spp., which are specialized on *Striga* spp., and their larvae feed within seed capsules, thus reducing seed production and spread of *Striga* (Parker and Riches 1993; Bashir 1987).

In West Africa, galling of fruits of *S. hermonthica* by *Smicronyx* (*Sm.*) is often very heavy and seed production can be reduced by 50–80 % or more (Greathead 1984; Parker and Riches 1993; Smith et al. 1993; Kroschel et al. 1995). The reduction in seed production from gall-forming (Fig. 26.1b) by *Smicronyx* spp. is often substantial, and these weevils have been promoted as potential biocontrol agents of *S. hermonthica* (Kroschel et al. 1995; Traoré et al. 1995). Soil cultivation, pesticide applications, and natural enemies keep their populations in check, thus further restricting the potential of *Smicronyx* to reduce the *Striga* seed bank. *Smicronyx* spp. is also the most common natural insect enemy of *S. gesnerioides*, but limited seed reduction occurs (Parker and Riches 1993).

Apparently, there has been no development of a mass-rearing biocontrol control program based on *Smicronyx* spp. The inoculative approach with *Smicronyx* would be difficult to implement, costly, and have minimal, if any, impact. The ability to prevent sufficient seed set is limited and will not be sufficient to lower the soil seed bank (Smith et al. 1993). An exceedingly high (>95 %) level of seed reduction is required for control and must be maintained from season to season for many years. In areas with low seed bank densities, *Sm. umbrinus* combined with other control strategies may have some impact on the seed bank and dispersal to un-infested regions (Smith et al. 1993; Smith and Webb 1996).

There has been one attempt with classical biological control of *Striga* using *Sm. albovariegatus* and *Eulocastra argentisparsa* Hamps. (*Lepidoptera*: Noctuidae)

(Greathead 1984; Bashir 1987). In 1974, the Commonwealth Institute of Biological Control (CIBC) collected these insect species in India and released them against *S. hermonthica* in Ethiopia. *Sm. albovariegatus* was released again in 1978 and was recovered the following year, but these insects failed to become established in Ethiopia (Tessema 2007).

26.3 Biocontrol of Parasitic Weeds with Microorganisms

The early growth stages of parasitic plant development, such as seed germination, host attachment, and tubercle development are key phases and ideal targets for successful management of these weeds. Thus, soil microorganisms and natural bioactive compounds interfering with those phases of parasite life cycle could result in attractive management strategies. Numerous fungi and bacteria can infect parasitic weeds, while other microbes may improve crop growth and deter parasitic attack. Soils that naturally suppress populations of parasitic weeds occur (Berner et al. 1995; Zermane et al. 2007). Vital and intensive interactions occur amongst the host plant, soil, and microorganisms in the rhizosphere of parasitic plants. Important biochemical interactions and exchanges of signal molecules occur between parasitic plants and soil microorganisms (Estabrook and Yoder 1998; Bouwmeester et al. 2007; Cardoso et al. 2011). Strigolactones are important signal molecules (see below and Chap. 10).

26.3.1 Role of the Rhizosphere in Parasitic Weed Demise

'*The rhizosphere is the playground and infection court where soilborne pathogens establish a parasitic relationship with the plant. However, the rhizosphere is also a battlefield where the complex rhizosphere community, both microflora and microfauna, interact with soilborne pathogens and influence the outcome of pathogen infection*' (Raaijmakers et al. 2009). Many rhizosphere microorganisms have been implicated in the biological control of parasitic weeds (Table 26.1).

Various *Pseudomonas*, *Azospirillum*, and other rhizobacteria disrupt *Striga*, *Orobanche* and *Phelipanche* seed germination and parasitic development. *Pseudomonas syringae* pathovar *glycinea* (*Psg*), an ethylene-producing rhizobacterium, was proposed as a biocontrol agent that could induce suicidal germination of *Striga* seeds (Berner et al. 1999). Ethylene does not, however, stimulate germination of *Orobanche* and *Phelipanche* seeds (see Sect. 22.4.1).

Several AM fungi reduce the incidence and effect of parasitic weed infestations. Colonization of cereal crop roots by AM fungi can reduce germination, subsequent attachment, emergence, and biomass of emerged *S. hermonthica* (Lendzemo et al. 2007). The formation of AM on cereals apparently leads to lowering of the production and the exudation of strigolactones, germination stimulants for *Striga*,

Table 26.1 Rhizosphere microorganisms and their effects of on parasitic weeds

Microorganism	Parasite	Action/response	Reference
Pseudomonas syringae pathovar glycinea (Psg)	*Striga hermonthica*, *S. aspera*, *S. gesnerioides*	Ethylene-producing bacteria promote seed germination in *Striga* species (suicidal germination)	Berner et al. (1999)
Psg with *Bradyrhizobium japonicum*	*S. hermonthica*	Seed death (suicidal germination) during non-host rotation	Ahonsi et al. (2003)
P. fluorescens	*Orobanche foetida*, *O. crenata*	Faba bean growth increased, parasite number and biomass decreased	Zermane et al. (2007)
P. fluorescens and *P. putida*	*S. hermonthica*	Seed germination inhibited	Ahonsi et al. (2002)
P. aeruginosa, *P. fluorescens*, *Bacillus atrophaeus*, *B. subtilis*	*Phelipanche aegyptiaca*, *O. cernua*	Radicle elongation inhibited	Barghouthi and Salman (2010)
Azospirillum brasilense	*P. aegyptiaca*	Inhibited seed germination and radicle elongation	Dadon et al. (2004)
A. brasilense	*S. hermonthica*	Seed germination and radicle elongation inhibited	Miché et al. (2000)
A. brasilense, *P. putida*, or combination of *A. amazonas* and *P. putida*	*S. hermonthica*	Inoculation with bacteria delayed emergence and reduced *Striga* incidence. Suppressed seed germination and haustorium development	Hassan et al. (2009)
A. brasilense; *P. putida* and other isolates	*S. hermonthica*	Inhibited germination by 40–85 %, haustorium initiation by 52–85 %, and attachment by 78–81 %	Hassan et al. (2011a)
Rhizobium strains	*O. crenata*	Decreased germination and number of tubercles	Mabrouk et al. (2007)
Glomus and *Paraglomus spp.* alone or with *Flavobacterium*, *Azotobacter, or Bacillus* sp.	*S. hermonthica*	AM fungi alone or in combination with plant growth-promoting bacteria (PGPR)-reduced *Striga* germination, seedling attachment, emergence, and delayed *Striga* emergence	Hassan et al. (2011b)
Glomus mosseae	*S. hermonthica*	Reduced damage, emergence reduced 62 %, and total dry matter yield of sorghum increased 30 %	Gweorgwo and Weber (2003)

(continued)

Table 26.1 (continued)

Microorganism	Parasite	Action/response	Reference
G. clarum, *Gigaspora margarita*	*S. hermonthica*	*Striga* shoot number reduced and cereal yield increased. *Striga* seed germination reduced by 88–97 %	Lendzemo et al. (2005, 2007)
G. mosseae, *G. intraradices*	*O. crenata*, *O. foetida*, *O. minor*, *P. aegyptiaca*	Colonized pea root exudates reduce parasite seed germination	Fernàndez-Aparicio et al. (2010)
Fusarium solani, *Macrophomina phaseolina*, *Alternaria alternata*, *Rhizoctonia solani*	*P. aegyptiaca*	*F. solani* caused superior disease symptoms and prevented damage to tomato plants	Dor and Hershenhorn (2009)
Fusarium solani	*S. hermonthica*	Culture filtrates inhibited germination	Ahmed et al. (2001)
Fusarium verticillioides	*P. aegyptiaca*, *P. ramosa*, *O. cumana*	Highly pathogenic but not to *O. crenata*	Dor et al. (2009)
Myrothecium verrucaria	*O. crenata*	Germination inhibition	El-Kassas et al. (2005)

and branching stimulants for AM fungi (Akiyama and Hayashi 2006). The reduced susceptibility of mycorrhized crop plants to parasitic attack is likely due to reduced levels of strigolactone in root exudates (Lendzemo et al. 2007) (see also Sects. 10.4.2).

The synthetic strigolactone, GR24, was recently shown to inhibit the radial growth and increase of hyphal branching of several root and foliar pathogens including *F. oxysporum* f. sp. *melonis*, *F. solani* f. sp. *mango*, *Sclerotinia sclerotiorum*, *Macrophomina phaseolina*, *Alternaria alternata*, *Colletotrichum acutatum*, and *Botrytis cinerea* (Dor et al. 2011). If strigolactones likewise inhibit the radial growth and/or increase branching of the *Fusarium* broomrape and witchweed bioherbicides, these bioherbicides would provide superior weed control when used with low-strigolactone-producing varieties.

26.3.2 Microbial Toxins

Many plant pathogens and non-pathogenic soil microbes produce phytotoxins that have been proposed as sources of natural or biorational herbicides (Strobel et al. 1991; Hoagland 2001). Microbial phytotoxins with the same modes of action as those of commercial herbicides and those with novel modes of action are of interest (Duke and Dayan 2011). Natural compounds, including fungal phytotoxins and amino acids, have been proposed for use in parasitic weed management (Vurro et al. 2009).

The potential of *Fusarium* isolates of *Striga*, *Orobanche*, and *Phelipanche* to produce phytotoxic metabolites that have bioherbicidal effects against different developmental stages of parasitic weeds has been investigated (Zonno and Vurro 1999, 2002; Abouzeid et al. 2004). 'These compounds inhibit seed germination or seedling elongation or, conversely, stimulate suicidal seed germination in the absence of the host' (Vurro et al. 2009).

Following the reports of *S. hermonthica* suppression by *F. nygamai* Burgess & Trimboli (Abbasher and Sauerborn 1992) and *F. oxysporum* (Schlecht) Snyd & Hans (Ciotola et al. 1995), the effects of known fungal toxins on germination of *S. hermonthica* seeds were examined in anticipation of finding bioactive compounds with potential as natural herbicides (Zonno and Vurro 1999). Two fungal toxins, T2 toxin and deoxynivalenol, completely inhibited seed germination of *S. hermonthica*, while other toxins, cytochalasin E, tenuazonic acid, and enniatin, were less inhibitory. In another study, toxins produced from *Fusarium* species were also assayed against *P. ramosa* seed germination (Zonno and Vurro 2002). Seven toxins, fusarenon X, nivalenol, T2 toxin, deoxynivalenol, HT-2, diacetoxyscirpenol, and neosolanioland, prevented all *P. ramosa* seeds from germinating.

Phytotoxic metabolites inhibitory to *P. ramosa* germination were also extracted from *F. compactum* (Wollenw.) and from *Myrothecium verrucaria* (Alb. & Schwein.) Ditmar (Abouzeid et al. 2004; Andolfi et al. 2005). The main metabolite produced by *F. compactum* was neosolaniol monoacetate, a well-known trichothecene. Active metabolites from *M. verrucaria*; verrucarins A, B, E, M, and L acetate; roridin A; isotrichoverrin B; and trichoverrol B are all trichothecenes.

Trichothecene compounds are the most active compounds that stop parasitic plant seeds from germinating, but they are also toxic to mammals (Zonno and Vurro 1999, 2002; Andolfi et al. 2005). The likelihood of government authorization to use these compounds is remote due to the risk to human and animal health. For example, *M. verrucaria* (IMI 361690) effectively controlled several weed species including kudzu (*Pueraria montana*), but mycotoxin production has prevented the registration of this bioherbicide by the US Environmental Protection Agency (US-EPA) (Abbas et al. 2002), even though macrocyclic trichothecenes were not detected in *P. montana* var. *lobata*-treated plants (Abbas et al. 2001). Abbas et al.'s (2002) concluding statement, '*Individual M. verrucaria toxins should be considered as possible biocontrol agents for kudzu, only if they exhibit substantial phytotoxicity coupled with low mammalian cytotoxicity*', is certainly applicable in the discovery of toxic metabolites for parasitic weed control.

Naturally occurring amino acids can be toxic to plants and overproduction of selected amino acids has improved biocontrol efficacy in several weed systems (Sands and Pilgeram 2009). Amino acids were surveyed to determine if any would differentially inhibit *P. ramosa* seed germination and not harm the parasite's host crop plant (Vurro et al. 2006). Methionine and arginine suppressed seed germination, while others encouraged suicidal germination of *P. ramosa* seeds. Vurro et al. (2006) have suggested '*Selection of an amino-acid/toxin-overproducing and -excreting strain of Fusarium should make the pathogen more effective than the wild strains at controlling broomrape infection …*'. Further studies are needed to determine the validity of this innovative approach.

26.3.3 Fungal Pathogens of Orobanche *and* Phelipanche *(Broomrape)*

Broomrapes are commonly diseased by fungi, and there were early attempts in the former Soviet Union, Hungary, China, and Iran to exploit pathogens including *F. lateritium* Nees: Fries, *F. solani* (Martius) Saccardo, *F. oxysporum*, and *Rhizoctonia solani* J.G. Kühn for broomrape biocontrol including the formulation and distribution of a 'Product F' in Russia (Parker and Riches 1993; Wan and Wang 1991 cited in Tessema 2007). Apparently, good results were obtained when Product F (*F. oxysporum*) was added into the planting holes of watermelon, but the technology has not withstood the test of time. Various fungi have continued to be isolated and tested for pathogenicity on broomrapes.

Two foliar pathogens, *Ulocladium atrum* (Preuss) Sacc. and *U. botrytis* Preuss (Pleosporaceae), have also been evaluated for broomrape control. Foliar suspensions of *U. atrum* were used experimentally against *O. crenata* parasitizing faba beans in Syria and destroyed some emerged shoots and underground tubercles (Linke et al. 1992). A strain of *U. botrytis* decreased *O. crenata* germination in vitro by 80 %, but it did not decrease the number of *O. crenata* shoots or tubercles in root chamber or pot experiments (Müller-Stöver and Kroschel 2005). *Ulocladium botrytis* was virulent on both *O. crenata* and *O. cumana*, but less virulent on *P. aegyptiaca*. The potential use of foliar applications of *Ulocladium* spp. is limited by the need for high humidity unless better water-holding formulations are developed. These results and studies with seed predators described above indicate that much more than 80 % reduction over several years is needed to provide control.

26.3.3.1 *Fusarium* spp. Attacking Broomrape (Table 26.2)

Initial encouraging results with *Fusarium* in Bulgaria (Bedi and Donchev 1991) spawned extensive *Orobanche* disease surveys in Europe and the Mediterranean. *Fusarium oxysporum* f. sp. *orthoceras* (Appel & Wollenw.) Bilay (FOO) was isolated from *O. cumana* parasitizing sunflowers. In small-scale field trials, broomrape emergence was reduced by 80 % and sunflower grain yields increased 18 times after fungus-colonized organic substrates were incorporated into the top 15 cm of

Table 26.2 *Fusarium* species investigated for the biological control of broomrapes

Fungal organism	Target parasite	Action/response	Reference
Fusarium oxysporum f. sp. *orthoceras* (FOO)	*Orobanche cumana*, *O. cernua*, *Phelipanche aegyptiaca*	Emergence deceased 80 %, grain yield increased 18×, 100 % disease severity, parasite biomass reduced 80 %, crop yield increased 5–11×. 89 % increase in dry matter of sunflower, compared to the fungus-free control	Bedi and Donchev (1991); Thomas et al. (1998); Shabana et al. (2002); Müller-Stöver et al. (2009b)
F. oxysporum (FOXY)	*P. aegyptiaca*, *P. ramosa*, *O. cernua*	Tubercle number and biomass decreased, 90 % control. Fusaric acid found in culture filtrates of *F. oxysporum*, but not in those of *F. arthrosporioides*	Amsellem et al. (2001a, b); Cohen et al. (2002a)
F. arthrosporioides (FARTH)	*P. aegyptiaca*, *P. ramosa*, *O. cernua*	Tubercle number and biomass decreased, 90 % control. Fusaric acid not found in culture filtrates of *F. arthrosporioides*	Amsellem et al. (2001a, b); Cohen et al. (2002a)
F. oxysporum (FT2), *F. solani*, *F. camptoceras*, *F. chlamydosporum*	*P. ramosa*	*F. oxysporum* and *F. solani* reduced number and biomass of emerging shoots by 50–70 %. Isolate FT2 regarded a strong candidate bioherbicide and a molecular marker was developed for FT2. *F. camptoceras* and *F. chlamydosporum* provided 50 % control	Boari and Vurro (2004); Abouzeid et al. (2004); Cipriani et al. (2009)
F. oxysporum (Foxy I), and (Foxy II)	*O. crenata*, *P. ramosa*	Germination decreased 40–80 %	Alla et al. (2008)
Fusarium sp.	*O. crenata*, *O. foetida*	Parasite emergence and biomass decreased 70–100 %	Boputiti et al. (2008)
F. oxysporum (FOG)	*P. ramosa*	Parasite number and biomass decreased 50–70 % for three consecutive years. Crop yield was not always increased	Müller-Stöver et al. (2009a); Kohlschmid et al. (2009)
F. verticillioides	*P. aegyptiaca*, *P. ramosa*, *O. cumana*	Highly pathogenic but not to *O. crenata*	Dor et al. (2009)
F. compactum	*O. crenata*, *P. ramosa*	Decreased germination	Abouzeid and El-Tarabilly (2010)

soil (Bedi and Donchev 1991; Bedi 1994). In a pot study, 90 % of the shoots were diseased, shoot density reduced by 67 %, and crop dry matter increased by 89 % after soil incorporation of 3×10^8 FOO conidia/kg (Thomas et al. 1998). Each developmental stage of *O. cumana* was susceptible to FOO attack. The fungus attacked seeds, seedlings, tubercles, and shoots in the soil resulting in parasite reduction and increased crop yield (Bedi and Donchev 1991; Bedi 1994; Amsellem et al. 2001b). *Orobanche cernua* plants parasitizing tobacco were also susceptible to the pathogen, but biotypes of *P. aegyptiaca* on tomato were not attacked by FOO.

When FOO was formulated in 'pesta' (encapsulation of chlamydospore-rich fungal biomass in a durum wheat-flour, kaolin, and sucrose matrix; Connick et al. 1991), close to 100 % disease severity and up to 80 % reduction in *Orobanche* biomass were recorded (Shabana et al. 2002). Sunflower seed yield increased by 5–11 times in comparison with fungus-free 'pesta' control treatments, and *O. cumana* emergence was reduced by 90 %. In a 3-year field trial in Bulgaria, the 'pesta' formulation of FOO reduced the number of emerged *Orobanche* shoots from 60 to 100 % illustrating variable control (Müller-Stöver et al. 2009b). Unfortunately, no data were provided on crop yield from this experiment.

Fusarium oxysporum E1d (FOXY) and a putative *F. arthrosporioides* E4a (FARTH) were isolated from diseased juvenile broomrape plants attacking melon in northern Israel (Amsellem et al. 2001b). Both strains infected *P. aegyptiaca*, *P. ramosa*, and *O. cernua*, but not *O. cumana*. Similar to the host range data of the FOO isolate, neither of these two *Fusarium* strains was pathogenic to the following tested crop plants: melons, potatoes, tomatoes, peppers, carrots, celery, chickpeas, and sunflowers (Amsellem et al. 2001b). When applied as seed, transplant, or post-transplant soil drench, fungal suspensions of FOXY and FARTH provided excellent control (90 %) of *P. aegyptiaca* in the greenhouse, but no mention of an effect on tomato yield was reported (Amsellem et al. 2001b). Formulated mycelia inocula were superior in performance to spore inocula when evaluated on *P. aegyptiaca* (Amsellem et al. 1999). FOXY and FARTH culture filtrates were each examined for the production of phytotoxins. These two isolates secrete non-fumonisin ceramide synthase inhibitors and do not secrete fumonisins FB1 and FB2 (Cohen et al. 2002b).

More than 50 pathogenic isolates were collected in Southern Italy from *P. ramosa*-infested fields of tomato, tobacco, and cauliflower, and the pathogenicity and virulence of their conidia and culture filtrates were assayed against broomrape (Abouzeid et al. 2004; Boari and Vurro 2004). *Fusarium oxysporum* (FT2), *F. solani*, *F. sambucinum* Fuckel, and *F. camptoceras* Wollenweber & Reinking all reduced the number and biomass of emerged broomrape shoots by 50–70 % as compared to the untreated controls (Boari and Vurro 2004). Isolate FT2 was host specific, highly virulent to *P. ramosa*, and considered a good bioherbicide candidate for the biological control of this weed. A molecular marker was developed using fluorescent amplified fragment length polymorphism (FAFLP) analysis to specifically identify this strain and track its fate after its release (Cipriani et al. 2009). The importance of the molecular marker is also relevant after field release to show non-liability when crops are attacked by other *Fusarium* sp. The present status of this isolate and its commercial potential are unclear but assumed to have been abandoned.

Several hundred *Fusarium* isolates were collected from *O. crenata* and *O. foetida* Poir. parasitizing faba bean in Tunisia (Boputiti et al. 2008). In pot experiments, two *Fusarium* sp. reduced the number of *O. crenata* and *O. foetida* by 70 % and 90 %, respectively, and their dry biomass by over 80 %. 'Pesta' formulations were superior and increased the reduction to near 100 %, but no mention of a positive or negative effect on crop yield was reported.

In Egypt, conidial suspension of two *F. oxysporum* isolates reduced *O. crenata* and *P. ramosa* germination in vitro by 76–80 %, in root chambers by 46–50 %, and in polyethylene bags by 40–55 % (Alla et al. 2008). Both isolates reduced attachment and number of tubercles and were reported to overcome the reduction in growth of faba bean and tomato (Alla et al. 2008).

Another *F. oxysporum* isolate (FOG) was discovered in south-western Germany from *P. ramosa* tubercles growing on tobacco roots (Müller-Stöver et al. 2009a). Both granular soil applications and conidial suspensions of FOG caused extensive mortality of *P. ramosa* in pot experiments. In field experiments over 3 years, results were inconsistent as *P. ramosa* shoot number and biomass were reduced by 50–70 % (Kohlschmid et al. 2009). However, a 50 % reduction in the number of *P. ramosa* shoots did not significantly increase crop yield.

The search for new pathogens continues in the Mediterranean area. *Fusarium verticillioides* (Sacc.) Nirenberg was isolated from the tubercles of *O. cumana* in Israel and shown to be highly pathogenic to *O. cumana*, *P. aegyptiaca*, and *P. ramosa*, but weakly pathogenic to *O. crenata* (Dor et al. 2009). Recently, 39 *Fusarium* isolates were obtained from newly emerged infected bean broomrape (*O. crenata*) and hemp broomrape (*P. ramosa*) collected from infested faba bean and tomato fields near Giza, Egypt (Abouzeid and El-Tarabilly 2010). One *F. oxysporum*, one *F. equiseti* (Corda) Sacc., and four *F. compactum* strongly inhibited broomrape seed germination, while one *F. compactum* isolate enhanced the growth of the infested crop.

26.3.3.2 Efforts to Improve Control Level of Broomrape Bioherbicide Candidates

As described above, numerous *Fusarium* species have been reported from broomrape and several have caused significant disease development on the parasitic plants in greenhouse and field trials. Results were variable and information on host plant response and crop yield was often not provided, suggesting that the level of control may not be an adequate amount for the farmers. Cohen et al. (2002a) concluded 'The organisms do not have sufficient virulence for field use (based on greenhouse testing), suggests that virulence should be transgenically enhanced or additional isolates sought'.

Several approaches have been proposed and experimented to enhance broomrape control. In one study, a mixture of pathogens provided modest to good improvement in the control of *O. cumana* and *P. aegyptiaca* (Dor et al. 2003). Others have suggested induced disease resistance may aid broomrape control by activation of immune responses before infection (Gonsior et al. 2004). In an effort to increase the efficacy of FOO, benzo (1,2,3) thiadiazole-7-carbothioic acid S-methyl ester (BTH), a product known to induce resistance against *O. cumana* in sunflower, was combined with FOO (Müller-Stöver et al. 2005). The combined treatment was not superior to FOO alone and thus BTH did not enhance the virulence of FOO. Enhancing the efficacy of biocontrol agents by fermentation, formulation, and application technologies has also been proposed (Sauerborn et al. 2007) but generally has not been followed up.

Several attempts to genetically enhance the virulence of FOXY and FARTH isolates to *P. aegyptiaca* have occurred. Both isolates were transformed with genes in the indole-3-acetic acid (IAA) pathway (Cohen et al. 2002b). Transgenic FOXY and FARTH overproducing IAA had enhanced virulence, as measured by reduction in the number and the size of broomrape shoots. FOXY and FARTH were also transformed with the *NEP*1 gene (Meir et al. 2009) that encodes a fungal toxin that successfully conferred hypervirulence when transformed into *F. oxysporum* f. sp. *erythroxyli* (Bailey et al. 2002) and into *Colletotrichum coccodes* (Wallr.) Hughes (Amsellem et al. 2002). None of the FOXY *NEP*1 transformants were hypervirulent, but the FARTH transformant killed *P. aegyptiaca* more rapidly than the wild type (Meir et al. 2009).The transformed *Fusarium* sp. has not been tested in the field because regulatory authorities have refused to grant a permit.

Except for the historical mention of Product F in Russia, no practical use has been made so far with these fungi for biocontrol of broomrape. Many *Fusarium* isolates have been extensively studied, but none are yet in wide-scale field testing and none appear to be proceeding towards commercialization. Perhaps superior virulent isolates will be discovered, perhaps best technologies will be developed and implemented to achieve best results, or perhaps new hypervirulent isolates will be transgenetically developed and receive regulatory acceptance (Gressel 2001).

26.3.4 Fungal Pathogens on Striga

Numerous fungal genera have representatives that are pathogenic to *Striga* including *Alternaria*, *Aspergillus*, *Bipolaris*, *Curvularia*, *Drechslera*, *Fusarium*, *Macrophomina*, *Paecilomyces*, *Phoma*, *Rhizopus*, and *Verticillium*. Widespread surveys have shown *F. oxysporum* to be the predominant soilborne pathogen on *S. hermonthica* in northern Ghana, Burkina Faso, Benin, Mali, Niger, and Nigeria (Ciotola et al. 1995; Abbasher et al. 1998; Marley et al. 1999; Yonli et al. 2010). Most surveys focused on *S. hermonthica* resulting in less information on pathogens of other *Striga* species. Several known *Fusarium* spp. and a new species, *F. brevicatenulatum* Nirenberg & O'Donnell, were isolated from diseased *S. asiatica* on upland rice in Madagascar (Nirenberg et al. 1998). *Fusarium* species were also the dominant pathogens isolated from *S. asiatica* in Malawi (Theu 2008).

Table 26.3 *Fusarium* species investigated for the biocontrol of witchweeds

Fungus	Target weed	Action/response	Reference
Fusarium equiseti	*Striga hermonthica*	'It may have some promise for reducing plant populations of *Striga* ...', '50 % mortality of *Striga* ...'	Zummo (1977); Kirk (1993)
F. nygamai	*S. hermonthica*	First pathogen to be evaluated as a biological control agent. Effective control of *Striga*. Mycotoxin fumonisin B_1 produced	Abbasher and Sauerborn (1992); Kroschel and Elzein (2004)
F. oxysporum (M12-4A)	*S. hermonthica*	400 % increase of sorghum dry weight. No mycotoxins produced. Chlamydospore powder seed coat technology	Ciotola et al. (1995, 2000); Savard et al. (1997)
F. oxysporum (Foxy 2)	*S. hermonthica*, *S. asiatica*	Reduced emergence 98 %. Increased sorghum 26 %. Pesta formulation implemented	Kroschel et al. (1996); Elzein and Kroschel (2004)
F. oxysporum (PSM-197)	*S. hermonthica*, *S. asiatica*, *S. gesnerioides*	Emergence of *S. hermonthica* reduced by 94 %, *S. asiatica* by 91 %, and *S. gesnerioides* by 82 %. Avirulent on *Alectra vogelii*	Marley et al. (1999, 2004, 2005); Marley and Shebayan (2005)
F. oxysporum, *F. equiseti*, *Fusarium* sp.	*S. hermonthica*	*Striga* emergence delayed 10 days and *Striga* biomass decreased by 47 %	Yonli et al. (2004, 2005, 2010)
F. nygamai, *F.* sp. 'Abuharaz'	*S. hermonthica*	Sorghum yield increased 80 %, less effective in the second year	Mohamed (2002); Zahran et al. (2008)

26.3.4.1 *Fusarium* spp. Attacking *Striga* (Table 26.3)

Biological control of *S. hermonthica* has gained considerable attention in recent years as a viable supplement to other control methods within an integrated approach. Following the first isolation and pathogenicity testing of *F. equiseti* from diseased *Striga* in 1977 (Zummo 1977), *Fusarium* species have received the greatest attention for the biological control of *S. hermonthica*. Intensive surveys on the occurrence of microorganisms pathogenic to *S. hermonthica* were conducted in Sudan (Abbasher and Sauerborn 1992), Burkina Faso, Mali, Niger (Abbasher et al. 1998; Ciotola et al. 1995; Yonli et al. 2004, 2010), Ghana (Kroschel et al. 1996; Abbasher et al. 1995), and Nigeria (Marley et al. 1999). The isolates attacked all growth stages of *Striga*, including seeds, germlings and seedlings, and flowering shoots, thus affecting the target prior to the onset of yield loss in addition to reducing the soil seed bank (Fig. 26.1c, d).

A fungal disease of *S. hermonthica* was observed in sorghum [*Sorghum bicolor* (L.) Moench]-infested fields in two regions of Sudan in 1986 (Kirk 1993). The fungus, *F. equiseti* (IMI Number 311922b), caused wilting of leaves and flowers followed by darkening of all the tissues destroying *Striga* plants before seed production. The sorghum crop was not affected. In a follow-up survey in Sudan, 28 fungal and two bacterial pathogens of *S. hermonthica* were isolated (Abbasher and Sauerborn 1992). One isolate, *F. nygamai* Burgess & Trimboli, was very virulent and reduced *Striga* emergence by up to 100 % when placed in potting soil prior to planting, and foliar applications of conidia killed emerged *Striga* plants. Unfortunately, the potential use of *F. nygamai* as a bioherbicide was compromised due to the production of fumonisin B_1, a powerful phytotoxin against *Striga* but also a carcinogenic mycotoxin toxic to mammals (Thiel et al. 1991; Kroschel and Elzein 2004).

Over 200 fungi were collected from over 100 diseased *S. hermonthica* shoots in Burkina Faso, Mali, and Niger (Ciotola et al. 1995). One isolate of *F. oxysporum* (M12-4A) from Mali, grown on sorghum straw and incorporated into potted soil, prevented all emergence of *S. hermonthica* and resulted in fourfold increase in sorghum yield (Fig. 26.1e). Isolate M12-4A applied as colonized, chopped, or ground sorghum straw resulted in a 60 % reduction of emerged *Striga* at 82 days after sowing and doubled sorghum biomass compared to the control (Diarra et al. 1996). Emergence of *S. hermonthica* was completely suppressed by applications of a chlamydospore powder in the planting hole or as a seed coating in the field (Ciotola et al. 2000). Only 80 g of the chlamydospore powder was needed to treat 1 ha.

Thirteen fungal species were isolated from diseased *S. hermonthica* in northern Ghana (Abbasher et al. 1995; Kroschel et al. 1996). The pathogenicity of 12 isolates, including *F. equiseti*, *F. equiseti* var. *bullatum*, *F. solani,* and *F. oxysporum*, was evaluated against *S. hermonthica* under controlled environmental conditions. All isolates were pathogenic, but one *F. oxysporum* (Foxy 2) was highly virulent, reducing the emergence of *S. hermonthica* by 98 % and increasing sorghum yield by 26 % (Kroschel et al. 1996).

Striga plants collected from a farmer's field near Samaru, Nigeria, led to the selection of another *F. oxysporum* isolate, PSM-197 (Marley et al. 1999; Marley and Shebayan 2005). Similar to the results reported for *F. oxysporum* (M12-14A) (Ciotola et al. 1995), emergence of *S. hermonthica* was completely inhibited when PSM-197 was grown on sorghum grain and incorporated into the soil. These surveys encompassed a substantial portion of Africa infested by *S. hermonthica* and displayed the abundance of *F. oxysporum* recovered from diseased *Striga* plants displaying vascular wilt symptoms.

26.3.4.2 *Fusarium oxysporum* f. sp. Strigae: A Bioherbicide Candidate Pathogen

Some *Fusarium* spp. have broad host ranges and produce mycotoxins, but many *Fusarium* spp. are specific at the genus or species level, leading to the *formae specialis* taxonomic classifications (Leslie and Summerell 2006). *Fusarium oxysporum* isolates M12-4A, PSM-197, and Foxy 2 are host restricted; they only infect plants in the genus *Striga* and do not cause disease on any tested crop plants (Ciotola et al. 1995; Kroschel et al. 1996; Marley et al. 1999; Elzein and Kroschel 2006b). Thus, these isolates can be considered to constitute the *formae speciales strigae* (Elzein et al. 2008). Contrary to the discovery of mycotoxins with *F. nygamai* described above, isolate M12-4A of *F. oxysporum* did not produce known mycotoxins, and hence, it does not constitute acknowledged health hazards to humans or livestock (Savard et al. 1997). The extent of mycotoxin production in isolates PSM-197 and Foxy 2 is presently being examined (Altus Viljoen, personal communication).

The genetic variability of *Striga* spp. is high (Mohameda et al. 2007), and this may impact the efficacy of the *F. oxysporum* bioherbicide. Genetic variability between and within individual populations or ecotypes of *Striga hermonthica* occurs. *Striga hermonthica*, an outcrossing species, demonstrates intra- and inter-population variation with differing degrees of virulence on host plants (see Chap. 19), which is likely to impact on *F. oxysporum*-bioherbicide efficacy (Elzein and Kroschel 2006b; Beed et al. 2007). Similar susceptibility/virulence variation amongst *Striga*/*Fusarium* populations can be expected. Venne et al. (2009) proposed that 'the efficacy of *F. oxysporum* f. sp. *strigae* isolates requires evaluation against multiple *Striga* populations from different hosts across varied environments'. Isolate Foxy 2 controlled both *S. hermonthica* and *S. asiatica* in greenhouse trials (Elzein and Kroschel 2004), whereas isolate PSM-197 was virulent on *S. asiatica*, *S. gesnerioides*, and *S. hermonthica*, but avirulent on the related *Alectra vogelii Benth* (Marley et al. 2005). Less is known about fungi on *Alectra*, but scientists from the Kenya Agricultural Research Institute (KARI) recently reported isolating fungi from dying *A. vogelii* plants displaying vascular wilt symptoms (Karanja et al. 2009).

Generally, *Fusarium* spp. isolates differ in their vegetative compatibility grouping (VCG) pattern in accordance with their host range (Leslie and Summerell 2006). There is a high degree of genetic similarity amongst the various isolates of *F. oxysporum* from *S. hermonthica* (Watson et al. 2007; Venne 2008; Elzein et al. 2008). The majority of *F. oxysporum* f. sp. *strigae* collected from Kenya, Niger, and Mali are in one VCG, while several isolates from Benin and Burkina Faso are in a second VCG group, and Foxy 2 from Ghana and PSM-197 from Nigeria are in a third VCG group (Venne 2008). The presence of a few VCG groups in a single *forma specialis* is consistent with data gathered from other *F. oxysporum* subspecies (Leslie and Summerell 2006; Watson et al. 2007). Random amplified polymorphic DNA (RAPD) assays have identified markers restricted to a set of *F. oxysporum* strains isolated from *Striga*. Two SCAR primers (FUN001 and FUN002) amplified a single band of 157 bp in most isolates tested from *Striga*, which included M12-4A from Mali, and several isolates from Benin and Burkina Faso, but one isolate of *F. oxysporum* from Burkina Faso, isolate Foxy 2 from Ghana, and isolate PSM 197 from Nigeria were not amplified by the SCAR primers. '*PCR-based assays confirm the VCG results, indicating F. oxysporum isolates from*

Striga are genetically similar suggesting co-existence of F. oxysporum f. sp. strigae with its host across the Sahel and the Savanna' (Beed et al. 2007).

Genetic characterization of these isolates shows similarities by virtue of unique DNA sequences that enabled them to be classified as a new *forma specialis* (f. sp. *strigae*). The genetic difference from known crop *F. oxysporum* pathogens suggests that they are biosafe and could be accepted as bioherbicides. Further research is required to develop molecular detection tools to confirm the relatedness of *F. oxysporum* f. sp. *strigae* isolates, to monitor the spread and persistence of this pathogen in the soil environment, and to distinguish the bioherbicide from crop pathogens. Diversity amongst the different *F. oxysporum*-bioherbicide isolates also needs to be examined relative to *S. hermonthica* diversity in the regions.

Fusarium oxysporum f. sp. *strigae* 'acts on the sorghum host like other non-pathogenic Fusarium strains by colonizing the root surface and the cortex of roots' (Elzein et al. 2010). This is in contrast with pathogenic *Fusarium* strains that invade the vascular system and cause typical disease symptoms. Hyphae grew in the intercellular space of the root cortex, penetrated parenchyma cells, and lysed cell walls and cytoplasm. Endodermal cells were not penetrated and hyphae were never found in the central cylinder of the sorghum root. Ndambi et al. (2011) concluded 'two mechanisms were identified by which Foxy 2 controls *S. hermonthica*; (1) complete digestion of *S. hermonthica* seedlings inside the host and (2) clogging of vessels of emerged *S. hermonthica* plants by hyphae contributing to wilting and subsequent death' of the parasite.

26.3.5 Formulation and Field Effectiveness of Fusarium-Based Bioherbicide for Striga Control

Various methods have been used to produce *Fusarium* inoculum for greenhouse and field trials. Inoculum production was based on simple, low-cost methods using locally available agricultural by-products including cereal grains, sorghum glumes, and sorghum straw (Ciotola et al. 2000; Elzein et al. 2006; Marley et al. 2004). Granular formulations, such as 'pesta', have been developed and are effective against *Striga* when placed directly into planting holes (Elzein and Kroschel 2006a). However, this formulation is expensive to produce and labour intensive to apply. A preferred field application is the application of a dry chlamydospore powder formulation of the fungus directly onto cereal seed using gum arabic as an adhesive (Ciotola et al. 2000). The use of gum arabic increases the rate of mycelial development and enhanced sporulation of *F. oxysporum f. sp. strigae*, while increasing the germination rate of the sorghum seeds.

Seed coating and 'pesta' formulations of various *F. oxysporum* isolates have been field evaluated for the control of *S. hermonthica*. Fifteen *F. oxysporum* isolates from diseased *S. hermonthica* were evaluated for the control of *Striga* in sorghum fields in Burkina Faso (Yonli et al. 2005). These 15 isolates were equally effective with *Striga* infestations reduced by 50 % and sorghum yields increased by 50 %. Field experiments conducted at Gezira, Sudan, over two consecutive seasons, examined the efficacy of two Sudanese *Fusarium* isolates, *F. nygamai* and *Fusarium* sp. 'Abuharaz', formulated as 'pesta' granules. Treatments reduced parasite shoot densities and increased sorghum yield by 80 % in the first year, but were less effective in the second season (Zahran et al. 2008).

The efficacy of 'pesta' granules of Foxy 2 and PSM-197 isolates, in combination with *Striga*-resistant and susceptible sorghum and maize cultivars, was tested under field conditions at two locations in Nigeria in 2003 (Schuab et al. 2006). Both Foxy 2 and PSM-197 were equally effective in controlling *Striga* on both susceptible and resistant maize and sorghum cultivars tested. Isolate PSM-197 was co-applied with a *Striga*-resistant sorghum cultivar and *Striga*-tolerant landrace in on-farm trials in Nigeria (Marley et al. 2004; Marley and Shebayan 2005). When the bioherbicide was applied with the resistant variety, *Striga* counts were reduced by over 90 %, crop vigour increased, and sorghum yields were 50 % higher, whereas the bioherbicide alone increased *Striga*-tolerant landrace cultivar yields by 20–40 % (Table 26.4).

Three isolates, M12-4A from Mali, Foxy 2 from Ghana, and PSM-197 from Nigeria, were similarly effective in *Striga* suppression in laboratory, pot studies, and field trials. These isolates attacked all growth stages of *Striga* and reduced the number of juvenile and flowering *Striga* plants. Repeated field trials performed in Nigeria, Burkina Faso, and Bénin compared the effects of M12-4A, Foxy 2, PSM-197, and other isolates originating from Bénin and Burkina Faso on *Striga* in *Striga*-resistant and *Striga*-susceptible varieties of sorghum and maize (Beed et al. 2007; Venne et al. 2009). Isolates PSM-197 and Foxy 2 were superior to M12-4A in suppressing witchweed under the range of field conditions tested. *Striga* emergence was reduced by 90 % when the biocontrol was used on a *Striga*-resistant maize line (Venne et al. 2009). The fungal treatments were coated onto the crop seed using locally available gum arabic or applied as kaolin-based 'pesta' granules into planting holes. With the smaller sorghum seeds, the granular formulation was more effective, but the granular formulation is more costly and difficult to distribute to farmers. When combined with *Striga*-resistant germplasm, the seed-coating method may offer the most cost-effective way of delivering the technology to the farmers.

When *Striga*-resistant varieties are used in combination with *F. oxysporum* as a biocontrol agent, the basis of the variety's resistance or tolerance should be chosen to favour the use of the fungus. The synthetic strigolactone GR24 was shown to inhibit the growth and increase of hyphal branching of several root and foliar pathogens including a *F. oxysporum* (Dor et al. 2011). If strigolactones inhibit the growth of *F. oxysporum f. sp. strigae* bioherbicides, these bioherbicides would provide superior weed control when used with low-strigolactone-producing varieties. Sorghum root exudates have been shown to strongly inhibited M12-4A chlamydospore germ tube elongation (Ciotola et al. 2000).

Seventy-five per cent or more of the farmers in sub-Saharan Africa (SSA) are subsistence farmers, without access to improved crop seed. A strategy is needed to

Table 26.4 *Striga* control by *Fusarium* mycoherbicide in sorghum fields

		Striga count (plot[b])		Grain yield (tonne/ha)	
Sorghum cultivar	*Fusarium* treatment[a]	Barhim village	Dutsen-Ma village	Barhim village	Dutsen-Ma village
SAMSORG 41	Treated	8.2	12.6	2.1	3.1
	Untreated	127	165	1.0	1.1
Farmers local	Treated	10	19.4	2.0	2.6
	Untreated	139	201.6	0.6	1.3
Mean		71.1	99.7	1.4	2.0
S.E.		10.95	14.83	0.12	0.15

Average *Striga hermonthica* shoot count and grain yield of two sorghum cultivars (SAMSORG 41and Farmers local) at harvest on nine farmer's fields each at two villages in the Nigerian savannah (from Marley and Shebayan 2005, with permission)
[a]Treatment with *Fusarium oxysporum* isolate PSM-197 mycoherbicide
[b]Plot size 9 m × 10 m = 25 rows, 10 m long

bring the *Fusarium* bioherbicide strategy to the majority of farmers while waiting for certified seed systems to be established. An inoculation production system was developed for *F. oxysporum* f. sp. *strigae* M12-4A utilizing a liquid suspension of finely ground sorghum straw as the substrate, fashioned on a cottage industry model (Ciotola et al. 2000). The material was produced, dried, ground, and stored at room temperature. Gum arabic was used to stick the bioherbicide powder, predominately chlamydospores (1×10^7 g^{-1}), onto sorghum seeds prior to planting. The concept was tested through training village-level producers of *Striga* bioherbicide in four villages in Mali (Bastiani 2002; Watson et al. 2007). When this was tested on farm, the production strategy was constrained by contamination of preparation utensils. The production of the bioherbicide could be carried out regionally by local entrepreneurs or farmer cooperatives with scientific capacity and facilities, but quality control would best be attained through production of dry powder inoculum for seed coating or a 'pesta' formulation at one or more central facilities with shipment to other locations. Feasibility and costs of this approach have yet to be determined. Further development of improved and certified seed production in *Striga*-infested regions would not only improve crop production but would certainly aid in the delivery of *F. oxysporum* f. sp. *strigae* bioherbicides. Most seed companies have seed-coating capabilities and several have experience in coating sorghum and maize seed with *F. oxysporum* f. sp. *strigae*.

26.4 Path to Commercialization of a *Striga* Bioherbicide

Steps towards the development and commercialization of a bioherbicide product for *Striga* control are ongoing in Kenya. The Africa Enterprise Challenge Fund (AECF) has funded the project 'Biological control of Striga—a weed of maize, millet and sorghum crops' to Real IPM Company (k) Ltd, Thika, Kenya. The objective of the 3-year project is to commercialize a bioherbicide product for *Striga* control in maize, millet, and sorghum in Kenya. Real IPM is using the Foxy 2 isolate of *F. oxysporum* f. sp. *strigae* and has developed a two-package seed-dressing system for small-scale farmers to control *Striga*. The use of the Gro-Plus seed priming and StopStriga involves soaking farmer-saved seeds overnight in a nutrient solution of Gro-Plus and the following morning, drain off the remaining solution and then apply the StopStriga dry powder and mix to coat the moistened seeds with fungus (RIUtv 2010).

A multimillion dollar project, 'Achieving sustainable *Striga* control for poor farmers in Africa (ISMA)', funded by the Bill and Melinda Gates Foundation, was launched by the International Institute of Tropical Agriculture (IITA) and partners in 2011. The project will implement and evaluate approaches including the deployment of biocontrol in Nigeria and Kenya. The project will link with Real IPM's commercial development of the Foxy 2 bioherbicide in Kenya and will advance the use of PSM-197 in Nigeria. Several challenges are ahead in the commercialization process: ownership and intellectual property rights need to be addressed and molecular tools are required to confirm the indigenous existence of *F. oxysporum* f. sp. *strigae* and to monitor the bioherbicide fungus in the environment. Registration is required to enable governmental approval for wide-scale deployment of a commercial bioherbicide and must be performed separately in each country. The registration process includes costly human health and safety testing and environmental toxicology. Large-scale mass production has not yet been achieved nor optimized. Critical studies are under way and there is optimism that if regulatory approval of a *Striga* bioherbicide is achieved, there will be another tool to be incorporated into Integrated Striga Management (ISM) programs for farmers. The future will determine costs and benefits of the bioherbicide compared to other controls. As Beed et al. (2007) have stated, *'clearly, a bioherbicide will only be adopted if field efficacy is proven to farmers/policy makers, and will only provide significant value if integrated with other techniques for control of Striga'*.

26.5 Conclusions and Future Possibilities

The principal obstacle in the long-term management of broomrape- and witchweed-infested fields is the durable seed bank that remains viable for several decades. The effectiveness of *Phytomyza* and *Smicronyx* to prevent seed set is often limited and, on their own, will not be enough to lower the soil seed bank significantly or control these parasitic weeds. Nonetheless, when other technologies reduce parasitic weed populations to low levels, these seed-feeding insects will naturally contribute to reduce seed production and reduce seed dispersal.

Isolates of the soilborne fungus, *Fusarium* spp., have the greatest potential as biocontrol agents for broomrape and witchweed. Broomrape has yet to benefit from this technology as comprehensive field evaluation has yet to be implemented; meanwhile, *Striga* bioherbicides are in the development phase. The genetic variability of *Striga* spp. is high and this may impact the long-term efficacy of the *F. oxysporum* bioherbicide. Plants can evolve resistance to pathogens and different *Fusarium* isolates are likely to be adapted to the different host populations and varied environmental conditions that exist across SSA. The efficacy of *F. oxysporum* f. sp. *strigae* isolates requires evaluation against multiple *Striga* populations from different hosts across varied environments. *Striga hermonthica* demonstrates intra- and interpopulation variations with differing degrees of virulence on host plants which is also likely to have impact on *F. oxysporum*-bioherbicide efficacy.

There is a need to have an improved understanding of the rhizosphere to promote beneficial rhizosphere conditions, so the proliferation and persistence of a biocontrol agent such as *F. oxysporum* is favoured. This requires an understanding of the agronomic and ISM practices that may favour proper rhizosphere conditions, and this in turn can only be achieved through improved knowledge of the rhizosphere. The manipulation of soil microbial communities may be of considerable value when integrated with other control interventions, in particular biological control with soilborne fungi such as *Fusarium* species. Perhaps the pathogenic *F. oxysporum* Fo47 (Olivain and Alabouvette 1997) can be jointly applied with the *Striga* bioherbicide to replace the fungicide in seed treatment as it has with the broomrape FOXY (J. Gressel, personal communication). It has also been proposed that an imazapyr (acetolactate synthase)-resistant *F. oxysporum* f. sp. *strigae* mutant could be jointly applied as a seed dressing with imazapyr to ALS-resistant maize seeds to effectively reduce the amount of herbicide used and extend the duration of protection from *Striga*, and it may delay evolution of resistance to the herbicide and the pathogen. More research is essential to understand the impacts of these interactions and the influence of other components of ISM and variance in the *Striga* seed bank.

The prospects for the integration of effective biocontrol agents into ISM systems are encouraging. Effective control of *Striga* has been demonstrated in the field using a safe, environmentally benign organism that can be readily grown, stored, formulated, and deployed. A commercial bioherbicide is hopefully imminent, but access to this bioherbicide by the vast majority of subsistence farmers will be distant unless there are significant improvements in seed production and distribution systems.

References

Abbas HK, Tak H, Boyette CD, Shier WT, Jarvis BB (2001) Macrocyclic trichothecenes are undetectable in kudzu (*Pueraria montana*) plants treated with a high-producing isolate of *Myrothecium verrucaria*. Phytochemistry 58:269–276

Abbas HK, Johnson BB, Shier WT, Tak H, Jarvis BB, Boyette CD (2002) Phytotoxicity and mammalian cytotoxicity of macrocyclic trichothecene from *Myrothecium verrucaria*. Phytochemistry 59:309–313

Abbasher AA, Sauerborn J (1992) *Fusarium nygamai* a potential bioherbicide for *Striga hermonthica* control in sorghum. Biol Control 2:291–296

Abbasher AA, Kroschel J, Sauerborn J (1995) Micro-organisms of *Striga hermonthica* in northern Ghana with potential as biocontrol agents. Biocontrol Sci Technol 5:157–161

Abbasher AA, Hess DE, Sauerborn J (1998) Fungal pathogens for biological control of *Striga hermonthica* on sorghum and pearl millet in West Africa. Afr Crop Sci J 6:179–188

Abouzeid MA, El-Tarabilly KA (2010) *Fusarium* spp. suppress germination and parasitic establishment of bean and hemp broomrapes. Phytopathol Mediterr 49:51–64

Abouzeid MA, Boari A, Zonno MC, Vurro M, Evidente A (2004) Toxicity profiles of potential biocontrol agents of *Orobanche ramosa*. Weed Sci 52:326–332

Ahmed NE, Sugimoto Y, Babiker AGT, Mohamed OE, Ma Y, Inanaga S, Nakajima H (2001) Effects of *Fusarium solani* isolates and metabolites on *Striga* germination. Weed Sci 49:354–358

Ahonsi MO, Berner DK, Emechebe AM, Lagoke ST (2002) Selection of rhizobacterial strains for suppression of germination of *Striga hermonthica* (Del.) Benth. seeds. Biol Control 24:143–152

Ahonsi MO, Berner DK, Emechebe AM, Lagoke ST, Sanginga N (2003) Potential of ethylene-producing *Pseudomonas* in combination with effective N^2-fixing *Bradyrhizobial* strains as supplements to legume rotation for *Striga hermonthica* control. Biol Control 28:1–10

Akiyama K, Hayashi H (2006) Strigolactones: chemical signals for fungal symbionts and parasitic weeds in plant roots. Ann Bot 97:925–931

Alla MMN, Shabana YM, Serag MM, Hassan NM, El-Hawary MM (2008) Granular formulation of *Fusarium oxysporum* for biological control of faba bean and tomato *Orobanche*. Pest Manag Sci 64:1237–1249

Amsellem Z, Zidack NK, Quimby PC Jr, Gressel J (1999) Long term dry preservation of active mycelia of two mycoherbicidal organisms. Crop Prot 18:643–649

Amsellem Z, Barghouthi S, Cohen B, Goldwasser Y, Gressel J, Hornok L, Kerenyi Z, Kleifeld Y, Klein O, Kroschel J, Sauerborn J, Müller-Stöver D, Thomas H, Vurro M, Zonno MC (2001a) Recent advances in the biocontrol of *Orobanche* (broomrape) species. Biocontrol 46:211–228

Amsellem Z, Kleifeld Y, Kerenyi Z, Hornok L, Goldwasser Y, Gressel J (2001b) Isolation, identification and activity of mycoherbicidal pathogens from juvenile broomrape plants. Biol Control 21:274–284

Amsellem Z, Cohen BA, Gressel J (2002) Engineering hypervirulence in an inundative mycoherbicidal fungus for efficient weed control. Nat Biotechnol 20:1035–1039

Andolfi A, Boari A, Evidente A, Vurro M (2005) Metabolites inhibiting germination of *Orobanche ramosa* seeds produced by *Myrothecium verrucaria* and *Fusarium compactum*. J Agric Food Chem 53:1598–1603

Bailey KL (2010) Canadian innovations in microbial biopesticides. Can J Plant Pathol 32:113–121

Bailey BA, Apel-Birkhold PC, Luster DG (2002) Expression of *NEP*1 by *Fusarium oxysporum* f. sp. *erythroxyli* after gene replacement and overexpression using PEG-mediated transformation. Phytopathology 92:833–841

Barghouthi S, Salman M (2010) Bacterial inhibition of *Orobanche aegyptiaca* and *Orobanche cernua* radical elongation. Biocontrol Sci Technol 20:423–435

Barton J (2005) Bioherbicides: all in a day's work . . . for a super hero. What's New in Biological Control of Weeds? Issue 32:4–6. Manaaki Whenua, Landcare Research, Lincoln, New Zealand

Bashir MO (1987) The potential for biocontrol of witchweeds. In: Musselman LJ (ed) Parasitic weeds in agriculture, *Striga*, vol I. CRC, Boca Raton, FL, pp 183–206

Bastiani C (2002) Seed coating with *Fusarium oxysporum* M12-4A for the biocontrol of *Striga hermonthica*. MSc Thesis, McGill University, Montreal, QC, Canada, p 72

Bedi JS (1994) Further studies on control of sunflower broomrape with *Fusarium oxysporum* f. sp. *orthoceras* a potential mycoherbicide. In: Pieterse AH, Verkleij JAH, ter Borg SJ (eds) Proceedings of the 3rd international workshop on *Orobanche* and related *Striga* research. Royal Tropical Institute, Amsterdam, The Netherlands, pp 539–544

Bedi JS, Donchev N (1991) Results on mycoherbicide control of sunflower broomrape (*Orobanche cumana* Wall.) under glasshouse and field conditions. In: Ransom JK, Musselman LJ, Worsham AT, Parker C (eds) Proceedings of the 5th international symposium on parasitic weeds. CIMMYT, Nairobi, Kenya, pp 76–82

Beed FD, Hallett SG, Venne J, Watson AK (2007) Biocontrol using *Fusarium oxysporum*, a critical component of integrated *Striga* management. Chapter 21. In: Ejeta G, Gressel J (eds) Integrating new technologies for *Striga* control: ending the witch-hunt. World Scientific, Hackensack, NJ, pp 283–300

Berner DK, Kling JG, Singh BB (1995) *Striga* research and control – a perspective from Africa. Plant Dis 79:652–660

Berner DK, Schaad NW, Völksch B (1999) Use of ethylene-producing bacteria for stimulation of *Striga* spp. seed germination. Biol Control 15:274–282

Boari A, Vurro M (2004) Evaluation of *Fusarium* spp. and other fungi as biological control agents of broomrape (*Orobanche ramosa*). Biol Control 30:212–219

Boputiti MZ, Souissi T, Kharrat M (2008) Evaluation of *Fusarium* as a potential biological control against *Orobanche* on faba bean in Tunisia. In: Julien MH, Sforza R, Bon MC, Evans HC, Hatcher PE, Hinz HL, Rector BG (eds) Proceedings of the XII international symposium on biological control of weeds. CAB International, Wallingford, UK, pp 238–244

Bouwmeester HJ, Roux C, Lopez-Raez JA, Bécard G (2007) Rhizosphere communication of plants, parasitic plants and AM fungi. Trends Plant Sci 12:225–230

Cardoso C, Ruyter-Spira C, Bouwmeester HJ (2011) Strigolactones and root infestation by plant-parasitic *Striga*, *Orobanche* and *Phelipanche* spp. Plant Sci 180:414–420

Ciotola M, Watson AK, Hallett SG (1995) Discovery of an isolate of *Fusarium oxysporum* with potential to control *Striga hermonthica* in Africa. Weed Res 35:303–309

Ciotola M, DiTommaso A, Watson AK (2000) Chlamydospore production, inoculation methods and pathogenicity of *Fusarium oxysporum* M12-4A, a biocontrol for *Striga hermonthica*. Biocontrol Sci Technol 10:129–145

Cipriani MG, Stea G, Moretti A, Altomare C, Mulè G, Vurro M (2009) Development of a PCR-based assay for the detection of *Fusarium oxysporum* strain FT2, a potential mycoherbicide of *Orobanche ramosa*. Biol Control 50:78–84

Cohen BA, Amsellem Z, Lev-Yadun S, Gressel J (2002a) Infection of tubercles of the parasitic weed *Orobanche aegyptiaca* by mycoherbicidal *Fusarium* species. Ann Bot 90:567–578

Cohen B, Amsellem Z, Maor R, Sharon A, Gressel J (2002b) Transgenically enhanced expression of indole-3-acetic acid (IAA) confers hypervirulence to plant pathogens. Phytopathology 92:590–596

Connick WJ Jr, Boyette CD, McAlpine JR (1991) Formulation of mycoherbicides using a pasta like process. Biol Control 1:281–287

Crawley MJ (2000) Seed predators and plant population dynamics. In: Fenner M (ed) Seeds: the ecology and regeneration in plant communities. CAB International, Wallingford, UK, pp 167–182

Dadon T, Nun NB, Mayer AM (2004) A factor from *Azospirillum brasilense* inhibits germination and radicle growth of *Orobanche aegyptiaca*. Israel Plant Sci 52:83–86

Diarra C, Ciotola M, Hallett SG, Hess DE, Watson AK (1996) Field efficacy of *Fusarium oxysporum* for the control of *Striga hermonthica*. Nuisibles Pest Pragas 4:257–263

Dor E, Vurro M, Hershenhorn J (2003) The efficacy of a mixture of fungi to control Egyptian and sunflower broomrape. http://cost849.ba.cnr.it/athens%20abstract.PDF. Accessed March 2013

Dor E, Hershenhorn J (2009) Evaluation of the pathogenicity of microorganisms isolated from Egyptian broomrape (*Orobanche aegyptiaca*) in Israel. Weed Biol Manag 9:200–208

Dor E, Hershenhorn J, Andolfi A, Cimmino A, Evidente A (2009) *Fusarium verticillioides* as a new pathogen of the parasitic weed *Orobanche* spp. Phytoparasitica 37:361–370

Dor E, Joel DM, Kapulnik Y, Koltai H, Hershenhorn J (2011) The synthetic strigolactone GR24 influences the growth pattern of phytopathogenic fungi. Planta 234:419–427

Duke SO, Dayan FE (2011) Modes of action of microbially-produced phytotoxins. Toxins (Basel) 8:1038–1064

Eilenberg J, Hajek A, Lomer C (2001) Suggestions for unifying the terminology in biological control. Biocontrol Sci Technol 46:387–400

El-Kassas R, Karam El-Din Z, Beale MH, Ward JL, Strange RN (2005) Bioassay-led isolation of *Myrothecium verrucaria* and verrucarin A as germination inhibitors of *Orobanche crenata*. Weed Res 45:212–219

Elzein A, Kroschel J (2004) *Fusarium oxysporum* "Foxy 2" shows potential to control both *Striga hermonthica* and *S. asiatica*. Weed Res 44:433–438

Elzein A, Kroschel J (2006a) Development and efficacy of granular formulation of *Fusarium oxysporum* "Foxy 2" for *Striga* control: an essential step towards practical field application in Africa. J Plant Dis Prot 20:889–905

Elzein A, Kroschel J (2006b) Host range studies of *Fusarium oxysporum* Foxy 2: an evidence for a new *forma specialis* and its implications for *Striga* control. J Plant Dis Prot 20:875–887

Elzein A, Kroschel J, Leth V (2006) Seed treatment technology: an attractive delivery system for controlling root parasitic weed *Striga* with mycoherbicide. Biocontrol Sci Technol 16:3–26

Elzein A, Brändle F, Cadisch G, Kroschel J, Marley P, Thines M (2008) *Fusarium oxysporum* strains as potential *Striga* mycoherbicides: molecular characterization and evidence for a new *forma specialis*. Open Mycol J 2:89–93

Elzein A, Heller A, Ndambi B, De Mol M, Kroschel J, Cadisch G (2010) Cytological investigations on colonization of sorghum roots by the mycoherbicide *Fusarium oxysporum* f. sp. *strigae* and its implications for *Striga* control using a seed treatment delivery system. Biol Control 53:249–257

Estabrook EM, Yoder JI (1998) Plant-plant communications: rhizosphere signaling between parasitic angiosperms and their hosts. Plant Phys 116:1–7

Fernàndez-Aparicio M, Garcia-Garrido JM, Ocampo JA, Rubiales D (2010) Colonization of field pea roots by arbuscular mycorrhizal fungi reduce *Orobanche* and *Phelipanche* species seed germination. Weed Res 50:262–268

Garren JM, Strauss SY (2009) Population-level compensation by an invasive thistle thwarts biological control from seed predators. Ecol Appl 19:709–721

Gonsior G, Buschmann H, Szinicz G, Spring O, Sauerborn J (2004) Induced resistance: an innovative approach to manage branched broomrape (*Orobanche ramosa*) in hemp and tobacco. Weed Sci 52:1050–1053

Greathead DJ (1984) The natural enemies of *Striga* spp. and the prospects for their utilization as biological control agents. In: Ayensu ES, Doggett H, Keynes RD, Marton-Lefevre J, Musselman LJ, Parker C, Pickering A (eds) Striga, biology and control. ICSU Press, Paris and The International Development Research Centre, Ottawa, Canada, pp 133–160

Gressel J (2001) Potential failsafe mechanisms against the spread and introgression of transgenic hypervirulent biocontrol fungi. Trends Biotechnol 19:149–154

Gweorgwo NA, Weber HC (2003) Arbuscular mycorrhizal fungi-parasite-host interaction for the control of *Striga hermonthica* (Del.) Benth. in sorghum [*Sorghum bicolor* (L.) Moench]. Mycorrhiza 13:277–281

Hassan MM, Gani MEA, Babiker AGE (2009) Management of *Striga hermonthica* in sorghum using soil rhizosphere bacteria and host plant resistance. Int J Agric Biol 11:367–373

Hassan MM, Gani MEA, Babiker AGT (2011a) Effects of bacterial strains and isolates on *in situ* germination, subsequent developmental stage of *Striga hermonthica* onto sorghum roots. Adv Environ Biol 5:3263–3269

Hassan MM, Abdelhalim TS, Yagoub SO, Osman AG, Gani MEA, Babiker AGE (2011b) Effects of arbuscular mycorrhiza fungi (AMF), plant growth promoting bacteria (PGPR) and interaction on *Striga hermonthica* management in sorghum. Int J Agric Res Rev 1:107–115

Hoagland RE (2001) Microbial allelochemicals and pathogens as bioherbicidal agents. Weed Technol 15:835–857

Karanja J, Nzioki SM, Nguluu S, Mugwika M, Khamisi R (2009) Screening cowpea for *Alectra vogelii* resistance in the Arid semi-Arid lands (ASALs) of Kenya. http://www.kari.org/kasal/docs/katumani_kasal_achievement_screening%20cowpea.pdf. Accessed Feb 2013

Kirk AA (1993) A fungal pathogen with potential for control of *Striga hermonthica* (Scrophulariaceae). Entomophaga 38:359–460

Klein O, Kroschel J (2002) Biological control of *Orobanche* spp. with *Phytomyza orobanchia*, a review. Biocontrol 47:245–277

Kohlschmid E, Sauerborn J, Müller-Stöver D (2009) Impact of *Fusarium oxysporum* on the holoparasitic weed *Phelipanche ramosa*: biocontrol efficacy under field-grown conditions. Weed Res 49(Suppl 1):56–65

Kroschel J, Elzein A (2004) Bioherbicidal effect of fumonisin B_1, a phytotoxic metabolite naturally produced by *Fusarium nygamai*, on parasitic weeds of the genus *Striga*. Biocontrol Sci Technol 14:117–128

Kroschel J, Abbasher AA, Sauerborn J (1995) Herbivores of *Striga hermonthica* in Northern Ghana and approaches to their use as biocontrol agents. Biocontrol Sci Technol 5:163–164

Kroschel J, Hundt A, Abbasher AA, Sauerborn J (1996) Pathogenicity of fungi collected in Northern Ghana to *Striga hermonthica*. Weed Res 36:515–520

Lendzemo VW, Kuyperb TW, Kropff MJ, van Ast A (2005) Field inoculation with arbuscular mycorrhizal fungi reduces *Striga hermonthica* performance on cereal crops and has the potential to contribute to integrated *Striga* management. Field Crops Res 91:51–61

Lendzemo VW, Kuyper TW, Matusova R, Bouwmeester HJ, van Ast A (2007) Colonization by arbuscular mycorrhizal fungi of sorghum leads to reduced germination and subsequent attachment and emergence of *Striga hermonthica*. Plant Signal Behav 2:28–62

Leslie JF, Summerell BA (2006) Fusarium laboratory manual. First edition. Blackwell, Ames, IA, p 388

Linke KH, Vorlaender C, Saxena MC (1990) Occurrence and impact of *Phytomyza orobanchia* (Diptera: Agromyzidae) on *Orobanche crenata* (Orobanchaceae) in Syria. Entomophaga 35:633–639

Linke KH, Scheibel C, Saxena MC, Sauerborn J (1992) Fungi occurring on *Orobanche* spp. and their preliminary evaluation for *Orobanche* control. Trop Pest Manag 38:127–130

Mabrouk Y, Zourgui L, Sifi B, Belhadj O (2007) The potential of *Rhizobium* strains for biological control of *Orobanche crenata*. Biologia 62:139–143

Marley PS, Shebayan JAY (2005) Field assessment of *Fusarium oxysporum* based mycoherbicide for control of *Striga hermonthica* in Nigeria. Biocontrol 50:389–399

Marley PS, Ahmed SM, Shebayan JAY, Lagoke STO (1999) Isolation of *Fusarium oxysporum* with potential for biocontrol of the witchweed (*Striga hermonthica*) in the Nigerian Savanna. Biocontrol Sci Technol 9:159–163

Marley PS, Aba DA, Shebayan JAY, Musa R, Sanni A (2004) Integrated management of *Striga hermonthica* in sorghum using a mycoherbicide and host plant resistance in the Nigerian Sudano-Sahelian savanna. Weed Res 44:157–162

Marley PS, Kroschel J, Elzein A (2005) Host specificity of *Fusarium oxysporum* Schlecht. (isolate PSM 197), a potential mycoherbicide for controlling *Striga* spp. in West Africa. Weed Res 45:407–412

McFadyen R (1998) Biological control of weeds. Ann Rev Entomol 43:369–393

Meir S, Amsellem Z, Al-Ahmad H, Safran E, Gressel J (2009) Transforming a NEP1 toxin gene into two *Fusarium* spp. to enhance mycoherbicide activity on *Orobanche* – failure and success. Pest Manag Sci 65:588–595

Miché L, Bouillant M-L, Rohr R, Sallé G, Bally R (2000) Physiological and cytological studies on the inhibition of *Striga* seed germination by the plant growth-promoting bacterium *Azospirillum brasilense*. Eur J Plant Pathol 106:347–351

Mohamed EAA (2002) Evaluation of *Fusarium* sp. Abu Haraz as a biocontrol for *Striga hermonthica* (Del). Benth. on *Sorghum bicolor* (L.). MSc Thesis, University of Gezira, Medani, Sudan, p 91

Mohameda KI, Bolin JF, Musselman LJ, Peterson AT (2007) Genetic diversity of *Striga* and implications for control and modeling future distributions. Chapter 6. In: Ejeta G, Gressel J (eds) Integrating new technologies for *Striga* control: ending the witch-hunt. World Scientific, Hackensack, NJ, pp 283–300

Müller-Stöver D, Kroschel J (2005) The potential of *Ulocladium botrytis* for biological control of *Orobanche* spp. Biol Control 33:301–306

Müller-Stöver D, Buschmann H, Sauerborn J (2005) Increasing control reliability of *Orobanche cumana* through integration of a biocontrol agent with a resistance-inducing chemical. Eur J Plant Pathol 111:193–202

Müller-Stöver D, Kohlschmid E, Sauerborn J (2009a) A novel strain of *Fusarium oxysporum* from Germany and its potential for biocontrol of *Orobanche ramosa*. Weed Res 49(Suppl 1):175–182

Müller-Stöver D, Batchvarova R, Kohlschmid E, Sauerborn J (2009b) Mycoherbicidal management of *Orobanche cumana*: observations from three years of field experiments. In: Rubiales D, Westwood J, Uledag A (eds) Proceedings of the 10th international world congress of parasitic plants, 8–12 June 2009, Kusadasi, Turkey, p 86

Ndambi B, Cadisch G, Elzein A, Heller A (2011) Colonization and control of *Striga hermonthica* by mycoherbicide component: an anatomical study. Biol Control 58:149–159

Nirenberg HI, O'Donnell K, Kroschel J, Andrianaivo AP, Frank JM, Mubatanhema W (1998) Two new species of *Fusarium*: *Fusarium brevicatenulatum* from the noxious weed *Striga asiatica* in Madagascar and *Fusarium pseudoanthophilum* from *Zea mays* in Zimbabwe. Mycologica 90:459–464

Norambuena H (2003) Control biológico de malezas en Chile: experiencias para la implementación rápida de proyectos. Manejo Integrado de Plagas y Agroecología (Costa Rica) 69:96–98

Olivain C, Alabouvette C (1997) Colonization of tomato root by a nonpathogenic strain of *Fusarium oxysporum*. New Phytol 137:481–494

Parker C, Riches CR (1993) Parasitic weeds of the world: biology and control. CAB International, Wallingford, UK, 332 p

Raaijmakers JM, Paulitz TC, Steinberg C, Alabouvette C, Moënne-Loccoz Y (2009) The rhizosphere: a playground and battlefield for soilborne pathogens and beneficial microorganisms. Plant Soil 321:341–361

RIUtv (2010) StopStriga and Gro-Plus Seed priming and bio-pesticides. http://www.researchintouse.com/bestbets/bb34stopstriga83riutv38.html. Accessed Dec 2012

Sands DC, Pilgeram AL (2009) Methods for selecting hypervirulent biocontrol agents of weeds: why and how. Pest Manag Sci 65:581–587

Sauerborn J, Müller-Stöver D, Hershenhorn J (2007) The role of biological control in managing parasitic weeds. Crop Prot 26:246–254

Savard ME, Miller JD, Ciotola M, Watson AK (1997) Secondary metabolites produced by a strain of *Fusarium oxysporum* used for *Striga* control in West Africa. Biocontrol Sci Technol 7:61–64

Schuab B, Marley M, Elzein A, Kroschel J (2006) Field evaluation of an integrated *Striga hermonthica* management in Sub-Saharan Africa: synergy between *Striga*-mycoherbicides (biocontrol) and sorghum and maize resistant varieties. J Plant Dis Prot 20:691–699

Shabana YM, Müller-Stöver D, Sauerborn J (2002) Granular Pesta formulation of *Fusarium oxysporum* f. sp. *orthoceras* for biological control of sunflower broomrape: efficacy and shelf-life. Biol Control 26:189–201

Shalaby FF, Ibrahim HMM, Hassanein EE (2004) Natural biocontroling activity of *Phytomyza orobanchia* (Kalt.) against *Orobanche crenata* and increasing its beneficial role by field release of the fly adults. Egypt Biol Pest Control 14:242–249

Singh SP (1989) Biological suppression of weeds. Biological Control Centre, NCIPM, Faridabad. Tech Bull 1:27

Smith MC, Webb M (1996) Estimation of the seed bank of *Striga* spp. [Scrophulariaceae] in Malian fields, and the implications for a model of biocontrol of *Striga hermonthica*. Weed Res 36:85–92

Smith MC, Holt J, Webb M (1993) Population model of the parasitic weed *Striga hermonthica* (Scrophulariaceae) to investigate the potential of *Smicronyx umbrinus* (Coleoptera: Curculionidae) for biological control in Mali. Crop Prot 12:471–476

Strobel GA, Kenfield D, Bunkers G, Sugawara F, Clardy J (1991) Phytotoxins as potential herbicides. Experientia 47:819–826

Tessema T (2007) The prospects of biological control of weeds in Ethiopia. Ethiop J Weed Manag 1:63–78

Theu MPKJ (2008) Biological control of witchweed (*Striga*) in maize (*Zea mays*) in Malawi. In: Plant protection progress report, 2007/2008 season, Department of Agricultural Research Services Planning and Review Meeting, Andrews Hotel, Mangochi, 14–20 September 2008, pp 90–103

Thiel PG, Marasas WF, Sydenham EW, Shephard GS, Gelderblom WC, Nieuwenhuis JJ (1991) Survey of fumonisin production by *Fusarium* species. Environ Microbiol 57:1089–1093

Thomas H, Sauerborn J, Müller-Stöver D, Ziegler A, Bedi JS, Kroschel J (1998) The potential of *Fusarium oxysporum* f. sp. *orthoceras* as a biological control agent for *Orobanche cumana* in sunflower. Biol Control 13:41–48

Traoré D, Vincent C, Stewart RK (1995) Life history of *Smicronyx guineanus* and *Sm. umbrinus* (Col.: Curculionidae) on *Striga hermonthica* (Scrophulariaceae). Biocontrol Sci Technol 40:211–221

van Mourik TA, Bianch FJJA, van der Werf W, Stomph TJ (2008) Long-term management of *Striga hermonthica*: strategy evaluation with a spatio-temporal population model. Weed Res 48:329–339

Venne J (2008) Molecular characterization and virulence analysis of *Fusarium oxysporum* strains used in biological control against *Striga hermonthica*. MSc Thesis, McGill University, Montreal, QC, Canada, p 93

Venne J, Beed F, Avocanh A, Watson A (2009) Integrating *Fusarium oxysporum* f. sp. *strigae* into cereal cropping systems in Africa. Pest Manag Sci 65:572–580

Vurro M, Boari A, Pilgeram AL, Sands DC (2006) Exogenous amino acids inhibit seed germination and tubercle formation by *Orobanche ramosa* (broomrape): potential application for management of parasitic weeds. Biol Control 36:258–265

Vurro M, Boari A, Evidente A, Andolfi A, Zermane N (2009) Natural metabolites for parasitic weed management. Pest Manag Sci 65:566–571

Wan FH, Wang R (1991) Achievement of biological control in the world and its prospect in China (in Chinese). Chin J Biol Control 7:81–87

Watson AK, Gressel J, Sands D, Hallett S, Vurro M, Beed F (2007) *Fusarium oxysporum* f. sp. *strigae*, Athletes Foot or Achilles Heel? In: Vurro M, Gressel J (eds) Novel biotechnologies for biocontrol agent enhancement and management. Springer, Dordrecht, pp 213–222

Yandoc-Ables CB, Rosskopf EN, Charudattan R (2006) Plant pathogens at work: progress and possibilities for weed biocontrol. Part 1: Classical vs. bioherbicidal approach. APSnet Feature Story Aug. 2006. http://www.apsnet.org/publications/apsnetfeatures/Pages/WeedBiocontrolPart1.aspx. Accessed March 2013

Yonli D, Traoré H, Hess DE, Abbasher AA, Boussim IJ (2004) Effect of growth medium and method of application of *Fusarium oxysporum* on infestation of sorghum by *Striga hermonthica* in Burkina Faso. Biocontrol Sci Technol 14:417–421

Yonli D, Traoré H, Hess DE, Abbasher AA, Sérémé P, Sankara P (2005) Biological control of witchweed in fields of Burkina Faso using isolates of *Fusarium oxysporum*. Afr Crop Sci 13:41–47

Yonli D, Traoré H, Sérémé P, Hess DE, Sankara P (2010) Pathogenicity of *Fusarium* isolates to *Striga hermonthica* in Burkina Faso. Pak J Biol Sci 13:201–208

Zahran E, Sauerborn J, Abbasher AA, Ahmed EA, Mohukker RI, Karlovsky P, Mohamed EA, Müller-Stöver D (2008) "Pesta" and alginate delivery systems of Fusarium spp. for biological control of Striga hermonthica (Del.) Benth. under Sudanese field conditions. Biocontrol 44:160–168

Zermane N, Souissi T, Kroschel J, Sikora R (2007) Biocontrol of broomrape (*Orobanche crenata* Forsk. and *Orobanche foetida* Poir.) by *Pseudomonas fluorescens* isolate Bf7-9 from the faba bean rhizosphere. Biocontrol Sci Technol 17:483–497

Zonno MC, Vurro M (1999) Effect of fungal toxins on germination of *Striga hermonthica* seeds. Weed Res 39:15–20

Zonno MC, Vurro M (2002) Inhibition of germination of *Orobanche ramosa* seeds by *Fusarium* toxins. Phytoparasitica 30:519–524

Zummo N (1977) Diseases of giant witchweed, *Striga hermonthica*, in West Africa. Plant Dis Rep 61:428–430

Index

D.M. Joel et al. (eds.), *Parasitic Orobanchaceae*,
DOI 10.1007/978-3-642-38146-1, © Springer-Verlag Berlin Heidelberg 2013

Q

R

MIX
Papier aus verantwortungsvollen Quellen
Paper from responsible sources
FSC® C105338

If you have any concerns about our products,
you can contact us on
ProductSafety@springernature.com

In case Publisher is established outside the EU,
the EU authorized representative is:
Springer Nature Customer Service Center GmbH
Europaplatz 3, 69115 Heidelberg, Germany

Printed by Libri Plureos GmbH
in Hamburg, Germany